Microbes in Microbial Communities

Raghvendra Pratap Singh •
Geetanjali Manchanda • Kaushik Bhattacharjee •
Hovik Panosyan
Editors

Microbes in Microbial Communities

Ecological and Applied Perspectives

Springer

Editors
Raghvendra Pratap Singh
Department of Research and Development
Azoth Biotech Pvt. Ltd.
Noida, Uttar Pradesh, India

Geetanjali Manchanda
Botany and Environmental Studies
DAV University
Jalandhar, Punjab, India

Kaushik Bhattacharjee
Division of Life Sciences
Institute of Advanced Study in Science and Technology
Guwahati, Assam, India

Hovik Panosyan
Department of Biochemistry, Microbiology and Biotechnology
Yerevan State University
Yerevan, Armenia

ISBN 978-981-16-5616-3 ISBN 978-981-16-5617-0 (eBook)
https://doi.org/10.1007/978-981-16-5617-0

This Springer imprint is published by the registered company Springer Nature Singapore Pte Ltd.
The registered company address is: 152 Beach Road, #21-01/04 Gateway East, Singapore 189721, Singapore

Preface

Microbial communities are ubiquitously present in the environment. In order to establish, maintain, adapt, and survive, microorganisms interact with each other and their hosts. The communities thus manifest multi-trophic interactions that result in high diversity. These interactions which may be synergistic or antagonistic shape the microbial communities and are responsible for the evolution and adaptations of the microorganisms. One of the mechanisms used by the bacteria for interactions is quorum sensing. It is a type of cell–cell communication system that plays an important role in cooperative and competitive microbial interactions. An understanding of quorum sensing mechanisms can provide vital clues about the functioning of microbial communities. Further, the interactions can be explored for potential uses in medical and agriculture applications.

This book explores different aspects of microbial communities as unique entities to study genetic diversity, ecological interactions between microbes and their adaptation mechanisms. It also covers the biotechnological prospects of microbes involved in various microbial communities and their potential use in agricultural, medical, and industrial applications.

The book brings together an outstanding team of scientists with interest and expertise in different aspects of microbial community ecology, organismal biology, biochemistry, and biotechnology. The topics addressed include but are not limited to diversity and microbial ecology, microbe–environment, microbe–microbe and microbe–other organism interactions, adaptation and evolution, element cycling, and biotechnological applications of microbes or their communities obtained from various ecosystems.

Within the book's 16 chapters are several studies focusing on microbial abundance in extreme environments, strategies of adaptation, and approaches in microbe-assisted bioremediation for a sustainable clean environment. Several chapters focus on synergistic and antagonistic interactions among microbes and try to answer the question how such interactions can be used as game changers in agriculture and health sectors. Two chapters describe bacterial community composition dynamics and quorum sensing in rhizosphere. A dedicated chapter discusses selected aspects of lichen microbiome, its diversity, biological role, and biotechnological applications. Two chapters describe the diversity and ecology of urinary and gut microbiome, respectively, and the role of these microbiomes in next-generation therapeutics. Antibacterial secondary metabolites produced by microbes from different ecosystems and their potential in biotechnology are also discussed in this book. There is a chapter in which the ecology and abundance of benzoate-degrading bacteria in industrial waste are discussed.

We hope you will enjoy the diverse and exciting research described in this book.

Noida, Uttar Pradesh, India — Raghvendra Pratap Singh
Jalandhar, Punjab, India — Geetanjali Manchanda
Guwahati, Assam, India — Kaushik Bhattacharjee
Yerevan, Armenia — Hovik Panosyan

Contents

Editors and Contributors

About the Editors

Raghvendra Pratap Singh is an eminent scientist in Microbial Biotechnology. He has received his Ph.D. from Gurukul Kangri University, India, and is presently working in Azoth Biotech Pvt. Ltd. as a Managing Director. His research contributions relate to the area of Cordyceps genomics, plant–microbes interaction and microbial genomics. He is a member of Aquatic Biodiversity Society and also has a certification from Food and Drug Association of India. He has received a number of awards from various scientific agencies and societies such as Young Scientist award by ABA-2017, SERB-DST grant, Chinese Postdoctoral Grant, DBT travel grant, etc. He has participated as speaker in several national and international scientific meets and conferences such as FEMS, Myxo-2016, AMI, and PTPB-2014. Dr. Singh has actively contributed a sundry of international research articles, book chapters, general articles, etc. in various microbiology and biotechnology journals and has two patents. He has deposited more than 400 gene sequences, six WGS, and three metagenomes at NCBI. He has authored 6 books, including a prestigious textbook that was published by the Indian Council of Agriculture Research, Government of India.

Geetanjali Manchanda is working as Head of the Botany Department in DAV University, Jalandhar, India. She received her M.Sc. degree from Delhi University and Ph.D. from Panjab University, Chandigarh. She has extensively worked on plant–microbe interactions in stressed and contaminated environments with a special focus on mycorrhizae for the fortification of various crops. She has received prestigious research grants from DST, India, and IFS Sweden. She has contributed immensely to the scientific community by publishing research papers and book chapters. She had recently authored a book on the use of omics technology for microbiology that has been published by ICAR, New Delhi, India.

Kaushik Bhattacharjee earned his Doctor of Philosophy (Ph.D.) from the North-Eastern Hill University of India in the Microbiology Laboratory by conducting research in the interdisciplinary field of Microbial diversity from extreme environments, Pharmaceutical microbiology, and Medicinal chemistry. He took his post-doctoral trainings in the Department of Botany at North-Eastern Hill University of India and IASST, Guwahati, India. He has so far contributed over 20 publications in referred journals of high repute and published about half a dozen of book chapters. He also serves as an editorial board member and invited journal reviewer for many highly reputed journals from publishing groups like Springer Nature, Elsevier, PLoS, Taylor & Francis, etc. He has been awarded the Outstanding Reviewer Award from Elsevier for the year 2018. He also serves as a Certified Mentor at Publons Academy, Clarivate Analytics, United States. His broad area of research interests include environmental microbiology and pharmaceutical microbiology.

Hovik Panosyan graduated in Biology from Yerevan State University (YSU) in 1999. He received his Ph.D. in microbiology from the Institute of Botany of NAS of Armenia in 2003. He has been a faculty member at YSU since 2002 and was promoted to Associate Professor in 2011. His main research area is microbial ecology and biology of extremophilic microbes. He has been awarded numerous research fellowships and awards including FEBS Short-Term Fellowship (2009 and 2004), FEMS Research Fellowship (2009), NFSAT (2011), and DAAD (2013) and has participated in international research together with USA, European, and Asian partners. He is currently coordinator and leader of international research and educational programs, as well as ISME ambassador of Armenia. He had work experience at the University of Bergen (Norway), LMU Munich (Germany), University of Nevada, Las Vegas (USA), and Institute of Biomolecular Chemistry Naples (Italy). Hovik Panosyan is actively engaged in studying the microbial community of extreme environments (terrestrial geothermal springs, alkali-saline soils, subterranean salt deposits, copper and molybdenum mines distributed on the territory of Armenia) based on culture-dependent and molecular techniques. He has published more than 60 research papers in peer-reviewed journals, four books, and 25 chapters.

Contributors

Mian Nabeel Anwar Environmental Engineering Department, University of Alberta, Edmonton, AB, Canada

Gajender Kumar Aseri Amity Institute of Microbial Technology, Amity University, Jaipur, Rajasthan, India

Aishiki Banerjee Department of Environmental Studies, Siksha-Bhavana, Visva-Bharati, Santiniketan, India

Mustapha Barakate Laboratory of Microbial Biotechnologies, Agrosciences & Environment, Faculty of Sciences Semlalia, Cadi Ayyad University, Marrakech, Morocco
Mohammed IV Polytechnic University (UM6P), AGBT, AgroBiosciences, Benguerir, Morocco

Mustapha Barakate Laboratory of Microbial Biotechnologies, Agrosciences & Environment, Faculty of Sciences Semlalia, Cadi Ayyad University, Marrakech, Morocco
Mohammed IV Polytechnic University (UM6P), AGBT, AgroBiosciences, Benguerir, Morocco

Dina Barman Microbial Ecology Laboratory, Department of Botany, Gauhati University, Guwahati, India

Aparajita Basu Department of Microbiology, University of Calcutta, Kolkata, India

Noureddine Bouras Laboratoire de Biologie des Systèmes Microbiens (LBSM), Ecole Normale Supérieure de Kouba, Alger, Algeria
Département de Biologie, Faculté des Sciences de la Nature et de la Vie et Sciences de la Terre, Université de Ghardaia, Ghardaïa, Algeria

Santanu Das Molecular Biology and Microbial Biotechnology Laboratory, Life Sciences Division, Institute of Advanced Study in Science and Technology (IASST), Guwahati, Assam, India
Department of Molecular Biology and Biotechnology, Cotton University, Guwahati, Assam, India

Sridevi Jagavati Department of Microbiology, Indira Priyadarshini Government Degree College for Women, Hyderabad, Telangana, India

Dhruva Kumar Jha Microbial Ecology Laboratory, Department of Botany, Gauhati University, Guwahati, India

Tulsi K. Joishy Molecular Biology and Microbial Biotechnology Laboratory, Life Sciences Division, Institute of Advanced Study in Science and Technology (IASST), Guwahati, Assam, India
Department of Molecular Biology and Biotechnology, Cotton University, Guwahati, Assam, India

Mojibur R. Khan Molecular Biology and Microbial Biotechnology Laboratory, Life Sciences Division, Institute of Advanced Study in Science and Technology (IASST), Guwahati, Assam, India

Neeraj Khare Amity Institute of Microbial Technology, Amity University, Jaipur, Rajasthan, India

Shweena Krishnani Amity Institute of Microbial Technology, Amity University, Jaipur, Rajasthan, India

Zhi Feng Li State Key Laboratory of Microbial Technology, Shandong University China, Jinan, China

Abha Manapure Department of Microbiology, S K Porwal College, Nagpur, Maharashtra, India

Geetanjali Manchanda Department of Botany, DAV University, Jalandhar, Punjab, India

Armine Margaryan Department of Biochemistry, Microbiology, and Biotechnology, Faculty of Biology, Yerevan State University, Yerevan, Armenia

Kunal Meena Amity Institute of Microbial Technology, Amity University, Jaipur, Rajasthan, India

Lotfi Mellouli Laboratory of Microorganisms and Enzymatic Biotechnology and Biomolecules, Center of Biotechnology of Sfax, University of Sfax, Sfax, Tunisia

Pooja Misra Crop Protection Division, CSIR-Central Institute of Medicinal Aromatic Plants, Lucknow, India

Goutam Mukherjee Department of Microbiology, University of Calcutta, Kolkata, India

Ashish Kumar Nayak Microbial Genomics and Diagnostics Lab, Microbiology and Plant Pathology Division, Regional Plant Resource Centre, Bhubaneswar, Odisha, India

Arun Pandiyan Food Safety Division, National Institute of Nutrition – ICMR, Hyderabad, Telangana, India

Hovik Panosyan Department of Biochemistry, Microbiology and Biotechnology, Yerevan State University, Yerevan, Armenia

Mukesh Poddar Department of Microbiology, Sikkim University, Gangtok, India

Rekha Punchappady-Devasya Yenepoya Research Centre, Yenepoya Deemed to be University, Mangalore, India

Alok R. Rai Department of Microbiology, S K Porwal College, Nagpur, Maharashtra, India

Manohari Rathi Centre for Environmental Science and Technology, School of Environment and Earth Sciences, Central University of Punjab, Bathinda, India

A. Vimala Rodhe Department of Microbiology, Silver Jubilee Govt. College, Kurnool, Andhra Pradesh, India

Mousumi Saha Department of Microbiology, University of Calcutta, Kolkata, India

Saurav Saha Department of Zoology, North-Eastern Hill University, Shillong, Meghalaya, India

R. R. Sargsyan Department of Biochemistry, Microbiology and Biotechnology, Faculty of Biology, Yerevan State University, Yerevan, Armenia

Arvind Saroj Department of Horticulture, Molecular Plant Breeding, North East Agriculture University, Harbin, China

Sreedevi Sarsan Department of Microbiology, St. Pious X Degree & PG College, Hyderabad, Telangana, India

Deepansh Sharma Amity Institute of Microbial Technology, Amity University, Jaipur, Rajasthan, India

Alok Kumar Sil Department of Microbiology, University of Calcutta, Kolkata, India

Deepti Singh Amity Institute of Microbial Technology, Amity University, Jaipur, Rajasthan, India

Dipti Singh VBS Purvanchal University, Jaunpur, Uttar Pradesh, India

Raghvendra Pratap Singh Department of Research and Development, Azoth Biotech Pvt. Ltd., Noida, Uttar Pradesh, India

Slim Smaoui Laboratory of Microorganisms and Enzymatic Biotechnology and Biomolecules, Center of Biotechnology of Sfax, University of Sfax, Sfax, Tunisia

Jagdip Singh Sohal Amity Institute of Microbial Technology, Amity University, Jaipur, Rajasthan, India

Atul Kumar Srivastava Research and Development Department, Uttaranchal University, Dehradun, Uttarakhand, India

Mangesh Suryavanshi Yenepoya Research Centre, Yenepoya Deemed to be University, Mangalore, India

A. Tsurykau Department of Botany and Plant Physiology, Faculty of Biology, Francisk Skorina Gomel State University, Gomel, Belarus
Department of Ecology, Botany and Nature Protection, Institute of Natural Sciences, Samara National Research University, Samara, Russia

Vineet Amity Institute of Microbial Technology, Amity University, Jaipur, Rajasthan, India

Amine Yekkour Laboratoire de Biologie des Systèmes Microbiens (LBSM), Ecole Normale Supérieure de Kouba, Alger, Algeria

Chapter 1
Synergistic Interactions Among Microbial Communities

Sreedevi Sarsan, Arun Pandiyan, A. Vimala Rodhe, and Sridevi Jagavati

Abstract Microorganisms often coexist with each other in close proximity such as micro-colonies or biofilms and are rare to be obtained as pure cultures from the environment. Hence, there is always a likelihood of microbe–microbe interactions among these communities, which can either be positive or negative. Various factors such as physical, chemical, biological and genetics regulate such interactions and the molecular mechanisms involved in these interactions between microorganisms. One of the most important positive microbial interactions is synergism. Microbial synergism is defined as the microbial interaction in which both or all the microbial population involved gets benefitted, by supporting each other's growth and proliferation. These cooperative systems are ubiquitous in nature and are involved in various beneficial activities such as driving various biogeochemical processes, enhancing soil biomass and nutrients, promoting plant growth, degradation of food in the colon, waste water treatment, medicine and food industry. Therefore, in the present chapter we explore the different types of microbial interactions and cooperation between communities. Further, we also discuss the chemical basis of synergism and the factors which influences the synergistic process such as environment, substrate, etc. Finally, this chapter emphasizes the potential applications and future prospects of microbial synergism in the field of medicine, food, agriculture and environment.

Keywords Microbial interactions · Microbial synergism · Cocultivation · Biofilms · Microbial consortium

S. Sarsan (✉)
Department of Microbiology, St. Pious X Degree & PG College, Hyderabad, Telangana, India

A. Pandiyan
Food Safety Division, National Institute of Nutrition – ICMR, Hyderabad, Telangana, India

A. V. Rodhe
Department of Microbiology, Silver Jubilee Govt. College, Kurnool, Andhra Pradesh, India

S. Jagavati
Department of Microbiology, Indira Priyadarshini Government Degree College for Women, Hyderabad, Telangana, India

R. P. Singh et al. (eds.), *Microbes in Microbial Communities*,
https://doi.org/10.1007/978-981-16-5617-0_1

1.1 Introduction

Microorganisms are omnipresent and play significant role in human health, industry, disease, animals and plants. Biological interactions include the effects produced upon each other by the organisms in a community. Microbial interactions are ubiquitous, unique and diverse in operating biological community. Microbial interactions hold an important practical relevance to various fields such as bioremediation, forestry, biotechnology, agriculture, food processing and environmental conservation (Frey-Klett et al. 2011). Major advantages from these microbial interactions include host colonization, biofilm formation and enhanced competition between same organisms (Singh et al. 2016; Tshikantwa et al. 2018).

Many types of interactions are known to exist in nature. Microbes interact with plants, animals, humans, other microbes and various environmental factors. These interactions are thus basically of two types and include abiotic and biotic interactions. In Abiotic interaction, there is no biological component interacting with the microbes. Abiotic factors include nonliving parts of the environment factors like sunlight, oxygen, water, pH and temperature which will have a major impact on living organisms. Interactions in which the bio component is involved are called as biotic interactions. Biotic factors include all the living organisms of an ecosystem like plants, animals and microorganisms and their interactions among themselves. Dynamic interactions among microorganisms have resulted as a consequence of a long evolutionary history in which many of the biological components adjust to one another over time. These microbes usually appear as complex interactive networks in natural ecosystem rather than existing as single species (Tecon and Or 2017). Therefore, microorganisms have established intricate communication systems via secretion of chemicals thus allowing intraspecies and interspecies interaction (Scherlach and Hertweck 2018). Such delicately balanced population of microorganisms each influencing and interacting with other members of the population forms a microbial ecosystem. Table 1.1 explains clearly about various terminologies used in this chapter.

The biotic interactions between any two populations are classified based on whether any one or both populations of a particular association are affected in a positive or negative manner (Fig. 1.1). The figure depicts different types of microbial interactions, with 0 denoting no effect, – denoting negative effect and + denoting positive effect among the interacting species. There are six categories of ecological interactions:

1. Mutualism: positive–positive interactions
2. Competition: negative–negative interactions
3. Antagonism: positive–negative interactions
4. Commensalism: positive–neutral interactions
5. Amensalism: negative–neutral interactions
6. Neutralism: neutral–neutral interactions

Table 1.1 Terminologies related to ecosystem and interactions

Term	Definition	References
Biotic interaction	Interactions occurring among two living organisms.	García-Callejas et al. (2018)
Productivity	Increase in cell numbers or biomass over time that reflecting the efficacy with which organisms rise to biomass.	Schnurr et al. (2016)
Synergistic interactions	It leads to an improved fitness of the cooperating individuals relative to single cultures.	Ellis et al. (2015)
Antagonism	Interaction between two organisms in which one profits at the cost of the other (in terms of fitness); the behaviour is a resulting strategy.	Sorci and Garnier (2019)
Competition	Negative interaction amongst the organism resulting from overlying resource requirements or chemical warfare, resultantly reduction in fitness of the interacting organisms.	Martin et al. (2016)
Mutualism	Interaction between organisms of varied species that benefited both of them.	Grover (2008)
Amensalism	Interaction between individuals of different species in which the acting organism experiences no benefit or detriment and the recipient organism experiences a negative outcome.	Pacheco and Segrè (2019)
Community	A group of potentially interacting species that co-occur	Nemergut et al. (2013)

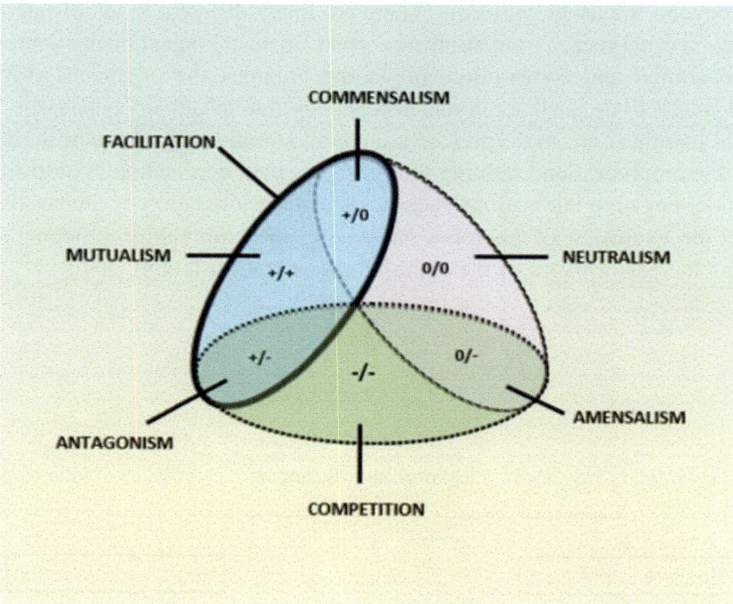

Fig. 1.1 Types of microbial interaction based on outcome of interaction. (Source: Adapted from Zélé et al. 2018)

These microbial interactions can happen between different species, same species, or even among microorganisms belonging to different families and genera. These interactions may be positive, negative or neutral. The term positive interaction can be defined as an "encounter between organisms that benefit at least one of the participants and cause harm to neither" while negative interaction can be defined as "an encounter between microorganisms that are detrimental to one of the participants" (Stachowicz et al. 2007).

1.2 Types of Microbial Interactions

There are many kinds of interactions among microorganisms which are immensely significant in the ecosystem functioning. This type of interactions existing among organisms can be either beneficial or detrimental to the host. Based on the positive or negative outcome of an interaction, there are different classes of microbial interactions. However, for a long time, microbial interactions were considered to be naturally inhibitory. But, recent innovations in microbiological research indicate that in a given natural environment, different microorganisms produce different products upon interactions thus proving to have wider applications and useful aspects beyond the usual antibiosis. There are many different kinds of interactions including synergism, amensalism, parasitism, antagonism, predation, protocooperation and competition, prevailing amongst the organisms (Fig. 1.2). Positive interactions such as commensalism or mutualism or synergism among microbial members are more prevalent and can significantly augment the productivity of the bioprocess and stability thus ensuring their application in industries. On the flipside, negative interactions such as parasitism, predation or amensalism that result in the exclusion of microbes sabotaging the community structure thereby disrupting the performance of the entire bioprocess (Ghosh et al. 2016).

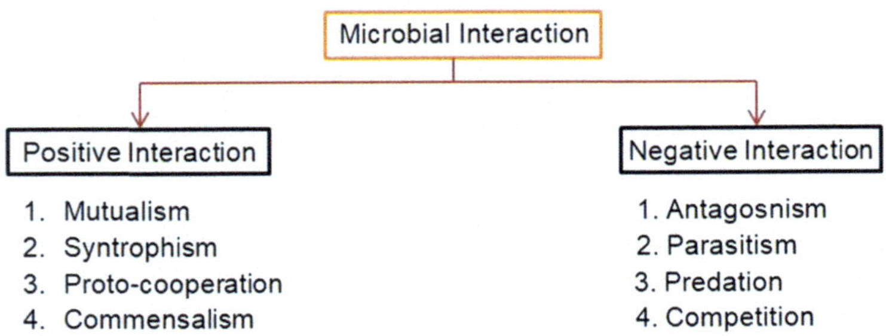

Fig. 1.2 Types of microbial interactions

1.2.1 *Positive Interactions*

The most common type of cooperative interactions seen among microbes are those of mutually beneficial. These positive interactions which are beneficial to both the participating species plays a key role in both evolutionary and ecological practices (Hernandez et al. 2019). There are many forms of positive associations benefitting the partners in association.

1.2.1.1 Mutualism

Among all the positive interactions, mutualisms has important part in determining the way different groups are organized and their performance (Traveset and Richardson 2014). Mutualism or symbiosis can be defined as an compulsory association where the mutualist and host are dependent on each other and signifies the interaction between the organisms, each getting benefited from the association. There exists a distinguishing feature between mutualism and its related term "symbiosis" which means the co-living of species within close proximity not necessarily benefitting each other. In mutualism, two communities interact with each other benefitting both the members accordingly (Fig. 1.3). There exists a common interest among the two interacting communities. Mutualism is a very specific type of relationship is which the members of association are consistent and other species of organisms cannot replace the existing organisms. In mutualism a specific, intimate physical contact is required between the organisms of association which allows them to carry out all activities as a single organism. Mutualism can be best explained by taking the example of Lichens. Lichens are an association including the biotic components fungi and algae belong to certain genera. The fungal component is termed as mycobiont and algal component is termed as phycobiont. The phycobiont may be a member belonging to green algae and cyanobacteria.

Microorganisms secrete a wide range of molecules into the environment making the nutrients available to the organisms in vicinity. They include amino acids, fermentation products like acetate, and electrons in the form of hydrogen or other

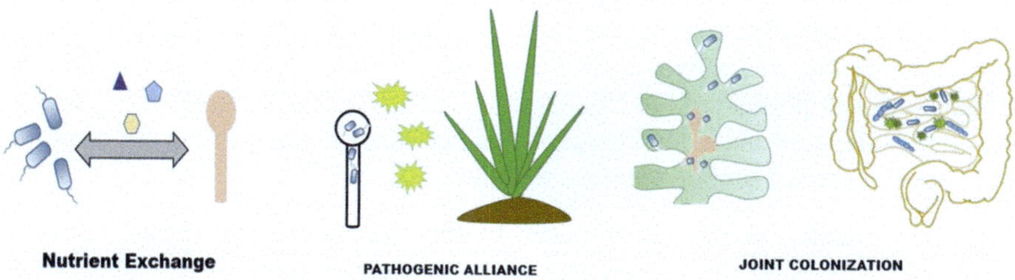

Fig. 1.3 Different forms of microbial mutualism. (**a**) Shared metabolism (**b**) Pathogenic association of bacteria and fungi (**c**) Joint colonization of niches. (Source: Adapted from Scherlach and Hertweck 2018)

molecules. In such cross-feeding interaction, one species benefits without affecting the other (Hillesland et al. 2011). For example, bacteria produce certain metabolites called siderophores during mutualistic microbial interactions. These molecules are exclusively secreted by the cell into the exterior environment transported via plasma membrane. Based on the need of one organism with other, mutualists may be obligatory or facultative. Facultative mutualism is that in which one organism didn't requires other for growth or reproduction but one species gets benefit from the other partner such as an increase in growth or capability. Obligate mutualists are those which require the interaction of the their partner for growth, reproduction and other life cycle functions (Mittelbach and Vannette 2017). Mutualisms also vary in specificity and are either species-specific or guild-specific. In species-specific based mutualisms, a precise partner is obligatory to fulfil activities that facilitate the mutualistic exchange. Whereas in guild-specific mutualism which is a more generalized and diffuse type, organisms within a league of functionally similar species will interact and benefit from each other (Crowley and Cox 2011) The evolution of mutualism in ecological communities is driven by the competitive process of natural selection (Darwin 1859). Long-term mutualisms may end when one of the partner ceases to require or to benefit from the other partner (Leigh 2010).

1.2.1.2 Syntrophism

The term "syntrophism" is defined as a cooperation in which the two organisms depend on each other to complete a metabolic activity and simple addition of a co-substrate or any nutrient will not suffice the mutual dependence (Schink and Stams 2006). Syntrophism can also be defined as "obligate mutualistic metabolism" (Morris et al. 2013). They are also considered as facultative mutualists as their survival depends on certain factors such as disease suppression and host organism and also environmental conditions (Singh et al. 2020). However, some of the classical definitions for syntrophism are given in Table 1.2. Syntrophy is a widespread phenomenon, because all organisms are interdependent in order to make sure the life does not stagnate. In Syntrophism, the growth of one of the organisms is dependent on the nutrients/substrates released by another organism. Syntrophism is an association in which both the associated organisms are benefitted from each other. Some of the syntrophic activities include: (1) Production of degradative or hydrolytic enzymes by soil microbes like *Arthrobacter* and *Streptomyces*, acting in conjunction with pesticides. (2) Utilizing toxic end products for their own metabolism, (3) chemotactic interaction between algal population and bacteria. Some of the examples of syntrophism are mentioned below:

1. In food industry, yoghurt is produced by the process of syntrophism, with the mutual cooperation of *Lactobacillus delbruecki* subsp. *bulgaricus* and *Streptococcus thermophillus* (Azam et al. 2017). This interaction involves the *L. bulgaricus*, which produces proteases from cell wall, in return for nutrients from *S. thermophillus* (Sieuwerts et al. 2008).

Table 1.2 Classical definitions and illustrations of syntrophy

Definition	References
Cooperations in which two organisms depend on each other to accomplish a particular metabolic activity and addition of a co-substrate or any nutrient will not overcome their mutual dependence.	Schink (1997)
Strongly coupled mutualistic interaction which is essentially proving its role in global carbon cycling under anaerobic environment.	McInerney et al. (2009)
The low concentration of degradation end products (hydrogen/formate/acetate) is essential to bring about the degradation of fatty acid and other compounds which is a thermodynamically interdependent	McInerney et al. (2011)
A nutritional condition where a substrate cannot be catabolised by either of the organism, rather, it is brought about by the joint metabolic capabilities of two or more organisms	Stams and Plugge (2009)
Associations where two organisms rely on each other for energy purposes to accomplish a fermentation process together, which none of them can do on their own	Schink (2002)

2. The bacteria like *Enterococcus faecalis* and *Lactobacillus arobinosus* are capable of growing together but not individually in minimal medium. The phenomenon of synergism is shown by two organisms: *E. faecalis* and *L. arobinosus*. *E. faecalis* requires folic acid produced by *L. arobinosus* which is dependent on phenylalanine formed by *E. faecalis*.
3. Some of the other applications or examples for syntrophism includes the biodegradation of aromatic and polyaromatic compounds under anaerobic condition (Berdugo-Clavijo et al. 2012; Fuchs et al. 2011), degradation of oil (Jones et al. 2008) and amino acid degradation (Schink and Stams 2006).

1.2.1.3 Protocooperation

This is a synergistic interaction/relationship in which both the organisms which are associated are mutually benefited from each other. The relationship between the organisms in protocooperation is similar to mutualism except that they are not obligatory as in mutualism. The synergistic association of *Chromatium* and *Desulfovibrio* involves protocooperation between the sulphur cycle and carbon cycle. Similarly, an interaction between *Cellulomonas*, a cellulolytic bacteria and N_2-fixing bacteria is a good example of protocooperation activity.

1.2.1.4 Commensalism

Commensalism is defined as the interaction between two microbes in which commensal gets profited, while the host remains unaffected, i.e. neither has beneficial nor harmful effect (Mathis and Bronstein 2020). This interaction is unidirectional and if the host and commensal are separated, the commensal is capable of surviving.

Table 1.3 Reported cases of commensalism in different biosystems

Interacting species partners	Interaction mechanism	Reference
Host for Mouse: *B. thetaiotaomicron*	Secretion of specific nutrients by mouse host for selective uses by bacteria	Hooper et al. (2000)
G. candidum, P. camemberti	*G. candidum* Presence boost the growth and metabolite production by *P. camemberti*	Aziza and Amrane (2006)
Pseudomonas sp. AS1: *A. oleivorans* DR1	Naphthalene degradation to salicylate by *Pseudomonas* sp. AS1 which favour the growth of *A. oleivorans* DR1 in naphthalene comprising medium	Seo et al. (2012)
Chlamydomonas reinhardtii: Stenotrophomonas sp. or *Pseudomonas* sp.	During hydrogen production, the hydrogenase enzyme of the algae is activated due to anoxic habitat which is created by bacterial species through respiration	Li et al. (2013)
Rhodotorula mucilaginosa R30: *Acidithiobacillus* sp.	During tannery sludge bioleaching process, Dissolved organic matter (DOM) which is repressive to *Acidithiobacillus* sp. And is degraded through heterotrophic *R. mucilaginosa* R30	Wang et al. (2010)

Commensalism does not involve physiological interaction or dependency between the host and the commensal. In commensalism, the two partners can survive independently and yet the commensal may be able to feed on elements/nutrients ingested by the host because of their spatial proximity. There are many examples of commensalism existing in different biosystems (Table 1.3). Commensalism may involve different mechanisms and are of different types.

1. Phoresy is a type of commensalism in which one microbe called as the phoretic is transported by other microbe mechanically—the host, without exercising nutritional or developmental consequences on the host (Houck and O'Connor 1991).
2. Inquilinism is a type of commensalism in which one of the species is used as a platform or cavity for the living condition of the recipient microbe.
3. Chemical commensalism is another type which is most often associated with two bacterial species but may not be always the case. In this type, one bacterium break down a chemical which is invaluable to the other species, and produces a metabolite that will be used as an energy source by the beneficiary second species (Veiga 2016).

Some of the representative examples of Commensalism are given below:

1. *E. coli* (*host*)—Non-pathogenic facultative anaerobe, in the intestinal tract of human, utilizes molecular oxygen and declines the Oxygen concentration. This creates a conducive environment where obligate anaerobes such as *Bacteroides* (commensal) start growing.
2. Commensalism between *Nitrosomonas* (host) and *Nitrobacter* (commensal) in the process of Nitrification: Ammonia is oxidized into Nitrite by *Nitrosomonas*

and this nitrite is used as a substrate by *Nitrobacter* for energy and thus oxidizes it into Nitrate.

3. In Swiss cheeses, the Lactic Acid Bacteria (LAB) produces lactic acid which used by propionic acid bacteria (Mounier et al. 2008). Similarly, in surface-ripened cheese, the lactic acid produced by LAB is being metabolized by *Geotrichum candidum and Debaryomyces hansenii* (Mounier et al. 2005).

1.2.2 Negative Interactions

Negative interactions are a type of interaction between the two microbial populations, where one population of organisms either attacks or inhibits the other organisms for the survival and food source. Negative interactions include many types such as antagonism, parasitism, predation and competition.

1.2.2.1 Antagonism

Antagonism or Ammensalism is a negative relationship in which two types of microbial populations exist in which one of the populations secretes substances that may be inhibitory or lethal to the other population (Prajakta et al. 2019; Singh et al. 2019). The Population which secretes or produces inhibitory compounds are benefited i.e. they are not affected or may exist in competition and successfully survive in the environment, at the same time inhibiting other population. This type of phenomenon where one of the populations is inhibited by biochemical molecules is called as antibiosis. Various types of antagonistic interactions existing in bacteria include inhibition of growth by antibiotics and/or hydrolytic enzymes, inhibition of adjacent cells. Their antagonistic activity may be limited to killing within the species or may also be able to kill between species from different genera, families and orders (García-Bayona and Comstock 2018).

Examples of Antagonism

1. Many normal flora organisms like Lactic acid bacteria (LAB) produce lactic acid in genital tract of females which is antagonistic to many pathogens like the fungus, *Candida albicans*.
2. Many pathogenic bacteria invading the skin are inhibited by several fatty acids produced by normal flora of skin (Nipa 2015)
3. *Thiobacillus thioxidans* produces sulphuric acid as a result of sulphur oxidation. This leads to lowering of pH in the medium and thus inhibits other microorganisms especially most bacteria.

Antagonism involves killing of competing organism with the release of extracellular compounds such as antibiotics or toxins. Earlier studies have shown that bacteria produce large group of antibacterial peptides and proteins by using the secretion systems to deliver these toxins to competing cells. There are diverse types of antimicrobial peptides and proteins produced by bacteria which vary in their structures and functions, cellular targets, mechanisms of action and spatial range. The most common type is "Bacteriocin" which are defined as a varied group of peptides or antibacterial toxins produced by bacteria (Schaechter 2009). Bacteriocins are ubiquitous and produced by almost all major groups of eubacteria and archaebacteria (Besse et al. 2015). On the contrary, bacteriocins characteristically have a narrow-spectrum of activity, targets and kills thoroughly related members of the producing strain (Cotter et al. 2013). Bacteriocins employ different mechanisms to kill the bacterial cells such as via pore-formation, degradation of peptidoglycan, cell wall synthesis and protein synthesis inhibition, nuclease activity and gyrase inhibition (Heng et al. 2007). These toxins produced by microbiota helps in evading pathogens and thus find potential applications in medical, agricultural and other industrial sectors (Pérez-García et al. 2011). A myriad of natural bioproducts produced by some fungi and bacteria are potential sources of important antibiotics and therapeutic drugs-like anticancer agents, immune suppressors, cholesterol controlling drugs, anaesthetics etc. these are used routinely to treat specific diseases and other medical conditions (Medema and Fischbach 2015).

1.2.2.2 Parasitism

Parasitism is the relationship between two different organism types whereby one organism (parasite) gets the benefit and acquires the required nutrition from the other (host) in the interaction/association by inducing damage to it, which gets harmed. The relationship between host and parasite may be physical or metabolic and usually exists for an extended period of time. Some parasites living external to the host and are known as ectoparasites while many live inside the host cell and called as endoparasites.

Examples of Parasitism

1. Endoparasite: Most common microbial parasitism in nature is represented by bacteriophages, viruses infecting bacteria. Viruses may be defined as acellular, obligate, intracellular parasites and host specific. Viruses are parasites to a variety of hosts like protozoa, cyanobacteria, bacteria, algae, fungi etc.
2. Ectoparasite *Bdellovibrio*: Many G –ve bacteria are infected or parasitised by *Bdellovibrio*

1.2.2.3 Predation

It is association among two microbes where one organism (predator) attacks or engulfs other organism (prey) usually resulting in death of the prey. This predator-prey interaction is of short duration. Bacterial predators produce an extensive range of secondary metabolites and degradative enzymes in order to lyse or kill other organisms (Jurkevitch 2006; Berleman et al. 2008). Bacterial predation also perform a significant role in advance of human pathogens (Erken et al. 2013). Based on predator-prey interaction, there are three groups among bacterial predators (Pérez et al. 2016).

1. Epibiotic predation: Predators consume the preys of outside before dividing into offspring cells and they remains have close proximity to the prey cell envelope
2. Endobiotic predation: Individual predatory cell causes direct invasion to modify prey cell wall by secretion of hydrolytic enzymes and penetrate into cytoplasm or the periplasmic space.
3. Group attack: The minimum colonies of predators are vital to kill and uptake the prey cells by generating the hydrolytic enzymes and metabolites.

Based on the differences in killing methods used these may be further subdivided in each group.

Examples of Predation

1. Protozoans–Bacteria interaction. The protozoans being larger microorganisms eat up or engulf bacterial cells present in soil. This declines the numbers of bacteria and thus leads to optimal levels of soil bacterial populations.
2. *Daptobacter*, *Vamparococcus*, *Bdellovibrio*, are common examples of bacteria acting as predators, which can feed on a variety of bacterial populations.

1.2.2.4 Competition

This type of interaction is a negative interaction where two microbial populations compete with each other. During this interaction, both the populations are affected negatively threatening their growth and survival in that environment. The organisms compete with each other for resources like nutrition, space, etc. which results in lower maximum density or growth rate of the microbial populations. They compete for carbon and nitrogen source, phosphorus, vitamins, growth factors and many other growth-limiting resources. Competition inhibits the occupancy of the same ecological niche by both the microbial populations, as one type will get successful, while the other gets eliminated. Competition can be exemplified by the organisms *Paramecium aurelia* and *Paramecium caudatum* both of which feed on the same bacterial population. *Paramecium aurelia* shows better growth rate than

Paramecium caudatum. In competition between microbial populations, bacteria employ several antimicrobial mechanisms such as secretion of compounds to kill or damage the neighbouring target cells (Hibbing et al. 2010). In highly competitive environments, the genetic persistence and survival of a species depends on their fight for resources (Bauer et al. 2018). The competitive phenotypes in such environment evolved in such a way to outcompete and displace their neighbouring species, in the environment of scarce nutrition and limited space. There are two types of competition, one is exploitative competition in which one of the participant species consume the resources and the other is Interference competition in which the cells invade and damage one another (Ghoul and Mitri 2016).

1.2.3 Microbial Synergism

Microbial synergism is an important positive interaction. Synergism is defined as the interaction in which two participating species supports each other by creating favourable conditions for growth and proliferation. Synergistic relationship occurs when two organisms grow better together than apart or when the by-products of one organism enhances the survival of another. In this relationship, members of an association receive benefits that exceed those that would result if each lived by itself. The cooperation between the microorganisms may be of many different ways and mutual association among the partners may differ from only minimal support to total mutual dependency (Fig. 1.4). Some of them include: (1) production of compounds accomplished in cooperation rather than individual organism as in syntropy, (2) affinity of microorganisms to each other as seen in the microbial consortium like bacterial cells attached to algal cell surfaces regarding chemotactic interactions among them, (3) capability shown by certain microorganisms to derive

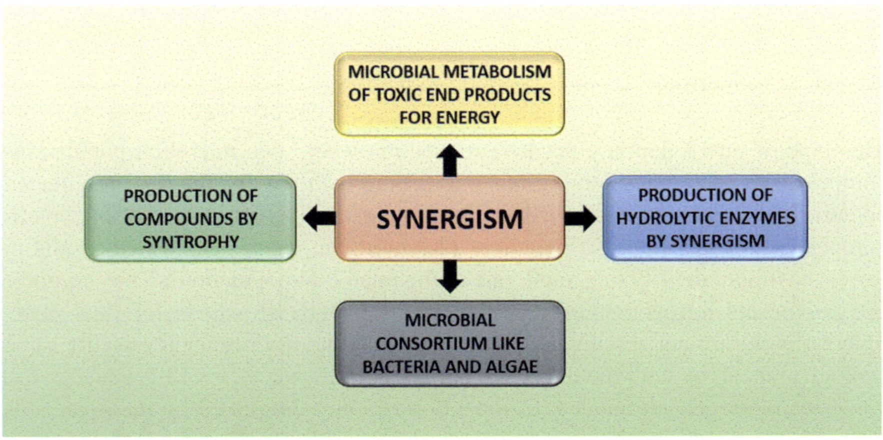

Fig. 1.4 Different mechanisms of Microbial Synergism

their energy from metabolism of toxic end products. (4) Synthesis of hydrolytic enzymes in the soil environment, for example by *Arthrobacter* and *Streptomyces*, show a combined activity of degradation of organophosphate pesticide and diazinon.

Many cases have been reported where one microorganism secretes metabolites such as vitamins or amino acid precursors, that may be beneficial to another organism that is deficient in specific pathway synthesis. Sometimes the partner in association might synthesize the particular compound individually, yet it depends on the other partner only to save its biosynthetic energy. There may be more strong types of cooperation and mutual interdependence observed among partners. In metabiosis type of cooperation, the latter partner gets profits from the former one in an interdependence metabolic pathway, but the former members may have insignificant or negligible advantage from the later partners. Also, the synergism includes strict syntrophic associations, where one organism depends on other for their growth and energy purposes and carry out a fermentation process together but neither of them can live on their own as seen in typical syntrophic associations (Schink 2002).

1.3 Chemical Basis of Synergism

Microorganisms which may include bacteria and fungi most significantly, function as populations of same organisms and as groups of different microorganisms. The evolution of such microbes happened due to a dynamic interaction between inorganic components and other higher organisms. They play a very important and beneficial role in the environment. The microbial interactions pave the way to working of the ecosystems, or self-controlling biological groups and their physical environment. These interactions also help in understanding the processes such as their role in environment and disease development. The important aspect of the understanding of microbial interactions is that the microbes constantly encounter their environments and bring about changes in the ecosystems to a greater extent. In nature, degradation of compounds is accomplished by complex microbial communities that work synergistically and efficiently. For instance, in nature, many bacterial and fungal species are capable of hydrolysing the lignocellulosic biomass via synergistic relationship through production of enzymes cellulases and xylanases (Tsegaye et al. 2018, 2019).

Microbial interactions are regulated by various chemical, biological, physical and genetic factors (Tshikantwa et al. 2018). The effects of these various factors are described in Table 1.4. The nature and the chemical intricacy of the substrate and nutrients usually affects the type of the interactions among microbes. A complex lignocellulose substrate may promote and encourage synergistic growth and positive interactions, while a simple substrate such as glucose supports negative interactions and competition among microbes. The type of substrate and the level of interaction were observed to be strongly related. The more complex the substrate like wheat straw, the stronger the synergistic relationship (Cortes-Tolalpa et al. 2017). In

Table 1.4 Factors affecting microbial interactions

S. No.	Factors	Effects and examples
1.	Physical factors	
	Salt concentration	Variation in salt concentrations influences the spoilage by microbes.
		Accumulation of spoilage yeasts in low salt concentrations as seen in baceman spoilage
	Temperature	Influences the actions of interacting microbial enzymes. Enzymes show enhanced microbial activity at optimum temperature. E.g. Japanese *shoyu* (soy sauce) production
	pH	Effects the capability of microbes to endure and live in a variety of environmental niches. Favourable optimum acidic condition causes LAB to thrive more during fermented milk products manufacture.
	Dissolved Oxygen	Concentration of dissolved oxygen is an important feature during microbial interactions within a community and is directly related to microbial growth.
		Growth of coryneform bacteria in baceman preparation favoured by high O_2 concentration
2.	Nutrients	Specific nutrients such as water, nitrogen, vitamins, and minerals and energy source are required to conduct several metabolic actions thus affecting the growth of diverse microorganisms. There exists a direct relationship between the concentration of nutrient and attached bacterial cell numbers. Thus, an increase in nutrient results in rise in number of cells.
3.	Chemical factors	Secondary metabolite produced by microorganisms play the crucial role in mediating complex microbial interactions. The interaction at molecular level amongst different microorganisms has resulted in a variety of chemical diversity which in turn lead to interplay of various microbes at the molecular level.
4.	Biologic factors	In cell-cell adhesion the interactions among microbial cells occur by adhering to surfaces or a metabolic substrate. Microbes also respond to environmental stimuli such as transfer of proteins and small molecules.
5.	Genetic factors	Provision of understanding of chemical signalling, horizontal gene transfer, motility, chemotaxis, pathogenesis, microbial viability and persistence.

contrast to this, when glucose alone was used there was no synergistic relationship found. A huge group of organisms show an inherent complexity of being diverse in their chemistry by producing varied enzymes. This diversity is desirable to degrade efficiently the various complex substrates or naturally occurring polymers into monomers. The degradation by these microbial communities in nature is brought about in a dynamic and time-dependent manner. The organisms which carry out degradation show dynamic interactions amongst themselves and show more activity when working in coordination or combination with each other than without help. This method is identified as synergistic growth often seen in microbial communities and closely linked with enzymatic activities (enzymatic synergism) (Cragg et al. 2015; Van Dyk and Pletschke 2012). Thus synergism in microbial consortia is of considerable importance, as projected in the recent research studies (Deng and Wang 2016; Mitri and Richard Foster 2013).

1.3.1 Cocultures

Any complex substrate in nature can be degraded or depolymerised by coexisting microorganisms which work more efficiently by synergistic metabolic activities. The organisms follow the process of niche partitioning to bring about the final result. The niche partitioning is signified by metabolic complementarity. One such classical example of complementarity is reported where rice straw substrate was hydrolysed and degraded for the production of bioethanol and hydrogen using *Bacillus* and *Clostridium* cocultures (Chang et al. 2008). Synergism between native bacteria is known to be significant in the hydrolysis of lignocellulosic substrates (Deng and Wang 2016). In the studies carried out by (Cortes-Tolalpa et al. 2017), the degradation potential of both monocultures and mixed cultures was examined and the results obtained with cocultures were observed to be more when compared to single cultures. The studies were performed to determine the enzymatic activities of *Citrobacter freundii/Sphingobacterium multivorum* biculture, the studies showed increased production of the degradative enzymes, cellobiohydrolases, mannosidases, and xylosidases. Metabolic complementarity was also reported in cocultures of *Trichoderma reesei* and *E. coli* in the degradation of pretreated corn stover and final optimal production of isobutanol (Minty et al. 2013). *Trichoderma reesei* synthesized the cellulolytic enzymes for conversion of corn stover into its constituent sugars. These sugars were further fermented by *E. coli* into isobutanol. From the above examples it is evident that the synergistic pairs or cocultures drive important chemical transformations by exchanging key metabolites or by niche partitioning especially when the independent organisms are lacking in particular metabolic pathways. The chemistry of transformations also depends upon the complexity of the substrate or the raw material source available for the varied kinds of microorganisms in a niche.

1.3.2 Biofilms

Biofilms are another classical example of microbial consortia. The development of biofilms is a well strategized process and specific controlled gene expression (McDougald et al. 2012). Microorganisms attach themselves to a biotic or abiotic surface and develop biofilm, by generating extracellular polymeric substances (EPS) (Donlan 2002). The organisms participated in the biofilm are protected from metals, antibiotics desiccation, biocides, ultraviolet (UV) rays and host defence system (Flemming and Wingender 2010). In the evolutionary perspective, biofilm is an important example on how microbial interaction helps the organisms to thrive under

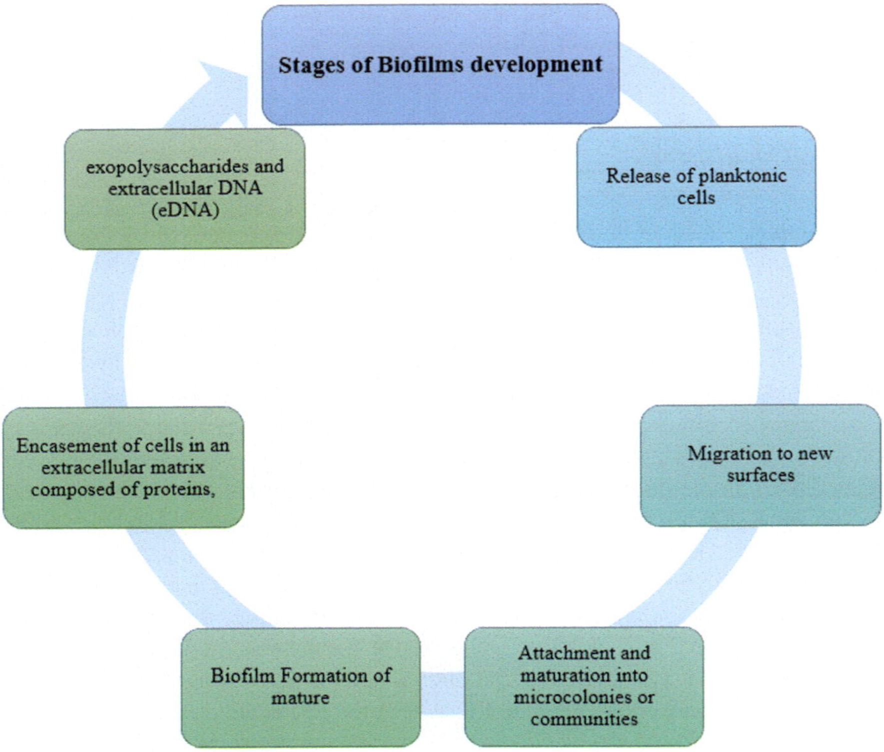

Fig. 1.5 Stages of biofilm development

extreme conditions by increasing their resistance against UV radiation (de Carvalho 2017), high temperature and pH (Harrison et al. 2007; Hoštacká et al. 2010), high salinity (Kim and Chong 2017), poor nutrition (Marsden et al. 2017) and different antibiotics (Hathroubi et al. 2017) through mutual cooperation. However, in food industry, these mixed species biofilms can cause equipment damage, food spoilage and also food borne diseases (Yuan et al. 2019).

Biofilms are highly structured and organized microbial communities which are closely connected to each other and produces the extracellular polymeric substance (EPS). The biofilm development consists of different stages such as Initial adhesion, Microcolony formation, Biofilm maturation and Dispersal (Figs. 1.5 and 1.6). The components of mature biofilms include both microbial cells and EPS matrix which is a scaffold for holding water, microbial cells, polysaccharides, proteins, DNA, RNA, ions in varied percentages and various biologically active molecules like cell communication signals (Flemming and Wingender 2010). For the microbial communities, production of biofilms is of great advantage to overcome and protect themselves from severe environments, lenience to physical and chemical stress leading to metabolic cooperation and community-coordination in the fine-tuning of gene expression. Biofilm mode of growth helps with transitory multicellular behaviour

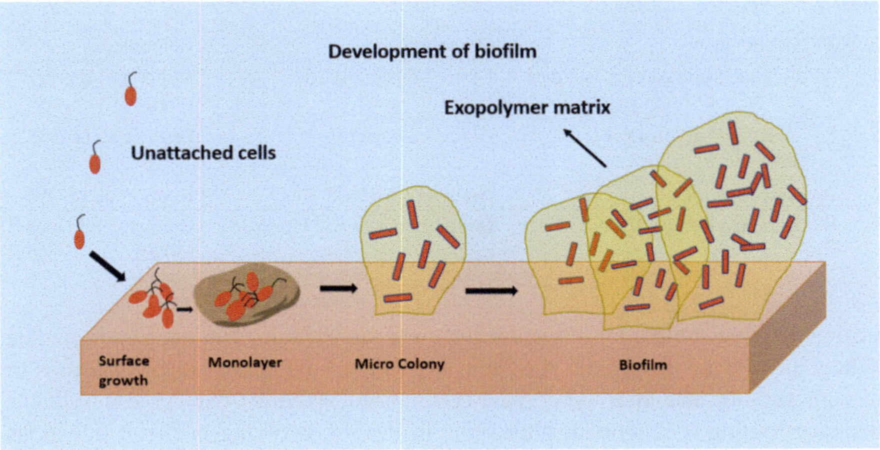

Fig. 1.6 Development of biofilm

causing an upsurge in local concentration of nutrients. The matrix of biofilms performs the role of digestion system. It accumulates digestive and degrading enzymes that can degrade various matrix components including complex nutrients and other substances. Degradation helps the products to be in close proximity to the microbial cells, further facilitating uptake of nutrients. Antimicrobial resistance is also reported. The matrix also has a significant role in shielding the cells against antimicrobials, toxins, and predators. Biofilm pathogens can tolerate host defence system. Biofilm microorganisms display collective and coordinated behaviour as they are adapted to their physiology and stress responses (Van Houdt and Michiels 2010; Zupančič et al. 2018). Opportunity of genetic material exchange is also observed to be on the higher side in biofilms.

1.3.3 Quorum Sensing

Wide range of chemical diversity has been observed among different microorganisms due to molecular level interactions. Regulation of interactions among microbes occurs even at molecular level and Quorum sensing is a unique physiological mechanism which is efficiently used by many bacteria. Several bacteria can show a good control over their synergistic activities and carry out the required specific physiological/cellular functions by the mechanism of quorum sensing (QS). The bacteria interact with each other through the sensing and exchanging of signals or chemical information. This is evident by secretions, sensing and response to small chemical signal molecules which can diffuse into the surroundings. These interactions are very useful to the organisms in the process of colonization, biofilms, increased adjustment and competitiveness in the changing environment. The quorum

Table 1.5 Types of autoinducers synthesized by bacteria

S. No.	Autoinducer	Organism	Reference
1.	Acylated homoserine lactones	Gram -ve bacteria	Miller and Bassler (2001)
2.	2-heptyl-3-hydroxy-4-quinolone (PQS)	Gram -ve bacteria	Pesci et al. (1999)
3.	Oligopeptides	Gram +ve bacteria	Rojas et al. (2019)
4.	Furanosyl borate diester	Both Gram +ve and Gram -ve bacteria	Bhuiyan and Noor (2020)

sensing activities hold a lot of significance in understanding virulence and possible pathogenic effects of bacteria. An inclusive study of the molecular mechanisms of quorum sensing and their synergistic coordinated activities may give a valuable evidence leading to important alterations in dealing evolving bacterial infections (Li and Tian 2016).

Autoinducers are an important group of chemical molecules which can act as signal molecules (Table 1.5). During the active bacterial growth, the cells synthesize these autoinducers which are involved in quorum sensing. These molecules act as mediator signalling molecules (Bouyahya et al. 2017) and they aid in communication and signalling between intra and interspecies bacterial communities (Rao and Kumavath 2020). As a response to alterations in cell population density, the autoinducers are synthesized. As the quorum sensing bacterial cell density increases, there is an increase in concentrations of autoinducers. Some of the unique and well-studied autoinducers are ACL (*N*-acyl homoserine lactone) from G −ve bacteria like *E. coli* (Lowery et al. 2008; Miller and Bassler 2001), oligopeptide-based autoinducers in gram-positive bacteria. Gram −ve and Gram +ve bacteria may also use other common auto inducers apart from Lactones such as autoinducer 2 also known as borate furanosyl (Bouyahya et al. 2017). An extensive study of how QS mechanisms work in bacteria for signalling has to be done to get a more understanding of the microbe-microbe interaction and further application of the same.

1.4 Potential Applications of Synergistic Interactions Among Microbial Communities

Many early studies were made by scientists to understand and explain the applications of Synergism. Synergism is better understood by the experiments conducted by (Sears and Putnam 1923) who were the first ones to study the concept of synergism. They have worked with many pairs of microorganisms and stated that the microbial pairs when present together produced gas utilizing the carbohydrate present in the nutrient medium and the gas was not produced in the absence of any one of the organisms or when they are grown individually in the nutrient medium (Table 1.6).

Table 1.6 Bacterial pairs that produce gas in synergism association

Carbohydrate	Organisms
Lactose	*Staphylococcus aureus* + *Salmonella schottmuelleri*
	S. faecalis + *S. schottmuelleri*
	S. faecalis + *S. choleraesuis*
	S. aureus + *P. vulgaris*
	S. faecalis + *P. vulgaris*
	S. faecalis + *S. paratyphi*
Sucrose	*S. aureus* + *E. coli*
	S. faecalis + *E. coli*
	S. equinus + *S. schottmuelleri*
	S. equinus + *S. paratyphi*
	S. aureus + *S. paratyphi*
Mannitol	*S. aureus* + *P. vulgaris*
	S. faecalis + *P. vulgaris*
	S. pyogenes + *P. vulgaris*
	Shigella paradysenteriae + *P. vulgaris*
	Eberthella typhosa + *P. vulgaris*

Source: Adapted from AJ Salle (1943)

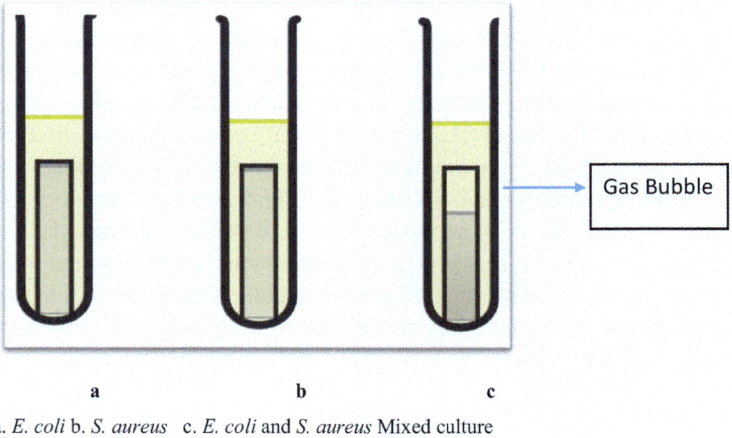

Fig. 1.7 Bacterial synergism in sucrose broth. (**a**) *E. coli* (**b**) *S. aureus* (**c**) *E. coli* and *S. aureus* mixed culture

Their explanation to this concept is that one of the organisms produced acid and the other organism has produced the gas. The acid forming organism has utilized the carbohydrate and released an intermediate product of the metabolism which acts as a substance for the second organism to release the gas. Example of sucrose utilization by *S. aureus* and *E. coli* is revealed in Fig. 1.7.

In furtherance to their studies, there were studies conducted on two bacteria namely, *Eberthella typhosa* and *Proteus morganii* which in synergistic association produced gas from mannitol. It is observed that *E. typhosa* produced intermediate compound from mannitol that was found to be stable at 100 °C. When the sample was heated to inactivate the culture of *E. typhosa* and later inoculated with the culture of *P. morganii*, the intermediate compound was fermented to produce gas similar to the one produced from combination of the two organisms when grown together (Castellani 1926; Graham 1932). Further it is reported that in presence of calcium carbonate, the gas produced was greater than before and is known to neutralize the action of acid produced by *Eberthella typhosa* which is why this organism remained actively growing and continued to ferment carbohydrate for long time. The greatest importance of synergism is perhaps observed in the field of bacteriological examination of water as seen in many research studies (Atkinson 1935; Atkinson and Wood 1938). More significant studies were taken up on these bacterial associations that are of common occurrence so that the role played by them in their natural environment is well understood.

The reports from recent research studies about the applications of synergistic microbial interactions in the fields of food, health, drug, agriculture and environment support the early studies conducted on synergism. The latest developments in research have proved that there are many classes of microbes within our surroundings which are capable of producing a variety of products through interactions with each other. This paves the way to an extensive range of potentially valuable applications. However, the interactions among microbes in nature have many deleterious effects also. There are many examples of microbial interactions in nature within prokaryotes or eukaryotes and also between prokaryotes and eukaryotes. The associations among bacteria and fungi are abundantly found in many environment including humans (Rashid et al. 2016). Microorganisms existing in synergistic associations produce different products that are beneficial to mankind in health, medicine, food, agriculture, and environment (Fig. 1.8) (Tshikantwa et al. 2018). The applications include many processes such as described in Table 1.7. Some of them include such as the synergistic association among gut microbiota influencing human health, the association among environmental microbes helping in regulating the sustainability of ecosystem, and various industrial processes making use of this microbial association in the production of valuable products (Zaccaria et al. 2017).

1.4.1 Role of Microbial Synergism in Health and Disease

1.4.1.1 The Human Microbiota in Health

The microorganisms that exist in human body are known as human microbiota and the number of microbiota is estimated to be approximately 10^{13} to 10^{14} microbial cells including protozoa, eukaryotic organisms, archaea, viruses and predominantly bacteria (Rocha 2016). They reside symbiotically on various parts of human body like skin, oral cavity, respiratory tract, gastrointestinal tract and urogenital system (Lloyd-Price et al. 2016). The gut of humans has been considered as an essential

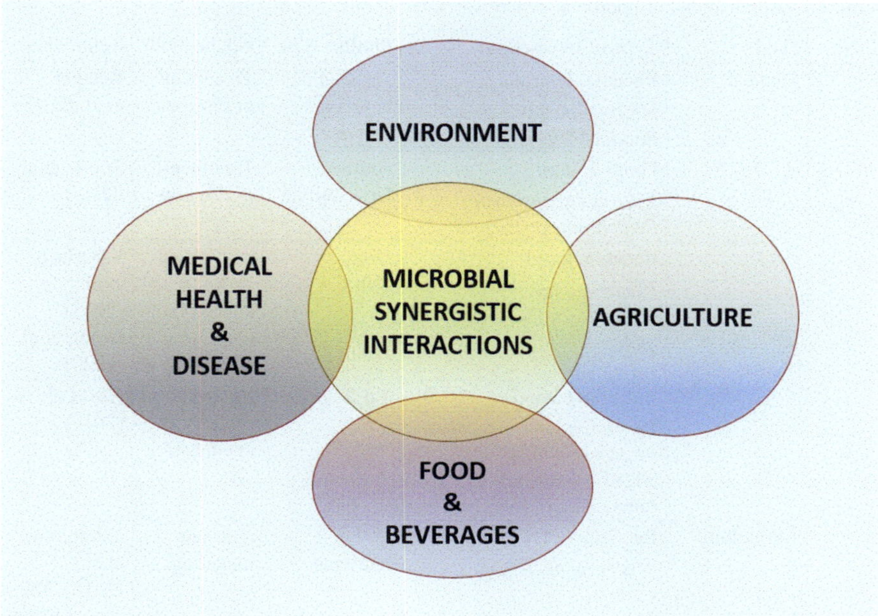

Fig. 1.8 Applications of synergistic interactions

organ, which is inhabited by more than 100 trillion symbiotic microorganisms which constitutes the gut microbiota. The gastrointestinal system has diverse microbiota comprising mainly of bacteria belongs to *Firmicutes*, *Bacteroidetes* and *Actinobacteria* phyla (Tap et al. 2009).

The microbiota of gut is in symbiotic association with the host and this relationship is controlled and steadied by the interactions of a complex network between it. Various interactions include metabolic, immune and neuroendocrine that are potentially facilitated by metabolites secreted by microbes. These metabolites act as signalling molecules and exhibit pleiotropic effects such as in controlling the host neuro-immune-inflammatory responses that could physiologically link gastrointestinal system with other systems (Cani 2018). The chief role of microbiota of gut on the host and the significant metabolites produced by them which are involved in maintaining host wellbeing are described in Table 1.8. Gut microbiome such as *Bacteroides*, *Bifidobacterium* and *Lactobacillus* species helps in metabolizing dietary fibres (xyloglucans) commonly found in vegetables and other foods that are indigestible by the stomach as well as small intestine. The gut microbiota also supplies essential nutrients and also prevent the colonization by opportunistic pathogens and thus finally aid in the formation of intestinal architecture. Many Studies have elucidated that the gut microbiota synthesize essential nutrients and vitamins and are involved in maintaining lipid and protein homeostasis (Goh and Klaenhammer 2015). The normal microbiota of gut releases fatty acids such as propionic acid, acetic acid and butyric acids. These SCFAs serve as source of energy

Table 1.7 Applications of microbial synergistic interactions

S. No.	Field	Type of interaction	Product type	Application	References
1.	Medical	Bacteria–Bacteria *Micromonospora* sp. and *Rhodococcus* sp.	Antibiotic keyicin	Treatment of diseases	Netzker et al. (2018)
		Fungi–Fungi *Xylaria cubensis* and *Penicillium restrictum*	Antibiotic Griseofulvin	Treatment of diseases	Stierle et al. (2017)
		Bacteria–fungi *Pseudomonas aeruginosa–Candida albicans*	Drug 5MPCA	Antifungal agent	Scherlach et al. (2013), Doing et al. (2020)
		Bacteria–bacteria *Propionibacterium freudenreichii* DSM 20271–*Lactobacillus brevis* ATCC 14869	Vit B12	Essential product synthesis	Xie et al. (2019)
2.	Agriculture	Bacteria–fungi *S. cerevisiae–R. etli*	Soil nutrients	Nutrient recycling	Tshikantwa et al. (2018), Xie et al. (2018)
3.	Environment	Bacteria–bacteria: *Brachymonas denitrificans B79, Comamonas denitrificans 110, Acinetobacter calcoaceticus ATCC23055* and *Aeromonas hydrophila L6*	Clean water	Waste water treatment	Tiwari and Lata (2018)
4.	Food and Beverage production	Fungi–Bacteria: *Pichia kudriazevii YS201* and *Bacillus subtilis BS38*	Alcohol	Production of alcoholic beverages	Ouattara et al. (2020)
		Bacteria: Bacteria *Lactobacillus delbrueckii* subsp. *Bulgaricus*, *Streptococcus salivarius* subsp. *Thermophilus*	Yogurt	Production of yogurt	Bocchi et al. (2020)
		Fungi–fungi: *Penicillium roqueforti–Rhizopus*	Surface-ripened cheese	Cheese production	Bocchi et al. (2020)
		Microbial consortia (Fungi–Fungi–Bacteria): *R. oryzae*, *S. cerevisiae* and *L. plantarum*	Wine	Wine production	Ly et al. (2019)

Table 1.8 Functions of key metabolites of gut microbiota in host

S. No.	Metabolites	Functions	Reference
1.	Short Chain Fatty Acids (SCFAs), Acetate, Butyrate, Propionate	Regulate host metabolic pathways	De Vadder et al. (2014)
2.	Indole derivatives	Energy homeostasis	Venkatesh et al. (2014)
3.	Bile acid metabolites	Activate host nuclear receptors and cell signalling pathways	Wahlström et al. (2016)
4.	Choline metabolites	Modulate lipid metabolism and glucose homeostasis	Wang et al. (2011)
5.	Vitamin B1, B2, B3, B6, B5, B7	Energy production and red blood cell formation	Forster et al. (2017), Lerner et al. (2017)
6.	Polyamines, Spermidine, Putrescine and Spermine	Control high proliferation rate of intestinal epithelial cells	Johnson et al. (2015), Rooks and Garrett (2016)

to the host intestinal epithelium and gets absorbed in the colon. They also perform variety of functions such as regulating gut motility, glucose homeostasis, inflammation and energy harvesting (Cani et al. 2013). The gut microbiota is known to deliver folates, vitamin B2, B7, B12, vitamin K and other vitamins to the host. Further, gut bacteria are also involved in the stimulation of cellular and humoral immunity. Some of the gut bacteria have novel potential health benefits. E.g.: *Faecalibacterium prausnitzii* can be used in the treatment of Irritable bowel syndrome (IBS) and Inflammatory bowel disease (IBD). Similarly, *Akkermansia muciniphila* of gut has a significant role in improving metabolic health of the hosts. There are many potential beneficial bacteria or probiotics in the gut such as *Lactobacillus* and *Bifidobacterium* that help in preventing or treatment of certain diseases (Valdes et al. 2018).

1.4.1.2 The Human Microbiota in Disease

The microbiome imbalance is referred as dysbiosis and impairs the normal functioning of gut microbiota and results in functional disease. The pathogens colonize the intestinal mucosa, causing a strong response of inflammation, thereby resulting in the disturbance of the gut bacteria leads to dysbiosis (Braga et al. 2016). Dysbiosis in gut microbiome is induced because of infectious disease and their treatment, consumption of antibiotics, dietary component, physical and psychological stress and various other factors. Several studies showed the intimate relationship between dysbiosis of the microbiota and infection. It has been revealed that the host infection is not only associated with the microbiome but also with the viruses. The intestinal microbiome is significantly changed in the case of patients with *Clostridium difficile* infection (CDI). This change in the microbiota is associated with the development of viral diseases including human immunodeficiency virus (HIV), hepatitis-B virus (HBV) and other diseases (Fig. 1.9) (Kho and Lal 2018).

1.4.1.3 Antimicrobials Production for Treatment of Diseases

Microorganisms are always in close proximity with other organisms in the environment and coexist by exchanging and sharing metabolites through chemical signals. Bipartite microbial cultivations/ co cultivations provide an attractive and potential reason to identify new compounds such as secondary metabolites including antibiotics with biological activity. Cocultivations of microorganisms simulate ecological interactions which are not present in microbial monocultures. However, multiple species interactions are essential, to simulate natural ecological conditions.

Microorganisms are known to excrete huge varieties of compounds that affect other microorganisms by inducing silent gene clusters. It has been proved that these products that are produced by monocultures can be increased by cocultivation of microorganisms. Further it is possible to exploit maximum amount of natural products, only through complex microbial consortia (Stierle et al. 2017). Cocultivation is more effective in stimulating the production of substances such as bioactive compounds, that are usually not produced in pure cultures. Thus, Cocultivation not only intends to open new avenues to their production, but also provides understanding of the natural role of the bioactive molecules and also the regulation of their development (Netzker et al. 2018).

Emergence of multi drug resistance (MDR) among bacteria and with no new drug discoveries made in the last couple of years is a major concern for the treatment of infection diseases in future. Considering the role of antibiotics, the analysis of microbial consortia in their ecological context forms the basis for the novel antibiotics discovery. As the frequency of new antibiotics discovery by microorganisms is considerably shrinking, attempts have been made to produce new metabolites using cocultures. In the screening of antibiotic producing strain of *Streptomyces tanashiensis*, it is observed that the desferrioxamine E produced by *Streptomyces griseus* growing in its close proximity has facilitated *S. tanashiensis* in compensating certain deficiency and resulted in promotion of its vegetative and productive growth phase (Ueda and Beppu 2017).

Several novel secondary metabolites have isolated from coculturing of microorganisms that is having new mode of action. Coculture of *A. nidulans* and the bacterium *S. rapamycinicus* leads to production of archetypal polyketide orsellinic acid due to the silent fungal gene cluster activation. Fumicyclines were also discovered as a result of mixed cultivation of *streptomycete species* and *A. fumigatus*. Mixed cultures of *Micromonospora* sp. with *Rhodococcus* sp. resulted in the identification of the antibiotic compound keyicin, which was found to be active against gram-positive bacteria selectively (Netzker et al. 2018). *P. fuscum* and *P. camembertii/clavigerum* when cocultivated, has led to the discovery of the new macrolide antibiotic called berkeleylactone with different mode of action (Zipperer et al. 2016). Human microbiota is another potent source for unknown secondary

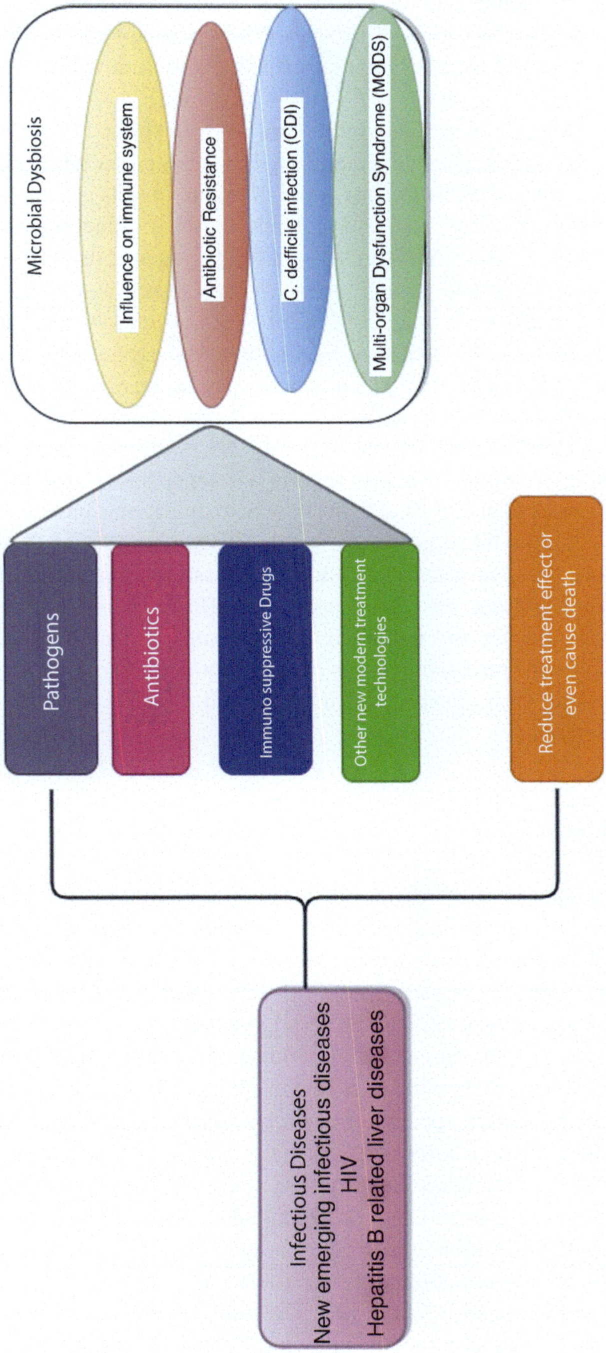

Fig. 1.9 Effect of Infectious diseases on microbiota leading to dysbiosis. (Source: Adapted from Kho and Lal 2018)

metabolites. A ribosomally synthesised thiopeptide known as Lactocillin, as well as lugdunin, the non-ribosomal peptide produced by human commensals, prevent the growth of neighbouring pathogens. This shows their participation in structuring of the human microbiota and their potential as effective antibiotics (Fig. 1.4) (Netzker et al. 2018).

The interactions among bacteria and fungi are not well studied in the background of human health. However, the interaction between, ubiquitous bacterium *Pseudomonas aeruginosa* and *Candida albicans*, a polymorphic fungus is a classic example of microbial interaction. This association leads to the production of 5-methyl-phenazine-carboxylic acid (5-MPCA) an antifungal agent. In this association, *C. albicans* produces ethanol which helps in the secretion of the phenazine 5-methyl-phenazine-carboxylic acid (5-MPCA) by *P. aeruginosa* and those phenazines cause an increase in fermentative metabolism and production of ethanol by *C. albicans* (Fourie and Pohl 2019). The interactions between *S. aureus* and *Candida albicans* is also an example of synergistic interactions (Shirtliff et al. 2009). The yeast *Cryptococcus neoformans* and the bacterium *Klebsiella aerogenes* coinfection is another example in which melanin acts as virulence factor for the pathogen *C. neoformans*. (Frases et al. 2006). Alkynyl-bis benzimidazoles have a very limited effect on Gram-negative bacteria but are outstanding inhibitors of Gram-positive bacteria. Therefore there is a need to study the possible synergistic associations between microorganisms and resolving barriers for effectiveness on G−ve bacteria, the outer membrane barrier and multidrug efflux pumps (Chamberlin et al. 2019). WHO also prioritises the research to focus on development of drugs that target multi drug-resistant (MDR) Gram-positive bacteria as well as those targeting the MDR Gram-negative pathogens.

1.4.1.4 Lethal Synergism

The synergistic association of microorganisms is not always beneficial. Not all human microbiota is involved in giving health benefits but some are also known to induce inflammation under certain conditions. *Lactobacillus* spp. Are known to cause rheumatoid arthritis by activating Toll-like receptor (TLR 2) and TLR4 which results in the increase of T_H1 and T_H17 activity and decreases T_{Reg}-cells function (Thaiss et al. 2016). Sometimes microbial associations may become lethal meaning that the coinfection with two organisms results in increased mortality whereas infections by single species of microbes (monomicrobial infections) are nonlethal. This phenomenon is termed as lethal synergism. *Candida albicans* and *Staphylococcus* species (*S. epidermidis* and *S. aureus*) have the ability to form obstinate biofilms in the host and also on abiotic surfaces such as medical devices. Interactions within these biofilm communities may lead to drug tolerance, increased virulence and immune evasion thus causing difficulty in treatment of infections. The most widely studied association is between *Candida albicans* and *Staphylococcus* species, which causes lethal infection allowing enhanced mortality (Kong et al. 2015; Esher et al. 2019; Todd et al. 2019). Adding to lethal killing, *C. albicans* shows high level vancomycin resistance to *S. aureus*. It is done by the formation of drug permeability barrier by releasing a carbohydrate-dense extracellular matrix during biofilm growth (Harriott and Noverr 2009; Carolus et al. 2019). In recent studies it is also established that alpha-toxin is essential for the lethal synergy caused by *S. aureus* in lung coinfection caused by Gram-negative opportunists (Cohen 2016).

1.4.2 Role of Microbial Synergism in Environment

Human actions like farming and agriculture, exploring mineral resources and other improvements of industrialization have been disturbing the natural ecosystem such as soil, water, and air which are the media of survival for life (Ul-Islam et al. 2016). The accumulation of toxic metals like mercury, arsenic, cadmium, lead, etc. have adverse effects on growth of plants (Hassan et al. 2017). Crop yields are directly influenced and affected by high concentrations of heavy metals beyond limits (Xiong et al. 2014). The interactions between microorganisms and their metabolic activities such as mobilization of metals, their transformation and detoxification help in removal of heavy metal contamination and are significant in the remediation of soils (Chen et al. 2015; Tiwari and Lata 2018). Many fungal species are capable of dealing with heavy metals. Examples are *Aspergillus*, *Penicillium*, *Trichoderma and Mucor* (Oladipo et al. 2018). These fungi possess different functional groups like carboxyl, phosphates, amines present as cell wall structure, that can bind and opsonize the metals (Ullah et al. 2017). Some plants–microbe interactions are known to help in control of heavy metal accumulation in plants.

Iron-oxidizing bacteria *Ferrimicrobium acidiphilum* and *Acidithiobacillus thioxidans* play a role in pyrite leaching when used as pure and mixed cultures. *Fm. acidiphilum* oxidizes ferrous iron to ferric iron which can attack the pyrite mineral but in this process the organism needs organic carbon to grow which is provided by *At. thiooxidans* that fix CO_2, which in turn depends on the former organism for its energy source i.e. reduced sulphur obtained from pyrite. Thus neither of the bacteria can grow as pure cultures in organic carbon-free media using pyrite as a source of energy, but they can interact synergistically to form a microbial consortium and grow successfully (Schaechter 2009). Synthetic community of microbes consisting of *Dehalococcoide*, *Desulfovibrio* and *Methanosarcina* was successfully studied in PCB dechlorination which can be used as biomarkers to evaluate and monitor the potential of PCB dechlorination at bioremediation sites. Further biostimulation of the sites with these three specific microbes may help in improving the efficiency of remediation (Shanquan Wang et al. 2019).

1.4.3 Role of Microbial Synergism in Agriculture

Plant-associated microbiota are significant in the growth and development of plants and provides protection from various biotic and abiotic stresses including plant pathogens (Hussain et al. 2018). The benefits of synergistic interactions among different microorganisms aids in improving the crop yields. There are different associations involving microbiota and plant groups in different combinations (Luo et al. 2018). The Plant growth-promoting microorganisms that are associated with crop plants include both Bacteria and Fungi. Plant growth-promoting bacteria (PGPBs) and plant growth-promoting fungi (PGPFs) are used in agriculture since many years and are the most promising for future prospects. Many beneficial microbiota gathers in the regions of phyllosphere, rhizosphere, and endosphere regions of plants and act as symbionts for plant roots E.g.: Arbuscular mycorrhizal fungi (AMFs), *Rhizobium* spp and *Frankia* spp. (Tian et al. 2020). The synergistic association between Arbuscular mycorrhizal fungi AMF *R. irregularis* and the PGPB *P. putida* help in improvement of the growth and also provide protection against pathogens of the wheat plants (Senapati et al. 2019). The crop yields of rice and wheat are improved by application of AMFs which help in resisting drought. Apart from improving the yield of cotton and soybean, AMFs are known to provide resistance to soyabean against *Macrophomina phaseolina* pathogen as well as they also help in promoting transfer of nitrogen from soil to plants (Spagnoletti et al. 2017). Medicinal plants *W. somnifera* and *H. niger* have shown an improvement in the alkaloid contents even under limited water conditions when Plant growth-promoting bacteria such as species of *Azospirillum*, *Pseudomonas*, *Azotobacter* and *Bacillus* were applied to the fields (Mathur et al. 2019). It is reported that the absorption of nitrogen by rice plants *Oryza sativa indica* and *Oryza sativa japonica* is associated with NRT1.1B gene, which encodes nitrate transporter and that is associated with microbiota of rhizosphere and nitrogen usage in rice fields (Zhang et al. 2019).

Studies on microbial synergism between *R. irregularis* and *B. amyloliquefaciens* showed a maximum rise in shoot weight and also improved efficiency of photosynthesis in plants with dual inoculation. Wherein, *B. amyloliquefaciens* helped in facilitating the colonization of arbuscular mycorrhizal fungus in all the tested plants and thus aided plant growth and proliferation (Nurmi et al. 2016; Tian et al. 2020).

1.4.4 Role of Microbial Synergism in Food Applications

Microbial interaction among bacteria, yeasts and fungi are significant in the production of several foods by fermentation (Scherlach et al. 2013). This joint action of microorganisms help to improve aroma, flavour, texture and other characteristics and also enhances the shelf life of fermented foods to a great extent (Dalié et al. 2010). Mixed cultures of fungi and bacteria are used in variety of fermentations including

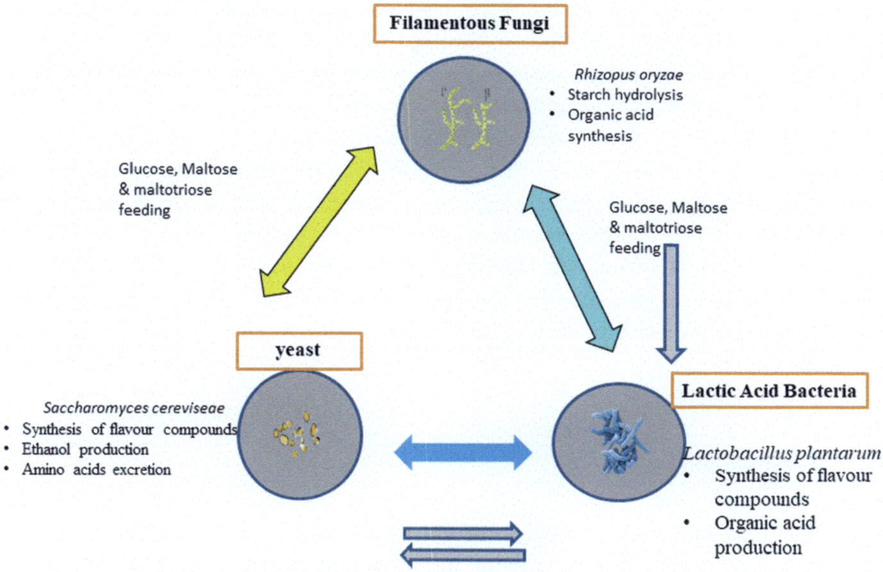

Fig. 1.10 Microbial consortia in wine production/fermentation

alcoholic beverages and dairy products (Bordet et al. 2020; Bocchi et al. 2020). Lactic acid bacteria (LAB), especially *Lactobacillus sanfranciscensis* and yeasts such as *Saccharomyces exiguous* or *Candida humilis* exhibit synergistic interaction, in the fermentation of sourdough. Cocultures of yeast and Lactic Acid Bacilli (LAB) generated an advanced quantity of lactic acid during sourdough fermentation than monocultures (Teleky et al. 2020).

Probiotics are known as the live microorganisms that have health benefit in or on the host. Probiotics can be given in moderate quantities as food supplements or through fermented food, such as traditional dairy products. They are known to increase the nutritional and functional value of food by promoting the amount and availability of nutrients and bioactive compounds such as organic acids, exopolysaccharides, and conjugated linoleic acid that originate from the joint action of microbial metabolism (Di Cagno et al. 2016). The probiotic microorganisms *Lactobacillus* spp. and *Bifidobacterium* spp. used in food fermentations interact synergistically with the substrate resulting in the improved beneficial features and this paves the way for designing functional foods for the specific requirements that happens during the transition from milk diet to solid foods on the human microbiota (Bocchi et al. 2020).

The requirement of joint metabolism for an effective fermentation is revealed in wine production experiments using microbial consortia of three microbial strains, *Saccharomyces cerevisiae*, *Rhizopus oryzae* and *Lactobacillus plantarum*. Modifying the ratio between these three species had shown significant effect on production of ethanol and organic acid and the profile of volatile compounds of the wine when compared to the wine produced by individual organisms (Fig. 1.10). It was observed that when three microbial strains were used in equal ratio during inoculation, it resulted in ethanol yield and flavour similar to that of mono cultures. This study paves the way for our understanding of synergistic interactions among prokaryote and eukaryotes with potential application in the area of food biotechnology (Ly et al. 2019). On the contrary, food fermentations with mixed culture of bacteria and fungi may also result in contamination of food leading to deleterious effects on human health. Consumption of Tempe bongkrek, a fermented dish made using fungus *Rhizopus oligosporus* causes food-intoxication in humans resulting in many number of deaths annually (Lackner and Hertweck 2011). The intoxication is known to be caused by the toxins produced by bacterium *Burkholderia cocovenenans* namely toxoflavin and bongkrekic acid (Li et al. 2019).

1.5 Conclusion

Interspecies interactions especially synergistic effect between different types of microorganisms is increasingly growing interest among researchers. The interaction between the microorganisms is of many different types and the mutual relationship between the partners is varied and may range from only marginal support to absolute mutual dependence. These microbial interactions may possibly act like "bio-engines" in producing various important metabolites such as volatile organic compounds. The cooperation among partners involved in these synergistic processes is strengthened due to closer proximity between the partner cells. There are many instances of this kind and are widespread in nature and still need to be unravelled in the future. The modern molecular and metagenomic approaches may facilitate scientists to study further and analyse the complete microbial diversity and their genetic capability to carry out the active metabolic pathways prevailing in a specified environment. These will hopefully facilitate and allow researchers to obtain the concealed information on various metabolites and microbial partners in situ. The knowledge and information about different microbial interactions and the possible benefits gives an opportunity for producing artificial microbiomes that can be efficiently applied to solve critical health issues, remediation of heavy metals, improving agricultural productivity and for producing novel fermentative products etc.

References

Atkinson N (1935) Synergic gas production by bacteria. Aust J Exp Biol Med Sci 13(2):67
Atkinson N, Wood E (1938) The false positive reaction in the presumptive test for bact. coli in water. Aust J Exp Biol Med Sci 16(1):115
Azam M, Mohsin M, Ijaz H, Tulain UR, Ashraf MA, Fayyaz A, Kamran Q (2017) Lactic acid bacteria in traditional fermented Asian foods. Pak J Pharm Sci 30(5):1803

Aziza M, Amrane A (2006) Commensalism during submerged mixed culture of Geotrichum candidum and Penicillium camembertii on glutamate and lactate. Process Biochem 41 (12):2452–2457

Bauer MA, Kainz K, Carmona-Gutierrez D, Madeo F (2018) Microbial wars: competition in ecological niches and within the microbiome. Microbial Cell 5(5):215

Berdugo-Clavijo C, Dong X, Soh J, Sensen CW, Gieg LM (2012) Methanogenic biodegradation of two-ringed polycyclic aromatic hydrocarbons. FEMS Microbiol Ecol 81(1):124–133

Berleman JE, Scott J, Chumley T, Kirby JR (2008) Predataxis behavior in Myxococcus xanthus. Proc Natl Acad Sci 105(44):17127–17132

Besse A, Peduzzi J, Rebuffat S, Carre-Mlouka A (2015) Antimicrobial peptides and proteins in the face of extremes: lessons from archaeocins. Biochimie 118:344–355

Bhuiyan AA, Noor R (2020) General perspectives on water and fluid borne microorganisms in Bangladesh. In: Applied microbiology: theory & technology. Universal Wiser Publisher, Singapore

Bocchi S, Sagheddu V, Elli M, Lim C, Morelli L (2020) The synergistic interaction between probiotics and food affects their beneficial features. Adv Nutri Food Sci ANAFS-165

Bordet F, Joran A, Klein G, Roullier-Gall C, Alexandre H (2020) Yeast–yeast interactions: mechanisms, methodologies and impact on composition. Microorganisms 8(4):600

Bouyahya A, Dakka N, Et-Touys A, Abrini J, Bakri Y (2017) Medicinal plant products targeting quorum sensing for combating bacterial infections. Asian Pac J Trop Med 10(8):729–743

Braga RM, Dourado MN, Araújo WL (2016) Microbial interactions: ecology in a molecular perspective. Braz J Microbiol 47:86–98

Cani PD (2018) Human gut microbiome: hopes, threats and promises. Gut 67(9):1716–1725

Cani PD, Everard A, Duparc T (2013) Gut microbiota, enteroendocrine functions and metabolism. Curr Opin Pharmacol 13(6):935–940

Carolus H, Van Dyck K, Van Dijck P (2019) Candida albicans and Staphylococcus species: a threatening twosome. Front Microbiol 10:2162

de Carvalho CC (2017) Biofilms: microbial strategies for surviving UV exposure. In: Ultraviolet light in human health, diseases and environment. Springer, New York, NY, pp 233–239

Castellani A (1926) Fermentation phenomena when different species of micro-organisms are in close association. Proc Soc Exp Biol Med 23(6):481–488

Chamberlin J, Story S, Ranjan N, Chesser G, Arya DP (2019) Gram-negative synergy and mechanism of action of alkynyl bisbenzimidazoles. Sci Rep 9(1):1–19

Chang J-J, Chou C-H, Ho C-Y, Chen W-E, Lay J-J, Huang C-C (2008) Syntrophic co-culture of aerobic Bacillus and anaerobic Clostridium for bio-fuels and bio-hydrogen production. Int J Hydrog Energy 33(19):5137–5146

Chen M, Xu P, Zeng G, Yang C, Huang D, Zhang J (2015) Bioremediation of soils contaminated with polycyclic aromatic hydrocarbons, petroleum, pesticides, chlorophenols and heavy metals by composting: applications, microbes and future research needs. Biotechnol Adv 33 (6):745–755

Cohen J (2016) Vaginal microbiome affects HIV risk. Science 353:331

Cortes-Tolalpa L, Salles JF, van Elsas JD (2017) Bacterial synergism in lignocellulose biomass degradation–complementary roles of degraders as influenced by complexity of the carbon source. Front Microbiol 8:1628

Cotter PD, Ross RP, Hill C (2013) Bacteriocins—a viable alternative to antibiotics? Nat Rev Microbiol 11(2):95–105

Cragg SM, Beckham GT, Bruce NC, Bugg TD, Distel DL, Dupree P et al (2015) Lignocellulose degradation mechanisms across the tree of life. Curr Opin Chem Biol 29:108–119

Crowley PH, Cox JJ (2011) Intraguild mutualism. Trends Ecol Evol 26(12):627–633

Dalié D, Deschamps A, Richard-Forget F (2010) Lactic acid bacteria–potential for control of mould growth and mycotoxins: a review. Food Control 21(4):370–380

Darwin C (1859) The origin of species by means of natural selection. John Murray, London

De Vadder F, Kovatcheva-Datchary P, Goncalves D, Vinera J, Zitoun C, Duchampt A et al (2014) Microbiota-generated metabolites promote metabolic benefits via gut-brain neural circuits. Cell 156(1–2):84–96
Deng Y-J, Wang SY (2016) Synergistic growth in bacteria depends on substrate complexity. J Microbiol 54(1):23–30
Di Cagno R, Filannino P, Gobbetti M (2016) Novel fermented fruit and vegetable-based products. In: Novel food fermentation technologies. Springer, New York, NY, pp 279–291
Doing G, Koeppen K, Occhipinti P, Hogan DA (2020) Conditional antagonism in co-cultures of Pseudomonas aeruginosa and Candida albicans: an intersection of ethanol and phosphate signaling distilled from dual-seq transcriptomics. BioRxiv
Donlan RM (2002) Biofilms: microbial life on surfaces. Emerg Infect Dis 8(9):881
Ellis CN, Traverse CC, Mayo-Smith L, Buskirk SW, Cooper VS (2015) Character displacement and the evolution of niche complementarity in a model biofilm community. Evolution 69 (2):283–293
Erken M, Lutz C, McDougald D (2013) The rise of pathogens: predation as a factor driving the evolution of human pathogens in the environment. Microb Ecol 65(4):860–868
Esher SK, Fidel PL, Noverr MC (2019) Candida/staphylococcal polymicrobial intra-abdominal infection: pathogenesis and perspectives for a novel form of trained innate immunity. J Fungi 5 (2):37
Flemming H-C, Wingender J (2010) The biofilm matrix. Nat Rev Microbiol 8(9):623–633
Forster VJ, McDonnell A, Theobald R, McKay JA (2017) Effect of methotrexate/vitamin B12 on DNA methylation as a potential factor in leukemia treatment-related neurotoxicity. Epigenomics 9(9):1205–1218
Fourie R, Pohl CH (2019) Beyond antagonism: the interaction between Candida species and Pseudomonas aeruginosa. J Fungi 5(2):34
Frases S, Chaskes S, Dadachova E, Casadevall A (2006) Induction by Klebsiella aerogenes of a melanin-like pigment in Cryptococcus neoformans. Appl Environ Microbiol 72(2):1542–1550
Frey-Klett P, Burlinson P, Deveau A, Barret M, Tarkka M, Sarniguet A (2011) Bacterial-fungal interactions: hyphens between agricultural, clinical, environmental, and food microbiologists. Microbiol Mol Biol Rev 75(4):583–609
Fuchs G, Boll M, Heider J (2011) Microbial degradation of aromatic compounds—from one strategy to four. Nat Rev Microbiol 9(11):803–816
García-Bayona L, Comstock LE (2018) Bacterial antagonism in host-associated microbial communities. Science 361(6408):eaat2456
García-Callejas D, Molowny-Horas R, Araújo MB (2018) The effect of multiple biotic interaction types on species persistence. Ecology 99(10):2327–2337
Ghosh S, Chowdhury R, Bhattacharya P (2016) Mixed consortia in bioprocesses: role of microbial interactions. Appl Microbiol Biotechnol 100(10):4283–4295
Ghoul M, Mitri S (2016) The ecology and evolution of microbial competition. Trends Microbiol 24 (10):833–845
Goh YJ, Klaenhammer TR (2015) Genetic mechanisms of prebiotic oligosaccharide metabolism in probiotic microbes. Annu Rev Food Sci Technol 6:137–156
Graham JG (1932) Bacterial synergism—the formation by B. typhosus or B. coli anaerogenes from Mannitol of an Intermediate Substance from which Morgan's Bacillus produces Gas. Epidemiol Infect 32(3):385–395
Grover J (2008) Population and community interactions
Harriott MM, Noverr MC (2009) Candida albicans and Staphylococcus aureus form polymicrobial biofilms: effects on antimicrobial resistance. Antimicrob Agents Chemother 53(9):3914–3922
Harrison JJ, Ceri H, Turner RJ (2007) Multimetal resistance and tolerance in microbial biofilms. Nat Rev Microbiol 5(12):928–938
Hassan TU, Bano A, Naz I (2017) Alleviation of heavy metals toxicity by the application of plant growth promoting rhizobacteria and effects on wheat grown in saline sodic field. Int J Phytoremed 19(6):522–529

Hathroubi S, Mekni MA, Domenico P, Nguyen D, Jacques M (2017) Biofilms: microbial shelters against antibiotics. Microb Drug Resist 23(2):147–156
Heng NC, Wescombe PA, Burton JP, Jack RW, Tagg JR (2007) The diversity of bacteriocins in gram-positive bacteria. In: Bacteriocins. Springer, New York, NY, pp 45–92
Hernandez D, Kiesewetter KN, Palakurty S, Stinchcombe JR, Afkhami ME (2019) Synergism and symbioses: unpacking complex mutualistic species interactions using transcriptomic approaches. In: The model legume Medicago truncatula. John Wiley & Sons, Hoboken, NJ, pp 1045–1054
Hibbing ME, Fuqua C, Parsek MR, Peterson SB (2010) Bacterial competition: surviving and thriving in the microbial jungle. Nat Rev Microbiol 8(1):15–25
Hillesland KL, Stahl DA, Harcombe W (2011) Measuring the costs of microbial mutualism. Microbe 6(10):427
Hooper LV, Falk PG, Gordon JI (2000) Analyzing the molecular foundations of commensalism in the mouse intestine. Curr Opin Microbiol 3(1):79–85
Hoštacká A, Čižnár I, Štefkovičová M (2010) Temperature and pH affect the production of bacterial biofilm. Folia Microbiol 55(1):75–78
Houck M, O'Connor B (1991) Ecological and evolutionary significance of phoresy in the Astigmata. Annu Rev Entomol 36(1):611–636
Hussain SS, Mehnaz S, Siddique KH (2018) Harnessing the plant microbiome for improved abiotic stress tolerance. In: Plant microbiome: stress response. Springer, New York, NY, pp 21–43
Johnson CH, Dejea CM, Edler D, Hoang LT, Santidrian AF, Felding BH et al (2015) Metabolism links bacterial biofilms and colon carcinogenesis. Cell Metab 21(6):891–897
Jones D, Head I, Gray N, Adams J, Rowan A, Aitken C et al (2008) Crude-oil biodegradation via methanogenesis in subsurface petroleum reservoirs. Nature 451(7175):176–180
Jurkevitch E (2006) Predatory prokaryotes: biology, ecology and evolution, vol 4. Springer Science & Business Media, New York, NY
Kho Z, Lal S (2018) The human gut microbiome-a potential controller of wellness and disease. Front Microbiol 9:1835. https://doi.org/10.3389/fmicb.2018.01835. PubMed PMID: 30154767
Kim LH, Chong TH (2017) Physiological responses of salinity-stressed Vibrio sp. and the effect on the biofilm formation on a nanofiltration membrane. Environ Sci Technol 51(3):1249–1258
Kong EF, Kucharíková S, Van Dijck P, Peters BM, Shirtliff ME, Jabra-Rizk MA (2015) Clinical implications of oral candidiasis: host tissue damage and disseminated bacterial disease. Infect Immun 83(2):604–613
Lackner G, Hertweck C (2011) Impact of endofungal bacteria on infection biology, food safety, and drug development. PLoS Pathog 7(6):e1002096
Leigh E Jr (2010) The evolution of mutualism. J Evol Biol 23(12):2507–2528
Lerner A, Neidhöfer S, Matthias T (2017) The gut microbiome feelings of the brain: a perspective for non-microbiologists. Microorganisms 5(4):66
Li YH, Tian XL (2016) Quorum sensing and bacterial social interactions in biofilms: bacterial cooperation and competition. In: Stress and environmental regulation of gene expression and adaptation in bacteria. John Wiley & Sons, Inc, Hoboken, NJ, pp 1195–1205
Li X, Huang S, Yu J, Wang Q, Wu S (2013) Improvement of hydrogen production of Chlamydomonas reinhardtii by co-cultivation with isolated bacteria. Int J Hydrog Energy 38 (25):10779–10787
Li J, Zhou L, Long C, Fang L, Chen Q, Chen Q et al (2019) An investigation of bongkrekic acid poisoning caused by consumption of a nonfermented rice noodle product without noticeable signs of spoilage. J Food Prot 82(10):1650–1654
Lloyd-Price J, Abu-Ali G, Huttenhower C (2016) The healthy human microbiome. Genome Med 8: 51
Lowery CA, Dickerson TJ, Janda KD (2008) Interspecies and interkingdom communication mediated by bacterial quorum sensing. Chem Soc Rev 37(7):1337–1346

Luo S, Tian L, Chang C, Wang S, Zhang J, Zhou X et al (2018) Grass and maize vegetation systems restore saline-sodic soils in the Songnen Plain of northeast China. Land Degrad Dev 29 (4):1107–1119
Ly S, Bajoul Kakahi F, Mith H, Phat C, Fifani B, Kenne T et al (2019) Engineering synthetic microbial communities through a selective biofilm cultivation device for the production of fermented beverages. Microorganisms 7(7):206
Marsden AE, Grudzinski K, Ondrey JM, DeLoney-Marino CR, Visick KL (2017) Impact of salt and nutrient content on biofilm formation by Vibrio fischeri. PLoS One 12(1):e0169521
Martin M, Hölscher T, Dragoš A, Cooper VS, Kovács ÁT (2016) Laboratory evolution of microbial interactions in bacterial biofilms. J Bacteriol 198(19):2564–2571
Mathis KA, Bronstein JL (2020) Our current understanding of commensalism. Annu Rev Ecol Evol Syst 51:167
Mathur S, Tomar RS, Jajoo A (2019) Arbuscular mycorrhizal fungi (AMF) protects photosynthetic apparatus of wheat under drought stress. Photosynth Res 139(1–3):227–238
McDougald D, Rice SA, Barraud N, Steinberg PD, Kjelleberg S (2012) Should we stay or should we go: mechanisms and ecological consequences for biofilm dispersal. Nat Rev Microbiol 10 (1):39–50
McInerney MJ, Sieber JR, Gunsalus RP (2009) Syntrophy in anaerobic global carbon cycles. Curr Opin Biotechnol 20(6):623–632
McInerney M, Sieber J, Gunsalus R (2011) Genomic sequences reveal systems required to produce hydrogen and formate, plus other hallmarks of the syntrophic lifestyle. Microbe Mag 6:479–485
Medema MH, Fischbach MA (2015) Computational approaches to natural product discovery. Nat Chem Biol 11(9):639
Miller MB, Bassler BL (2001) Quorum sensing in bacteria. Annu Rev Microbiol 55(1):165–199
Minty JJ, Singer ME, Scholz SA, Bae C-H, Ahn J-H, Foster CE et al (2013) Design and characterization of synthetic fungal-bacterial consortia for direct production of isobutanol from cellulosic biomass. Proc Natl Acad Sci 110(36):14592–14597
Mitri S, Richard Foster K (2013) The genotypic view of social interactions in microbial communities. Annu Rev Genet 47:247–273
Mittelbach M, Vannette RL (2017) Mutualism in yeasts. In: Yeasts in natural ecosystems: ecology. Springer, New York, NY, pp 155–178
Morris BE, Henneberger R, Huber H, Moissl-Eichinger C (2013) Microbial syntrophy: interaction for the common good. FEMS Microbiol Rev 37(3):384–406
Mounier J, Gelsomino R, Goerges S, Vancanneyt M, Vandemeulebroecke K, Hoste B et al (2005) Surface microflora of four smear-ripened cheeses. Appl Environ Microbiol 71(11):6489–6500
Mounier J, Monnet C, Vallaeys T, Arditi R, Sarthou A-S, Hélias A, Irlinger F (2008) Microbial interactions within a cheese microbial community. Appl Environ Microbiol 74(1):172–181
Nemergut DR, Schmidt SK, Fukami T, O'Neill SP, Bilinski TM, Stanish LF et al (2013) Patterns and processes of microbial community assembly. Microbiol Mol Biol Rev 77(3):342–356
Netzker T, Flak M, Krespach MK, Stroe MC, Weber J, Schroeckh V, Brakhage AA (2018) Microbial interactions trigger the production of antibiotics. Curr Opin Microbiol 45:117–123
Nipa KK (2015) Human skin bacteria: innocuous symbiotic association, pathogenic action and antibiotic effect. Plant Environ Dev 4(1):6–13
Nurmi V, Votsis A, Perrels A, Lehvävirta S (2016) Green roof cost-benefit analysis: special emphasis on scenic benefits. J Benefit Cost Anal 7(3):488–522
Oladipo OG, Awotoye OO, Olayinka A, Bezuidenhout CC, Maboeta MS (2018) Heavy metal tolerance traits of filamentous fungi isolated from gold and gemstone mining sites. Braz J Microbiol 49(1):29–37
Ouattara HG, Elias RJ, Dudley EG (2020) Microbial synergy between Pichia kudriazevii YS201 and Bacillus subtilis BS38 improves pulp degradation and aroma production in cocoa pulp simulation medium. Heliyon 6(1):e03269
Pacheco AR, Segrè D (2019) A multidimensional perspective on microbial interactions. FEMS Microbiol Lett 366(11):fnz125

Pérez J, Moraleda-Muñoz A, Marcos-Torres FJ, Muñoz-Dorado J (2016) Bacterial predation: 75 years and counting! Environ Microbiol 18(3):766–779

Pérez-García A, Romero D, De Vicente A (2011) Plant protection and growth stimulation by microorganisms: biotechnological applications of Bacilli in agriculture. Curr Opin Biotechnol 22(2):187–193

Pesci EC, Milbank JB, Pearson JP, McKnight S, Kende AS, Greenberg EP, Iglewski BH (1999) Quinolone signaling in the cell-to-cell communication system of Pseudomonas aeruginosa. Proc Natl Acad Sci 96(20):11229–11234

Prajakta BM, Suvarna PP, Raghvendra SP, Alok RR (2019) Potential biocontrol and superlative plant growth promoting activity of indigenous Bacillus mojavensis PB-35 (R11) of soybean (Glycine max) rhizosphere. SN Appl Sci 1(10):1–1

Rao TE, Kumavath R (2020) Role of quorum sensing in microbial infections and biofilm formation model organisms for microbial pathogenesis. In: Biofilm formation and antimicrobial drug discovery. Springer, New York, NY, pp 61–78

Rashid MI, Mujawar LH, Shahzad T, Almeelbi T, Ismail IM, Oves M (2016) Bacteria and fungi can contribute to nutrients bioavailability and aggregate formation in degraded soils. Microbiol Res 183:26–41

Rocha EP (2016) Using sex to cure the genome. PLoS Biol 14(3):e1002417

Rojas F, Silvester E, Young J, Milne R, Tettey M, Houston DR et al (2019) Oligopeptide signaling through TbGPR89 drives trypanosome quorum sensing. Cell 176(1–2):306–317.e316

Rooks MG, Garrett WS (2016) Gut microbiota, metabolites and host immunity. Nat Rev Immunol 16(6):341–352

Salle AJ (1943) Fundamental principles of bacteriology, AJ Salle, BS, MS, PHD, 6th impression. McGraw-Hill book Company, Inc

Schaechter M (2009) Encyclopedia of microbiology. Academic Press, New York, NY

Scherlach K, Hertweck C (2018) Mediators of mutualistic microbe–microbe interactions. Nat Prod Rep 35(4):303–308

Scherlach K, Graupner K, Hertweck C (2013) Molecular bacteria-fungi interactions: effects on environment, food, and medicine. Annu Rev Microbiol 67:375–397

Schink B (1997) Energetics of syntrophic cooperation in methanogenic degradation. Microbiol Mol Biol Rev 61(2):262–280

Schink B (2002) Synergistic interactions in the microbial world. Antonie Van Leeuwenhoek 81 (1–4):257–261

Schink B, Stams AJ (2006) Syntrophism among prokaryotes. Prokaryotes 2:309–335

Schnurr PJ, Molenda O, Edwards E, Espie GS, Allen DG (2016) Improved biomass productivity in algal biofilms through synergistic interactions between photon flux density and carbon dioxide concentration. Bioresour Technol 219:72–79

Sears H, Putnam JJ (1923) Gas production by bacteria in symbiosis. J Infect Dis 32:270–279

Senapati N, Stratonovitch P, Paul MJ, Semenov MA (2019) Drought tolerance during reproductive development is important for increasing wheat yield potential under climate change in Europe. J Exp Bot 70(9):2549–2560

Seo H, Kim J, Jung J, Jin HM, Jeon CO, Park W (2012) Complexity of cell–cell interactions between Pseudomonas sp. AS1 and Acinetobacter oleivorans DR1: metabolic commensalism, biofilm formation and quorum quenching. Res Microbiol 163(3):173–181

Shirtliff ME, Peters BM, Jabra-Rizk MA (2009) Cross-kingdom interactions: Candida albicans and bacteria. FEMS Microbiol Lett 299(1):1–8

Sieuwerts S, De Bok FA, Hugenholtz J, van Hylckama Vlieg JE (2008) Unraveling microbial interactions in food fermentations: from classical to genomics approaches. Appl Environ Microbiol 74(16):4997–5007

Singh RP, Manchanda G, Singh RN, Srivastava AK, Dubey RC (2016) Selection of alkalotolerant and symbiotically efficient chickpea nodulating rhizobia from North-West Indo Gangetic Plains. J Basic Microbiol 56:14–25. https://doi.org/10.1002/jobm.201500267

Singh RP, Manchanda G, Maurya IK, Maheshwari NK, Tiwari PK et al (2019) Streptomyces from rotten wheat straw endowed the high plant growth potential traits and agro-active compounds. Biocatal Agric Biotechnol 17:507–513. https://doi.org/10.1016/j.bcab.2019.01.014
Singh S, Kumar V, Dhanjal DS, Singh J (2020) Biological control agents: diversity, ecological significances, and biotechnological applications. In: Natural bioactive products in sustainable agriculture. Springer, New York, NY, pp 31–44
Sorci G, Garnier S (2019) Evolutionary ecology: evolution of parasitism. Elsevier, Amsterdam
Spagnoletti F, Carmona M, Gómez NET, Chiocchio V, Lavado RS (2017) Arbuscular mycorrhiza reduces the negative effects of M. phaseolina on soybean plants in arsenic-contaminated soils. Appl Soil Ecol 121:41–47
Stachowicz JJ, Bruno JF, Duffy JE (2007) Understanding the effects of marine biodiversity on communities and ecosystems. Annu Rev Ecol Evol Syst 38:739–766
Stams AJ, Plugge CM (2009) Electron transfer in syntrophic communities of anaerobic bacteria and archaea. Nat Rev Microbiol 7(8):568–577
Stierle AA, Stierle DB, Decato D, Priestley ND, Alverson JB, Hoody J et al (2017) The berkeleylactones, antibiotic macrolides from fungal coculture. J Nat Prod 80(4):1150–1160
Tap J, Mondot S, Levenez F, Pelletier E, Caron C, Furet J et al (2009) Towards the human intestinal microbiota phylogenetic core. Environ Microbiol 11:2574–2584
Tecon R, Or D (2017) Biophysical processes supporting the diversity of microbial life in soil. FEMS Microbiol Rev 41(5):599–623
Teleky B-E, Martău AG, Ranga F, Chețan F, Vodnar DC (2020) Exploitation of lactic acid bacteria and Baker's yeast as single or multiple starter cultures of wheat flour dough enriched with soy flour. Biomolecules 10(5):778
Thaiss CA, Zmora N, Levy M, Elinav E (2016) The microbiome and innate immunity. Nature 535 (7610):65–74
Tian L, Lin X, Tian J, Ji L, Chen Y, Tran L-SP, Tian C (2020) Research advances of beneficial microbiota associated with crop plants. Int J Mol Sci 21(5):1792
Tiwari S, Lata C (2018) Heavy metal stress, signaling, and tolerance due to plant-associated microbes: an overview. Front Plant Sci 9:452
Todd OA, Fidel PL, Harro JM, Hilliard JJ, Tkaczyk C, Sellman BR et al (2019) Candida albicans augments Staphylococcus aureus virulence by engaging the staphylococcal agr quorum sensing system. MBio 10(3):e00910
Traveset A, Richardson DM (2014) Mutualistic interactions and biological invasions. Annu Rev Ecol Evol Syst 45:89–113
Tsegaye B, Balomajumder C, Roy P (2018) Biodegradation of wheat straw by Ochrobactrum oryzae BMP03 and Bacillus sp. BMP01 bacteria to enhance biofuel production by increasing total reducing sugars yield. Environ Sci Pollut Res 25(30):30585–30596
Tsegaye B, Balomajumder C, Roy P (2019) Isolation and characterization of novel lignolytic, cellulolytic, and hemicellulolytic bacteria from wood-feeding termite Cryptotermes brevis. Int Microbiol 22(1):29–39
Tshikantwa TS, Ullah MW, He F, Yang G (2018) Current trends and potential applications of microbial interactions for human welfare. Front Microbiol 9:1156
Ueda K, Beppu T (2017) Antibiotics in microbial coculture. J Antibiot 70(4):361–365
Ul-Islam M, Wajid Ullah M, Khan S, Kamal T, Ul-Islam S, Shah N, Kon Park J (2016) Recent advancement in cellulose based nanocomposite for addressing environmental challenges. Rec Pat Nanotechnol 10(3):169–180
Ullah MW, Shi Z, Shi X, Zeng D, Li S, Yang G (2017) Microbes as structural templates in biofabrication: study of surface chemistry and applications. ACS Sustain Chem Eng 5 (12):11163–11175
Valdes AM, Walter J, Segal E, Spector TD (2018) Role of the gut microbiota in nutrition and health. BMJ 361:k2179

Van Dyk J, Pletschke B (2012) A review of lignocellulose bioconversion using enzymatic hydrolysis and synergistic cooperation between enzymes—factors affecting enzymes, conversion and synergy. Biotechnol Adv 30(6):1458–1480

Van Houdt R, Michiels C (2010) Biofilm formation and the food industry, a focus on the bacterial outer surface. J Appl Microbiol 109(4):1117–1131

Veiga J (2016) Commensalism, amensalism, and synnecrosis. Elsevier, Amsterdam

Venkatesh M, Mukherjee S, Wang H, Li H, Sun K, Benechet AP et al (2014) Symbiotic bacterial metabolites regulate gastrointestinal barrier function via the xenobiotic sensor PXR and Toll-like receptor 4. Immunity 41(2):296–310

Wahlström A, Sayin SI, Marschall H-U, Bäckhed F (2016) Intestinal crosstalk between bile acids and microbiota and its impact on host metabolism. Cell Metab 24(1):41–50

Wang S, Zheng G, Zhou L (2010) Heterotrophic microorganism Rhodotorula mucilaginosa R30 improves tannery sludge bioleaching through elevating dissolved CO2 and extracellular polymeric substances levels in bioleach solution as well as scavenging toxic DOM to Acidithiobacillus species. Water Res 44(18):5423–5431

Wang Z, Klipfell E, Bennett BJ, Koeth R, Levison BS, DuGar B et al (2011) Gut flora metabolism of phosphatidylcholine promotes cardiovascular disease. Nature 472(7341):57–63

Wang S, Chen C, Zhao S, He J (2019) Microbial synergistic interactions for reductive dechlorination of polychlorinated biphenyls. Sci Total Environ 666:368–376

Xie L, Lehvävirta S, Timonen S, Kasurinen J, Niemikapee J, Valkonen JP (2018) Species-specific synergistic effects of two plant growth—promoting microbes on green roof plant biomass and photosynthetic efficiency. PLoS One 13(12):e0209432

Xie C, Coda R, Chamlagain B, Varmanen P (2019) Co-fermentation of Propionibacterium freudenreichii and Lactobacillus brevis in Wheat Bran for in situ Production of Vitamin B12. Front Microbiol 10:1541

Xiong T, Leveque T, Shahid M, Foucault Y, Mombo S, Dumat C (2014) Lead and cadmium phytoavailability and human bioaccessibility for vegetables exposed to soil or atmospheric pollution by process ultrafine particles. J Environ Qual 43(5):1593–1600

Yuan L, Hansen MF, Røder HL, Wang N, Burmølle M, He G (2019) Mixed-species biofilms in the food industry: current knowledge and novel control strategies. Crit Rev Food Sci Nutr 60:2277–2293

Zaccaria M, Dedrick S, Momeni B (2017) Modeling microbial communities: a call for collaboration between experimentalists and theorists. Processes 5(4):53

Zélé F, Magalhães S, Kéfi S, Duncan AB (2018) Ecology and evolution of facilitation among symbionts. Nat Commun 9(1):1–12

Zhang J, Liu Y-X, Zhang N, Hu B, Jin T, Xu H et al (2019) NRT1. 1B is associated with root microbiota composition and nitrogen use in field-grown rice. Nat Biotechnol 37(6):676–684

Zipperer A, Konnerth MC, Laux C, Berscheid A, Janek D, Weidenmaier C et al (2016) Human commensals producing a novel antibiotic impair pathogen colonization. Nature 535 (7613):511–516

Zupančič J, Raghupathi PK, Houf K, Burmølle M, Sørensen SJ, Gunde-Cimerman N (2018) Synergistic interactions in microbial biofilms facilitate the establishment of opportunistic pathogenic fungi in household dishwashers. Front Microbiol 9:21

Chapter 2
Actinobacterial Secondary Metabolites from Maghrebian Ecosystems: An Overview of Half-Century of Investigation

Amine Yekkour, Noureddine Bouras, Slim Smaoui, Lotfi Mellouli, and Mustapha Barakate

This book chapter is dedicated to the late Dr Nasserdine Sabaou (1956–2019): professor in microbiology at the École Normale Supérieure de Kouba, Alger (Algeria) *and founder of the* Laboratoire de Biologie des Systèmes Microbiens (LBSM) *As a pioneer scientist, he devoted his life to the exploration of Saharan inhabiting actinobacterial diversity and their involved metabolites. At present, the valuable works of Professor N Sabaou and collaborators are the source of the most important literature on these topics and permitted to discover several novel species, genera and metabolites.*

A. Yekkour (✉)
Laboratoire de Biologie des Systèmes Microbiens (LBSM): Ecole Normale Supérieure de Kouba, Alger, Algeria

Institut National de la Recherche Agronomique d'Algérie, Alger, Algeria

N. Bouras
Laboratoire de Biologie des Systèmes Microbiens (LBSM): Ecole Normale Supérieure de Kouba, Alger, Algeria

Département de Biologie, Faculté des Sciences de la Nature et de la Vie et Sciences de la Terre, Université de Ghardaia, Ghardaïa, Algeria

S. Smaoui · L. Mellouli
Laboratory of Microorganisms and Enzymatic Biotechnology and Biomolecules, Center of Biotechnology of Sfax, University of Sfax, Sfax, Tunisia

M. Barakate
Laboratory of Microbial Biotechnologies, Agrosciences & Environment, Faculty of Sciences Semlalia, Cadi Ayyad University, Marrakech, Morocco

Mohammed IV Polytechnic University (UM6P): AGBT, AgroBiosciences, Benguerir, Morocco

R. P. Singh et al. (eds.), *Microbes in Microbial Communities*,
https://doi.org/10.1007/978-981-16-5617-0_2

Dr Nasserdine Sabaou (1956–2019)

Abstract Because of their remarkable ability to provide a broad range of bioactive compounds, *Actinobaceria* have attracted special interest members of this bacterial group, in particular those belonging to the genus *Streptomyces*, are considered as the most important producers of bioactive molecules and have led to the discovery of a large number of valuable compounds with broad range of applications (therapeutic, industrial, agricultural, etc.) Maghreb countries (Northwest Africa) present a highly diversified area in term of ecosystems coexistence that potentially facilitate the development of specialized microbial metabolisms adapted for survival to specific conditions. Here we presented a review on the achievements from the exploration of Maghrebian-originated actinobacteria ability to produce active secondary metabolites.

As a whole, 152 active compounds were reported, with diversified chemical structures Amongst these molecules, 31 were described for the first time Most of the determined structures are produced by strains that belong to *Streptomyces* and *Saccharothrix* genera involved in arid/semiarid ecosystems, especially from the Algerian Sahara.

Keywords Actinobacteria · Maghreb · Secondary metabolites · Antibiotic · Saharan ecosystem

2.1 Introduction

Actinomycetes are a group of filamentous Gram-positive bacteria with a percentage of guanine–cytosine higher than 55%. These bacteria are widely spread in nature, including ecological niches, with the existence of various lifestyles possibilities such as commensals, symbionts, and even plant–animal pathogens (Ait Barka et al. 2015).

Because of their ability to provide a broad range of bioactive compounds, *Actinobaceria* have attracted much interest. Members of this group are considered as the most important producers of bioactive molecules with potential benefit to human health, industrial processes and agriculture (Subhashini et al. 2017; Solecka et al. 2012).

At the globe scale, Maghreb countries (Northwest Africa) represent a restricted area in term of ecosystems coexistence (Le Houérou 1990). Despite those the major central and south parts are desert (Sahara), the great bioclimatic range (from humid to hyperarid) and the variety of geomorphic situations make these ecosystems highly diversified and potentially results in the development of specialized microbial metabolisms adapted for survival in such conditions (Sabaou et al. 1998).

This book chapter aimed to present a review on the achievements from the exploration of Maghrebian-originated actinobacteria ability to produce active secondary metabolites and revealed the remarkable potential in term of diversity in both chemical structures and antibiosis capacities.

2.2 Importance of Actinobacteria and Their Secondary Metabolites

The term "secondary metabolites" was defined more than a century ago in opposition to "primary metabolites," which are involved in physiological functions essential for development, reproduction, and cell function of the organism (Hesketh et al. 2002).

In this sense, secondary metabolites are organic compounds that are highly structurally diverse and produced in very low amounts by several organisms, including plants, fungi, and bacteria, but are not strictly mandatory for the maintenance and reproduction of their producers. This does not imply an "auxiliary importance" of secondary metabolites in contrast to importance of primary metabolites.

Secondary metabolites produced by microorganisms appear to be secreted in connection to their surrounding living context to exert various biological effects in an attempt to control biotic interactions and enhance the potential of an organism to endure specific conditions. Thus, secondary metabolites can be regarded as carriers of chemical communication within the microbiota for "relational functions" and the field of ecological chemistry address the role of these secreted metabolites in an ecosystem (Sandegren and Andersson 2009; Nguyen et al. 2012; Prajakta et al. 2019).

The remarkable variable chemical structures of these compounds and the putatively related biological activities has led biologists and chemists to assess the potential (medical, industrial process, agriculture) value of new metabolites rather than determining their function in nature.

The improvement of analytical methods for secondary metabolite isolation and characterization in the last decades has shown the huge diversity of microbial-derived secondary compounds. Neither in silico drug design nor synthetic chemistry has outperformed microorganisms as a source of innovative chemical structures.

Amongst microorganisms, actinobacteria are particularly interesting for their amazing capacity to produce secondary metabolites with diverse chemical structures and diverse biological activities including antivirals, antiparasitics, immunostimulants, immunosuppressants, cytostatic and antitumor (Watve et al. 2001; Demain 2006; Solecka et al. 2012; Li et al. 2017; Takahashi and Nakashima 2018). In fact, about 45% of the microbial-derived molecules documented (Solecka et al. 2012) and 70% of marketed active molecules (Solanki and Kahanna 2008) are of actinobacterial origin. However, actinobacteria, particularly those belonging to the genus *Streptomyces*, are recognized for their great ability to produce antibacterial and antifungal antibiotics. Indeed, approximately, two-third of natural antibiotics were isolated from Actinomycetes (Newman et al. 2003) and about 80% of antibiotic producing actinobacteria belong to the genus *Streptomyces* (Demain 2006; Demain and Sanchez 2009).

The increasing incidence of multidrug resistance in pathogenic microorganisms necessitates the use of antibiotic compounds expressing differential levels of toxicity and side effects (Berdy 2005; Messai et al. 2008; Fair and Tor 2014; Subhashini and Singh 2014; Li and Webster 2018). Thus, it is essential to maintain a framework of research for the isolation of new microbial-derived antibiotics in the hope of finding new effective and less toxic compounds in order to control pathogenic microorganisms along with complementary strategies of preparation of synthetic/semisynthetic antibiotics and search for novel targets within the microbial pathogen.

By postulating involvement of particular adapted metabolisms, one of the strategies for enhancing the likelihood of obtaining particularly interesting isolates and secondary metabolites is to analyse uncommon/niche habitats, such extreme environments (Sabaou et al. 1998).

2.3 Diversity of Maghreb Ecosystems

Maghreb refers to the Northwest African region which totalizing more than 6 million km^2. With a Pacific Ocean and Mediterranean Sea sides and a major central desert part, called Sahara, as well as a contrasted elevations and precipitations, the Maghreb has an impressive bioclimatic variation going from humid to hyperarid (Le Houérou 1990). The related variety of geomorphic situations (alluvial lands in sea costs, mountainous terrain, high lands, oasis, rock and send desert) make these ecosystems highly diversified and potentially results in the development of specialized microbial metabolisms adapted for survival in such conditions (Sabaou et al. 1998).

2.4 General Process for Secondary Metabolites Production, Purification, and Chemical Structure Elucidation

Amongst the isolated actinobacterial strains, isolates that exhibit interesting antimicrobial activity, through targeted in vitro testing assay, are selected to investigate the active compounds involved (Fig. 2.1).

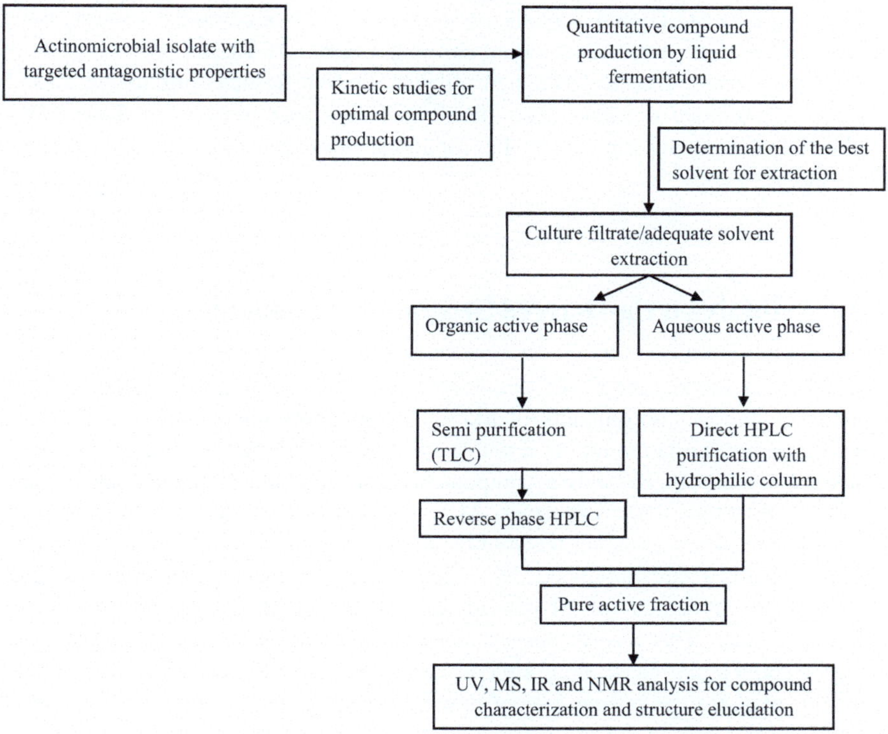

Fig. 2.1 Flow diagram showing the general process followed in the production, purification and chemical structure elucidation of active secondary metabolites derived from actinobacteria

Generally, the process of characterizing new secondary metabolites starts with empiric search of optimal conditions (mainly feeding and incubation parameters in complex or semisynthetic medium such as ISP2[1] and SSM[2], respectively) for compound production and secretion in liquid fermentation experiments (kinetic studies). Then, the culture filtrate is extracted with an adequate solvent (solvent that gave the most interesting extraction depending to the nature of the extractible compounds) using a standard liquid/liquid extraction procedure. In particular cases where a strain is only able to grow, or produce active compound in solid culture media (solid-state fermentation), the containing medium is cut into small pieces then subjected to solvent extraction (Badji et al. 2005; Lahoum et al. 2019). For purification, the active phase, mainly organic, is subjected to successive chromatographic analysis, from partial purification processes, such as silica gel thin layer

[1] International *Streptomyces* Project medium 2 (per liter of distilled water): 4 g yeast extract, 10 g malt extract, 4 g glucose and 20 g agar (Shirling and Gottlieb 1966).

[2] Semi-synthetic medium (per liter of distilled water): 10 g D-glucose, 2 g $(NH_4)_2SO_4$, 2 g NaCl, 5 g KH_2PO_4, 1 g K_2HPO_4, 2 g $MgSO_4 \cdot 7H_2O$, 5 g $CaCO_3$ and 2 g yeast extract (Bouras et al. 2006).

chromatography (TLC) and Sephadex LH20 columns, to a complete separation by reverse phase high performance liquid chromatography (HPLC).

The pure active fraction, obtained after successive re-injections in the HLPC system, is submitted to mass spectrometry (MS) and spectroscopic analysis (UV–visible, IR, and various forms of NMR) for compound characterization and structure elucidation.

2.5 Actinobacteria-Derived Secondary Metabolites

The Table 2.1 showed the list of secondary metabolites derived from actinobacteria inhabiting several environments of Maghreb, especially soil habitat, as gathered from the scientific literature. As a whole, studies of such Actinobacerial potential permitted to identify 152 active compounds, with diversified chemical structures, produced. Amongst these molecules, 31 were described for the first time; and are therefore considered as novel metabolites (Table 2.1, Fig. 2.2). Most of the determined structures are produced by strains that belong to *Streptomyces* and *Saccharothrix* genera involved in arid/semiarid ecosystems, especially from the Algerian Sahara. This result indicates that, through the ability to contain adapted actinobacterial strain, these environments constitute an ecological niche with significant potential for the prospection of new active metabolites. Moreover, extended feeding experiments in the culture of some particular strains, such as *Saccharothrix algeriensis* NRRL B-24137, have successfully led to the induction of new compounds. In fact, *Saccharothrix algeriensis* NRRL B-24137, a rare *actinobacterium* isolated from a palm grove soil in Southern Algeria (Zitouni et al. 2004) was shown to produce several dithiolopyrrolones depending on the incorporated precursors available in the culture medium. The selection of appropriate sources of carbon and nitrogen was shown to induce the biosynthesis of 13 new dithiolopyrrolones (Table 2.1).

The number of actinobacteria-derived molecules from Algeria was 75, 68 from Tunisia and nine from Morocco, as indicated in Table 2.1. As far as we know, no reports are available regarding the characterization of actinobacterial-produced compounds from other remaining countries.

2.6 Conclusion

Actinobacteria inhabiting Maghreb environments possess an outstanding ability to produce diversified active secondary metabolites, particularly those originated from arid Saharan environment, but remain scarcely explored when compared to huge areas under Saharan condition and the diversity of niche habitats (adapted plants, oasis soils, hyper saline soils, rocky plateau, etc.) that exist.

Table 2.1 List of secondary metabolites derived from actinobacteria isolated from different countries in Maghreb

No.	Compound name [Chemical group]	Producer microorganism	Microbial isolation origine	Molecular weight (g/mol)	Target	Reference
1.	Thiolutin (Acetyl-pyrrothine) [Dithiolopyrrolone]	*Saccharothrix algeriensis* NRRL B-24137	Palm grove soil, Adrar, Algeria (Zitouni et al. 2004)	228	Gram+ bacteria	Lamari et al. (2002), Bouras et al. (2006)
					Microfungi	
					Yeasts	
2.	Iso-butyryl-pyrrothine [Dithiolopyrrolone]	*Saccharothrix algeriensis* NRRL B-24137	Palm grove soil, Adrar, Algeria (Zitouni et al. 2004)	256	Gram+ bacteria	Lamari et al. (2002), Bouras et al. (2006)
					Microfungi	
					Yeasts	
3.	Butanoyl-pyrrothine[a] [Dithiolopyrrolone]	*Saccharothrix algeriensis* NRRL B-24137	Palm grove soil, Adrar, Algeria (Zitouni et al. 2004)	256	Gram+ bacteria	Lamari et al. (2002), Bouras et al. (2006)
					Microfungi	
					Yeasts	
4.	Senecioyl-pyrrothine[a] [Dithiolopyrrolone]	*Saccharothrix algeriensis* NRRL B-24137	Palm grove soil, Adrar, Algeria (Zitouni et al. 2004)	268	Gram+ bacteria	Lamari et al. (2002), Bouras et al. (2006)
					Microfungi	
					Yeasts	
5.	Tigloyl-pyrrothine[a] [Dithiolopyrrolone]	*Saccharothrix algeriensis* NRRL B-24137	Palm grove soil, Adrar, Algeria (Zitouni et al. 2004)	268	Gram+ bacteria	Lamari et al. (2002), Bouras et al. (2006)
					Microfungi	
					Yeasts	
6.	Propionylpyrrothine (aureothricin) [Dithiolopyrrolone]	*Saccharothrix algeriensis* NRRL B-24137	Palm grove soil, Adrar, Algeria (Zitouni et al. 2004)	242	Gram+ bacteria	Merrouche et al. (2010)
					Microfungi	
					Yeasts	
7.	Valeryl-pyrrothine[a] [Dithiolopyrrolone]	*Saccharothrix algeriensis* NRRL B-24137	Palm grove soil, Adrar, Algeria (Zitouni et al. 2004)	270	Gram+ bacteria	Bouras et al. (2008), Merrouche et al. (2010)
					Microfungi	
					Yeasts	
8.	Isovaleryl-pyrrothine[a] [Dithiolopyrrolone]	*Saccharothrix algeriensis* NRRL B-24137	Palm grove soil, Adrar, Algeria (Zitouni et al. 2004)	270	Gram+ bacteria	Merrouche et al. (2010)
					Microfungi	
					Yeasts	

(continued)

Table 2.1 (continued)

No.	Compound name [Chemical group]	Producer microorganism	Microbial isolation origine	Molecular weight (g/mol)	Target	Reference
9.	Formyl-pyrrothine[a] [Dithiolopyrrolone]	*Saccharothrix algeriensis* NRRL B-24137	Palm grove soil, Adrar, Algeria (Zitouni et al. 2004)	214	Gram+ bacteria Microfungi Yeasts	Bouras et al. (2008), Merrouche et al. (2010)
10.	Crotonyl-pyrrothine[a] [Dithiolopyrrolone]	*Saccharothrix algeriensis* NRRL B-24137	Palm grove soil, Adrar, Algeria (Zitouni et al. 2004)	254	Gram+ bacteria Microfungi Yeasts	Merrouche et al. (2011)
11.	Sorbyl-pyrrothine[a] [Dithiolopyrrolone]	*Saccharothrix algeriensis* NRRL B-24137	Palm grove soil, Adrar, Algeria (Zitouni et al. 2004)	280	Gram+ bacteria Microfungi Yeasts	Merrouche et al. (2011)
12.	2-Hexonyl-pyrrothine[a] [Dithiolopyrrolone]	*Saccharothrix algeriensis* NRRL B-24137	Palm grove soil, Adrar, Algeria (Zitouni et al. 2004)	282	Gram+ bacteria Microfungi Yeasts	Merrouche et al. (2011)
13.	2-Methyl-3-pentenyl-pyrrothine[a] [Dithiolopyrrolone]	*Saccharothrix algeriensis* NRRL B-24137	Palm grove soil, Adrar, Algeria (Zitouni et al. 2004)	282	Gram+ bacteria Microfungi AntiYeasts	Merrouche et al. (2011)
14.	Benzoyl-pyrrothine[a] [Dithiolopyrrolone]	*Saccharothrix algeriensis* NRRL B-24137	Palm grove soil, Adrar, Algeria (Zitouni et al. 2004)	290	Gram+ bacteria Microfungi Yeasts	Bouras et al. (2008), Merrouche et al. (2019a)
15.	Iso-hexanoyl-pyrrothine[a] [Dithiolopyrrolone]	*Saccharothrix algeriensis* NRRL B-24137	Palm grove soil, Adrar, Algeria (Zitouni et al. 2004)	284	Gram+ bacteria Microfungi Yeasts	Merrouche et al. (2019b)
16.	Benzoyl-holothine (Demethyl-benzoyl-pyrrothine)[a] [Dithiolopyrrolone]	*Saccharothrix algeriensis* NRRL B-24137	Palm grove soil, Adrar, Algeria (Zitouni et al. 2004)	276	Gram+ bacteria Microfungi Yeasts	Bouras et al. (2008)

17.	Holomycin[b] [Dithiolopyrrolone]	*Saccharothrix algeriensis* NRRL B-24137	Palm grove soil, Adrar, Algeria (Zitouni et al. 2004)	214	Gram+ bacteria	Merrouche et al. (2020)
					Microfungi	
					Yeasts	
18.	Mutactimycin C[b] [Anthracyclin]	*Saccharothrix* SA 103	Saharan soil of Hoggar, Tamanrasset, Southern Algeria	630	Gram + bacteria	Zitouni et al. (2004)
19.	Mutactimycin PR[a,b] [Anthracyclin]	*Saccharothrix* SA 103	Saharan soil of Hoggar, Tamanrasset, Southern Algeria	662	Gram+ bacteria	Zitouni et al. (2004)
20.	Saquayamycines A [Angucyclin]	*Streptomyces* sp. PAL114	Palm grove Soil, Béni-isguen, Ghardaïa, Algeria	820	Gram+ bacteria	Aouiche et al. (2014)
					Microfungi	
					Yeasts	
21.	Saquayamycines C [Angucyclin]	*Streptomyces* sp. PAL114	Palm grove Soil, Béni-isguen, Ghardaïa, Algeria	824	Gram+ bacteria	Aouiche et al. (2014)
					Microfungi	
					Yeasts	
22.	Vineomycin A1 [Angucyclin]	*Streptomyces* sp. PAL114	Palm grove Soil, Béni-isguen, Ghardaïa, Algeria	934	Gram+ bacteria	Aouiche et al. (2015)
					Microfungi	
					Yeasts	
23.	Chaetoglobosin A [Alkaloid]	*Streptomyces* sp. PAL114	Palm grove Soil, Béni-isguen, Ghardaïa, Algeria	528	Gram+ bacteria	Aouiche et al. (2015)
					Microfungi	
					Yeasts	
24.	Mzabimycin A[a] [Angucyclin]	*Streptomyces* sp. PAL114	Palm grove Soil, Béni-isguen, Ghardaïa, Algeria	1089	Gram+ bacteria	Tata et al. (2019)
25.	Mzabimycin B[a] [Angucyclin]	*Streptomyces* sp. PAL114	Palm grove Soil, Béni-isguen, Ghardaïa, Algeria	1121	Gram+ bacteria	Tata et al. (2019)

(continued)

Table 2.1 (continued)

No.	Compound name [Chemical group]	Producer microorganism	Microbial isolation origine	Molecular weight (g/mol)	Target	Reference
26.	Pigment-like antibiotic R2[a] [Angucyclinone]	*Streptosporangium* sp. Sg3	Saharan soil, Adrar, Algeria	476	Gram+ bacteria	Boudjella et al. (2010)
27.	Pigment-like antibiotic R1[c] [Quinone-anthracycline]	*Streptosporangium* sp. Sg3	Saharan soil, Adrar, Algeria	462	Gram+ bacteria	Boudjella et al. (2007)
28.	Pigment-like biotic R3[c] [Quinone-anthracycline]	*Streptosporangium* sp. Sg3	Saharan soil, Adrar, Algeria	490	Gram+ bacteria	Boudjella et al. (2007)
29.	Pigment-like biotic HP17[c] [Naphthoquinone]	*Spirillospora* sp.719	Palm grove Soil beni Abbes	nd	Gram+ bacteria	Hacène and Lefebvre (1996)
30.	Pigment-like antibiotic AH17[c] [Quinone]	*Spirillospora* sp.719	Palm grove Soil beni Abbes	nd	Microfungi	Hacène and Lefebvre (1995)
31.	6-Deoxy-8-*O*-methylrabelomycin[b] [Angucyclinone]	*Nocardiopsis* sp. HR-4	Salt lake soil (Sebkha), Ain Salah, Algeria	336	Gram+ bacteria	Boukhalfa et al. (2017)
32.	(−)-7-Deoxy-8-*O*-methyltetrangomycin[a,b] [Angucyclinone]	*Nocardiopsis* sp. HR-4	Salt lake soil (Sebkha), Ain Salah, Algeria	337	Gram+ bacteria	Boukhalfa et al. (2017)
33.	Kenalactam A[a] [Polyene macrolactam]	*Nocardiopsis* sp. CG3	Saltpan soil, Kenadsa, Bechar, Algeria	367	nd	Messaoudi et al. (2019)
34.	Kenalactam B[a] [Polyene macrolactam]	*Nocardiopsis* sp. CG3	Saltpan soil, Kenadsa, Bechar, Algeria	367	nd	Messaoudi et al. (2019)
35.	Kenalactam C[a] [Polyene macrolactam]	*Nocardiopsis* sp. CG3	Saltpan soil, Kenadsa, Bechar, Algeria	458.8	*Staphylococcus*	Messaoudi et al. (2019)
					Microfungi	
					Yeasts	
					Virus (hepatitis C)	
					Cytotoxic	
36.	Kenalactam D[a] [Polyene macrolactam]	*Nocardiopsis* sp. CG3	Saltpan soil, Kenadsa, Bechar, Algeria	498	*Staphylococcus*	Messaoudi et al. (2019)
					Virus (hepatitis C)	
					Cytotoxic	

37.	Kenalactam E[a] [Polyene macrolactam]	*Nocardiopsis* sp. CG3	Saltpan soil, Kenadsa, Bechar, Algeria	538.7	*Staphylococcus*	Messaoudi et al. (2019)
					Virus (hepatitis C)	
					Cytotoxic	
38.	Oligomycin A [Macrolide]	*Streptomyces* sp. HG29	Saharan soil, Hoggr, Tamanrasset, Algeria	790	Microfungi	Khebizi et al. (2018)
39.	Oligomycin E [Macrolide]	*Streptomyces* sp. HG29	Saharan soil, Hoggr, Tamanrasset, Algeria	820	Microfungi	Khebizi et al. (2018)
40.	Cyanogriside I[a,b] [Bipyridine]	*Saccharothrix xinjiangensis* ABH26	Saharan soil, Adrar, Algeria	580	Gram+ bacteria	Lahoum et al. (2019)
41.	Cyanogriside J[a,b] [Bipyridine]	*Saccharothrix xinjiangensis* ABH26	Saharan soil, Adrar, Algeria	378	Gram+ bacteria	Lahoum et al. (2019)
42.	Caerulomycin A[b] (isomer E and isomer Z) [Bipyridine]	*Saccharothrix xinjiangensis* ABH26	Saharan soil, Adrar, Algeria	231	Gram+ bacteria	Lahoum et al. (2019)
					Microfungi	
43.	Caerulomycin F[b] [Bipyridine]	*Saccharothrix xinjiangensis* ABH26	Saharan soil, Adrar, Algeria	218	Gram+ bacteria	Lahoum et al. (2019)
					Microfungi	
					Yeasts	
44.	Caerulomycinonitrile[b] [Bipyridine]	*Saccharothrix xinjiangensis* ABH26	Saharan soil, Adrar, Algeria	213	Gram+ bacteria	Lahoum et al. (2019)
45.	Chloramphenicol[b] [Aromatic]	*Saccharothrix* sp. PAL54	Palm grove Soil, Béni-isguen, Ghardaïa, Algeria	322	nd	Aouiche et al. (2012)
46.	Compound 104A1[c] [Aromatic]	*Actinomadura* sp. AC104	Palm grove Soil, beni Abbes, Algeria	312	Microfungi	Badji et al. (2006)
					Yeasts	
47.	Compound 104A2[c] [Aromatic]	*Actinomadura* sp. AC104	Palm grove Soil, beni Abbes, Algeria	326	Microfungi	Badji et al. (2006)
					Yeasts	

(continued)

Table 2.1 (continued)

No.	Compound name [Chemical group]	Producer microorganism	Microbial isolation origine	Molecular weight (g/mol)	Target	Reference
48.	Compound 104A3[c] [Aromatic]	*Actinomadura* sp. AC104	Palm grove Soil, beni Abbes, Algeria	340	Microfungi	Badji et al. (2006)
					Yeasts	
49.	Compound 104A4[c] [Aromatic]	*Actinomadura* sp. AC104	Palm grove Soil, beni Abbes, Algeria	354	Microfungi	Badji et al. (2006)
					Yeasts	
50.	Compound 94A1[c] [Aromatic]	*Nonomuraea* sp. NM94	Palm grove Soil, beni Abbes, Algeria	298	Gram+ bacteria	Badji et al. (2007)
					Microfungi	
					Yeasts	
51.	Compound 94A2[c] [Aromatic]	*Nonomuraea* sp. NM94	Palm grove Soil, beni Abbes, Algeria	312	Gram+ bacteria	Badji et al. (2007)
					Microfungi	
					Yeasts	
52.	Compound 94A3[c] [Aromatic]	*Nonomuraea* sp. NM94	Palm grove Soil, beni Abbes, Algeria	326	Gram+ bacteria	Badji et al. (2007)
					Microfungi	
					Yeasts	
53.	Compound 94A4[c] [Aromatic]	*Nonomuraea* sp. NM94	Palm grove Soil, beni Abbes, Algeria	340	Gram+ bacteria	Badji et al. (2007)
					Microfungi	
					Yeasts	
54.	Compound 94A5[c] [Aromatic]	*Nonomuraea* sp. NM94	Palm grove Soil, beni Abbes, Algeria	340	Gram+ bacteria	Badji et al. (2007)
					Microfungi	
					Yeasts	
55.	Di-(2-ethylhexyl) phthalate [Phthalate]	*Streptomyces* sp. G60	Palm grove, Metlili, Ghardaïa, Algeria	390.5	*Staphylococcus* (methicillin-resistant)	Driche et al. (2015)
56.	Compound AH97-F1[c] [Aminoglycosidic]	*Actinoalloteichus* sp. AH97	Saline Saharan soil, Hoggar, Tamanrasset, Algeria	850	nd	Boudjelal et al. (2011)

57.	Dioctyl phthalate (compound AH97-F2) [phthalate]	*Actinoalloteichus* sp. AH97	Saline Saharan soil, Hoggar, Tamanrasset, Algeria	390	nd	Boudjelal et al. (2011)
58.	2,4-Di-tert-butylphenol [Alkylphenol]	*Streptomyces* sp. G61	Palm grove, Metlili, Ghardaïa, Algeria	206	Microfungi	Belghit et al. (2016)
					Yeasts	
59.	4-Hydroxy-4-{9-hydroxy-9-[2-(1-hydroxy-6-methyl-hepta-2,4-dienylidene)-3,6-dioxo-cyclohexylidene]-nona-1,3,5,7-tetraenyl}-6-oxa-bicyclo [3.1.0]hex-2-ene-2-carboxylic acid[a] (antibiotic A4) [epoxylated]	*Saccharothrix* sp. SA198	Saharan soil, Tamanrasset, southern Algeria	606	Gram+ bacteria	Boubetra et al. (2013)
					Gram− bacteria	
					Microfungi	
60.	4-Hydroxy-4-{9-hydroxy-9-[2-(1-hydroxy-6-methyl-octa-2,4-dienylidene)-3,6-dioxo-cyclohexylidene]-nona-1,3,5,7-tetraenyl}-6-oxa-bicyclo [3.1.0]hex-2-ene-2-carboxylic acid[a] (antibiotic A5) [Epoxylated]	*Saccharothrix* sp. SA198	Saharan soil, Tamanrasset, southern Algeria	520	Gram+ bacteria	Boubetra et al. (2013)
					Gram− bacteria	
					Microfungi	
61.	5-[(5E,7E,11E)-2,10-Dihydroxy-9,11-dimethyl-5,7,11-tri-decatrien-1-yl]-2-hydroxy-2-(1-hydr-oxyethyl)-4-methyl-3 (2H)-furanone (compound AT37-1) [Furanone]	*Streptomyces* sp. AT37	Saharan soil, Adrar, Algeria	394	*Staphylococcus* (methicillin-resistant) biofilm	Driche et al. (2017)
62.	Actinomycin D [Peptidic with cyclic structure]	*Streptomyces* sp. IA1	Saharan soil, Ain amenas, Algeria	1254	Microfungi	Toumatia et al. (2015)
		Streptomyces sp. GSBNT10	Saharan soil, Beni Abbes, Bechar, Algeria	1254	Gram+ bacteria	Djinni et al. (2019)
					Gram− bacteria	
					Microfungi	
63.	Compound BJ1-B [b,c] [Aminoglycosidic]	*Streptosporangium* sp. Sg163.	Palm grove soil, Adrar, Algeria	340	Yeasts	Boudjella et al. (2014)

(continued)

Table 2.1 (continued)

No.	Compound name [Chemical group]	Producer microorganism	Microbial isolation origine	Molecular weight (g/mol)	Target	Reference
64.	Compound BJ1-G [b,c] [Aminoglycosidic]	*Streptosporangium* sp. Sg163	Palm grove soil, Adrar, Algeria	358	Yeasts	Boudjella et al. (2014)
65.	Compound JP2[b,c] [Aminoglycosidic]	*Streptosporangium* sp. Sg163	Palm grove soil, Adrar, Algeria	458	Yeasts	Boudjella et al. (2014)
66.	Compound BJ2-8 [b,c] [Aminoglycosidic]	*Streptosporangium* sp. Sg163	Palm grove soil, Adrar, Algeria	396	Yeasts	Boudjella et al. (2014)
67.	2-amino-*N*-(2-amino-3-phenylpropanoyl)-*N*-hydroxy-3-phenylpropanamide[a] [Hydroxamic acid-containing]	*Streptomyces* sp. WAB9	Saharan soil, Bechar, Algeria	327	Gram+ bacteria	Yekkour et al. (2015)
					Gram− bacteria	
					Microfungi	
					Yeasts	
68.	Streptazolin	*Streptomyces* sp. SRC3	Oued (river) sediment, Ziama Mansouria, Jijel, Algeria	207	Gram+ bacteria	Djinni et al. (2018)
					Microfungi	
					Yeasts	
69.	Abierixin [Polyether]	*Streptomyces* sp. SF10	Semiarid soil, Chélia, Khenchela, Algeria	725	Nd	Leulmi et al. (2019)
70.	Nigericin [Polyether]	*Streptomyces* sp. SF10	Semiarid soil, Chélia, Khenchela, Algeria	724	Nd	Leulmi et al. (2019)
71.	Epinigericin [Polyether]	*Streptomyces* sp. SF10	Semiarid soil, Chélia, Khenchela, Algeria	724	Nd	Leulmi et al. (2019)
72.	Grisorixin methyl ester[a] [Polyether]	*Streptomyces* sp. SF10	Semiarid soil, Chélia, Khenchela, Algeria	722	Cytotoxic to glio-blastoma cancer stem	Leulmi et al. (2019)
73.	Spectinabilin [Polyketide]	*Streptomyces* sp. V_{002}	Forest soil, El Ogbane, Saida, Algeria	477.5	Gram+ bacteria	Gacem et al. (2020)
					Gram− bacteria	
					Antioxydant	

74.	Undecylprodigiosin [Prodigiosin derivative red pigment]	*Streptomyces* sp. V_{002}	Forest soil, El Ogbane, Saida, Algeria	393.5	Gram+ bacteria	Gacem et al. (2020)
					Antioxydant	
75.	Metacycloprodigiosin [Prodigiosin derivative red pigment]	*Streptomyces* sp. V_{002}	Forest soil, El Ogbane, Saida, Algeria	391.5	Gram+ bacteria	Gacem et al. (2020)
					Antioxydant	
76.	Cyclo-(Leu-Pro) [Cyclo dipeptide diketopiperazine]	*Streptomyces* sp. TN638	Industrial waste soil, Rades, Tunisia	210	Gram+ bacteria	Kaaniche et al. (2020)
					Gram− bacteria	
77.	Cyclo-(Val-Pro) [Cyclo dipeptide diketopiperazine]	*Streptomyces* sp. TN638	Industrial waste soil, Rades, Tunisia	196	Gram+ bacteria	Kaaniche et al. (2020)
					Gram− bacteria	
78.	Cyclo-(Phe-Pro) [Cyclo dipeptide diketopiperazine]	*Streptomyces* sp. TN638	Industrial waste soil, Rades, Tunisia	244	Gram+ bacteria	Kaaniche et al. (2020)
					Gram− bacteria	
79.	Nonactin [Macrotetrolide]	*Streptomyces* sp. TN638	Industrial waste soil, Rades, Tunisia	737	Gram+ bacteria	Kaaniche et al. (2020)
					Gram− bacteria	
80.	Monactin [Macrotetrolide]	*Streptomyces* sp. TN638	Industrial waste soil, Rades, Tunisia	751	Gram+ bacteria	Kaaniche et al. (2020)
					Gram− bacteria	
81.	Dinactin [Macrotetrolide]	*Streptomyces* sp. TN638	Industrial waste soil, Rades, Tunisia	765	Gram+ bacteria	Kaaniche et al. (2020)
					Gram− bacteria	
82.	Trinactin [Macrotetrolide]	*Streptomyces* sp. TN638	Industrial waste soil, Rades, Tunisia	307	Gram+ bacteria	Kaaniche et al. (2020)
					Gram− bacteria	
83.	3-Phenylpyrazin-2(1H)-one[a]	*Streptomyces* sp. TN82	Saharan soil, Tunisia	172	Gram+ bacteria	El Euch et al. (2018)
					Gram− bacteria	
84.	3-*O*-Methylviridicatin[b] [Alkaloid]	*Streptomyces* sp. TN82	Saharan soil, Tunisia	251	Gram+ bacteria	El Euch et al. (2018)
					Gram− bacteria	
85.	Cyclo-(Leu-Pro) [Cyclo dipeptide diketopiperazine]	*Streptomyces* sp. TN82	Saharan soil, Tunisia	210	Nd	El Euch et al. (2018)
86.	Anthranilic acid [Aminio acid]	*Streptomyces* sp. TN82	Saharan soil, Tunisia	137	Nd	El Euch et al. (2018)
87.	Indole-3-carbaldehyde	*Streptomyces* sp. TN82	Saharan soil, Tunisia	145	Nd	El Euch et al. (2018)

(continued)

Table 2.1 (continued)

No.	Compound name [Chemical group]	Producer microorganism	Microbial isolation origine	Molecular weight (g/mol)	Target	Reference
88.	*p*-Hydroxybenzoic acid [Phenolic derivative]	*Streptomyces* sp. TN82	Saharan soil, Tunisia	138	Nd	El Euch et al. (2018)
89.	2,4-Bis (1,1-dimethylethyl) phenol	*Streptomyces* sp. TN272	Arid soil, South Tunisia	278.5	Nd	Elleuch et al. (2012)
90.	1-Hexadecene [Alkene]	*Streptomyces* sp. TN272	Arid soil, South Tunisia	224	Nd	Elleuch et al. (2012)
91.	5-Octadecen [Alkene]	*Streptomyces* sp. TN272	Arid soil, South Tunisia	252.5	nd	Elleuch et al. (2012)
92.	Cis-cyclo (L-prolyl-L-valyl) [Cyclo peptide diketopiperazine]	*Streptomyces* sp. TN272	Arid soil, South Tunisia	196	nd	Elleuch et al. (2012)
93.	Hexadecanoic acid [Fatty acids]	*Streptomyces* sp. TN272	Arid soil, South Tunisia	256	nd	Elleuch et al. (2012)
94.	Cyclo (leucyl-prolyl) [Cyclo peptide diketopiperazine]	*Streptomyces* sp. TN272	Arid soil, South Tunisia	210	nd	Elleuch et al. (2012)
95.	Cis-cyclo(phenyl-prolyl) [Cyclo peptide diketopiperazine]	*Streptomyces* sp. TN272	Arid soil, South Tunisia	244	nd	Elleuch et al. (2012)
96.	3(Z)-tetradecene [Alkene]	*Streptomyces* sp. TN272	Arid soil, South Tunisia	196	nd	Elleuch et al. (2012)
97.	Trans1,10-dimethyl-trans-9-decal	*Streptomyces* sp. TN272	Arid soil, South Tunisia	182	nd	Elleuch et al. (2012)
98.	1-Nonadecene [Alkene]	*Streptomyces* sp. TN272	Arid soil, South Tunisia	266.5	nd	Elleuch et al. (2012)
99.	7-Hydroxy-tetradecanoic acid [Hydroxy fatty acids]	*Streptomyces* sp. TN272	Arid soil, South Tunisia	244	nd	Elleuch et al. (2012)
100.	7-Hydroxy-pentadecanoic acid [Hydroxy fatty acids]	*Streptomyces* sp. TN272	Arid soil, South Tunisia	272	nd	Elleuch et al. (2012)

101.	9-Hydroxy-hexadecanoic acid [Hydroxy fatty acids]	*Streptomyces* sp. TN272	Arid soil, South Tunisia	272	nd	Elleuch et al. (2012)
102.	9-hydroxy-heptadecanoic acid [Hydroxy fatty acids]	*Streptomyces* sp. TN272	Arid soil, South Tunisia	286	nd	Elleuch et al. (2012)
103.	lichenysin-G9a [Cyclic lipopeptide]	*Streptomyces* sp. TN272	Arid soil, South Tunisia	1035	nd	Elleuch et al. (2012)
104.	lichenysin-G9b [Cyclic lipopeptide]	*Streptomyces* sp. TN272	Arid soil, South Tunisia	1049	nd	Elleuch et al. (2012)
105.	lichenysin-G8b [Cyclic lipopeptide]	*Streptomyces* sp. TN272	Arid soil, South Tunisia	1078	nd	Elleuch et al. (2012)
106.	lichenysin-G8a [Cyclic lipopeptide]	*Streptomyces* sp. TN272	Arid soil, South Tunisia	662	nd	Elleuch et al. (2012)
107.	Compound 18[c] [Cyclic lipopeptide]	*Streptomyces* sp. TN272	Arid soil, South Tunisia	662	nd	Elleuch et al. (2012)
108.	*N*-[2-(1H-indol-3-yl)-2 oxo-ethyl] acet-amide [Alkaloid]	*Streptomyces* TN256	Saharan soil, Tunisia	216	Gram+ bacteria	Smaoui et al. (2012)
					Gram− bacteria	
					Microfungi	
109.	Di-(2-ethylhexyl) phthalate [Phthalate]	*Streptomyces* TN256	Saharan soil, Tunisia	390	Gram+ bacteria	Smaoui et al. (2012)
					Gram− bacteria	
					Microfungi	
110.	1-Nonadecene [Alkene]	*Streptomyces* TN256	Saharan soil, Tunisia	266	Gram+ bacteria	Smaoui et al. (2012)
					Gram− bacteria	
					Microfungi	
111.	Cyclo (L-Pro-LTyr) [Cyclo dipeptide diketopiperazine]	*Streptomyces* TN256	Saharan soil, Tunisia	260	Gram+ bacteria	Smaoui et al. (2012)
					Gram− bacteria	
					Microfungi	
112.	1-Acetyl-β-carboline [Alkaloid]	*Streptomyces* sp. TN262	Arid soil, South Tunisia	210	nd	Elleuch et al. (2010)

(continued)

Table 2.1 (continued)

No.	Compound name [Chemical group]	Producer microorganism	Microbial isolation origine	Molecular weight (g/mol)	Target	Reference
113.	Tryptophol [Aromatic alcohol]	*Streptomyces* sp. TN262	Arid soil, South Tunisia	161	Gram+ bacteria	Elleuch et al. (2010)
					Gram− bacteria	
					Microfungi	
114.	Cineromycin B [Macrolide]	*Streptomyces* sp. TN262	Arid soil, South Tunisia	294	Gram+ bacteria	Elleuch et al. (2010)
115.	2,3-dihydrocineromycin B [Macrolide]	*Streptomyces* sp. TN262	Arid soil, South Tunisia	296	Gram+ bacteria	Elleuch et al. (2010)
116.	Cyclo-(tyrosyl prolyl) [Cyclo dipeptide diketopiperazine]	*Streptomyces* sp. TN262	Arid soil, South Tunisia	828	nd	Elleuch et al. (2010)
117.	3-(hydroxyacetyl)-indole	*Streptomyces* sp. TN262	Arid soil, South Tunisia	175	nd	Elleuch et al. (2010)
118.	Brevianamide F [Alkaloid]	*Streptomyces* sp. TN262	Arid soil, South Tunisia	283	nd	Elleuch et al. (2010)
119.	Cis-cyclo-(L-prolyl-L-leucyl) [Alkaloid]	*Streptomyces* sp. TN262	Arid soil, South Tunisia	210	nd	Elleuch et al. (2010)
120.	Benzophenone [Alkaloid]	*Streptomyces* sp. TN262	Arid soil, South Tunisia	182	nd	Elleuch et al. (2010)
121.	*N*-butyl-benzenesulfonamide	*Streptomyces* sp. TN262	Arid soil, South Tunisia	213	nd	Elleuch et al. (2010)
122.	Hexanedioic acid-bis-(2-ethylhexyl) ester[a] [Polyether]	*Streptomyces* sp. TN262	Arid soil, South Tunisia	370	nd	Elleuch et al. (2010)
123.	Brevianamide F [Alkaloid]	*Streptomyces* sp. TN605	Arid soil, South Tunisia	283	Gram+ bacteria	Mehdi et al. (2009), Amar et al. (2012)
124.	Tryptophol [aromatic alcohol]	*Streptomyces* sp. TN605	Arid soil, South Tunisia	161	Gram+ bacteria	Amar et al. (2012)
					Yeasts	

125.	Cyclo (L-Leu, L-Pro) [Cyclo dipeptide diketopiperazine]	*Streptomyces* sp. TN605	Arid soil, South Tunisia	210	Gram+ bacteria	Mehdi et al. (2006), Amar et al. (2012)
					Gram− bacteria	
					Microfungi	
126.	2-Hydroxyphenylacetic acid [Aromatic]	*Streptomyces* sp. TN605	Arid soil, South Tunisia	152	nd	Amar et al. (2012)
127.	Cyclo (L-Leu-L-Arg) [Cyclo dipeptide diketopiperazine]	*Streptomyces* sp. TN17	Oasis soil, South Tunisia	269	Gram+ bacteria	Smaoui et al. (2011)
					Gram− bacteria	
					Microfungi	
128.	Di-(2-ethylhexyl) phthalate [Phthalate]	*Streptomyces* sp. TN17	Oasis soil, South Tunisia	390	Gram+ bacteria	Smaoui et al. (2011)
					Microfungi	
129.	Cyclo 1-(2-(cyclopentanecarbonyl-3-phenyl-propionyl)-pyrrolidine-2-carboxylic acid (1-carbamoyl-propyl)-amide [Cyclic tetrapeptide]	*Streptomyces* sp. TN17	Oasis soil, South Tunisia	488	Gram+ bacteria	Smaoui et al. (2011)
					Microfungi	
130.	brevianamide F[b] [Alkaloid]	*Streptomyces* sp. TN58	Tunisian soil	283	Gram+ bacteria	Mehdi et al. (2009)
					Microfungi	
131.	*N*-acetyltryptamine [β-carbolines]	*Streptomyces* sp. TN58	Tunisian soil	202	Microfungi	Mehdi et al. (2009)
132.	Thiazolidomycin [Thiazolidine]	*Streptomyces* sp. TN58	Tunisian soil	217	Gram+ bacteria	Mehdi et al. (2009)
					Gram− bacteria	
					Microfungi	
133.	1-*O*-(2-Aminobenzoyl)-L-rhamnoside [Rhamnopyranosides]	*Streptomyces* sp. TN58	Tunisian soil	283	*Staphylococcus*	Mehdi et al. (2009)
134.	4-Hydroxybenzoyl-L-rhamnopyranosid [Rhamnopyranosides]	*Streptomyces* sp. TN58	Tunisian soil	284	*Staphylococcus*	Mehdi et al. (2009)
135.	Cis-cyclo (Leucyl-Prolyl) [Cyclo dipeptide diketopiperazine]	*Streptomyces* sp. TN97	Oasis soil, Tunisia	210	Gram+ bacteria	Mehdi et al. (2006)
					Gram− bacteria	
					Microfungi	

(continued)

Table 2.1 (continued)

No.	Compound name [Chemical group]	Producer microorganism	Microbial isolation origine	Molecular weight (g/mol)	Target	Reference
136.	Cis-cyclo (L-phenyl, L-prolyl) [Cyclo dipeptide diketopiperazine]	*Streptomyces* sp. TN97	Oasis soil, Tunisia	244	Microfungi	Mehdi et al. (2006)
137.	6,8-dihydroxy-3-methylisocoumarin [Isocoumarins]	*Streptomyces* sp. TN97	Oasis soil, Tunisia	192	nd	Mehdi et al. (2006)
138.	*N*-acetyl-tyramine	*Streptomyces* sp. TN97	Oasis soil, Tunisia	179	Gram+ bacteria	Mehdi et al. (2006)
					Microfungi	
					Yeasts	
139.	Irumamycin [Macrolide]	*Streptomyces* sp. US80	Oasis soil, Tunisia	763	Gram+ bacteria	Fguira et al. (2005)
					Microfungi	
					Yeasts	
140.	X-14952B [Macrolide]	*Streptomyces* sp. US80	Oasis soil, Tunisia	155	Gram+ bacteria	Fguira et al. (2005)
					Microfungi	
					Yeasts	
141.	17-hydroxy-venturicidin A [Macrolide]	*Streptomyces* sp. US80	Oasis soil, Tunisia	765	Gram+ bacteria	Fguira et al. (2005)
					Microfungi	
					Yeasts	
142.	Cyclo (L-phe, L-pro)[b] [Cyclo dipeptide diketopiperazine]	*Streptomyces* sp. US24	Tunisian soil	244	nd	Mehd et al. (2004)
143.	3-indoleethanol[b] [Aromatic alcohol]	*Streptomyces* sp. US24	Tunisian soil	161	nd	Mehd et al. (2004)
144.	Novonestmycin A [Nonpolyenic Macrolide]	*Streptomyces* sp. Z26	Moroccan Rhizospheric soil	1229	Microfungi	Nafis et al. (2018a)
					Yeasts	

145.	Novonestmycin B [Nonpolyenic Macrolide]	*Streptomyces* sp. Z26	Moroccan rhizospheric soil	1343	Microfungi	Nafis et al. (2018a)
					Yeasts	
146.	Antimycin A19 [Nonpolyenic lactone]	*Streptomyces* sp. AS25	Moroccan rhizospheric soil	548	Microfungi	Nafis et al. (2018b)
					Yeasts	
147.	Ymycine[a]	*Streptomyces* sp. BS46	Moroccan rhizospheric soil	441	Gram+ bacteria (methicillin-resistant)	Ouhdouch and Barakate (2006)
					Microfungi	
					Yeasts	
148.	Streptochlorin [Indole alkaloid]	*Streptomyces* sp. S37	Moroccan rhizospheric soil	218	Microfungi	Couillerot et al. (2014)
149.	Nigericin [Carboxylic polyether]	*Streptomyces* sp. S37	Moroccan rhizospheric soil	725	Microfungi	Couillerot et al. (2014)
150.	Piericidin A1 [α-pyridone]	*Streptomyces* sp. S37	Moroccan rhizospheric soil	415	Microfungi	Couillerot et al. (2014)
151.	Isochainin [Macrolide]	*Streptomyces* sp. AP1	Moroccan rhizospheric soil	610	Microfungi	Bouizgarne et al. (2006)
152.	Viridomycin like D2 [iron chelating green pigment]	*Streptomyces griseus*	Moroccan phosphate mines	506	Gram-positive bacteria and Microfungi	Hamdali et al. (2021)

nd not determined

[a]Novel compound (Cf. Fig. 2.2 for molecular structure details as referenced according there numbering in this table)

[b]Antibiotic group or compound-first documented production from the reported bacterial genus

[c]Partially characterized compound

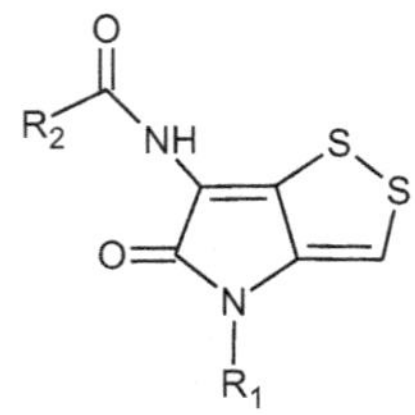

R1= CH_3 (pyrrothin group)

R2=	$(CH_2)_2–CH_3$	Butanoyl-pyrrothine
	$CH=C(CH_3)_2$	Senecioyl-pyrrothine
	$C(CH_3)=CH(CH_3)$	Tigloyl-pyrrothine
	$CH_2CH_2CH_2CH_3$	Valeryl-pyrrothine
	$CH(CH_3)(CH_2CH_3)$	Isovaleryl-pyrrothine
	H	Formyl-pyrrothine
	$CH=CH(CH_3)$	Crotonyl-pyrrothine
	$CH=CH–CH=CH(CH_3)$	Sorbyl-pyrrothine
	$CH=CH–CH_2CH_2CH_3$	2-Hexonyl-pyrrothine
	$CH=C(CH_3)(CH_2CH_3)$	2-Methyl-3-pentenyl-pyrrothine
	C_6H_6	Benzoyl-pyrrothine

R1= H (holothin group)

R2=	CH_2CH_5	Demethyl-benzoyl-pyrrothine (Benzoyl-holothine)

Dithiolopyrrolones *(1-17)*

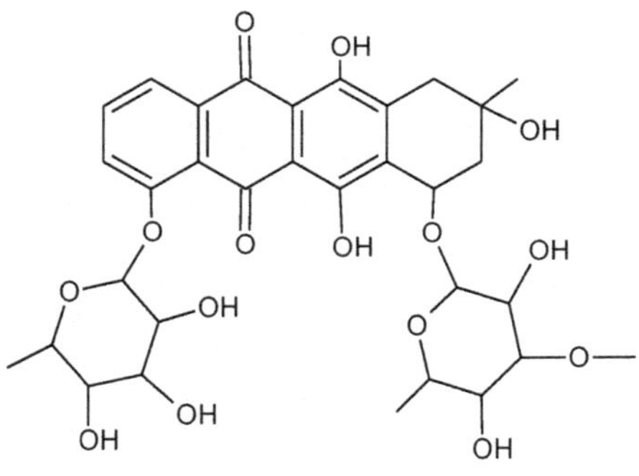

Mutamycin PR *(19)*

Fig. 2.2 Molecular structure of the novel compounds derived from actinobacteria inhabiting Maghreb environments Italicized numbers referred to the corresponding molecule order in Table 2.1

The success of the postulated strategy of enhancing the potential of obtaining interesting secondary metabolites from the analysis of adapted microorganisms to extreme conditions encourage fostering the intensification efforts of microbial

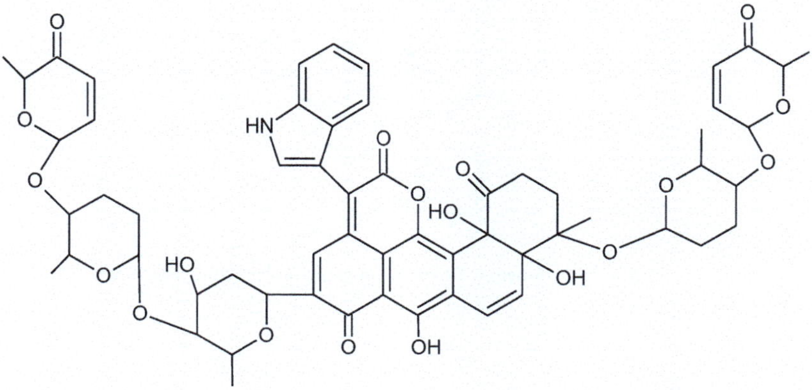

Mzabimycin A *(24)*

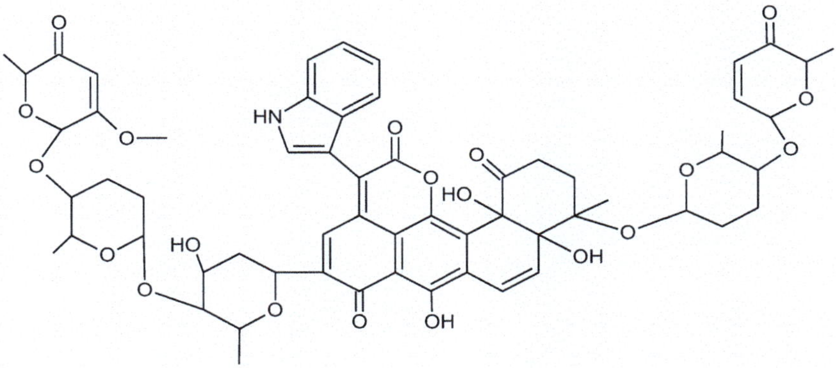

Mzabimycin B *(25)*

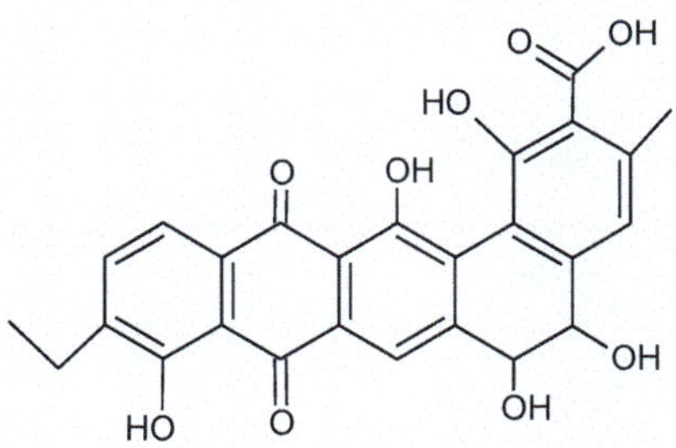

Angucyclinone R2 *(26)*

Fig. 2.2 (continued)

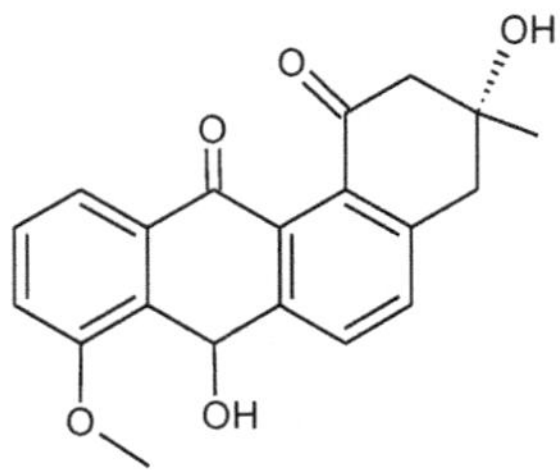

(−)-7-deoxo-6-deoxy-7-hydroxy-8-O-methylrabelomycin *(32)*

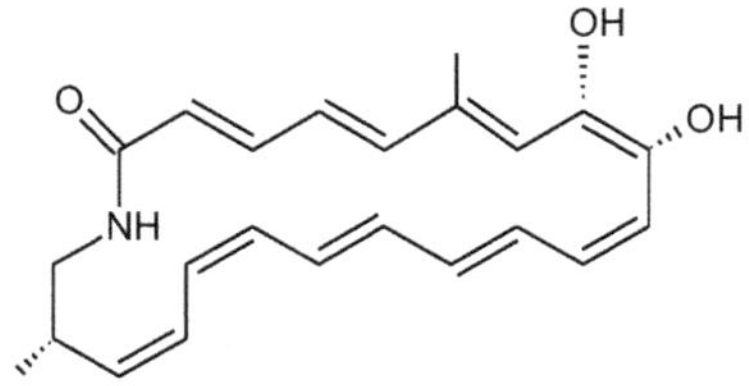

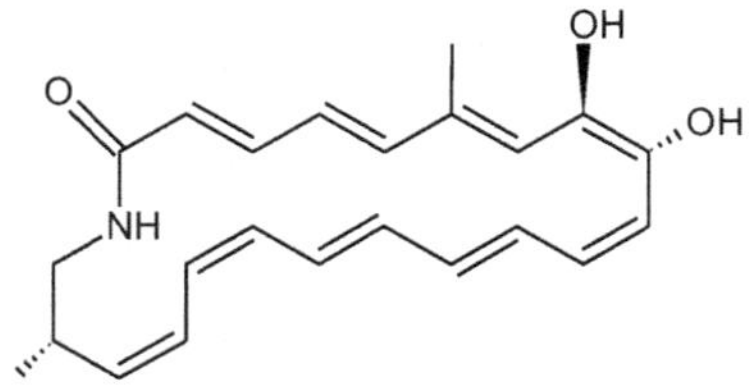

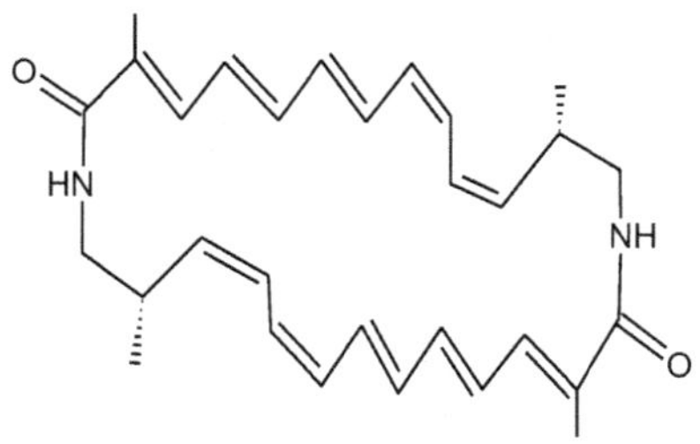

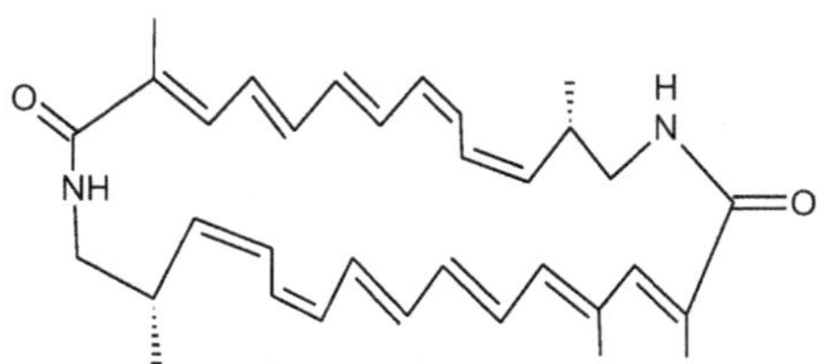

Fig. 2.2 (continued)

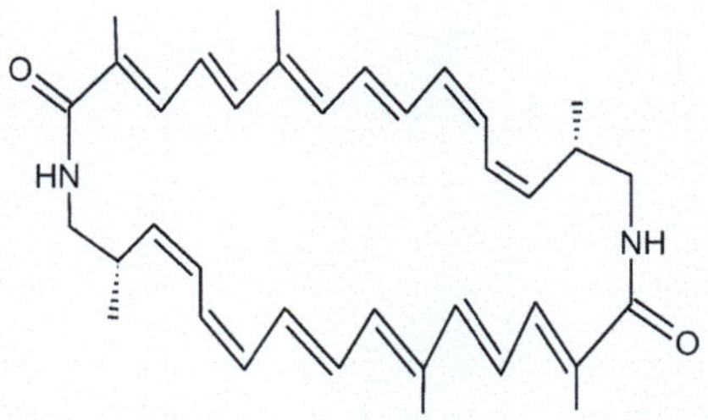

Kenalactams A–E *(33-37)*

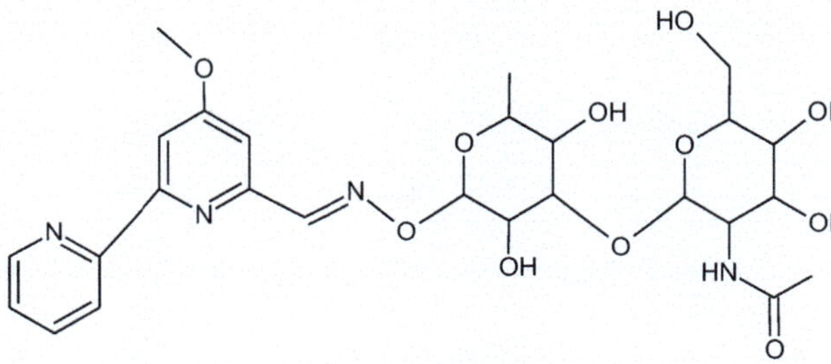

Cyanogriside I *(40)*

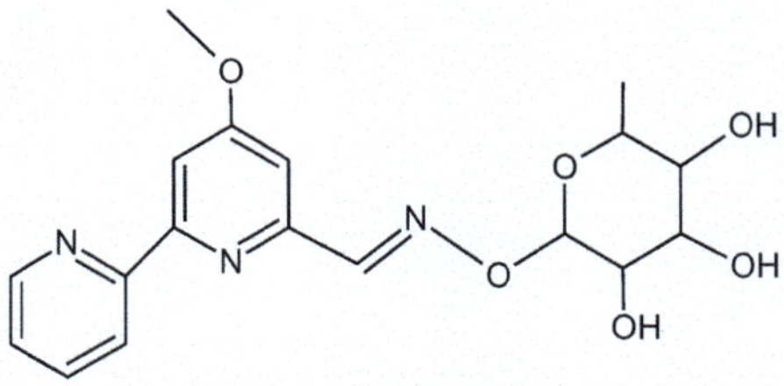

Cyanogriside J *(41)*

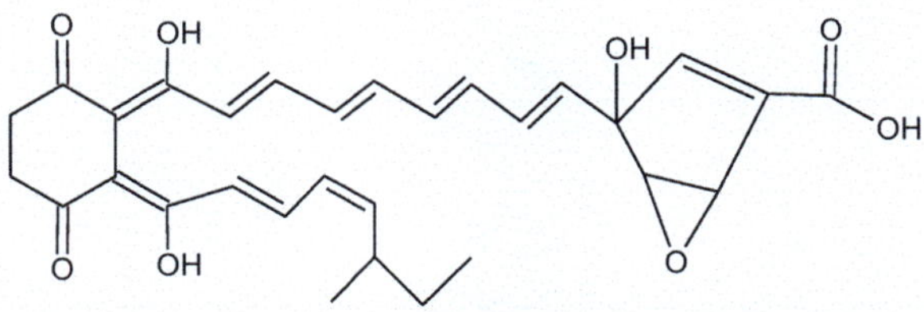

4-hydroxy-4-{9-hydroxy-9-[2-(1-hydroxy-6-methyl-hepta- 2,4-dienylidene)-3,6-dioxo-cyclohexylidene]-nona-1,3,5,7- tetraenyl}-6-oxa-bicyclo[310]hex-2-ene-2-carboxylic acid (antibiotic A4) *(59)*

Fig. 2.2 (continued)

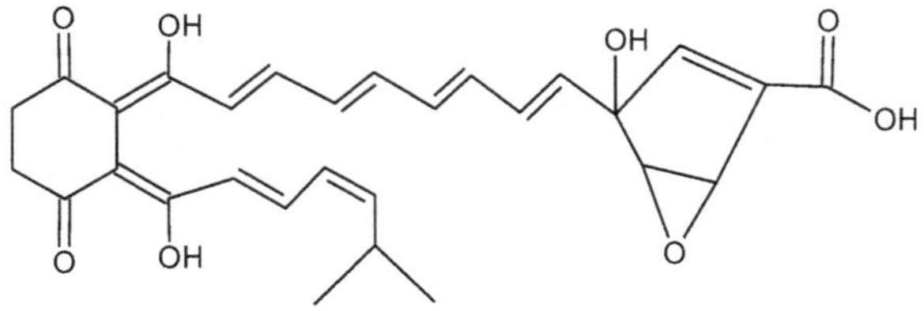

4-hydroxy-4-{9-hydroxy-9-[2-(1-hydroxy-6-methyl-octa- 2,4-dienylidene)-3,6-dioxo-cyclohexylidene]-nona-1,3,5,7-tetraenyl}-6-oxa-bicyclo[310]hex-2-ene-2-carboxylic acid (antibiotic A5) *(60)*

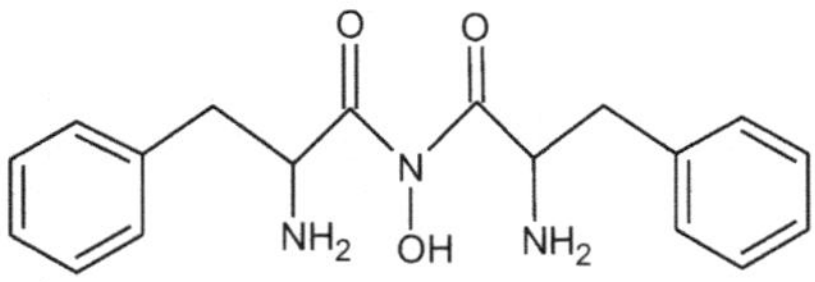

2-amino-N-(2-amino-3-phenylpropanoyl)-N-hydroxy-3-phenylpropanamide *(67)*

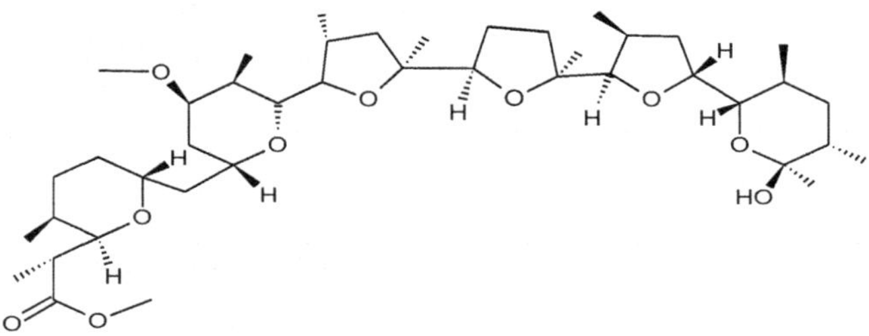

Grisorixin methyl ester *(72)*

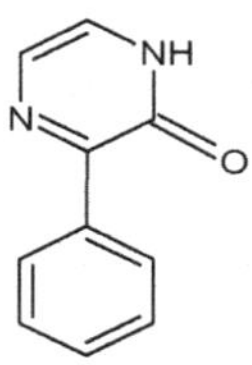

3-phenylpyrazin-2(1H)-one *(83)*

Fig. 2.2 (continued)

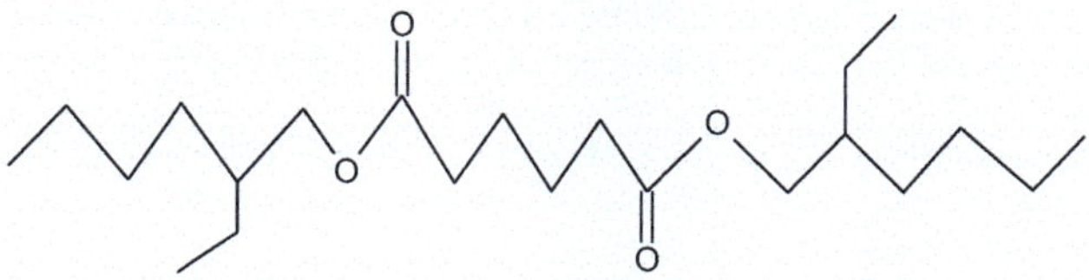

Hexanedioic acid-bis-(2-ethylhexyl) ester *(122)*

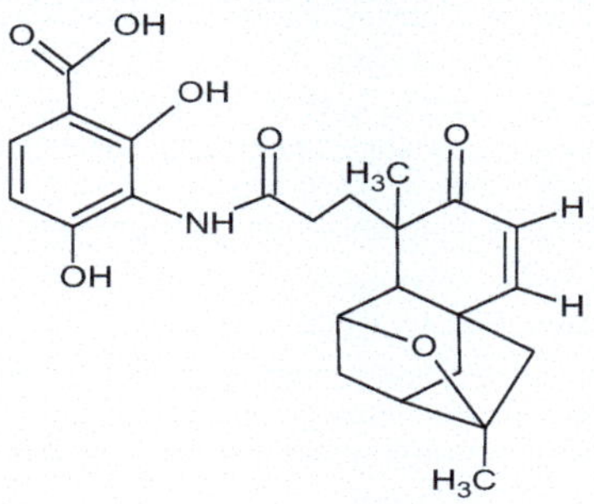

Ymycine *(147)*

Fig. 2.2 (continued)

screening on such versatile environments to expect future finding of bioactive compounds with potential benefit in human health, industrial process and agriculture.

References

Ait Barka E, Vatsa P, Sanchez L, Gaveau-Vaillant N, Jacquard C, Meier-Kolthoff JP, Klenk H-P, Clément C, Ouhdouch Y, van Wezel GP (2015) Taxonomy, physiology, and natural products of actinobacteria. Microbiol Mol Biol Rev 80:1–43

Amar MB, Elleuch L, Abd-Alla HI, Najah S, Chakchouk A, Damak M, Ben Salem R, Mohamed Shaaba M, Mellouli L (2012) The new *Streptomyces* sp TN605 strain secretes simultaneously three active compounds and a high level of the interesting pharmaceutical industry intermediate: 2–hydroxyphenylacetic acid. Bull Environ Pharmacol Life Sci 1:48–56

Aouiche A, Sabaou N, Meklat A, Zitouni A, Bijani C, Mathieu F, Lebrihi A (2012) *Saccharothrix* sp PAL54, a new chloramphenicol–producing strain isolated from a Saharan soil. World J Microbiol Biotechnol 28(3):943–951

Aouiche A, Bijani C, Zitouni A, Mathieu F, Sabaou N (2014) Antimicrobial activity of saquayamycins produced by *Streptomyces* spp PAL114 isolated from a Saharan soil. J Med Mycol 24(2):e17–e23

Aouiche A, Meklat A, Bijani C, Zitouni A, Sabaou N, Mathieu F (2015) Production of vineomycin A1 and chaetoglobosin A by *Streptomyces* sp PAL114. Ann Microbiol 65(3):1351–1359

Badji B, Riba A, Mathieu F, Lebrihi A, Sabaou N (2005) Antifungal activity of a saharan Actinomadura strain against various pathogenic and toxinogenic fungi. J Med Mycol 15 (4):211–219

Badji B, Zitouni A, Mathieu F, Lebrihi A, Sabaou N (2006) Antimicrobial compounds produced by *Actinomadura* sp AC104 isolated from an Algerian Saharan soil. Can J Microbiol 52 (4):373–382

Badji B, Mostefaoui A, Sabaou N, Lebrihi A, Mathieu F, Seguin E, Tillequin F (2007) Isolation and partial characterization of antimicrobial compounds from a new strain *Nonomuraea* sp NM94. J Ind Microbiol Biotechnol 34(6):403–412

Belghit S, Driche EH, Bijani C, Zitouni A, Sabaou N, Badji B, Mathieu F (2016) Activity of 2, 4–di–tert–butylphenol produced by a strain of *Streptomyces mutabilis* isolated from a Saharan soil against *Candida albicans* and other pathogenic fungi. J Med Mycol 26(2):160–169

Berdy J (2005) Bioactive microbial metabolites. J Antibiot 58(1):1–26

Boubetra D, Sabaou N, Zitouni A, Bijani C, Lebrihi A, Mathieu F (2013) Taxonomy and chemical characterization of new antibiotics produced by *Saccharothrix* SA198 isolated from a Saharan soil. Microbiol Res 168(4):223–230

Boudjelal F, Zitouni A, Mathieu F, Lebrihi A, Sabaou N (2011) Taxonomic study and partial characterization of antimicrobial compounds from a moderately halophilic strain of the genus *Actinoalloteichus*. Braz J Microbiol 42(3):835–845

Boudjella H, Bouti K, Zitouni A, Mathieu F, Lebrihi A, Sabaou N (2007) Isolation and partial characterization of pigment-like antibiotics produced by a new strain of *Streptosporangium* isolated from an Algerian soil. J Appl Microbiol 103(1):228–236

Boudjella H, Zitouni A, Coppel Y, Mathieu F, Monje MC, Sabaou N, Lebrihi A (2010) Antibiotic R2, a new angucyclinone compound from *Streptosporangium* sp Sg3. J Antibiot 63 (12):709–711

Boudjella H, Lamari L, Bouti K, Sabaou N (2014) Activité antilevurienne d'une souche d'actinobactérie appartenant au genre *streptosporangium* et isolée d'un sol saharien. Algerian J Ari Environ 4(2):3–18

Bouizgarne B, El Hadrami I, Ouhdouch Y (2006) Novel production of isochainin by a strain of *streptomyces* sp isolated from rhizosphere soil of the indigenous Moroccan plant *Argania Spinosa* L. World J Microbiol Biotechnol 22(5):423–429

Boukhalfa RYH, Eveno Y, Karama S, Selama O, Lauga B, Duran R, Hacène H, Eparvier V (2017) Isolation, purification and chemical characterization of a new angucyclinone compound produced by a new halotolerant *Nocardiopsis* sp HR–4 strain. World J Microbiol Biotechnol 33 (6):126

Bouras N, Mathieu F, Sabaou N, Lebrihi A (2006) Effect of amino acids containing sulfur on dithiolopyrrolone antibiotic productions by *Saccharothrix algeriensis* NRRL B-24137. J Appl Microbiol 100(2):390–397

Bouras N, Merrouche R, Lamari L, Mathieu F, Sabaou N, Lebrihi A (2008) Precursor–directed biosynthesis of new dithiolopyrrolone analogs by *Saccharothrix algeriensis* NRRL B–24137. Process Biochem 43(11):1244–1252

Couillerot O, Loqman S, Toribio A, Hubert J, Gandner L, Nuzillard JM, Ouhdouch Y, Clement C, Ait Barka S, Renault JH (2014) Purification of antibiotics from the biocontrol agent *Streptomyces anulatus* S37 by centrifugal partition chromatography. J Chromatogr B 944:30–34

Demain AL (2006) From natural products discovery to commercialization: a success story. J Ind Microbiol Biotechnol 33(7):486–495

Demain AL, Sanchez S (2009) Microbial drug discovery: 80 years of progress. J Antibiot 62 (1):5–16

Djinni I, Djoudi W, Souagui S, Rabia F, Rahmouni S, Mancini I, Kecha M (2018) *Streptomyces thermoviolaceus* SRC3 strain as a novel source of the antibiotic adjuvant streptazolin: a statistical approach toward the optimized production. J Microbiol Methods 148:161–168

Djinni I, Defant A, Djoudi W, Chaouch FC, Souagui S, Kecha M, Mancini I (2019) Modeling improved production of the chemotherapeutic polypeptide actinomycin D by a novel *Streptomyces* sp strain from a Saharan soil. Heliyon 5(5):e01695

Driche EH, Belghit S, Bijani C, Zitouni A, Sabaou N, Mathieu F, Badji B (2015) A new *Streptomyces* strain isolated from Saharan soil produces di–(2–ethylhexyl) phthalate, a metabolite active against methicillin–resistant *Staphylococcus aureus*. Ann Microbiol 65 (3):1341–1350

Driche EH, Sabaou N, Bijani C, Zitouni A, Pont F, Mathieu F, Badji B (2017) *Streptomyces* sp AT37 isolated from a Saharan soil produces a furanone derivative active against multidrug–resistant *Staphylococcus aureus*. World J Microbiol Biotechnol 33(6):105

El Euch IZ, Frese M, Sewald N, Smaoui S, Shaaban M, Mellouli L (2018) Bioactive secondary metabolites from new terrestrial *Streptomyces* sp TN82 strain: isolation, structure elucidation and biological activity. Med Chem Res 27(4):1085–1092

Elleuch L, Shaaban M, Smaoui S, Mellouli L, Karray-Rebai I, Fguira LFB, Shaaban KA, Laatsch H (2010) Bioactive secondary metabolites from a new terrestrial *Streptomyces* sp TN262. Appl Biochem Biotechnol 162(2):579–593

Elleuch L, Shaaban KA, Abdel-Aziz MS, Chakchouk A, Nagia MM, Mellouli L, Shaaban M (2012) Cyclic lipopeptides and other bioactive secondary metabolites from a new terrestrial *Streptomyces* sp TN272. Afr J Microbiol Res 6(9):2202–2210

Fair RJ, Tor Y (2014) Antibiotics and bacterial resistance in the 21st Century. Perspect Medicin Chem 6(6):25–64

Fguira LFB, Fotso S, Ameur-Mehdi RB, Mellouli L, Laatsch H (2005) Purification and structure elucidation of antifungal and antibacterial activities of newly isolated *Streptomyces* sp strain US80. Res Microbiol 156(3):341–347

Gacem MA, Ould-El-Hadj-Khelil A, Boudjemaa B, Wink J (2020) Antimicrobial and antioxidant effects of a forest actinobacterium V_{002} as new producer of spectinabilin, undecylprodigiosin and metacycloprodigiosin. Curr Microbiol 77(10):2575–2583

Hacène H, Lefebvre G (1995) AH17, a new non–polyenic antifungal antibiotic produced by a strain of *Spirillospora*. Microbios 83(336):199–205

Hacène H, Lefebvre G (1996) HP17, a new pigment-like antibiotic produced by a new strain of *Spirillospora*. J Appl Bacteriol 80(5):565–569

Hamdali H, Lebrihi A, Monje MC, Benharref A, Hafidi M, Ouhdouch Y, Virolle MJ (2021) A molecule of the viridomycin family originating from a *Streptomyces griseus*-related strain has the ability to solubilize rock phosphate and to inhibit microbial growth. Antibiotics 10(1):72

Hesketh AR, Chandra G, Shaw AD, Rowland JJ, Kell DB, Bibb MJ, Chater KF (2002) Primary and secondary metabolism, and post–translational protein modifications, as portrayed by proteomic analysis of *Streptomyces coelicolor*. Mol Microbiol 46:917–932

Kaaniche F, Hamed A, Elleuch L, Chakchouk-Mtibaa A, Smaoui S, Karray-Rebai I, Koubaa I, Arcile G, Allouche N, Mellouli L (2020) Purification and characterization of seven bioactive compounds from the newly isolated *Streptomyces cavourensis* TN638 strain via solid–state fermentation. Microb Pathog 142:104106

Khebizi N, Boudjella H, Bijani C, Bouras N, Klenk HP, Pont F, Mathieu F, Sabaou N (2018) Oligomycins A and E, major bioactive secondary metabolites produced by *Streptomyces* sp strain HG29 isolated from a Saharan soil. J Med Mycol 28(1):150–160

Lahoum A, Sabaou N, Bijani C, Bouras N, Pont F, Snini SP, Mathieu F (2019) Antimicrobial activities of novel bipyridine compounds produced by a new strain of *Saccharothrix* isolated from Saharan soil. Saudi Pharm J 27(1):56–65

Lamari L, Zitouni A, Dob T, Sabaou N, Lebrihi A, Germain P, Seguin E, Tillequin F (2002) New Dithiolopyrrolone Antibiotics from *Saccharothrix* sp SA 233 II Physicochemical Properties and Structure Elucidation. J Antibiot 55(8):702–706

Le Houérou HN (1990) Recherches écoclimatique et biogéographique sur les zones arides. de l'Afrique du Nord CEPE/CNRS, Montpellier. 600 p

Leulmi N, Sighel D, Defant A, Khenaka K, Boulahrouf A, Mancini I (2019) Nigericin and grisorixin methyl ester from the Algerian soil-living *Streptomyces youssoufiensis* SF10 strain: a computational study on their epimeric structures and evaluation of glioblastoma stem cells growth inhibition. Natur P Res 33(2):266–273

Li B, Webster TJ (2018) Bacteria antibiotic resistance: new challenges and opportunities for implant–associated orthopaedic infections. J Orthop Res 36(1):22–32

Li ZF, Li P, Jing YG, Singh RP, Li YZ (2017) Isolation and characterization of the epothilone genecluster with flanks from high alkalotolerant strain *Sorangium cellulosum* (So0157-2). World J Microbiol Biotechnol 33(7):137. https://doi.org/10.1007/s11274-017-2301-y

Mehd RBA, Mellouli L, Chabchoub F, Fotso S, Bejar S (2004) Purification and structure elucidation of two biologically active molecules from a new isolated *Streptomyces* sp US24 strain. Chem Nat Compound 40(5):510–513

Mehdi RBA, Sioud S, Fourati Ben Fguira L, Bejar S, Mellouli L (2006) Purification and structure determination of four bioactive molecules from a newly isolated *Streptomyces* sp TN97 strain. Process Biochem 41(7):1506–1513

Mehdi RBA, Shaaban KA, Rebai IK, Smaoui S, Bejar S, Mellouli L (2009) Five naturally bioactive molecules including two rhamnopyranoside derivatives isolated from the *Streptomyces* sp strain TN58. Natur P Res 23(12):1095–1107

Merrouche R, Bouras N, Coppel Y, Mathieu F, Monje MC, Sabaou N, Lebrihi A (2010) Dithiolopyrrolone antibiotic formation induced by adding valeric acid to the culture broth of *Saccharothrix algeriensis*. J Nat Prod 73(6):1164–1166

Merrouche R, Bouras N, Coppel Y, Mathieu F, Sabaou N, Lebrihi A (2011) New dithiolopyrrolone antibiotics induced by adding sorbic acid to the culture medium of *Saccharothrix algeriensis* NRRL B–24137. FEMS Microbiol Lett 318(1):41–46

Merrouche R, Yekkour A, Coppel Y, Bouras N, Lamari L, Zitouni A, Mathieu F, Sabaou N (2019a) Effective biosynthesis of benzoyl-pyrrothine dithiolopyrrolone antibiotic by cinnamic acid-precursor addition in culture of *Saccharothrix algeriensis* NRRL B-24137. Lett Appl Microbiol 68(2):165–172

Merrouche R, Yekkour A, Coppel Y, Bouras N, Zitouni A, Lebrihi A, Mathieu F, Sabaou N (2019b) A new dithiolopyrrolone antibiotic triggered by a long fermentation of *Saccharothrix algeriensis* NRRL B-24137 in sorbic acid-amended medium. Lett Appl Microbiol 69 (4):294–301

Merrouche R, Yekkour A, Coppel Y, Bouras N, Zitouni A, Mathieu F, Sabaou N (2020) *Saccharothrix algeriensis* NRRL B–24137, the first non-*Streptomyces* actinobacterium, produces holomycin after cystine feeding. Arch Microbiol 202(9):2509–2516

Messai Y, Iabadence H, Benhassine T, Alouache S, Tazir M, Guatier V, Arlet G, Bakour R (2008) Prevalence and characterization of extended spectrum *b*–lactamases in *Klebsiella pneumoniae* in Algiers hospitals (Algeria). Pathol Biol 56(5):319–325

Messaoudi O, Sudarman E, Bendahou M, Jansen R, Stadler M, Wink J (2019) Kenalactams A-E, Polyene Macrolactams Isolated from Nocardiopsis CG3. J Nat Prod 82(5):1081–1088

Nafis A, Oubaha B, Elhidar N, Ortlieb N, Kulik A, Niedermeyer T, Hassani L, Barakate M (2018a) Novel production of two nonpolyenic antifungal macrolide derivatives by *Streptomyces* Z26 isolated from Moroccan rhizospheric soil. Online J Biol Sci 18:176–185

Nafis A, Elhidar N, Oubaha B, Samri SE, Niedermeyer T, Ouhdouch Y, Hassani L, Barakate M (2018b) Screening for non–polyenic antifungal produced by actinobacteria from moroccan habitats: assessment of antimycin A19 production by *Streptomyces albidoflavus* AS25. Int J Mol Cell Med 7(2):133

Newman DJ, Cragg GM, Snader KM (2003) Natural products as sources of new drugs over the period 1981–2002. J Nat Prod 66:1022–1037

Nguyen Q-T, Merlo ME, Medema MH, Jankevics A, Breitling R, Takano E (2012) Metabolomics methods for the synthetic biology of secondary metabolism. FEBS Lett 586:2177–2183

Ouhdouch Y, Barakate M (2006) Antibiotic chemical structure obtained from the strain *Streptomyces* sp bs46 and its use against methicillin–resistant bacteria. Patent WO2008032140 A1

Prajakta BM, Suvarna PP, Singh RP, Rai AR (2019) Potential biocontrol and superlative plant growth promoting activity of indigenous *Bacillus mojavensis* PB-35(R11) of soybean (*Glycine max*) rhizosphere. SN Appl Sci 1:1143. https://doi.org/10.1007/s42452-019-1149-1

Sabaou N, Boudjella H, Bennadji A, Mostefaoui A, Zitouni A, Lamari L, Bennadji H, Lefebvre G, Germain P (1998) Les sols des oasis du Sahara algérien, source d'actinomycètes rares producteurs d'antibiotiques. Séch 9:147–153

Sandegren L, Andersson DI (2009) Bacterial gene amplification: implications for the evolution of antibiotic resistance. Nat Rev Microbiol 7(8):7578–7588

Shirling EB, Gottlieb D (1966) Methods for characterization of *Streptomyces* species. Int J Syst Evol Microbiol 16:313–340

Smaoui S, Mellouli L, Lebrihi A, Coppel Y, Fguira LFB, Mathieu F (2011) Purification and structure elucidation of three naturally bioactive molecules from the new terrestrial *Streptomyces* sp TN17 strain. Nat Prod Res 25(8):806–814

Smaoui S, Mathieu F, Elleuch L, Coppel Y, Merlina G, Karray-Rebai I, Mellouli L (2012) Taxonomy, purification and chemical characterization of four bioactive compounds from new *Streptomyces* sp TN256 strain. World J Microbiol Biotechnol 28(3):793–804

Solanki R, Kahanna M (2008) Bioactive compounds from marine actinomycetes. Indian J Microbiol 48(4):410–431

Solecka J, Zajko J, Postek M, Rajnisz A (2012) Biologically active secondary metabolites from actinomycetes. Cent Eur J Biol 7:373–390

Subhashini DV, Singh RP (2014) Isolation of endophytic actinomycetes from roots and leaves of tobacco (*Nicotiana tabacum* L). Ann Plant Pro Sci 22(2):458–459

Subhashini DV, Singh RP, Manchanda G (2017) OMICS approaches: tools to unravel microbial systems. Directorate of Knowledge Management in Agriculture, Indian Council of Agricultural Research, New Delhi. ISBN: 9788171641703. https://books.google.com/books/about/OMICS_Approaches.html?id=vSaLtAEACAAJ

Takahashi Y, Nakashima T (2018) Actinomycetes, an inexhaustible source of naturally occurring antibiotics. Antibiotics 7(3):74

Tata S, Aouiche A, Bijani C, Bouras N, Pont F, Mathieu F, Sabaou N (2019) Mzabimycins A and B, novel intracellular angucycline antibiotics produced by *Streptomyces* sp PAL114 in synthetic medium containing L–tryptophan. Saudi Pharm J 27(7):907–913

Toumatia O, Yekkour A, Goudjal Y, Riba A, Coppel Y, Mathieu F, Sabaou N, Zitouni A (2015) Antifungal properties of an actinomycin D-producing strain, *Streptomyces* sp IA1, isolated from a Saharan soil. J Basic Microbiol 55(2):221–228

Watve MG, Tickoo R, Jog MM, Bhole BD (2001) How many antibiotics are produced by the genus *Streptomyces*? Arch Microbiol 176:386–390

Yekkour A, Meklat A, Bijani C, Toumatia O, Errakhi R, Lebrihi A, Mathieu F, Sabaou N (2015) A novel hydroxamic acid-containing antibiotic produced by a Saharan soil-living *Streptomyces* strain. Lett Appl Microbiol 60(6):589–596

Zitouni A, Boudjella H, Mathieu F, Sabaou N, Lebrihi A (2004) Mutactimycin PR, a New Anthracycline Antibiotic from *Saccharothrix* sp SA 103. J Antibiot 57(6):367–372

Chapter 3
Study of Potential Interrelationship Criteria of Microorganisms for Sustainable Diversity

Mousumi Saha, Goutam Mukherjee, Aparajita Basu, and Alok Kumar Sil

Abstract Modernization has brought humans as well as nature at the verge of petrifying situation. Toward restraining the sustainable developments of the world, wise and proper utilization of microbes and scientific consideration toward microbial diversity can be a new avenue for scientific exploration as microbes are gregarious organisms that can form very diverse microscopic communities and potentially responsible for nutrient cycling. Microbial communities are not only important for biogeochemical cycles but also essential for the food web, thus an elaborate knowledge of their dynamics will be critical in predicting the biosphere modules with a focus on future environmental conditions. Thus, structural analysis on microbial ecology of a specific area helps in better observation of intra or interdependency of different microbial species with respect to their interaction and trophic relationships. In this context, soil nitrogen (N) and carbon (C) cycle plays a pivotal role in maintaining various ecosystem functions like climate change, nutrient cycling diversity and so on, which are directly associated with human welfare and the microbes present in this community exhibit metabolic exchange capability for effective and economic resource utilization in the soil ecosystem. As a result, effective, economic, and wise resource utilization of microscopic community imparts the imperishable activities of soil productivity to balance other ecological functions. Besides this, soil microbes are also rendering significant role in bioremediation or biotransformation of xenobiotic or toxic compounds. In this chapter, we will consider the microbial diversity along with potential interrelationship of microbes in terrestrial and aquatic ecosystem that regulates its interaction and plays a vital role in improving bioremediation of toxic pollutants, regulating biogeochemical cycle and the foremost improvement in agriculture. Thus, the aim of our study is to co-relate the interdependence of microbes into convertible science of development for combating environmental pollution, climate change along with producing commercially important goods and services for human welfare.

Mousumi Saha and Goutam Mukherjee contributed equally with all other contributors.

M. Saha (✉) · G. Mukherjee · A. Basu · A. K. Sil
Department of Microbiology, University of Calcutta, Kolkata, India

R. P. Singh et al. (eds.), *Microbes in Microbial Communities*,
https://doi.org/10.1007/978-981-16-5617-0_3

Keywords Microbial community · Ecosystem function · Bioremediation · Microbial diversity · Biogeochemical cycle · Agroecosystem · Nutraceuticals

Abbreviations

C	Carbon
HGT	Horizontal gene transfer
N	Nitrogen
OTUs	Operational taxonomic units
P	Phosphorus
S	Sulfur
SD	Sustainable development
SOM	Soil organic matter

3.1 Introduction

Microorganisms play an integral and often unique role in the functioning and maintaining a sustainable ecosystem (Trivedi et al. 2016; Singh et al. 2016a, b). Microbes play the most significant role in food web, climate change, soil fertility, plant productivity, bioremediation, human welfare and in development of industrial products. Sustainable development (SD) meets the requirement of the present and at the same time it paves the way for a better future for the coming generation through the conservation of resources. The idea of SD emerged in 1987 with the Brundtland Report (United Nations General Assembly 1987). Soon after that "SD" becomes a very popular but debatable term. With the rapid development, humans have exposed themselves to a huge risk of health issues, difficult livelihood, and unfavorable environmental conditions for survival. Due to the limitless anthropogenic activity, soil and water are also getting contaminated (Balbus et al. 2013; Singh et al. 2020b). Microbial communities are one of the major driving forces related with various ecological functions, human and environmental welfare. Enormous diverse and incredible function of these communities is not well understood. This knowledge gap is the most fascinating fact and calls for the exigent problem in the present scenario regarding the study of microbial ecology. With advancement of research related to environmental microbiology, biotechnology and molecular biology, another discipline named ~~as~~ Biogeochemistry has emerged. It has assisted in appropriate explanation of microbial community with its orientation for N-, C-, P-, and S-cycle. It is very essential but critical to establish effective policies to explore and preserve microbial diversity for environmental security as well as for human wellbeing. Thus, to attain the goals of SD, microbes or microbial community can be considered as a bridge between economic, social and environmental sustainability (Fig. 3.1). Microbial diversity is the key player for a balance ecosystem (Wagg et al. 2014; Subhashini et al. 2017; Singh et al. 2020a). Due to the natural forces and/or anthropogenic activities, microbial community structure changes constantly and

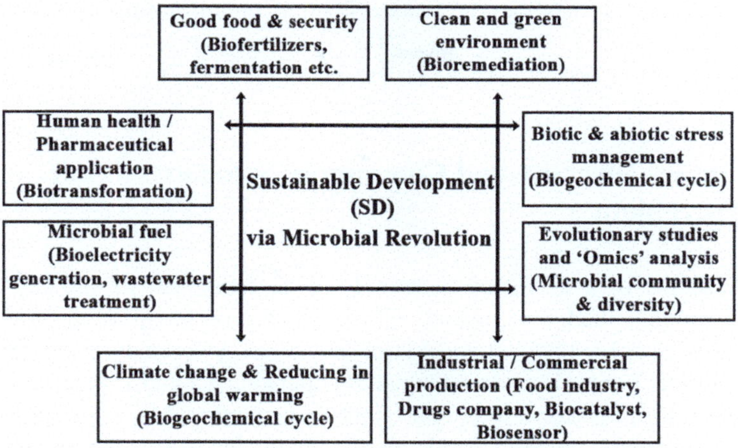

Fig. 3.1 Potential application of microbes and microbial enzymes in sustainable development (SD)

over time it gets slowly replaced with a new one resulting in microbial succession. Considering the sustainable ecosystem functioning, identification of structure and function of microbial community are critical challenge, but with the emergence of new tools and techniques help us resolving these issues. Human activities are not only responsible for environmental pollution, but also directly or indirectly associated with disturbing the biogeochemical cycles (Griggs et al. 2013; Yang et al. 2020a, b).

Microbial diversity can be utilized for various application with prime focus in SD. Managing, preserving, and proper utilization of microbial diversity are the areas which need attention by the policy makers. Little is known about the potential contribution of microbial diversity to the national economy, to wealth creation and to improvements in the quality of life. SD aims to meet the needs of the present generation without the affecting of natural resources unsympathetically. The exponential growth of human population and indiscriminate usage of resources create various wastes that are hazardous to the environment and pollute soil, water, air, etc. In this context, environmental and industrial sustainability goals can be achieved by microbial consortia that have the capability for bioremediation and biotransformation. In addition, microbes have beneficial roles in agriculture, food production, pest control, wastewater treatment, disease prevention, fermentations, antibiotic productions, vaccine productions, bioenergy/biofuel production, etc. (Vitorino and Bessa 2017). Considering this, the management of microbial diversity has an important role in SD. Modern biotechnology based industry has the huge potential to fabricate new products and processes by exploiting the capability of microbes in various ecosystem.

In the current chapter, efforts were targeted to elucidate the potential role of microorganisms in sustainable development and their interrelationship in the ecosystem functioning. Here, it is also tried to delineate functional and taxonomic frame of soil microbial community and its succession with N/C-cycle and it will also discuss the role of microbes in bioremediation, biotransformation and human welfare.

3.2 Species Diversity and Community Structure of Soil Microbiota

Characterization of a microbial community in an ecosystem by the abundance, diversity profiling and functionality of the species or group is the traditional approach to decipher the ecological structure. Strategy to analyze the microbial diversity differs from microbial ecologist to population ecologist. Previously, phylogenetic analysis of bacteria was done on the basis of morphology, metabolism and other physiological properties. All these culture-dependent approaches were followed by the botanist and zoologist, which was very difficult. Study of microbial diversity in a community is important to identify potential ecosystem players. Knowledge on this is very useful to determine the seasonal variation of microbial community. In this regard, microbial community directly participates in crop production, disease and pest control in farming and thus can be linked with agroecosystem sustainability (Fig. 3.2). For example, both gram positive and gram-negative bacteria like *Micrococcus*, *Staphylococcus*, *Corynebacterium*, *Brucellaceae*, *Burkholderiaceae*, *Xanthomonadaceae*, and many more were found

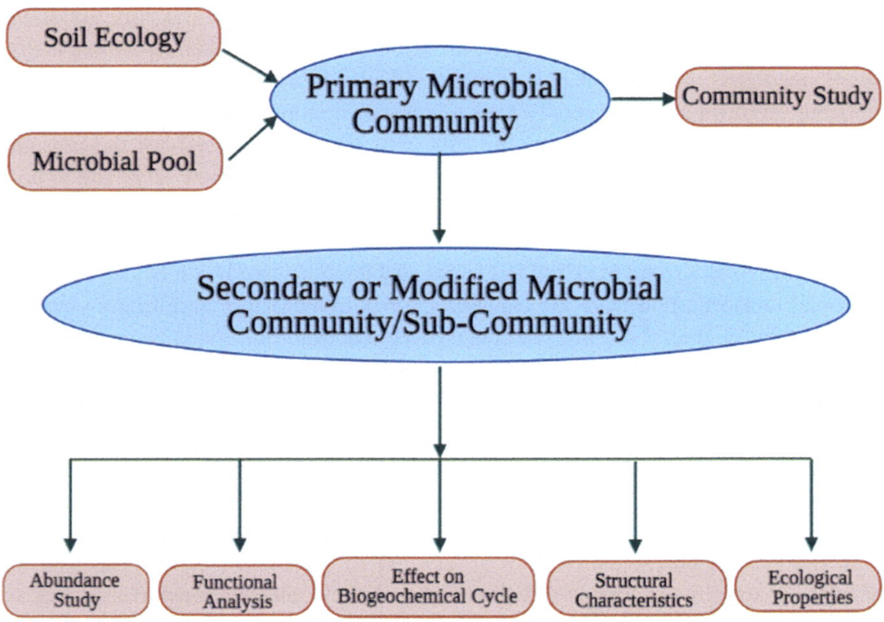

Fig. 3.2 Ecosystem functioning of microbial community

in rice ecosystem (Panizzon et al. 2015). Structure and function of bacterial community is affected by the plant metabolites, quality and depth of the soil (Wang et al. 2018; Singh et al. 2020a, b, c).

Microscopic observation or culture-based methods don't provide appropriate profiling of the diverse bacterial communities (Stewart 2012).

Obviously, culture-based methods are insufficient to determine the bacterial diversity from environment, because a typical soil sample contains thousands of individual taxa (often called "operational taxa"; OTU). Some estimates indicate that an individual soil sample may contain more than 100 taxa approximately (Stewart 2012; Portillo et al. 2013; Louca et al. 2019). This brings microbial ecologist to search the underlying concept of diversity with functional significance of microbes.

The theory of biological diversity is based on the idea of niche and survival banishment: if species compete with each other perfectly, they cannot coexist stably in one niche. Therefore, each species must have certain functional differences. However, niche-based theories are struggling with the huge habitat of plants, and plants are competing for limited resources (light, water, minerals). Zhang et al. (2017) states that around hundred species may be present in some environment. Different schemes have been proposed in order to explain the abnormal pattern of huge biodiversity. The neutral theory holds that when the adaptive changes between individuals are as large as the adaptive changes between species, species can coexist in the niche, but when the competition between individuals and the competition between species are so fierce, the species also can coexist. In various environments when the organism is fixed, dynamic changes allow coexistence of taxa with overlapping function; it may be suitable for many soils microbial clade conditions (Kneitel and Chase 2004). Diverse phylogeny is due to different microbial function and are related to resource allocation strategies. However, it is difficult to explain this difference in function among close related groups. Therefore, high-throughput sequencing methods are applied to sequence communities and their structures are analyzed according to ecologically significant populations (Koskella et al. 2017; Gallego et al. 2019; Yang et al. 2020a, b).

Soil parameters, pattern of usage and climate of the respective soil ecosystem contribute a lot toward microbial community of that specific soil. Similarly, presence of microbial community is directly linked with plant production. Thus, an interrelated study of soil science and microbiology is of great economic importance (Baliyarsingh et al. 2017). Microbial abundance in soil is effected by different soil properties, various environmental factors and interaction with other organisms. Changes in parameters like oxygen, moisture, and pH, can be selected for growth of fungi and some bacteria, while in other cases taxonomic orientation is the choice (Kneitel and Chase 2004; Koskella et al. 2017; Singh et al. 2016a, b). *Clostridium*, is an obligate anaerobe whereas the genus *Bacillus* are aerobic bacteria. These are the two genera which are closely arranged in Firmicutes on the basis of evolutionary relationship. Microorganisms are accomplished with different characteristics in order to survive in diverse ecosystem. Decomposer depends on enzymes to degrade plant polymer for their metabolism. Some rhizosphere bacteria rely on plant secretory products and adapt to the rhizosphere ecosystem.

Acidic bacteria appear to be stress-tolerant oligotrophic organisms, while Bacteroides and β-proteobacteria are trophic bacteria which need sufficient water (Compant et al. 2019).

Higher level of microbial phylum can be used for development of phylogenetic model which can explore the function of those families. Actinomycetes and Verrucomicrobia quickly adapts to hydration while Firmicutes response moderately though they maintain conserved rRNA in drought condition. However, the response of bacteria is different in class (Portillo et al. 2013; Louca et al. 2019). Effective repetition is usually seen at the species level (Barnes et al. 2020). In this case, HGT (Horizontal Gene Transfer) transmit gene from one microbe to other for various activities. Simple pathways rarely needs enzymes among related group where HGT is a regular phenomenon (Andam and Gogarten 2011; Anwar et al. 2019). HGT does not involved in rearrangement of metabolic function or their related gene as transfer of gene to highly diverse organism is not associated with HGT.

Denitrification is one of the most common processes among microbes because nitrogen can easily be modified. Maintaining the alternative physiology of aerobic respiration is easy. In contrast, sulfur reducing bacteria is unable to replace cytochrome sulfate reductase only by in the electron transport chain. The electron transmission system of sulfate reducing agent is completely different, so it is difficult for aerobic bacteria to become SO_4^{2-} reducing agent through HGT (Price et al. 2014).

β diversity is the usual soil habitats, Thus, it indicates that soil is a habitat of definite restricted microbial community. In this case, rhizosphere bulbs choose replicable bacteria on the basis of physicochemical properties of the rhizome sediments (Cordovez et al. 2019). Acid bacteria are in habitual in large soil aggregates while less common in internal microaggregates. Pore size of soil is also responsible for community variation. Abundance of microbes are related with reduction in water content and pore relatedness (Carson et al. 2010). Very scanty information regarding soil microbial community, its regulation and function is known, thus there is a huge scope of research in this important area.

3.3 Microbial Succession and Biogeochemical Cycles

Succession is one of the influential appellations with respect to augmentation for microbial community composition. Microbes play a crucial and decisive role to conduct innumerable reaction in the soil which regulates ecosystem and biogeochemical systems. Microbial communities are gaining major environmental importance as they are predominating components in regulating biogeochemical cycles and can be considered as the trailblazer of ecological succession over time (Konopka 2009; Fuhrman 2009). Knowledge of interconnecting microbial community and ecosystem function with successional study is very scare and inaccessible till date (Trivedi et al. 2016). But, culture-independent metagenomics approaches have emerged as a powerful tool for studying the identity; the function of dominant and

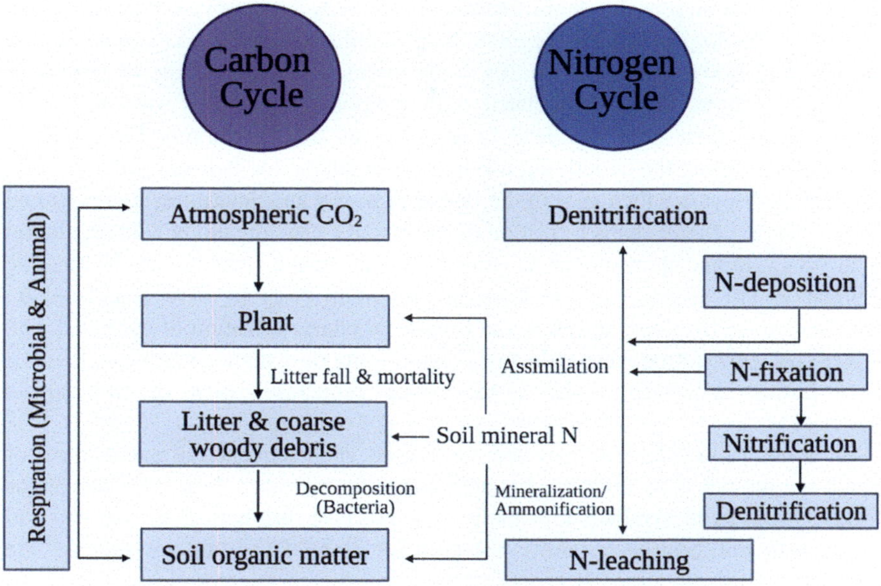

Fig. 3.3 Corelation between nitrogen and carbon cycle

complex microbial communities along with its role in different biogeochemical processes and better interpretation of taxonomic or functional models (Bulseco et al. 2020). A recent study showed a perfect correlation between plant diversity and microbial diversity in terrestrial ecosystem (Liu et al. 2020). Similarly, another study showed that the plant–microbe interaction in rhizospheric soil plays an intricate role in controlling the biogeochemical cycle and thus plays very important role to sustain biotic diversity in terrestrial ecosystem. In addition, this research also suggested that changes in an ecosystem, will definitely affect the microbial community and the microbial diversity, biogeochemical cycle with emphasis on Nitrogen (N)-cycle.

In nature, N exists in different forms and is the basic element of two important biomolecules i.e., protein and nucleic acid. N-cycle is the significant nutrient cycle as distinct group of microbes are involved in each single step like biological N fixation, mineralization/ammonification, nitrification, and denitrification, and plants are also involved in this cycle (Hayatsu et al. 2008). N changes its form into gaseous nitrogen (N_2), nitrogen oxide (NO), nitrogen dioxide (NO_2), ammonia (NH_3) or others and recycles through various ecosystem and microbes are the regulators of all these stages (Fig. 3.3). Biological N fixation can be considered as the first step where N_2 transforms into NO and NO_2. Some symbiont microbes or N fixers fix the N so that it can readily uptake or use by the plants.

Ammonification is the step where formation of NH_3 takes place, then it reacts with water present in the soil and forms NH_4. Nitrification is the next step in which different nitrifying bacteria convert ammonia into nitrites, NO_2^-, and nitrates,

NO_3^-. The last step is the denitrification where NO_3^- are recycle back to N_2 by denitrifiers. Biological N fixers regulate N availability and are the key player in the successional study (Huang et al. 2011). Microbes contribute to the dynamics of N-cycle which in turn is associated with temporal and spatial variation in N processes and their rates (Table 3.1) (Zhao et al. 2014). Microbial community study helps to study the underlying mechanism of interaction between soil microbes involved in N-cycle which govern the fate of N in soil and determine the functional link as well as genetic characterization of community structure along with regulation of the N-cycle and its response to environmental change (Isobe et al. 2020). The abundance and diversity of the bacteria present in N-cycle were elucidated by genetic characterization via qPCR and functional characterization of the genes like 16S rRNA, *nifH* (nitrogenase reductase), *amoA* (ammonia monooxygenease), *narG/napA* (nitrate reductase), *nosZ* (nitrous oxide reductase), *nirK* or *nirS* (nitrite reductase).

In carbon (C)-cycle, C is recycled and reused via a sequence of events most of which are regulated by both aerobic and anaerobic microbes. Soil microbes act as decomposers, plant symbionts, or pathogens which in turn help in C turnover and retention in soil. Soil ecosystem is the pool for global CO_2 emission and has the potential to mitigate atmosphere CO_2. Photosynthesis by plant removes CO_2 from their and is utilized for photosynthesis. Then C moves to the soil from plant and animal in the form of dead organisms, wood, and plant litters etc. C is then getting back in the atmosphere through microbial as well as animal respiration and by the utilization of fossil fuels. In some course of events, it gets exported in the ocean or remains sequestrated in soils (Table 3.1) (Zhao et al. 2014; Madsen 2011). Researches on C fixation, sequestration and degradation have provided us with a complete insight for the abundance and contribution of soil microbes in C-cycle (Fig. 3.3). It has also helped to understand the regulation and the involvement of enzymes in C turnover by the soil microbial community. The C fixation pathways include the C4 dicarboxylic acid cycle, Calvin cycle, reductive tricarboxylic acid cycle (rTCA) cycle, 3-hydroxypropionate cycle, reductive acetyl-CoA pathway, dicarboxylate-hydroxybutyrate cycle, and hydroxypropionate-hydroxybutyrate pathway (Berg et al. 2010). Further analysis of C fixation pathways has ~~have~~ carried out by studying the transcripts of key genes like *Ppc*, phosphoenolpyruvate carboxylase; *rbcL* and *S*, ribulose bisphosphate carboxylase large chain, *cbbL*; *porA*, *B*, *D* and *G*, pyruvate ferredoxin oxidoreductase subunits; *idh1*, isocitrate dehydrogenase; *korA*, *B*, *D* and *G*, 2-oxoglutarate/2-oxoacid ferredoxin oxidoreductase subunits; *accA*, *B*, *C* and *D*, acetyl-CoA carboxylase subunits. Thus, all these analyses demonstrate the presence of enzymes in soil, which are involved in CO_2 sequestration and soil respiration rate (Arai 2011).

Elaborate knowledge regarding microbial succession is of unique requirement in terms of soil ecology. Successional study with soil metagenomics analysis can be used to predict the shift in microbial community. Soil physicochemical properties and the stages and rate of litter decomposition also contribute toward the shift in structural and functional diversity of bacteria. Changes in the regulation of expression of C and N-cycle genes were observed with microbial succession (Zhong et al.

Table 3.1 The Biogeochemical cycles steps and responsible microbes and gene/enzymes

Cycle	Steps	Microbes/bacteria	Gene/enzyme
N	Nitrogen Fixation	N_2-fixers	Nitrogenase, *nif*
		1. Free-living nitrogen-fixing bacteria; Aerobic—Azotobacter, Beijemickia; Anaerobic—Clostridium, Rhodospirillum	
		2. Symbiotic nitrogen-fixing bacteria—Rhizobium	
		3. Cyanobacteria—Nostoc, Anabaena, Spirulina	
	Assimilation	Cyanobacteria, *Synechococcus elongatus* and *Anabaena* sp. PCC 7120	Nitrate assimilation nirA operon, glutamine synthetase, glutamate synthase
	Nitrification	Ammonia oxidising bacteria, *Nitrosomonas*, *Nitrosococus*	Ammonia monooxygenase, *amoA*, Hydroxylamine oxido-reductase, *hao*
		Nitrite oxidising bacteria, *Nitrobacter*	Nitrite oxidoreductase, *nxrA*
	Denitrification	Denitrifying bacteria *Pseudomonas*, *Thiobacillus*, *Paraccocus*, *Azospirillum*, *Bradyrhizobium*	Nitrate reductase, *narG*, *napA*; nitrite reductase, *nirK*, *nirS*; nitrous oxide reductase, *nosZ*
	Ammonification	Ammonifying bacteria, *Bacillus*, *Pseudomonas*, *Clostridium*, *Proteus*	Glutamate Dehydrogenase, *gdh*; urease, *ure*; alkaline metallopeptidase, *apr*
C	Photosynthesis (Carbon fixation)	Cyanobacteria, green and purple sulfur bacteria	Ribulose-1,5-bisphosphate carboxylase/oxygenase, *rubisco*; Phosphoenolpyruvate carboxylase, *pcc*
		Clostridium thermocellum, *Rhodopseudomonas palustris*	Carbon monoxide dehydrogenase/acetyl-CoA synthase (CODH/ACS)
	Respiration (Aerobic and anaerobic)	*Aerobic prokaryotes*; *Escherichia coli*, *Bacillus subtilis*	NADH dehydrogenase; cytochrome oxidase, *aa3*, *cbb3*
		Paracoccus denitrificans, *Acetobacterium woodii*	Nitrate reductase, nitrite reductase
	Decomposition	Actinomycetes, clostridia, bacilli, arthrobacters and pseudomonads are involved in decomposition	Genes or enzymes: *pectinase*, *endochitinase*, *exoglucanase*, *acetylglucosaminidase*
	Combustion	Burning of fossil fuels, release CO_2 in the environment and increases it amount in the atmosphere.	
	Methanogenesis	Methanogenic bacteria; *Methanosarcina barkeri*, *Methanobacterium thermoautotrophicum*	Formyl-MF dehydrogenase; Methyl-coenzyme M reduc tase, *mcrA*

(continued)

Table 3.1 (continued)

Cycle	Steps	Microbes/bacteria	Gene/enzyme
	Methane oxidation	Methanotrophs include *Methylococcaceae*, *Methylocystaceae*. *Methylococcus capsulatus*	Methane monooxygenase (MMO); Soluble methane monooxygenase (sMMO)
S	Sulfur oxidation	Sulfur oxidizers, *Thiobacillus thiooxidans*, *Acidithiobacillus ferrooxidans*	*soxXA*, *soxYZ*, *soxB*, *soxCD*
	Dissimilatory Sulfate reduction	Sulfate reducers, *Desulfotomaculum* bacteria, *Desulfovibrio desulfuricans*	*ATP sulfurlyase*, *sat/atpS*; *APS reductase apr/aps*
	Oxidation of hydrogen sulfide	Photosynthetic green and purple sulfur bacteria and some chemolithotrophs; *Chlorobium*, *Chromatium*	Sulfide quinone oxidoreductase, *sqr*; flavocytochrome c/sulfide dehydrogenases, *fccAB*
	Sulfur reduction	*Desulfuromonas*, *Desulfurella*	Sulfite reductase, *dsr*
	Assimilative sulfate reduction	Sulfate (SO_4) is reduced to organic sulfhydryl groups (R–SH)	
P	Mineralization, precipitation and utilization of phosphorus	*Bacillus subtilis*, *Arthobacter*, *Bartonella quintana*, *Mesorhizobium loti*, *Polaromonas naphthalenivorans*, *Campylobacter coli*	Phosphatase; Polyphosphate kinase, *ppk*; exopolyphosphatase, *ppx*

2018). Based on the metagenomic approach, it was also reported that litter decomposition in C-cycle is highly regulated by microbes like *Bacteroidetes*, *Helotiales*, *Acidobacteria*, and many other microbes during the course of succession (Herzog et al. 2019).

Sulfur (S), like N is another essential element that is constituent of amino acids and different co-factors. A diverse group of microbes are involved in the conversion of one form of S into others. The different steps in S-cycle are assimilative sulfate reduction, desulfurization, oxidation of hydrogen sulfide, oxidation of elemental S, dissimilative S reduction and dissimilative sulfate reduction (Table 3.1) (Rodriguez-Mora et al. 2016; Madsen 2011). The phosphorus (P) cycle is another important biogeochemical cycle, but it is slow. The steps of the P-cycle include weathering, absorption by plants and animals and decomposition (Table 3.1) (Richardson and Simpson 2011).

In soil, N/P ratio is very critical for plant productivity, but each nutrient is not the major requirement rather an association of both the cycles is crucial with the C plays a lead role in maintaining various ecosystem functions like climate change, nutrient cycling diversity, and so on, which are directly associated with human welfare. With industrial and agriculture progression, the amount of C in the environment increases which in turn results in an enhanced recycling of N as organic matter. Long term monitoring of microbial community has contributed significantly to improve the framework for growth and interaction between and among different groups of terrestrial and marine biomes with their environmental activities like C (greenhouse gas fluxes) and N fluxes, biomass production, climate changes and responses toward different stresses (biotic and abiotic) (Rousk and Bengtson 2014). Transcriptomic and metagenomics analysis has created various routes for the construction of community structure of the soil microbes based on its taxonomic composition, diversity, abundance, interaction, and their importance in soil ecology with respect to their role indifferent biogeochemical cycles (N, C, and other cycles). Identification of most abundant, active phylum level OTUs; phylogenetic orientation; gene expression and functional gene analysis are also feasible via metagenomics and metatranscriptomic data analysis. Metagenomics study has also helped in understanding the importance of gene expression recovered directly from the environmental samples. Culture-based method and ribotyping have the application to determine levels of ammonification, nitrification, and denitrification; C fixation, degradation, and sequestration capacity of the different microbial strains. Identification of the new or novel gene(s) present in microbes involved in various steps in N/C-cycle can be possible, which in turn help in developing more effective strategies for bioremediation or formulation of probiotics. Soil organic matter (SOM) is closely associated with the composition of the microbial community and it is thought that microbes are the major contributors toward ecological succession. Although microbial activities and biomass are related to biological N fixation, variation in concentration of C as well as C sequestration, methanogenesis, and sulfate reduction, soil-plant interaction is also another important component of these events (Rousk and Bengtson 2014). Thus, the study of plant microbe interaction is also important to fully understand the various biogeochemical cycles.

3.4 Ecosystem Function and Microbial Diversity

Microbes dominate the soil ecosystem. Various anthropogenic pressures are related with shift in microbial loads, but microbial abundance and function in soil is still unclear. More scientific attention is required to establish the interrelation among microbial community and functioning. Microbes plays a pivotal role in agriculture, pollution control, and others but its significance is argued (Carson et al. 2010; Maron et al. 2018). Microbial abundance is crucial factor in controlling climate change and soil fertility. Various groups of microbes are responsible for similar type of significant function in soil. For example, heterotrophs a vast group of microbes helps in decomposition of organic matter in soil (Carson et al. 2010; Yang et al. 2018; Cavicchioli et al. 2019; Cordovez et al. 2019). Few microbial species are corelated with the CO_2 released organic matter decomposition. Climate change has some impacts on soil microbial ecology, which in turns affect the soil carbon content (Cavicchioli et al. 2019).

Microbial decomposition and mineralization processes in soil helps in maintaining the versatility of ecosystems. These processes allow material and energy to be transferred between the ground and the ground. More and more experiments and observational studies have provided proof that biodiversity is directly related to ecological function. Any changes in ecosystem will affect the microbial function (Maron et al. 2018). Scarcity of information regarding microbial pool and its activity, biotic and abiotic stresses can hinder the activity of microbes and this will be a hurdle in the pathway for sustainable development policies (Cavicchioli et al. 2019).

Loss of biodiversity is a major area of research as it has some impact in improper ecological activity. Microbial diversity and ecological function is directly proportional, but its underlying mechanism varies between different ecological communities. Different terrestrial microbes depicted similar activity than aquatic communities (Miki et al. 2014). Microbes can be categorized under board or narrow type on the basis of different processes they perform in soil. Process like respiration or mineralization is conducted by board group of diverse soil microbial community. Thymidine and leucine incorporation, carbon mineralization, denitrification and nitrification process is not affected by loss of microbial community (Philippot et al. 2013; Cavicchioli et al. 2019).

Nutrient cycle is performed by versatile microbes present in both in terrestrial/aquatic ecosystems (Singh et al. 2014). Microbial nitrogen cycles are usually associated with high levels of plant diversity, thereby stimulating productivity (Philippot et al. 2013; Maron et al. 2018). Plant community can enhance the soil microbial community by supporting multiple litter qualities. Microbes promote decomposition and increase soil organic matter content (Miki et al. 2014). Different types of microorganisms are required in degradation of organic material. Microbes help in releasing soil nutrients for food and fiber production by other microbial community. Therefore, despite being widely ignored, microbes helps in maintaining versatility by altering nutrient supply and resource allocation, resulting in high material processing rates in terrestrial ecosystems (Miki et al. 2014; Carson et al. 2010). Various studies already hypothesized a direct relationship between microbial diversity and ecosystem, but their exact mechanisms were not established. To confirm this hypothesis a group of researchers followed the method of genotype-phenotype analysis with the traditional approaches. Then they finally concluded that using this mapping technique can analyze the microbial physiology and community responsible for ecological changes and the effect of human activities in ecosystem functioning (Morris et al. 2020). It is known that our ecosystem is at the verge of danger due to increase in anthropogenic activities which are detrimental to ecosystem. Thus, using modern molecular approaches like metagenomics, transcriptomic, next generation sequencing has proven to be a promising techniques for identifying the reasons behind the change in microbial ecology (Heintz-Buschart et al. 2020; Wang et al. 2019).

In this regard, plant, soil and microbes together regulate the ecology, nutrient cycling, and survival strategy (Liu et al. 2018; Cordovez et al. 2019). For example, different groups of bacteria are involved in decomposition of organic matter and conversion of N and phosphorous (P) into various inorganic forms (mineralization), for availability to plants (Maron et al. 2018). Abundance of microbes is associated with indigenous plant and thus have relationship with proper ecosystem functioning (Cavicchioli et al. 2019; Cordovez et al. 2019).

3.5 Bioremediation and Biotransformation

Soil is polluted by the persistence and excessive presence of hazardous materials like toxic compounds, radioactive materials, chemicals, disease-causing agents, heavy metals, insoluble aromatic contaminants, etc. which create a threat to the normal growth of plant and health of animals. In this regard, bioremediation is considered as the major non-excludable strategy of involving microbes for transformation or degradation of waste or contaminants. Bacteria involved in N- and P-cycle are significantly used as bio-fertilizers and bio-pesticides. They not only contribute a lot in the improvement of crop productivity and security but also in reduction of soil and water pollution; and increase soil fertility. The microbes that produce enzymes like oxidoreductase, hydrolase, oxygenase, laccase, protease, lipase, peroxidase, transferase, isomerase, cellulase, etc. are the most potent pollutant degraders, bio-transformers, suitable for bioremediation or highly used in commercial or industrial field. Bioremediation is also very effective in oil-contaminated sites (Abatenh et al. 2017). Microbial enzymes can co-metabolize, adsorbed, degrade, immobilize, detoxify, precipitate or change the oxidation state of the pollutants.

Phytoremediation, biostimulation, bioaugementation, bioventing, biopiles, composting, and bio-attenuation are alternative forms of in situ or ex situ bioremediation for removal of wastes or other toxic compounds from the environment. One of the biggest challenges is the field test of the microbial enzymes responsible for bioremediation, biotransformation, and biodegradation (Fig. 3.4).

Thus, rigorous and complete exploration of molecular mechanism, metabolic regulation, and downstream processing is required. Few researchers have developed microbial biosensors to detect the amount of contamination in a polluted site. Meta-transcriptomics study of the microbial community or diversity of the contaminated zone will render the scientists to understand the interaction of the introduced pollutant degrader with the inhabitant population of the microbes (Singh et al. 2017; Nostrand et al. 2012). Microbe-mediated remediation is very advantageous in aquatic and marine ecosystems too. It was found that several toxic and harmful chemical pollutants changed the water quality and thus distorting ecological balance by bioaccumulation and biomagnifications and thus caused several diseases of humans as well as other animals (Akpor and Muchie 2010). Extensive use of plastic and its associated materials are major sources of the environmental threat to lake, pond, river, ocean, and terrestrial ecosystems (Cole et al. 2011). In this context, several reports demonstrated that many bacterial species like, *Pseudomonas* sp., *Exeguobacterium* sp., *Bacillus* sp., *Rhizobium* sp. and many fungal species like, mycelial fungus were very much efficient in bioremediation of various forms of plastic. The degradation of multi-aromatic contaminations in the soil is carried by consortia of different bacteria (like *Pseudomonas* sp., *Acinetobacter* sp., and

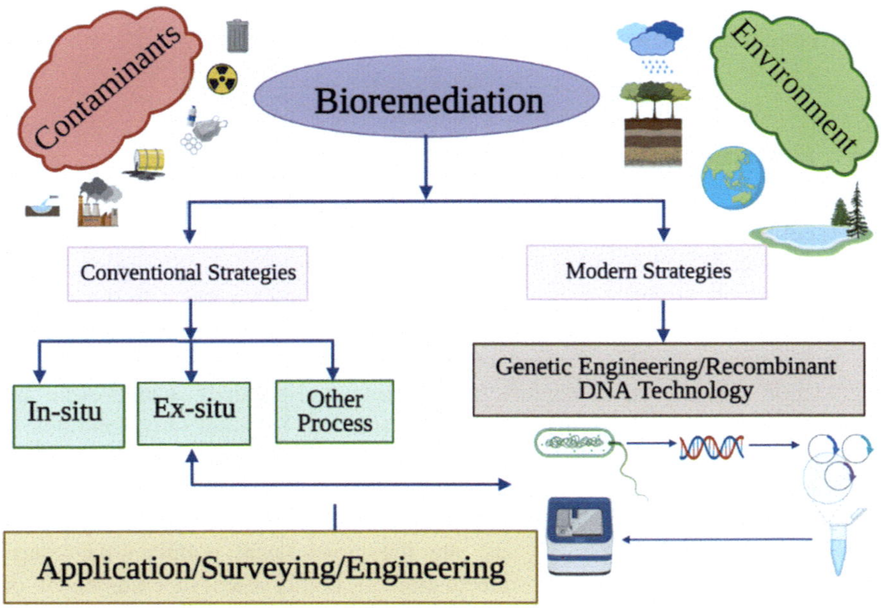

Fig. 3.4 Scheme of bioremediation

Rhodococcus sp.) (Yu et al. 2005). Organic maters and inorganic contaminants in the pond like NO_2, H_2S, and other harmful waste decomposed and mineralized by several photosynthetic bacteria. Apart from the conventional high-cost management systems some bacteria were found to reduce and detoxify the heavy metals in in-situ as well as ex-situ conditions (Marzan et al. 2017). Several heavy metal-reducing bacteria include *Pseudomonas* sp., *Bacillus* sp., *Citrobacter* sp., *Aerococcus* sp., and many more.

Besides, biotransformation is the technique of alteration of a chemical/compound (substrate) into structurally related products by using microbes or microbial enzymes as a biocatalyst. With the advent of new technology, this method has become vital in the industrial sector for the production of chemicals and pharmaceuticals due to its effectiveness and eco-friendly advantages. With this transformation tool rare and notable compounds have emerged containing unique, structural and functional attributes and enhanced pharmacokinetic properties. Types of biotransformation reactions include oxidation, isomerization, hydrolysis, reduction, etc. In biotransformation different types of biocatalyst are used like growing cell, immobilized cell, enzymes (oxidoreductase, lyase) or immobilized enzymes. Immobilized enzymes are frequently used for glucose isomerase, penicillin, acylase, etc., and different species of *Streptomyces*, *Corynebacterium*, *Nocardia*, are responsible for biotranformation of taxol, testosterone, cholesterol toluene and xylenes (Hegazy et al. 2015).

3.6 Microbes in Human Welfare

Improvement and modernization of biotechnological research provide a great impact on large scale industrial production of materials essential for life. The bacterial production of enzymes, vaccines, antibiotics, probiotics, vitamins, and coenzymes and biofuel occupied a great place of utilization in pharmaceutical, biotech, and bio-processing industries (Fig. 3.5) (Vitorino and Bessa 2017).

The industrial production of α-amylase, lipase, beta-lactamase, streptokinase, glutaminase, asparaginase, collagenase, proteases, and penicillinase by a diverse group of bacteria is very important (Singh et al. 2016a, b). Vaccines for several pathogenic diseases such as pneumonia, whooping cough, meningitis, tetanus, Q fever, plague, typhoid, anthrax, tuberculosis etc. are produced by several recombinant bacterial strains (Detmer and Glenting 2006). Bacteria produce antibiotics like gentamicin, bacitracin, polymyxin B (Chi and Holo 2018). Beside this, the probiotics and prebiotics are of major therapeutic importance in the food and health sector. Nutraceuticals is another major contribution of microbes toward food

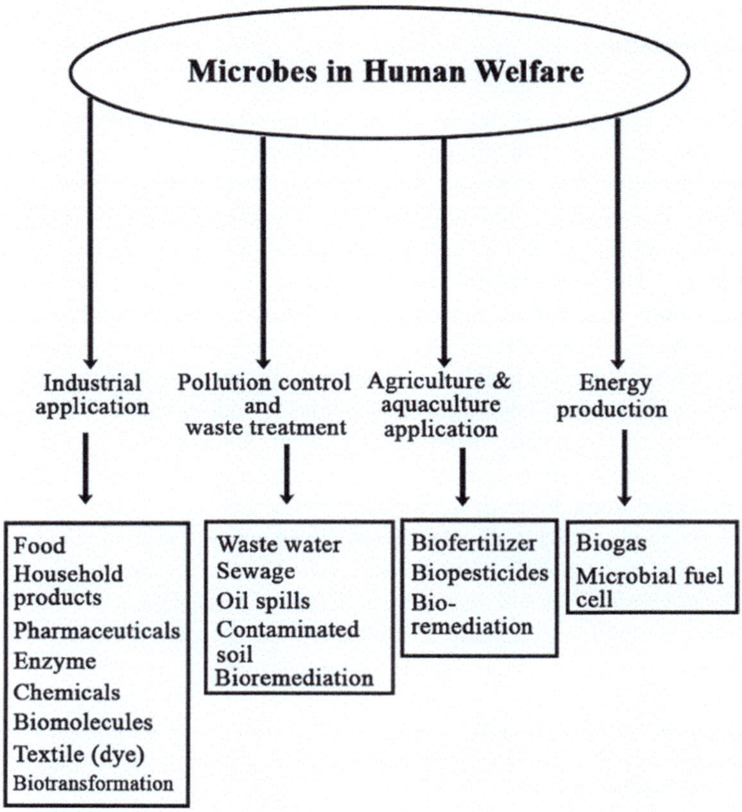

Fig. 3.5 Schematic representation of involvement of microbes in human development

industry, as these are future food with low cost but have more nutritive value. More scientific research and proper establishment of all this must be improve to achieve the goal of sustainable development. In this context, rice bran oil have some prebiotic effect (Panizzon et al. 2015). Probiotics are widely used in yogurts, cheese, ice cream, and other diary related food items as they can help to cure the gastrointestinal disease and bowel discomfort and also to boost the immune system. Several microorganisms are the natural sources of most vitamins. Enrichment of natural resources and biotechnological implications made it easy for the utilization of several bacteria in industrial scale for the productions of large quantities of vitamins. Production of biofuels by lignocellulosic biomass expected to lower the emission of greenhouse gases which ultimately reduce environmental pollution (Li et al. 2018). Butanol and ethanol are the most promising source of biofuel. Bio-hydrogen is the potential source of clean energy and may find a great future application (Su et al. 2018). Several widely used bacteria in industrial applications are *Bacillus*, *Pseudomonas*, *Enterococcus*, *Streptococcus*, *Erwinia*, *Acetobacter*, *Clostridium*, *Geobacter*, *Lactobacillus*, and many others (Singh et al. 2016a, b).

3.7 Conclusion

This article has imparted a sketch of the extensive perception of microbial communities together with its role in SD. It described different forms of microbial communities and their structural association with the soil ecosystem. All this information may provide the scientist or researcher with the knowledge of microbial functional diversity for improving N/C cycle and predicting ecosystem responses to environmental change. The beneficial role of microbes or microbial enzymes in terms of decontamination, degradation, bioconversion, detoxification, bioremediation, biotransformation, and other aspects like food security, production of commercially important items and human health have also been discussed that made a significant contribution toward biotechnological applications. In addition, this chapter is also tried to gather and conclude different aspects of SD goals like food security, production of commercially important items, human health and environmental protection related to microbes or microbial community.

Summarizing all, the vast abundance, easy availability, metabolic diversity, and easy manipulation made microorganisms a valuable tool in sustainable developments. The government should make new policies and awareness must be created for the microbial revolution to develop cost-effective technologies in health and industrial sectors.

Acknowledgment Authors would like to thank SERB-Department of Science and Technology (DST), Government of India and Council of Scientific and Industrial Research (CSIR), Government of India for the financial support.

References

Abatenh E, Gizaw B, Tsegaye Z, Wassie M (2017) The role of microorganisms in bioremediation - a review. Open J Environ Biol 2(1):038–046

Akpor OB, Muchie M (2010) Bioremediation of polluted wastewater influent: phosphorus and nitrogen removal. Sci Res Essays 5(21):3222–3230

Andam CP, Gogarten JP (2011) Biased gene transfer and its implications for the concept of lineage. Biol Direct 6:47. https://doi.org/10.1186/1745-6150-6-47

Anwar MN, Li ZF, Gong Y, Singh RP, Li YZ (2019) Omics studies revealed the factors involved in the formation of colony boundary in *Myxococcus xanthus*. Cell 8(6):530

Arai H (2011) Regulation and function of versatile aerobic and anaerobic respiratory metabolism in *Pseudomonas aeruginosa*. Front Microbiol 2:103. https://doi.org/10.3389/fmicb.2011.00103

Balbus JM, Boxall AB, Fenske RA, McKone TE, Zeise L (2013) Implications of global climate change for the assessment and management of human health risks of chemicals in the natural environment. Environ Toxicol Chem 32(1):62–78. https://doi.org/10.1002/etc.2046

Baliyarsingh B, Nayak SK, Mishra BB (2017) Soil microbial diversity: an ecophysiological study and role in plant productivity. In: Adhya TK, Mishra BB, Annapurna K, Verma DK, Kumar U (eds) Advances in soil microbiology: recent trends and future prospects. Springer, Singapore, pp 1–17. https://doi.org/10.1007/978-981-10-7380-9_1

Barnes EM, Carter EL, Lewis JD (2020) Predicting microbiome function across space is confounded by strain-level differences and functional redundancy across taxa. Front Microbiol 11:101. https://doi.org/10.3389/fmicb.2020.00101

Berg IA, Ramos-Vera WH, Petri A, Huber H, Fuchs G (2010) Study of the distribution of autotrophic CO_2 fixation cycles in Crenarchaeota. Microbiology 156(Pt 1):256–269. https://doi.org/10.1099/mic.0.034298-0

Bulseco AN, Vineis JH, Murphy AE, Spivak AC, Giblin AE, Tucker J, Bowen JL (2020) Metagenomics coupled with biogeochemical rates measurements provide evidence that nitrate addition stimulates respiration in salt marsh sediments. Limnol Oceanogr 65:S321–S339. https://doi.org/10.1002/lno.11326

Carson JK, Gonzalez-Quiñones V, Murphy DV, Hinz C, Shaw JA, Gleeson DB (2010) Low pore connectivity increases bacterial diversity in soil. Appl Environ Microbiol 76(12):3936–3942. https://doi.org/10.1128/AEM.03085-09

Cavicchioli R, Ripple WJ, Timmis KN et al (2019) Scientists' warning to humanity: microorganisms and climate change. Nat Rev Microbiol 17(9):569–586. https://doi.org/10.1038/s41579-019-0222-5

Chi H, Holo H (2018) Synergistic antimicrobial activity between the broad spectrum bacteriocingarvicin KS and nisin, farnesol and polymyxin B against gram-positive and gram-negative bacteria. Curr Microbiol 75(3):272–277

Cole M, Lindeque P, Halsband C, Galloway TS (2011) Microplastics as contaminants in the marine environment: a review. Mar Pollut Bull 62(12):2588–2597. https://doi.org/10.1016/j.marpolbul.2011.09.025

Compant S, Samad A, Faist H, Sessitsch A (2019) A review on the plant microbiome: ecology, functions, and emerging trends in microbial application. J Adv Res 19:29–37. https://doi.org/10.1016/j.jare.2019.03.004

Cordovez V, Dini-Andreote F, Carrión VJ, Raaijmakers JM (2019) Ecology and evolution of plant microbiomes. Annu Rev Microbiol 73:69–88. https://doi.org/10.1146/annurev-micro-090817-062524

Detmer A, Glenting J (2006) Live bacterial vaccines--a review and identification of potential hazards. Microb Cell Factories 5:23. https://doi.org/10.1186/1475-2859-5

Fuhrman JA (2009) Microbial community structure and its functional implications. Nature 459 (7244):193–199. https://doi.org/10.1038/nature08058

Gallego I, Venail P, Ibelings BW (2019) Size differences predict niche and relative fitness differences between phytoplankton species but not their coexistence. ISME J 13 (5):1133–1143. https://doi.org/10.1038/s41396-018-0330-7
Griggs D, Stafford-Smith M, Gaffney O et al (2013) Policy: sustainable development goals for people and planet. Nature 495(7441):305–307. https://doi.org/10.1038/495305a
Hayatsu M, Tago K, Saito M (2008) Various players in the nitrogen cycle: diversity and functions of the microorganisms involved in nitrification and denitrification. Soil Sci Plant Nutr 54 (1):33–45. https://doi.org/10.1111/j.1747-0765.2007.00195.x
Hegazy ME, Mohamed TA, ElShamy AI et al (2015) Microbial biotransformation as a tool for drug development based on natural products from mevalonic acid pathway: a review. J Adv Res 6 (1):17–33. https://doi.org/10.1016/j.jare.2014.11.009
Heintz-Buschart A, Guerra C, Djukic I, Cesarz S, Chatzinotas A, Patoine G, Sikorski J, Buscot F, Küsel K, Wegner C-E, Eisenhauer N (2020) Microbial diversity-ecosystem function relationships across environmental gradients. Res Ideas Outc 6:e52217. https://doi.org/10.3897/rio.6.e52217
Herzog C, Hartmann M, Frey B et al (2019) Microbial succession on decomposing root litter in a drought-prone Scots pine forest. ISME J 13:2346–2362. https://doi.org/10.1038/s41396-019-0436-6
Huang LN, Tang FZ, Song YS et al (2011) Biodiversity, abundance, and activity of nitrogen-fixing bacteria during primary succession on a copper mine tailings. FEMS Microbiol Ecol 78 (3):439–450. https://doi.org/10.1111/j.1574-6941.2011.01178.x
Isobe K, Ise Y, Kato H et al (2020) Consequences of microbial diversity in forest nitrogen cycling: diverse ammonifiers and specialized ammonia oxidizers. ISME J 14(1):12–25. https://doi.org/10.1038/s41396-019-0500-2
Kneitel JM, Chase JM (2004) Trade-offs in community ecology: linking spatial scales and species coexistence. Ecol Lett 7(1):69–80
Konopka A (2009) What is microbial community ecology? ISME J 3(11):1223–1230. https://doi.org/10.1038/ismej.2009.88
Koskella B, Hall LJ, Metcalf CJE (2017) The microbiome beyond the horizon of ecological and evolutionary theory. Nat Ecol Evol 1(11):1606–1615. https://doi.org/10.1038/s41559-017-0340-2
Li T, Zhang C, Yang KL, He J (2018) Unique genetic cassettes in a Thermoanaero bacterium contribute to simultaneous conversion of cellulose and monosugars into butanol. Sci Adv 4(3): e1701475
Liu YH, Wang ET, Jiao YS, Tian CF, Wang L, Wang ZJ, Guan JJ, Singh RP, Chen WX, Chen W (2018) Symbiotic characteristics of *Bradyrhizobium diazoefficiens* USDA 110 mutants associated with shrubby sophora (*Sophoraflavescens*) and soybean (*Glycine max*). Microbiol Res 214:19–27
Liu L, Zhu K, Wurzburger N, Zhang J (2020) Relationships between plant diversity and soil microbial diversity vary across taxonomic groups and spatial scales. Ecosphere 11(1):e02999. https://doi.org/10.1002/ecs2.2999
Louca S, Mazel F, Doebeli M, Parfrey LW (2019) A census-based estimate of Earth's bacterial and archaeal diversity. PLoS Biol 17(2):e3000106. https://doi.org/10.1371/journal.pbio.3000106
Madsen EL (2011) Microorganisms and their roles in fundamental biogeochemical cycles. Curr Opin Biotechnol 22(3):456–464. https://doi.org/10.1016/j.copbio.2011.01.008
Maron PA, Sarr A, Kaisermann A et al (2018) High microbial diversity promotes soil ecosystem functioning. Appl Environ Microbiol 84(9):e02738–e02717. https://doi.org/10.1128/AEM.02738-17
Marzan LW, Hossain M, Mina SA, Akter Y, Chowdhury AM (2017) Isolation and biochemical characterization of heavy-metal resistant bacteria from tannery effluent in Chittagong city, Bangladesh: bioremediation viewpoint. Egypt J Aquat Res 43(1):65–74

Miki T, Yokokawa T, Matsui K (2014) Biodiversity and multifunctionality in a microbial community: a novel theoretical approach to quantify functional redundancy. Proc R Soc B Biol Sci 281 (1776):20132498

Morris A, Meyer K, Bohannan B (2020) Linking microbial communities to ecosystem functions: what we can learn from genotype-phenotype mapping in organisms. Philos Trans R Soc Lond Ser B Biol Sci 375(1798):20190244. https://doi.org/10.1098/rstb.2019.0244

Nostrand JD, He Z, Zhou J (2012) Use of functional gene arrays for elucidating in situ biodegradation. Front Microbiol 3:339. https://doi.org/10.3389/fmicb.2012.00339

Panizzon J, Pilz J, Harry KN, Ramos R, Ziegler D, Fiuza L (2015) Microbial diversity: relevance and relationship between environmental conservation and human health. Braz Arch Biol Technol 58:137–145. https://doi.org/10.1590/S1516-8913201502821

Philippot L, Spor A, Hénault C et al (2013) Loss in microbial diversity affects nitrogen cycling in soil. ISME J 7(8):1609–1619. https://doi.org/10.1038/ismej.2013.34

Portillo MC, Leff JW, Lauber CL, Fierer N (2013) Cell size distributions of soil bacterial and archaeal taxa. Appl Environ Microbiol 79(24):7610–7617. https://doi.org/10.1128/AEM.02710-13

Price MN, Ray J, Wetmore KM et al (2014) The genetic basis of energy conservation in the sulfate-reducing bacterium *Desulfovibrio alaskensis* G20. Front Microbiol 5:577. https://doi.org/10.3389/fmicb.2014.00577

Richardson AE, Simpson RJ (2011) Soil microorganisms mediating phosphorus availability update on microbial phosphorus. Plant Physiol 156(3):989–996. https://doi.org/10.1104/pp.111.175448

Rodriguez-Mora MJ, Edgcomb VP, Taylor C, Scranton MI, Taylor GT, Chistoserdov AY (2016) The diversity of sulfide oxidation and sulfate reduction genes expressed by the Bacterial Communities of the Cariaco Basin, Venezuela. Open Microbiol J 10:140–149. https://doi.org/10.2174/1874285801610010140

Rousk J, Bengtson P (2014) Microbial regulation of global biogeochemical cycles. Front Microbiol 5:103. https://doi.org/10.3389/fmicb.2014.00103

Singh RP, Singh RN, Manchanda G, Tiwari PK, Srivastava AK et al (2014) Biofuels: a way ahead. Ind J Appl Res 4(1):3–6

Singh R, Kumar M, Mittal A, Mehta PK (2016a) Microbial enzymes: industrial progress in 21st century. 3. Biotech 6(2):174. https://doi.org/10.1007/s13205-016-0485-8

Singh RP, Manchanda G, Singh RN, Srivastava AK, Dubey RC (2016b) Selection of alkalotolerant and symbiotically efficient chickpea nodulating rhizobia from North-West Indo Gangetic Plains. J Basic Microbiol 56:14–25. https://doi.org/10.1002/jobm.201500267

Singh RP, Manchanda G, Li ZF, Rai AR (2017) Insight of proteomics and genomics in environmental bioremediation. In: Bhakta JN (ed) Handbook of research on inventive bioremediation techniques. IGI Global, Hershey, PA. https://doi.org/10.4018/978-1-5225-2325-3

Singh R, Manchanda G, Maurya I, Wei Y (eds) (2020a) Microbial versatility in varied environments. Springer, Singapore. https://doi.org/10.1007/978-981-15-3028-9

Singh RP, Handa R, Manchanda G (2020b) Nanoparticles in sustainable agriculture: an emerging opportunity. J Control Release 329:1234–1248. https://doi.org/10.1016/j.jconrel.2020.10.051

Singh RP, Manchanda G, Yang Y, Singh D, Srivastava AK et al (2020c) Deciphering the factors for nodulation and symbiosis of Mesorhizobium associated with *Cicer arietinum* in Northwest India. Sustainability 12:1–17

Stewart EJ (2012) Growing unculturable bacteria. J Bacteriol 194(16):4151–4160. https://doi.org/10.1128/JB.00345-12

Su X, Zhao W, Xia D (2018) The diversity of hydrogen-producing bacteria and methanogens within an *in-situ* coal seam. Biotechnol Biofuels 11:245. https://doi.org/10.1186/s13068-018-1237-2

Subhashini DV, Singh RP, Manchanda G (2017) OMICS approaches: tools to unravel microbial systems. Directorate of Knowledge Management in Agriculture, Indian Council of Agricultural Research, New Delhi. ISBN: 9788171641703. https://books.google.co.in/books?id=vSaLtAEACAAJ

Trivedi P, Delgado-Baquerizo M, Trivedi C et al (2016) Microbial regulation of the soil carbon cycle: evidence from gene-enzyme relationships. ISME J 10(11):2593–2604. https://doi.org/10.1038/ismej.2016.65

United Nations General Assembly (1987) Report of the world commission on environment and development: our common future. United Nations General Assembly, Development and International Co-operation: Environment, Oslo

Vitorino LC, Bessa LA (2017) Technological microbiology: development and applications. Front Microbiol 8:827. https://doi.org/10.3389/fmicb.2017.00827

Wagg C, Bender SF, Widmer F, van der Heijden MG (2014) Soil biodiversity and soil community composition determine ecosystem multifunctionality. Proc Natl Acad Sci U S A 111 (14):5266–5270. https://doi.org/10.1073/pnas.1320054111

Wang X, Wang Z, Jiang P et al (2018) Bacterial diversity and community structure in the rhizosphere of four *Ferula* species. Sci Rep 8:5345. https://doi.org/10.1038/s41598-018-22802-y

Wang JT, Egidi E, Li J et al (2019) Linking microbial diversity with ecosystem functioning through a trait framework. J Biosci 44:109. https://doi.org/10.1007/s12038-019-9928-9

Yang SH, Chen WH, Wang ET, Chen WF, Yan J, Han XZ, Tian CF, Sui XH, Singh RP, Jiang GM, Chen WX (2018) Rhizobial biogeography and inoculation application to soybean in four regions across China. J Appl Microbiol 125(3):853–866

Yang Y, Liu L, Singh RP, Meng C, Ma S, Jing C, Li Y, Zhang C (2020a) Nodule and root zone microbiota of salt-tolerant wild soybean in coastal sand and saline-alkali soil. Front Microbiol 11:2178. https://doi.org/10.3389/fmicb.2020.523142

Yang Y, Singh RP, Song D, Chen Q, Zheng X, Zhang C, Zhang M, Li Y (2020b) Synergistic effect of *Pseudomonas putida* II-2 and *Achromobacter* sp. QC36 for the effective biodegradation of the herbicide quinclorac. Ecotoxicol Environ Saf 188:109826

Yu SH, Ke L, Wong YS, Tam NF (2005) Degradation of polycyclic aromatic hydrocarbons by a bacterial consortium enriched from mangrove sediments. Environ Int 31(2):149–154. https://doi.org/10.1016/j.envint.2004.09.008

Zhang Z, Qu Y, Li S, Feng K, Wang S, Cai W, Liang Y, Li H, Xu M, Yin H, Deng Y (2017) Soil bacterial quantification approaches coupling with relative abundances reflecting the changes of taxa. Sci Rep 7(1):1

Zhao M, Xue K, Wang F et al (2014) Microbial mediation of biogeochemical cycles revealed by simulation of global changes with soil transplant and cropping. ISME J 8(10):2045–2055. https://doi.org/10.1038/ismej.2014.46

Zhong YY, Weiming S, Guan Z (2018) Decreased occurrence of carbon cycle functions in microbial communities along with long-term secondary succession. Soil Biol Biochem 123:207. https://doi.org/10.1016/j.soilbio.2018.05.017

Chapter 4
Antimicrobial Peptide and Toxin-Based Mutualism: Obligate Symbiotic Entomopathogenic Nematode—Bacterium Associations

Aishiki Banerjee and Saurav Saha

Abstract Entomopathogenic nematodes (EPN) *Steinernema* and *Heterorhabditis* live in an obligate symbiosis with the bacterial strains of *Xenorhabdus* and *Photorhabdus*, which form the Entomopathogenic Bacteria (EPB), respectively. Each symbiotic complex is a stable coevolutionary product. The EPN/EPB complex (*Steinernema/Xenorhabdus*; *Heterorhabditis/Photorhabdus*) produces numerous antimicrobial compounds and other natural products including antibiotic peptides. The third member of this mutualistic unit is the insect. The target insect is killed by the extremely strong protein-type toxin produced also by the EPB symbiont. The toxins are strain-specific, structurally different but each of them is highly toxic only to insects. These peptides provide monoxenic conditions for the EPN/EPB symbiotic complex to sustain under polyxenic (insect gut, soil) conditions. Each symbiotic complex produces a set of well-conserved AMP molecules with unique structure and target spectrum. Each obligate EPN/EPB symbiosis is a unique type of mutualism between the respective EPN and EPB strains. The type EPN—insect relation is parasitism, while the type of EPB-insect relation is pathogenicity. The chapter has been discussing all known EPN/EPB symbiotic associations in details; attempts of using EPB toxins as plant protection agents in transgenic plants; and the perspective of using antimicrobial peptides to overcome multidrug-resistant (MDR) plant pathogenic, veterinary and clinical pathogenic bacteria, oomycetes, fungi, and protozoa.

Keywords Bacteria · Bioactive compounds · Entomopathogen · Mutualism · Nematode · Symbiosis

---

A. Banerjee (✉)
Department of Environmental Studies, Siksha-Bhavana, Visva-Bharati, Santiniketan, India

S. Saha
Department of Zoology, North-Eastern Hill University, Shillong, Meghalaya, India

R. P. Singh et al. (eds.), *Microbes in Microbial Communities*,
https://doi.org/10.1007/978-981-16-5617-0_4

4.1 Introduction

Over the years, green developments and sustainable approaches have swathed in most aspects of our lives and the results are most profound in the field of agriculture. Shifting toward bio-diesel, biogas, bio fertilizers and bio pesticides the world has drifted from the use of complex industrial chemicals to environment friendly alternatives (Singh et al. 2014a, b, 2020a, b; Maheshwari et al. 2021). The increasing needs of the modern world have led to an increased research with impetus to the use of biopesticides and biofertilizers leading to greater yields and quality (Chitwood 2003; Fan and Rosegrant 2008; Donadio et al. 2012). Pests that cause a menace to the crop plants can be controlled organically using toxins produced by some entomopathogenic bacteria (EPB) and have become the key pesticides for commercial use to control pest populations (Forst et al. 1997; Ehlers 2001; Chattopadhyay et al. 2004). Host–microbe interactions ranging from symbiotic to pathogenic are omnipresent in various ecological niches in land and water (Bozhüyük et al. 2016). Various studies have been reported based on these interactions including symbiotic relationships between microbes and plants (Delaux and Schornack 2021), vertebrates and invertebrates and the multiple outcomes of these interactions have also been exploited. One such interaction with its ample usage as bio pesticide, has grabbed attention of the scientist community is the entomopathogenic microorganisms and its invertebrate host (Forst et al. 1997; Ogier et al. 2010; Dreyer et al. 2018). Microbial entomopathogens attack several insects, mites, or ticks *Clostridium*, *Bacillus*, *Paenibacillus* among the spore-formers and *Serratia*, *Pseudomonas*, *Providencia*, *Yersinia*, *Photorhabdus*, and *Xenorhabdus* among the non-spore-formers (Kalha et al. 2014). The entomopathogenic, motile, Gram-negative genera of *Xenorhabdus* and *Photorhabdus* belonging to the Gammaproteobacteria has been widely studied for their unique life cycle which involves a complex cycle within more than one host, having symbiotic relationship with nematodes of the genera *Steinernema* and *Heterorhabditis* respectively (Forst et al. 1997; Boemare 2002; Bozhüyük et al. 2016). Most validations of *Xenorhabdus* and *Photorhabdus* cultures have been isolated from soil-dwelling nematodes although the cells have been independently cultured under laboratory conditions (Fukruksa et al. 2017). Notably, free-living cellular entities of the bacteria have not been reported from any environmental samples (Poinar and Grewal 2012). Characteristically, a symbiotic bacterium is associated with a single species of the nematode while the same species of the bacteria can be hosted by a wide species range of the entomopathogenic nematode (Goodrich-Blair and Clarke 2007). The ratio of nematode species is more compared to bacterial species as 61 entomopathogenic nematodes (*Steinernema*) was researched to be associated with 26 species of bacteria (*Xenorhabdus*) thus, increasing the species diversity of the nematodes than the bacteria (Sajnaga et al. 2018). These studies prove that the symbiotic association is an indispensable part for the persistence of the bacteria in the soil. Also, the bacteria are intended to bring about efficient pathogenicity of the insect host facilitating the nematode to capably complete its life cycle.

In the pathogenic bacterial life cycle, the invading bacterium must adapt to the challenging environment inside the host species, combat the imminent host microbiota and dodge the different host immune responses (Sicard et al. 2004; Herbert and Goodrich-Blair 2007; Richards and Goodrich-Blair 2009; Nielsen-LeRoux et al. 2012). *Photorhabdus* and *Xenorhabdus* have evolved to fight the host environment challenges with the help of toxin proteins and peptides which eventually leads to the death of the host species (Singh et al. 2014a, b; Nielsen-LeRoux et al. 2012). Most entomopathogens directly attack the host and cause its death while the genera *Photorhabdus* and *Xenorhabdus* have a unique life cycle as its life is spent inside a nematode and is carried into the target insect through its primary host (Snyder et al. 2007; Goodrich-Blair and Clarke 2007; Poinar and Grewal 2012). These invertebrate nematodes have an obligatory dependency on their microbial symbionts for completing most of their life history traits (Poinar and Grewal 2012). A complex series of events take place in the insect-nematode-microbe interaction and to understand the underlying features, the whole paradigm needs to be ventured.

Antibiotics represent a revolutionary achievement in treating infectious diseases which has significantly advanced the health sector and increased life expectancy in the entire world. Numerous natural resources are being studied including plants, animals, and microorganisms. Manipulation of these natural products using several chemical and biotechnological tools, these resources generated newer compounds which act promising antimicrobial agents (Planson et al. 2011; Butler et al. 2013; Mbaveng et al. 2015; Li et al. 2017). The first bacteriocin recognized in 1925 (Gratia 1925), enabled the expansion of an entire research arena devoted to discover and identify potent antimicrobial compounds effective against various pathogenic bacterial, fungal and viral species (Akerey et al. 2009; Hassan et al. 2012; Torres et al. 2013; Prajakta et al. 2019), and even against natural resistant structures such as bacterial biofilms (Park et al. 2011; Yasir et al. 2018). The use of microbial metabolites for pest resistance dates back to 1950s when insecticidal sprays containing the Cry toxin from *Bacillus thuringiensis*, under the name of Thuricide™ was commercialized (de Maagd et al. 1999; Chattopadhyay et al. 2004). The "natural products" from the bacteria *Photorhabdus* and *Xenorhabdus* is under much research that targets a wide range of pests and also produces antimicrobial compounds which arise from a complex mutualistic behavior that can be further exploited for the use of mankind (Chaston et al. 2011; Bozhüyük et al. 2016).

In this chapter we specifically look into the antimicrobial peptide and toxin-based mutualism caused by the obligate symbiotic entomopathogenic nematode (EPN)—bacterium having monoxenic associations and their applications in modern science.

4.2 Mutualism Between EPN and EPB

Xenorhabdus and *Photorhabdus* are symbiotically associated with the two genera of entomopathogenic nematodes (EPNs) *Steinernema* and *Heterorhabditis*, respectively and these nematodes are distinctive because (1) they possess the ability to symbiotically carry bacteria introducing them in the body hemocoel of the insects or pests, (2) these rhabditid nematodes are readily grown and cultured on synthetic solid or liquid media (Chaston et al. 2011; Bozhüyük et al. 2016), and are pathogenic to a wide range of insects that are pests to a huge range of crop plants (Boemare et al. 1993; Ehlers 2001).

4.2.1 *The Complex Life Cycle*

The cycle (Fig. 4.1) begins with the free-living dauer stage of the nematodes living in soil, also known as Infective Juveniles (IJs) (Lacey et al. 2015).

These stages of IJs harbor the bacteria (*Xenorhabdus* and *Photorhabdus*) in the anterior part of its gut as symbionts and carry them into the target insect host by penetrating through mouth, spiracle, anus, or by inserting through the cuticle making their way to hemocoel (Gulcu et al. 2017; da Silva et al. 2020). After successful infection of the host, the nematodes release the bacterium into the insect hemolymph (Dowds and Peters 2002), where they multiply, and produce numerous metabolites

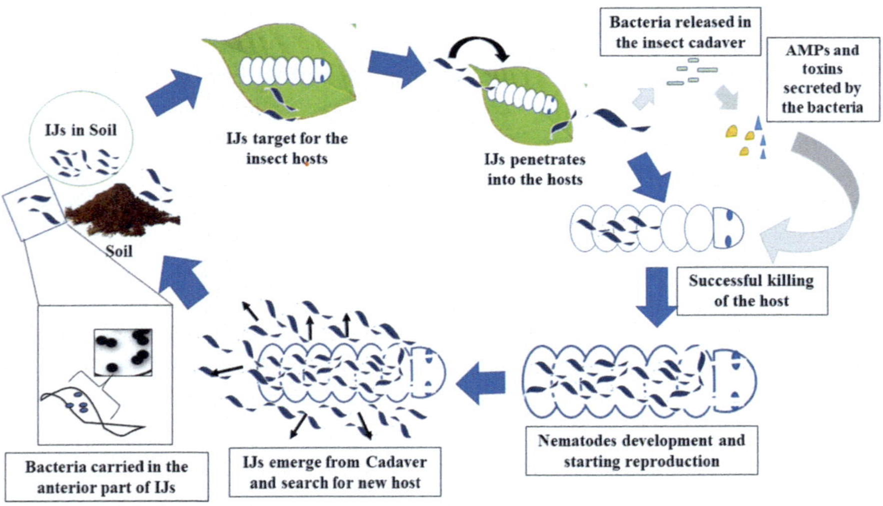

Fig. 4.1 The insect-nematode-bacteria life cycle with the infective juveniles (IJs) of the entomopathogenic nematodes (EPNs) infecting the target host and bringing about the series of changes with the successful infection and release of the symbiotic bacteria *Xenorhabdus* and *Photorhabdus* spp. in the host cadaver

or natural products that aid in the insect demise, degradation of tissues, protection from host immune responses (Richards and Goodrich-Blair 2010) and also importantly from the other competing microorganisms present in the imminent environment within the insect cadaver (Waterfield et al. 2009; Tobias et al. 2018). Thus, "offensive mutualism" is important for the adult nematodes, which mature inside the host, in killing the prey using the bacterial metabolites and protecting the insect carcass from predating organisms (Koppenhöfer and Gaugler 2009). The bacteria initiate replication inside the host and reach the stationary phase before the sexual reproduction of the nematode begins (Han and Ehlers 2000; Tobias et al. 2016, 2017). Optimum nematode reproduction occurs when the symbiotic bacterial population is maximum, which also acts as a food source for the EPN (Snyder et al. 2007). Later, the insect body is slowly degraded and used up for energy as the nematodes enter the reproductive phase (Fig. 4.1) and the new generation of EPNs leave the dead body in search of a new host (Forst and Clarke 2002). The life cycle of the bacteria *Xenorhabdus* and *Photorhabdus* are interweaved with the life cycle of their symbiotic IJs. After discharge of the bacteria inside the hemocoel, evading the host immune response, the bacteria start multiplying and producing factors that provide protection and assistance in nurturing the new cells of both the EPBs and EPNs from predation (Daborn et al. 2001; Eleftherianos et al. 2009). The IJ is the only free-living stage and as the energy source declines these new units leave the carcass to find new hosts employing a species-specific approach using artifices from ambush to scavenging (Adams et al. 2006; Ciche and Ensign 2003). Most importantly, the obligate mutualistic bacteria are reassociated with the newly produced nematodes before exiting the body of the host (Martens et al. 2003). Several natural products have been studied that help in maintenance of the symbiosis enabled by either of the enzymes polyketide synthesases (PKSs) or non-ribosomal peptide synthesases (NRPSs) (Crawford et al. 2012; Singh et al. 2015; Tobias et al. 2016).

Both the genus of bacteria is capable of producing certain extracellular enzymes like lipases, proteases and wide range of antibiotics as secondary metabolites and are believed to be secreted inside the hemolymph of the insect body when the cells reach the stationary phase of their growth cycle (Bode 2009; Pidot et al. 2014). These cells have been cultured in laboratory conditions and typically produce cytoplasmic inclusion bodies composed of intricate structure of crystalline proteins (Stock et al. 2017).

A distinctive variant of *Xenorhabdus* and *Photorhabdus* spp. is the phase II cells which characteristically do not exist as symbionts with the nematodes (Hazir et al. 2016) however, the phase I cells are the only forms of the two genera that can form symbiosis with the IJs of the nematodes. Specific isolates of *X. nematophila* was isolated from *Steinernema carpocapsae*; *X. hominickii* from *Steinernema monticolum* and *Photorhabdus temperate temperata* from *Heterorhabditis megidis* (Bird and Akhurst 1983; Park and Kim 2000; Kang et al. 2004). This interesting bacteria-nematode-insect life cycle has long been studied with the implications of using the antagonistic features of the bacteria as antibiotic compounds and pest control.

4.2.2 Biochemical Factors Involved in the Life Cycle: Toxin-Based Mutualism

From the life cycle of the three entities: bacteria > nematode > insect, it is understood that a complex biochemical chain with a number of compounds is involved behind each step of the cycle. With the entry of the IJs in the host, both nematode and bacteria start killing the host, but in most cases the bacteria are alone highly virulent and autonomously cause the death of the insect. On recognition of pathogens, the innate immunity of the insect triggers releasing several recognition receptors like, cytokine, nitric oxide, eicosanoids and biogenic monoamines (Gillespie et al. 1997; Lemaitre and Hoffmann 2007). Phospholipase A2 or PLA2 secretion initiates with eicosanoid activation and PLA2 catalyzes the synthesis of eicosanoids and subsequently synthesizes leukotrienes, prostaglandins, and epoxyeicosatrienoic acids (Seo et al. 2012; Park et al. 2003; Kim et al. 2018). The EPBs have evolved to synthesize a number of compounds that are potential immunosuppressant of insect immune system, also producing a variety of insecticidal toxins in the form of crystalline proteins competently killing the same insect, suggesting PLA2 inhibition a major tactic for *Xenorhabdus* and *Photorhabdus* pathogenesis (Hasan et al. 2019). Kim et al. (2018) have identified that the PLA2 suppression occurs within 3 h of injection into the insect cadaver (Kim et al. 2018). Several PLA2 inhibitors have been documented (Seo et al. 2012; Sadekuzzaman and Kim 2017), benzylidene acetone being the first to be identified from a liquid culture of *X. nematophila* (Ji et al. 2004).

Recent works in this regard indicated that toxin-mediated killing of the host is enhanced with the use of several immunosuppressive compounds. Reports of high molecular weight toxins secreted by *Photorhabdus luminescence* and *Xenorhabdus nematophila* showed an important cause for the insect death (Dunphy and Webster 1984; Bowen and Ensign 1998; da Silva et al. 2013). Studies on the bacteria-insect transaction revealed that *Photorhabdus* and *Xenorhabdus* enters the insect moth *Manduca sexta* through the anterior midgut region which is initially colonized and later spreads to the posterior end (da Silva et al. 2013). *Photorhabdus* sp. brings about a breakdown of the intestinal epithelial lining inhibiting the insects' ingestion (Owuama 2001; da Silva et al. 2013). According to scientific explanations pathogenicity brought about by the bacteria affects cell replication of the host and toxin production causes histological injury and septicemia (Owuama 2001). The toxin complexes represented as Tcs is produced both by *Xenorhabdus* and *Photorhabdus* species and are highly lethal on insects. Tcs produced by *Xenorhabdus* induce inhibition of eicosanoid synthesis of insect thus, interfering with the insect immune system (Dunphy and Webster 1984; Park and Kim 2000). Fabclavins produced by *Xenorhabdus budapestensis* are fusion compounds having antibiotic and insecticidal properties (Akhurst 1982; Sergeant et al. 2006; Bode 2009; Fuchs et al. 2014) while another eight set of insect immunosuppressing metabolites are synthesized by *Xenorhabdus nematophila* as reported by Eom et al. (2014). Four different types of insecticidal toxins have been reported by different species of *Photorhabdus* which

Table 4.1 Production of secondary antimicrobial compounds of *Xenorhabdus* spp. And *Photorhabdus* spp.

Species name	Secondary metabolic compounds	References
Xenorhabdus cabanillasii	Nemaucin, Cabanillasin, Rhabdopeptide	Gualtieri et al. (2009), Houard et al. (2013), Reimer et al. (2013)
Xenorhabdus doucetiae	Xenoamicin, Xenocoumacin, Xenorhabdin, Phenylethylamine, Tryptamide	Zhou et al. (2012), Bode et al. (2017)
Xenorhabdus budapestensis	Fabclavine, Bicornitun	Xiao et al. (2012), Fuchs et al. (2014), Tobias et al. (2017)
Xenorhabdus mauleonii	Xenoamicin, Xenocoumacin, Xenorhabdin	Zhou et al. (2012), Tobias et al. (2017)
Xenorhabdus kozodoii	Xenocoumacin	Tobias et al. (2017)
Xenorhabdus nematophila	Nematophin	Crawford et al. (2011), Reimer et al. (2014), Seo et al. (2012), Brachmann et al. (2012), Shi and Bode (2018), Hong et al. (2015), Reimer et al. (2013), Guo et al. (2017), Morales-Soto and Forst (2011), Morales-Soto et al. (2012), Singh and Banerjee (2008)
	Oxindole and Benzylideneacetone	
	Pristinamycin	
	Rhabduscin	
	Rhabdopeptide	
	Xenocoumacin	
	Xenorhabdicin	
	Xenocin	
Xenorhabdus szentirmaii	Fabclavine, Szentiamide	Grundmann et al. (2013), Fuchs et al. (2014), Tobias et al. (2017)
Photorhabdus luminescens	Leucine Responsive Protein (lrp)	Nollmann et al. (2015)

include the, Pvc (*Photorhabdus* virulence cassettes), Mcf (make caterpillars floppy), Pir (insect-related protein) and Tcs (toxin complexes) (Rodou et al. 2010). Pvc toxins are virulence factors which act in making the insects vulnerable and simultaneously Mcf assumes control causing apoptosis in the hemocoel (Forst and Nealson 1996; Jallouli et al. 2010) (Table 4.1).

Tcs toxin complex has a mode of action analogous to that of the δ-endotoxin of *B. thuringiensis* and aids in destroying epithelial cells of the insect midgut (Aktories et al. 2014). Certain species of *Xenorhabdus* is effective only in association with the nematode *Steinernema* forming the nemato-bacterial complex like the *X. innexi* which feeds mostly on crickets (Bonifassi et al. 1999; Sicard et al. 2003, 2005). *X. nematophila* and *P. luminescens*, infection has been studied in mosquito larva (*Aedes egypti*) and showed to cause larvae mortality, and interfered the development of pupae and adults (da Silva et al. 2020).

4.3 Insect Immune Response and the Antimicrobial Peptides

The humoral and cellular immune responses of the insect host combine an antibacterial defence mechanism to combat the bacterial invaders. It is well established that insects are also very resistant to the bacterial infections. They can produce an extensive range of proteins and peptides as a first line of defence against pathogens (Oñate-Garzón et al. 2017). We especially focus on a large group of AMPs that are present in various insects such as defensins, cecropins, attacins, lebocins, drosocin, dipterins, metchnikowin, ponericins, jelleines, apisimin, pyrrhocoricin, persucatusin, and melittin. Three groups of the insect AMPs have been classified according to the amino acid sequence and structures. (a) Defensins with 6–8 conserved cysteine residues, having a stabilizing array of three or four disulfide bridges and three domains consisting in a flexible amino-terminal loop; (b) Cecropins, the linear peptides with-helix but lack cysteine residues; and (c) peptides with an over representation of Proline and/or Glycine residues (Makarova et al. 2018). The most important insect AMPs are cecropins, drosocin, attacins, diptericins, defensins, ponericins, drosomycin, and metchnikowin. Though, more new peptides can still be discovered (Mylonakis et al. 2016; Zhang and Gallo 2016).

Defensins are small (~4 kDa) cationic AMPs with six conserved cysteine residues that form three intramolecular disulfide bridges, and they have been identified in nearly all living organisms (Zhu and Gao 2013). This peptide inhibits the Gram-positive bacteria and also fungi. It is particularly effective against the larvae of the bee pathogen *Paenibacillus*, which causes American foulbrood (Blikova et al. 2002). Cecropins are a family of cationic antimicrobial peptides of 31–39 residues first isolated from the immunized hemolymph of *H. cecropia* pupae (Hultmark et al. 1982; Steiner et al. 1981), and have been identified in lepidopteran, dipteran, and coleopteran insects. Cecropins are synthesized as secreted proteins and mature active cecropins are generated after removal of signal peptides (Table 4.2) (Hultmark et al. 1982). Cecropins depict a broad range of activity against Gram-negative and Gram-positive bacteria, as well as fungi (DeLucca et al. 1997; Ekengren and Hultmark 1999; Vizioli et al. 2000).

Attacins are synthesized as pre-proproteins containing a signal peptide, a pro-peptide (P domain), and an N-terminal attacin domain, followed by two glycine rich domains (G1 and G2 domains) (Hedengren et al. 2000). Attacins were first purified from the hemolymph of bacteria immunized *H. cecropia* pupae (Hultmark et al. 1983) and can inhibit growth of *E. coli* cells by directly targeting bacterial outer membrane to increase permeability (Engström et al. 1984), and also, they can inhibit synthesis of several bacterial outer membrane proteins, including OmpC, OmpF, OmpA, and LamB by binding to LPS even without entering the inner membrane or cytoplasm (Carlsson et al. 1991).

Melittin, the bee venom, causing cell lysis (Rady et al. 2017) has a strong antibacterial effect against a variety of bacteria (Leandro et al. 2015; Wu et al.

Table 4.2 Bacterial symbiont species associated with entomopathogenic nematodes

S. No.	Symbionts	Host nematodes	References
1	*Photorhabdus luminescens*	*Heterorhabditis bacteriophora*	Boemare et al. (1993)
2	*Xenorhabdus nematophilus*	*Steinernema carpocapsae*	Boemare and Akhurst (1988)
3	*Xenorhabdus poinarii*	*Steinernema glaseri*	Boemare and Akhurst (1988)
4	*Xenorhabdus bovienii*	*Steinernema feltiae*	Boemare and Akhurst (1988)
5	*Xenorhabdus bovienii*	*Steinernema affinis*	Boemare and Akhurst (1988)
6	*Xenorhabdus* sp.	*Steinernema scapterisci*	Aguillera et al. (1993)
7	*Xenorhabdus japonica*	*Steinernema kushidai*	Yamanaka et al. (1992)
8	*Xenorhabdus beddingii*	*Steinernema* sp.	Boemare and Akhurst (1988)
9	*Xenorhabdus* sp.	*Steinernema cubana*	Mráček et al. (1994)

Table 4.3 Different types of toxins and their functions of *Xenorhabdus* spp. and *Photorhabdus* spp.

Species name	Toxins	Functions	References
Xenorhabdus	Tcs	Induce immunosuppression in insects by inhibiting eicosanoid synthesis	Park and Kim (2000)
Xenorhabdus budapestensis	Fabclavins	It exhibits antibiotic and insecticide activities	Sergeant et al. (2006), Bode (2009), Fuchs et al. (2014)
Photorhabdus	Pvc (*Photorhabdus* virulence cassettes)	It destroys epithelial cells in the *M. sexta* and *G. mellonella*	Forst and Nealson (1996)
	Pir (insect-related protein)	It causes the death of larvae of *G. mellonella*, with high toxicity	Waterfield et al. (2005)
	Tcs (toxin complexes)	It destroys epithelial cells from the middle intestine of insects	Rodou et al. (2010)
	Mcf (make caterpillars floppy)	It promotes hemocytes apoptosis in the hemocoel	Jallouli et al. (2010)

2016; Picoli et al. 2017; Socarras et al. 2017) specifically against *Xanthomonas oryzae* pv. a destructive bacterial disease of rice, indicating that this peptide may have potential applications in plant protection (Shi et al. 2016). In several other insects, proline-rich antimicrobial peptides of 16–34 residues with different names have been identified (Table 4.3).

The immune response so far studied in insects has a clear picture of how the immune system and its components work against the pathogens. On nematode infection, host pattern recognition by the peptides is the first action of defence which is followed by PAMPs (pathogen-associated molecular patterns) mediated recognition of the antigen (Bettencourt et al. 2004; Buchon et al. 2014). This triggers the synthesis of insect antimicrobial compounds, initiating the immunogenic reactions and causing phagocytosis (Marmaras and Lampropoulou 2009). Synthesis of the peptide molecules occurs late in the reaction process which are secreted by the cells of the gut, lymph glands, ovaries and midgut several hours after the infection (Rosales and Vonnie 2017). Several insect models are being generated to study the mode of action of various bacterial pathogens. The Lepidopteran moth *Manduca sexta* has been thoroughly studied for its immunogenic pathways against bacterial infections and has shown to have various mechanism of resistance against bacteria (Silva et al. 2002). In a susceptibility test against the pathogenic P11-1 strain of *Pseudomonas aeruginosa*, higher resistance was shown by the larva of *Manduca sexta* than *Galleria mellonella* (Dunn and Drake 1983). On *Photorhabdus* infection of *M. sexta* the insect is mostly killed due the suppression of its hemocyte-mediated phagocytosis (Dowds and Peters 2002; Silva et al. 2002). Recent molecular expression studies have shown that the Cecropin encoding gene was inhibited by cells of *X. nematophila* in *Spodoptera exigua* (Duvic et al. 2012). Hwang et al. (2013) also reported a similar study where the AMP genes were inhibited in insects infected with *X. nematophila*. The modulation of the expression inhibition of the antimicrobial peptide genes varies with different EPN/EPB symbionts (Darsouei et al. 2017). Upregulation of the peptide genes of attacin and drosomycin has also been a method of insect killing. The nematode *Steinernema carpocapsae* and *Heterorhabditis bacteriophora* has been reported to induce the antimicrobial peptide genes and cause cell death using the insect immune system (Vega and Kaya 2012; Alvandi et al. 2017).

4.4 Insecticidal Activity and Pest Management

Reports on the insecticidal activity from earlier studies reveal significant contribution of the nematode-bacteria association (Lu et al. 2017; Chang et al. 2019). The EPNs can infect a broad range of insects which are predominantly pests of crop plants and are easily capable of multiplication in huge numbers bringing about effective pest control (Shapiro-Ilan and Gaugler 2002; Shapiro-Ilan et al. 2014). Application of EPNs along with the bacterial symbionts has been studied as harmless with no side-effects for the environment and human beings and other prevalent non-target organisms (Akhurst and Smith 2002; Ehlers 2005). *Photorhabdus* bacteria, when released in the insect hemolymph, independently have shown exceptionally high pathogenicity (Boemare et al. 1997). The nemato-bacterial complex *Xenorhabdus innexi* and *Steinernema scapterisci* have been effective killers of crickets. *X. nematophila* and *P. luminescens* cultures have shown larvicidal activity and have been tested on the mosquito larva (da Silva et al. 2020). However, an important feature of this activity is that there has been a few information on insect resistance to these virulent genera of bacteria (Carlsson et al. 1991; Srisailam et al.

2000; Lu et al. 2016). Most advantageously the mode of action of the toxins from *P. luminescens* is different from the much-complained BT toxin obtained from *B. thuringiensis* and may act as potent pesticides (Bărnuŀiu et al. 2011; Miyoshi et al. 2017). The virulence efficiency against the common wax moth *Galleria mellonella* have been studied in *Photorhabdus* and *Xenorhabdus* and has been shown that the toxins from the bacteria are alone lethal for the larval stage of the moth (Rahoo et al. 2011). Application of IJs of *S. feltiae* and *S. carpocapsae* against the universal pest of apple and pomegranate, codling moth of the Lepidoptera family (*Cydia pomonella*) provide a very good instance of the effective usage of EPNs for bio-control of pests (Lacey et al. 2006). Other pest targets of EPNs have been noted in the order Lepidoptera are the navel orangeworm, *Amyelois transitella* (Siegel et al. 2006); peach tree borer *Synanthedon pictipes* (Shapiro-Ilan et al. 2010); filbert moth, *Melissopus latiferreanus* (Chambers et al. 2010); and oriental fruit moth, *Grapholita molesta* (Riga et al. 2006). Lepidopteran cutworms of the genera *Agrotis*, *Amathes*, *Noctua*, *Peridroma*, *Prodenia* spp. could be suitably controlled by the use of EPNs (Shapiro-Ilan et al. 2002; Ebssa and Koppenhöfer 2011).

Transgenic studies in crop protection using *Bacillus thuringiensis* have been an extremely successful agricultural advancement but the antagonistic effect of resistance in the pest populations pose a threat to its future prospects. Hence, alternative bioprospection of novel compounds from unexplored sources and their deployment in pest control trials has been the current initiative of agriculturists. Expression studies of the toxic GroEL chaperones extracted from strains of *X. nematophila* has deliberated resistance in tomato and tobacco plants (Kumari et al. 2015a). Resistance in tomato and tobacco was also observed by the same research group by using pilin encoding gene of *X. nematophila* against the pest *Helicoverpa armigera* (Kumari et al. 2015b). Although transgenic plants are restrained in pest control applications, the novel insecticidal genes from the two rhabditid genera can show us light in the near future, helping to prevent massive crop losses due to pests.

4.5 Novel Natural Products, Their Applications and Impact on Future Research

The Gram-negative *Xenorhabdus* and *Photorhabdus* spp. of the Enterobacteriaceae synergistically produce many bioactive components with the symbiotic nematodes required for the propagation and reproduction of both the counterparts in the association (Dreyer et al. 2018).

4.5.1 Bioactive Compounds from Xenorhabdus *and* Photorhabdus *spp.*

Xenorhabdus spp. produces bioactive metabolites and compounds in its entire life cycle which have been largely ignored and research is still underrated in this domain. These bioactive compounds synthesized by *Xenorhabdus* spp. have a wide range of targets including bacteria, fungi, protozoa, and also inhibiting nematodes and insects (Webster et al. 2002). Mostly the enzymes NRPS (non-ribosomal peptide synthetases) and PKS (polyketide synthetases) initiate the synthesis of the diverse group of peptides that act as the precursors of various active compounds like the fabclavins (Fuchs et al. 2014), depsipeptides (Zhou et al. 2012; Kronenwerth et al. 2014), PAX (peptide-antimicrobial-*Xenorhabdus*) peptides (Fuchs et al. 2011), and xenocoumacins (Reimer 2018). Other important compounds like the indole-derivative compounds (Sundar and Chang 1993), benzylideneacetone (Ji et al. 2004), and bacteriocins like xenocins (Singh and Banerjee 2008) have been reported from different species of *Xenorhabdus* spp. Other unnamed peptides with broad spectrum antimicrobial property have also been reported by Xiao et al. (2012).

Photorhabdus spp. also has been studied for its antimicrobial bioactive compounds which are critical in maintaining a monoxenic paradigm inside the host bug and avert further infection of the larval body from surrounding pathogenic microbes and infestation by arthropods. Antibacterial compounds were first obtained by Paul et al. (1981) from pure cultures of *P. luminescens* and structurally characterized two antibiotic compounds. Only a few of these metabolites have been studied, isolated and tested for their biological activities while most of their large-scale usages are still in the budding stage (Bozhüyük et al. 2016). An exhaustive genome sequence study analyzing the biosynthesis gene clusters for the NRPS and PKS in seven species and subspecies of *Photorhabdus* spp. showed positive results along with 22 gene clusters all responsible for synthesis of different and novel natural products (Tobias et al. 2016). Rhabduscin, a rare compound synthesized both by *Xenorhabdus* and *Photorhabdus* spp. is chemically an isonitrile required by the bacteria to inhibit the enzyme phenoloxidase produced by the insect host (Crawford et al. 2012). Isopropyl stilbene (IPS) a derivative of dialkylresorcinol (DAR) involved in rhabduscin biosynthesis is an important compound in nematode biosynthesis and also shows antibacterial activity and are cytotoxic to eukaryotic cell lines (Buscató et al. 2013). It is also likely that these bacteria employ these natural products as signaling molecules and hence necessitates in-depth research regarding these compounds (Schöner et al. 2015; Bozhüyük et al. 2016). Specific strains of *P. luminescens* and *P. asymbiotica* have been found to produce several antifungal compounds like cepafungin and glidobactin along with their derivatives which are potent proteasome inhibitors and are involved in protection against fungi and other saprophytes (Stein et al. 2012; Theodore et al. 2012; Bozhüyük et al. 2016). *Photorhabdus* and *Xenorhabdus* strains commonly synthesize rhabdopeptides which show antiprotozoal and cytotoxic activity that are involved in the defence against saprophytes in the soil (Bode et al. 2012, 2015; Tobias et al. 2016; Zhao et al.

2018). Various other peptides photopyrones and other derivatives from DARs have quorum sensing properties (Kresovic et al. 2015). Transcinnamic acid from pure cultures of *P. luminescens* identified by Bock et al. (2014) proved effective against a pecan fungal pathogen, *Fusicladium effusum* exhibiting a clear zone at a concentration of 200 mg/mL. A recent novel toxic peptide Galtox, with unknown mode of action has been reported by Ahuja et al. (2021) by the Indian strains of *P. akhurstii*.

Identification and isolation of these natural products from *Photorhabdus* has been researched and an approach based on the whole genome sequence has been carried out by Bode and Müller (2005) which in short is known as OSMAC (one strain many compounds). This uses detailed genetic techniques under different growth conditions like promoter exchange, manipulation of regulatory proteins and heterologous expression and also chemical tools to identify and study various natural products and their synthesis process which has been comprehensively discussed in a review by Bozhüyük et al. (2016).

4.5.2 Usage of the Natural Products and Application-Based Studies

Almost certainly the major task in this area of natural product research is of recognizing the target compounds of specific natural bioactives from the bacteria and it's final on-field application. As per reports dated 2005, approximate of ten species of EPNs native to Europe and America, are marketed commercially as antipest agents for crop management (Grewal and Peters 2005). The natural compounds from *Xenorhabdus* and *Photorhabdus* spp. show multidimensional actions and cover an enormous paradigm applicable to various day to day usages of human kind. Although these tend to perform better in their indigenous environmental conditions which also alleviates the fear of the unwanted effect of the exotic EPN products (Ehlers 2005; Kumar et al. 2015). Thus, isolating native EPNs from different environmental niches is essential for better performance of the EPN/EPB symbionts. This also increases the chance of finding novel products with researchers contributing from various countries (Malan et al. 2006; Noosidum et al. 2010; Ganguly et al. 2010). One of the research group has conducted antibiotic synthesis from *Xenorhabdus bovienii* strain YL002; *X. nematophila* TB (Fang et al. 2010) and YL001 (Wang et al. 2008) in a larger scale, upregulating the process (Fang et al. 2012). Ullah et al. (2015) carried out a research where benzaldehyde produced by *Photorhabdus temperata* strain M1021 prove to be antimicrobial, insecticidal and antioxidant in nature. NP like trans cinnamic acid produced by the isolate *P. luminescens* offered a powerful replacement to the harsh chemical pesticides, reducing the menace of fungicide resistance, and also abating any adverse environmental impacts (Bock et al. 2014). Biological control of *Aedes aegypti* larvae using inhibitory compounds of *Xenorhabdus* and *Photorhabdus* spp. has shown great potential in Northern Thailand (Fukruksa et al. 2017). da Silva et al. (2020) in a recent review put forward the usage and research of insecticidal properties of *Xenorhabdus* and *Photorhabdus* in biologically controlling the extensive spread of mosquito vectors on an urgent basis.

Compounds obtained from a single strain of *Xenorhabdus* and *Photorhabdus* has been found to be insecticidal, nematocidal, antiprotozoal, and also target cancerous cells apart from being antibacterial and antifungal in nature. The possibility of discovering novel antimicrobial compounds is promising and methods need to be developed to produce such compounds at higher concentrations. The important hurdle in this production process is that these compounds are synthetized non-ribosomally and are not a single gene product (Dreyer et al. 2018).

4.6 Conclusion

It is evident that *Xenorhabdus* and *Photorhabdus* spp. genus of the endosymbiotic bacteria are an excellent source for novel antimicrobial metabolites and bioactive compounds. The antimicrobials from these bacteria have been ignored as antibiotic source and there is a dearth in their application-based studies (Pidot et al. 2014). A number of research has focused on the remarkable potential of these metabolites and explained their use both in vitro and in vivo (Barkai-Golan 2001; Böszörményi et al. 2009; Shapiro-Ilan et al. 2009; Fang et al. 2011). These compounds find their usages not only in the field of agriculture but also in multiple spheres encompassing various industries, the healthcare and food industries. Upliftment of agricultural science, medical science and veterinary medicine and the food and feedstuff industries can be made with application-based research on these antimicrobial peptides (AMPs) (Pidot et al. 2014; Dreyer et al. 2018).

Multi-drug resistance has become a major problem in medical treatment against various diseases caused by bacteria and the novel antimicrobial compounds from these endosymbiotic bacteria can provide a solution to the imminent problem. A single species of these bacteria produces a wide arsenal of antimicrobial compounds which surpass the number of the recognized antibiotics (Dreyer et al. 2018). Although research is required to upregulate and optimize the production technique of such useful compounds. It is apparent that production of these bioactive compounds is not an easy job and thorough in-depth studies need to be carried out to refine the process and obtain the desired product.

The insecticidal nature of these group of bacteria have been studied but latest research on its further use and its implications has not much been reported. It is also important to find out which strain of the bacterial isolate is more effective and has results without posing any kind of threats to the environment.

The endosymbiotic bacteria along with their nematode is a stunning example of nature's wonders and hidden mysteries which still needs much attention of the scientific community. Further research is also required to bring into light the diverse perspectives of the symbiosis and the number of beneficial impacts on the different spheres of life.

References

Adams BJ, Fodor A, Koppenhöfer HS, Stackebrandt E, Stock SP, Klein MG (2006) Reprint of Biodiversity and systematics of nematode–bacterium entomopathogens. Biol Control 37:32–49

Aguillera MM, Hodge NC, Stall RE, Smart GC Jr (1993) Bacterial symbionts of *Steinernemascapterisci*. J Invertebr Pathol 62:68–72

Ahuja A, Kushwah J, Mathur C, Chauhan K, Dutta TK, Somvanshi VS (2021) Identification of Galtox, a new protein toxin from Photorhabdus bacterial symbionts of *Heterorhabditis* nematodes. Toxicon 194:53–62

Akerey B, Le-Lay C, Fliss I, Subirade M, Rouabhia M (2009) *In vitro* efficacy of nisin Z against *Candida albicans* adhesion and transition following contact with normal human gingival cells. J Appl Microbiol 107:1298–1307

Akhurst RJ (1982) Antibiotic activity of *Xenorhabdus* spp. bacteria symbiotically associated with insect pathogenic nematodes of the families heterorhabditidae and steinernematidae. J Gen Microbiol 128:3061–3065

Akhurst R, Smith KI (2002) Regulation and safety. Entomopathog Nematol 31:311–332

Aktories K, Schmidt G, Lang AE (2014) *Photorhabdus luminescens* toxins TccC3 and TccC5: insecticidal ADP-ribosyltransferases that modify threonine and glutamine. Curr Top Microbiol Immunol 384:53–67

Alvandi J, Karimi J, Ghadamyari M, Sharifi M, Asoodeh A (2017) Physiological defense of the white grub, polyphylla adspersa motschulsky (Col., Scarabaeidae) against entomopathogenic nematodes. J Asia Pac Entomol 20:878–885

Barkai-Golan R (2001) Postharvest diseases of fruits and vegetables: development and control. Elsevier, Amsterdam

Bărnuţiu LI, Mărghitaş LA, Dezmirean DS, Bobiş O, Mihai CM, Pavel C (2011) Antimicrobial compounds of royal jelly. Bull UASVM Anim Sci Biotechnol 68:1–2

Bettencourt R, Tanji T, Yagi Y, Ip YT (2004) Toll and Toll-9 in *Drosophila* innate immune response. J Endotoxin Res 10:261–268

Bird AF, Akhurst RJ (1983) The nature of the intestinal vesicle in nematodes of the family Steinernematidae. Int J Parasitol 13:599–606

Blikova K, Hanes J, Nordhoff E, Saenger W, Klaudiny J, Simuth J (2002) Apisimin, a new serine-valine-rich peptide from honeybee (*Apis mellifera* L.) royal jelly: purification and molecular characterization. FEBS Lett 528:125–129

Bock CH, Shapiro-Ilan DI, Wedge DE, Cantrell CL (2014) Identification of the antifungal compound, trans-cinnamic acid, produced by *Photorhabdus luminescens*, a potential biopesticide against pecan scab. J Pest Sci 87:155–162

Bode HB (2009) Entomopathogenic bacteria as a source of secondary metabolites. Curr Opin Chem Biol 13:224–230

Bode HB, Müller R (2005) The impact of bacterial genomics on natural product research. Angew Chem Int Ed 44:6828–6846

Bode HB, Reimer D, Fuchs SW, Kirchner F, Dauth C, Kegler C, Lorenzen W, Brachmann AO, Grün P (2012) Determination of the absolute configuration of peptide natural products by using stable isotope labeling and mass spectrometry. Chem Eur J 18:2342–2348

Bode E, Brachmann AO, Kegler C, Simsek R, Dauth C, Zhou Q, Kaiser M, Klemmt P, Bode HB (2015) Simple "on-demand" production of bioactive natural products. ChemBioChem 16:1115–1119

Bode E, He Y, Vo TD, Schultz R, Kaiser M, Bode HB (2017) Biosynthesis and function of simple amides in *Xenorhabdus doucetiae*. Environ Microbiol 19:4564–4575

Boemare N (2002) Interactions between the partners of the entomopathogenic bacterium nematode complexes, Steinernema-*Xenorhabdus* and Heterorhabditis-*Photorhabdus*. Nematology 4:601–603

Boemare NE, Akhurst RJ (1988) Biochemical and physiological characterization of colony form variants in *Xenorhabdus* spp.(Enterobacteriaceae). Microbiology 134:751–761

Boemare NE, Akhurst RJ, Mourant RG (1993) DNA relatedness between *Xenorhabdus* spp. (Enterobacteriaceae), symbiotic bacteria of entomopathogenic nematodes, and a proposal to transfer *Xenorhabdus luminescens* to a new genus, *Photorhabdus* gen. nov. Int J Syst Evol Microbiol 43:249–255

Boemare NE, Givaldan A, Brehelin M, Laumond C (1997) Symbiosis and pathogenicity of nematode-bacterium complexes. Symbiosis 22:21–45

Bonifassi E, Fischer-Le Saux M, Boemare N, Lanois A, Laumond C, Smart G (1999) Gnotobiological study of infective juveniles and symbionts of *Steinernemascapterisci*: a model to clarify the concept of the natural occurrence of monoxenic associations in entomopathogenic nematodes. J Invertebr Pathol 74:164–172

Böszörményi E, Érsek T, Fodor A, Fodor AM, Földes LS, Hevesi M, Hogan JS, Katona Z, Klein MG, Kormány A, Pekár S (2009) Isolation and activity of *Xenorhabdus* antimicrobial compounds against the plant pathogens *Erwinia amylovora* and *Phytophthora nicotianae*. J Appl Microbiol 107:746–759

Bowen DJ, Ensign JC (1998) Purification and characterization of a high molecular-weight insecticidal protein complex produced by the entomopathogenic bacterium *Photorhabdus luminescens*. Appl Environ Microbiol 64:3029–3035

Bozhüyük KA, Zhou Q, Engel Y, Heinrich A, Pérez A, Bode HB (2016) Natural products from *Photorhabdus* and other entomopathogenic bacteria. Curr Top Microbiol Immunol 402:55–79

Brachmann AO, Reimer D, Lorenzen W, Augusto Alonso E, Kopp Y, Piel J, Bode HB (2012) Reciprocal cross talk between fatty acid and antibiotic biosynthesis in a nematode symbiont. Angew Chem 124:12252–12255

Buchon N, Silverman N, Cherry S (2014) Immunity in *Drosophila melanogaster*—from microbial recognition to whole-organism physiology. Nat Rev Immunol 14:796–810

Buscató EL, Büttner D, Brüggerhoff A, Klingler FM, Weber J, Scholz B, Živković A, Marschalek R, Stark H, Steinhilber D, Bode HB (2013) From a multipotent stilbene to soluble epoxide hydrolase inhibitors with antiproliferative properties. Chem Med Chem 8:919–923

Butler MS, Blaskovich MA, Cooper MA (2013) Antibiotics in the clinical pipeline in 2013. J Antibiot 66:571–591

Carlsson A, Engström P, Palva ET, Bennich H (1991) Attacin, an antibacterial protein from *Hyalophora cecropia*, inhibits synthesis of outer membrane proteins in *Escherichia coli* by interfering with *omp* gene transcription. Infect Immun 59:3040–3045

Chambers U, Bruck DJ, Olsen J, Walton VM (2010) Control of overwintering filbert worm (Lepidoptera: Tortricidae) larvae with *Steinernemacarpocapsae*. J Econ Entomol 103:416–422

Chang DZ, Serra L, Lu D, Mortazavi A, Dillman AR (2019) A core set of venom proteins is released by entomopathogenic nematodes in the genus Steinernema. PLoS Pathog 15:e1007626

Chaston JM, Suen G, Tucker SL, Andersen AW, Bhasin A, Bode E, Bode HB, Brachmann AO, Cowles CE, Cowles KN, Darby C (2011) The entomopathogenic bacterial endosymbionts *Xenorhabdus* and *Photorhabdus*: convergent lifestyles from divergent genomes. PLoS One 6: e27909

Chattopadhyay A, Bhatnagar NB, Bhatnagar R (2004) Bacterial insecticidal toxins. Crit Rev Microbiol 30:33–54

Chitwood DJ (2003) Research on plant-parasitic nematode biology conducted by the United States Department of Agriculture–Agricultural Research Service. Pest Manag Sci 59:748–753

Ciche TA, Ensign JC (2003) For the insect pathogen *Photorhabdus luminescens*, which end of a nematode is out? Appl Environ Microbiol 69:1890–1897

Crawford JM, Portmann C, Kontnik R, Walsh CT, Clardy J (2011) NRPS substrate promiscuity diversifies the xenematides. Org Lett 13:5144–5147

Crawford JM, Portmann C, Zhang X, Roeffaers MB, Clardy J (2012) Small molecule perimeter defense in entomopathogenic bacteria. PNAS 109:10821–10826

Da Silva OS, Prado GR, Da Silva JLR, Silva CE, Da Costa M, Heermann R (2013) Oral toxicity of *Photorhabdus luminescens* and *Xenorhabdus nematophila* (Enterobacteriaceae) against *Aedes aegypti* (Diptera: Culicidae). Parasitol Res 112:2891–2896

Daborn PJ, Waterfield N, Blight MA, Ffrench-Constant RH (2001) Measuring virulence factor expression by the pathogenic bacterium *Photorhabdusluminescens* in culture and during insect infection. J Bacteriol 183:5834–5839

Darsouei R, Karimi J, Ghadamyari M, Hosseini M (2017) Differential change patterns of main antimicrobial peptide genes during infection of entomopathogenic nematodes and their symbiotic bacteria. J Parasitol 103:349–358

Delaux PM, Schornack S (2021) Plant evolution driven by interactions with symbiotic and pathogenic microbes. Science 371:6531

Delucca AJ, Bland JM, Jacks TJ, Grimm C, Cleveland TE, Walsh TJ (1997) Fungicidal activity of cecropin A. Antimicrob Agents Chemother 41:481–483

Donadio SX, Monciardini P, Alduina R, Mazza P, Chiocchini C, Cavaletti C, Sosio M, Puglia AM (2012) Microbial technologies for the discovery of novel bioactive metabolites. J Biotechnol 99: 187–198

Dowds BC, Peters AR (2002) Virulence mechanisms. In: Entomopathogenic nematology. CABI International, Wallingford, pp 79–98

Dreyer J, Malan AP, Dicks LM (2018) Bacteria of the genus *Xenorhabdus*, a novel source of bioactive compounds. Front Microbiol 9:3177–3191

Dunn PE, Drake DR (1983) Fate of bacteria injected into naive and immunized larvae of the tobacco hornworm *Manduca sexta*. J Invertebr Pathol 41:77–85

Dunphy GB, Webster JM (1984) Interaction of *Xenorhabdus nematophilus* subsp. nematophilus with the haemolymph of *Galleria mellonella*. J Insect Physiol 30:883–889

Duvic B, Jouan V, Essa N, Girard PA, Pagès S, Abi Khattar Z, Volkoff NA, Givaudan A, Destoumieux-Garzon D, Escoubas JM (2012) Cecropins as a marker of Spodoptera frugiperda immunosuppression during entomopathogenic bacterial challenge. J Insect Physiol 58:881–888

Ebssa L, Koppenhöfer AM (2011) Efficacy and persistence of entomopathogenic nematodes for black cutworm control in turfgrass. Biocontrol Sci Tech 21:779–796

Ehlers RU (2001) Mass production of entomopathogenic nematodes for plant protection. Appl Microbiol Biotechnol 56:623–633

Ehlers RU (2005) Forum on safety and regulation. In: Grewal PS, Ehlers R-U, Shapiro-Ilan DI (eds) Nematodes as biological control agents. CABI Publishing, Wallingford, pp 107–114

Ekengren S, Hultmark D (1999) *Drosophila* cecropin as an antifungal agent. Insect Biochem Mol Biol 29:965–972

Eleftherianos I, Waterfield NR, Bone P, Boundy S, Ffrench-Constant RH, Reynolds SE (2009) A single locus from the entomopathogenic bacterium *Photorhabdus luminescens* inhibits activated *Manduca sexta* phenoloxidase. FEMS Microbiol Lett 293:170–176

Engström P, Carlsson A, Engström A, Tao ZJ, Bennich H (1984) The antibacterial effect of attacins from the silk moth *Hyalophora cecropia* is directed against the outer membrane of *Escherichia coli*. EMBO J 3:3347–3351

Eom S, Park Y, Kim Y (2014) Sequential immunosuppressive activities of bacterial secondary metabolites from the entomopahogenic bacterium *Xenorhabdus nematophila*. J Microbiol 52: 161–168

Fan SA, Rosegrant MW (2008) Investing in agriculture to overcome the world food crisis and reduce poverty and hunger. IFPRI (International Food Policy Research Institute), Washington, DC

Fang XL, Feng JT, Zhang WG, Wang YH, Zhang X (2010) Optimization of growth medium and fermentation conditions for improved antibiotic activity of *Xenorhabdus nematophila* TB using a statistical approach. Afr J Biotechnol 9:8068–8077

Fang XL, Li ZZ, Wang YH, Zhang X (2011) *In vitro* and *in vivo* antimicrobial activity of *Xenorhabdus bovienii* YL002 against *Phytophthora capsici* and *Botrytis cinerea*. J Appl Microbiol 111:145–154

Fang XL, Han LR, Cao XQ, Zhu MX, Zhang X, Wang YH (2012) Statistical optimization of process variables for antibiotic activity of *Xenorhabdus bovienii*. PLoS One 7:e38421

Forst ST, Clarke DA (2002) Bacteria-nematode symbiosis. Entomopathog Nematol 31:57–74

Forst S, Nealson K (1996) Molecular biology of the symbiotic-pathogenic bacteria *Xenorhabdus* spp. and *Photorhabdus* ssp. Microbiol Rev 60:21–43
Forst S, Dowds B, Boemare N, Stackebrandt E (1997) *Xenorhabdus* and *Photorhabdus* spp.: bugs that kill bugs. Annu Rev Microbiol 51:47–72
Fuchs SW, Proschak A, Jaskolla TW, Karas M, Bode HB (2011) Structure elucidation and biosynthesis of lysine-rich cyclic peptides in *Xenorhabdus nematophila*. Org Biomol Chem 9: 3130–3132
Fuchs SW, Grundmann F, Kurz M, Kaiser M, Bode HB (2014) Fabclavines: bioactive peptide–polyketide-polyamino hybrids from *Xenorhabdus*. ChemBioChem 15:512–516
Fukruksa C, Yimthin T, Suwannaroj M, Muangpat P, Tandhavanant S, Thanwisai A, Vitta A (2017) Isolation and identification of *Xenorhabdus* and *Photorhabdus* bacteria associated with entomopathogenic nematodes and their larvicidal activity against *Aedes aegypti*. Parasit Vectors 10:440–450
Ganguly S, Kumar S, Rathour KS (2010) Availability of biopesticidal nematodes to suit different agro-climatic region of India for commercialisation. Ind J Nematol 40:261–264
Gillespie JP, Kanost MR, Trenczek T (1997) Biological mediators of insect immunity. Annu Rev Entomol 42:611–643
Goodrich-Blair H, Clarke DJ (2007) Mutualism and pathogenesis in *Xenorhabdus* and *Photorhabdus*: two roads to the same destination. Mol Microbiol 64:260–268
Gratia A (1925) Sur un remarquableexempled'antagonisme entre deux souches de coilbacille. CR Seances Soc Biol Fil 93:1040–1041
Grewal PS, Peters A (2005) Formulation and quality. In: Grewal PS, Ehlers RU, Shapiro-Ilan D (eds) Nematodes as biocontrol agents. CABI Publishing, Wallingford, pp 79–90
Grundmann F, Kaiser M, Kurz M, Schiell M, Batzer A, Bode HB (2013) Structure determination of the bioactive depsipeptidexenobactin from *Xenorhabdus* sp. PB30.3. RSC Adv 3:22072–22077
Gualtieri M, Aumelas A, Thaler JO (2009) Identification of a new antimicrobial lysine-rich cyclolipopeptide family from *Xenorhabdus nematophila*. J Antibiot 62:295–302
Gulcu B, Cimen H, Raja RK, Hazir S (2017) Entomopathogenic nematodes and their mutualistic bacteria: their ecology and application as microbial control agents. Biopestic Int 13:79–112
Guo S, Zhang S, Fang X, Liu Q, Gao J, Bilal M, Wang Y, Zhang X (2017) Regulation of antimicrobial activity and xenocoumacins biosynthesis by pH in *Xenorhabdus nematophila*. Microb Cell Factories 16:1–4
Han R, Ehlers RU (2000) Pathogenicity, development, and reproduction of *Heterorhabditisbacteriophora* and *Steinernemacarpocapsae* under axenic *in vivo* conditions. J Invertebr Pathol 75:55–58
Hasan MA, Ahmed S, Mollah MM, Lee D, Kim Y (2019) Variation in pathogenicity of different strains of *Xenorhabdus nematophila*; Differential immunosuppressive activities and secondary metabolite production. J Invertebr Pathol 166:107221–107231
Hassan M, Kjos M, Nes IF, Diep DB, Lotfipour F (2012) Natural antimicrobial peptides from bacteria: characteristics and potential applications to fight against antibiotic resistance. J Appl Microbiol 113:723–736
Hazir S, Shapiro-Ilan DI, Bock CH, Hazir C, Leite LG, Hotchkiss MW (2016) Relative potency of culture supernatants of *Xenorhabdus* and *Photorhabdus* spp. on growth of some fungal phytopathogens. Eur J Plant Pathol 146:369–381
Hedengren M, Borge K, Hultmark D (2000) Expression and evolution of the *Drosophila* attacin/diptericin gene family. Biochem Biophys Res Commun 279:574–581
Herbert EE, Goodrich-Blair H (2007) Friend and foe: the two faces of *Xenorhabdus nematophila*. Nat Rev Microbiol 5:634–646
Hong SJ, Ullah I, Park GS, Jung BK, Choi J, Khan AR, Kim MC, Shin JH (2015) Draft genome sequence and annotation of the insect pathogenic bacterium *Xenorhabdusnematophila* strain C2-3, isolated from nematode *Steinernemacarpocapsae* in the Republic of Korea. Genome Announce 3:e01521–e01514

Houard J, Aumelas A, Noël T, Pages S, Givaudan A, Fitton-Ouhabi V, Villain-Guillot P, Gualtieri M (2013) Cabanillasin, a new antifungal metabolite, produced by entomopathogenic *Xenorhabdus cabanillasii* JM26. J Antibiot 66:617–620

Hultmark D, Engström A, Bennich H, Kapur R, Boman HG (1982) Insect immunity: isolation and structure of cecropin D and four minor antibacterial components from *Cecropia pupae*. Eur J Biochem 127:207–217

Hultmark D, Engström A, Andersson K, Steiner H, Bennich H, Boman HG (1983) Insect immunity. Attacins, a family of antibacterial proteins from *Hyalophora cecropia*. EMBO J 2:571–576

Hwang J, Park Y, Kim Y, Hwang J, Lee D (2013) An entomopathogenic bacterium, *Xenorhabdus nematophila*, suppresses expression of antimicrobial peptides controlled by Toll and Imd pathways by blocking eicosanoid biosynthesis. Arch Insect Biochem Physiol 83:151–169

Jallouli W, Zouari N, Jaoua S (2010) Involvement of oxidative stress and growth at high cell density in the viable but nonculturable state of *Photorhabdus temperata* ssp. temperata strain K122. Process Biochem 45:706–713

Ji D, Yi Y, Kang GH, Choi YH, Kim P, Baek NI, Kim Y (2004) Identification of an antibacterial compound, benzylideneacetone, from *Xenorhabdus nematophila* against major plant-pathogenic bacteria. FEMS Microbiol Lett 239:241–248

Kalha CS, Singh PP, Kang SS, Hunjan MS, Gupta V, Sharma R (2014) Entomopathogenic viruses and bacteria for insect-pest control. In: Integrated pest management. Academic Press, New York, NY, pp 225–244

Kang S, Han S, Kim Y (2004) Identification of an entomopathogenic bacterium, *Photorhabdustemperata* subsp. temperata, in Korea. J Asia Pac Entomol 7:331–337

Kim H, Choi D, Jung J, Kim Y (2018) Eicosanoid mediation of immune responses at early bacterial infection stage and its inhibition by *Photorhabdus temperata* subsp. temperata, an entomopathogenic bacterium. Arch Insect Biochem Physiol 99:e21502

Koppenhöfer HS, Gaugler R (2009) Entomopathogenic nematode and bacteria mutualism. In: Defensive mutualism in microbial symbiosis, vol 26, pp 99–116

Kresovic D, Schempp F, Cheikh-Ali Z, Bode HB (2015) A novel and widespread class of ketosynthase is responsible for the head-to-head condensation of two acyl moieties in bacterial pyrone biosynthesis. Beilstein J Org Chem 11:1412–1417

Kronenwerth M, Bozhüyük KA, Kahnt AS, Steinhilber D, Gaudriault S, Kaiser M, Bode HB (2014) Characterisation of Taxlllaids A–G; natural products from *Xenorhabdusindica*. Chem A Eur J 20:17478–17487

Kumar P, Ganguly S, Somvanshi VS (2015) Identification of virulent entomopathogenic nematode isolates from a countrywide survey in India. Int J Pest Manag 61:135–143

Kumari P, Mahapatro GK, Banerjee N, Sarin NB (2015a) Ectopic expression of GroEL from *Xenorhabdus nematophila* in tomato enhances resistance against *Helicoverpaarmigera* and salt and thermal stress. Transgenic Res 24:859–873

Kumari P, Mahapatro GK, Banerjee N, Sarin NB (2015b) A novel pilin subunit from *Xenorhabdus nematophila*, an insect pathogen, confers pest resistance in tobacco and tomato. Plant Cell Rep 34:1863–1872

Lacey LA, Arthurs SP, Unruh TR, Headrick H, Fritts R Jr (2006) Entomopathogenic nematodes for control of codling moth (Lepidoptera: Tortricidae) in apple and pear orchards: effect of nematode species and seasonal temperatures, adjuvants, application equipment, and post-application irrigation. Biol Control 37:214–223

Lacey LA, Grzywacz D, Shapiro-Ilan DI, Frutos R, Brownbridge M, Goettel MS (2015) Insect pathogens as biological control agents: back to the future. J Invertebr Pathol 132:1–41

Leandro LF, Mendes CA, Casemiro LA, Vinholis AH, Cunha WR, de Almeida R, Martins CH (2015) Antimicrobial activity of apitoxin, melittin and phospholipase A_2 of honey bee (*Apismellifera*) venom against oral pathogens. An Acad Bras Cienc 87:147–155

Lemaitre B, Hoffmann J (2007) The host defense of *Drosophila melanogaster*. Annu Rev Immunol 25:697–743

Li ZF, Li P, Jing YG, Singh RP, Li YZ (2017) Isolation and characterization of the epothilone gene cluster with flanks from high alkalotolerant strain *Sorangium cellulosum* (So0157-2). World J Microbiol Biotechnol 33(7):137. https://doi.org/10.1007/s11274-017-2301-y
Lu D, Geng T, Hou C, Huang Y, Qin G, Guo X (2016) *Bombyx mori* cecropin A has a high antifungal activity to entomopathogenic fungus *Beauveria bassiana*. Gene 583:29–35
Lu D, Sepulveda C, Dillman AR (2017) Infective juveniles of the entomopathogenic nematode *Steinernemascapterisci* are preferentially activated by cricket tissue. PLoS One 12:e0169410
de Maagd RA, Bosch D, Stiekema W (1999) *Bacillus thuringiensis* toxin-mediated insect resistance in plants. Trends Plant Sci 4:9–13
Maheshwari NK, Singh RP, Manchanda G, Dubey RC, Maheshwari DK (2021) Sunn Hemp (*Crotalaria juncea*) nodulating bacteria capable for high antagonistic potential and plant growth promotion attributes. J Microbiol Biotechnol Food Sci 10(3):385–389. https://doi.org/10.15414/jmbfs.2020.10.3.385-389
Makarova O, Johnston P, Rodriguez-Rojas A, El Shazely B, Morales JM, Rolff J (2018) Genomics of experimental adaptation of *Staphylococcus aureus* to a natural combination of insect antimicrobial peptides. Sci Rep 8:1–8
Malan AP, Nguyen KB, Addison MK (2006) Entomopathogenic nematodes (Steinernematidae and Heterorhabditidae) from the southwestern parts of South Africa. Afr Plant Prot 12:65–69
Marmaras VJ, Lampropoulou M (2009) Regulators and signalling in insect haemocyte immunity. Cell Signal 21:186–195
Martens EC, Heungens K, Goodrich-Blair H (2003) Early colonization events in the mutualistic association between *Steinernemacarpocapsae* nematodes and *Xenorhabdusnematophila* bacteria. J Bacteriol 185:3147–3154
Mbaveng AT, Sandjo LP, Tankeo SB, Ndifor AR, Pantaleon A, Nagdjui BT, Kuete V (2015) Antibacterial activity of nineteen selected natural products against multi-drug resistant Gram-negative phenotypes. SpringerPlus 4:1–9
Miyoshi N, Isogai E, Hiramatsu K, Sasaki T (2017) Activity of tick antimicrobial peptide from *Ixodes persulcatus* (persulcatusin) against cell membranes of drug-resistant *Staphylococcus aureus*. J Antibiot 70:142–146
Morales-Soto N, Forst SA (2011) The xnp1 P2-like tail synthesis gene cluster encodes xenorhabdicin and is required for interspecies competition. J Bacteriol 193:3624–3632
Morales-Soto N, Gaudriault S, Ogier JC, Thappeta KR, Forst S (2012) Comparative analysis of P2-type remnant prophage loci in *Xenorhabdus bovienii* and *Xenorhabdus nematophila* required for xenorhabdicin production. FEMS Microbiol Lett 333:69–76
Mráček Z, Hernández EA, Boëmare NE (1994) *Steinernemacubana* sp. n.(Nematoda: Rhabditida: Steinernematidae) and the preliminary characterization of its associated bacterium. J Invertebr Pathol 64:123–129
Mylonakis E, Podsiadlowski L, Muhammed M, Vilcinskas A (2016) Diversity, evolution and medical applications of insect antimicrobial peptides. Phil Trans R Soc Lond Ser B 371: 20150290–20150301
Nielsen-LeRoux C, Gaudriault S, Ramarao N, Lereclus D, Givaudan A (2012) How the insect pathogen bacteria *Bacillus thuringiensis* and *Xenorhabdus/Photorhabdus* occupy their hosts. Curr Opin Microbiol 15:220–231
Nollmann FI, Heinrich AK, Brachmann AO, Morisseau C, Mukherjee K, Casanova-Torres ÁM, Strobl F, Kleinhans D, Kinski S, Schultz K, Beeton ML (2015) A *Photorhabdus* natural product inhibits insect juvenile hormone epoxide hydrolase. ChemBioChem 16:766–771
Noosidum A, Hodson AK, Lewis EE, Chandrapatya A (2010) Characterization of new entomopathogenic nematodes from Thailand: foraging behavior and virulence to the greater wax moth, *G. mellonella* L. (Lepidoptera: Pyralidae). J Nematol 42:281–291
Ogier JC, Calteau A, Forst S, Goodrich-Blair H, Roche D, Rouy Z, Suen G, Zumbihl R, Givaudan A, Tailliez P, Médigue C (2010) Units of plasticity in bacterial genomes: new insight from the comparative genomics of two bacteria interacting with invertebrates, *Photorhabdus* and *Xenorhabdus*. BMC Genomics 11:1–21

Oñate-Garzón J, Manrique-Moreno M, Trier S, Leidy C, Torres R, Patiño E (2017) Antimicrobial activity and interactions of cationic peptides derived from *Galleria mellonella* cecropin D-like peptide with model membranes. J Antibiot 70:238–245

Owuama CI (2001) Entomopathogenic symbiotic bacteria, *Xenorhabdus* and *Photorhabdus* of nematodes. World J Microbiol Biotechnol 17:505–515

Park Y, Kim Y (2000) Eicosanoids rescue *Spodoptera exigua* infected with *Xenorhabdusnematophilus*, the symbiotic bacteria to the entomopathogenic nematode *Steinernemacarpocapsae*. J Insect Physiol 46:1469–1476

Park Y, Kim Y, Putnam SM, Stanley DW (2003) The bacterium *Xenorhabdus nematophilus* depresses nodulation reactions to infection by inhibiting eicosanoid biosynthesis in tobacco hornworms, *Manduca sexta*. Arch Insect Biochem Physiol 52:71–80

Park SC, Park Y, Hahm KS (2011) The role of antimicrobial peptides in preventing multidrug-resistant bacterial infections and biofilm formation. Int J Mol Sci 12:5971–5992

Paul VJ, Frautschy S, Fenical W, Nealson KH (1981) Antibiotics in microbial ecology. J Chem Ecol 7:589–597

Picoli T, Peter CM, Zani JL, Waller SB, Lopes MG, Boesche KN, Vargas GDÁ, Hübner SO, Fischer G (2017) Melittin and its potential in the destruction and inhibition of the biofilm formation by *Staphylococcus aureus*, *Escherichia coli* and *Pseudomonas aeruginosa* isolated from bovine milk. Microb Pathog 112:57–62

Pidot SJ, Coyne S, Kloss F, Hertweck C (2014) Antibiotics from neglected bacterial sources. Int J Med Microbiol 304:14–22

Planson AG, Carbonell P, Grigoras I, Faulon JL (2011) Engineering antibiotic production and overcoming bacterial resistance. Biotechnol J 6:812–825

Poinar GO Jr, Grewal PS (2012) History of entomopathogenic nematology. J Nematol 44:153–161

Prajakta BM, Suvarna PP, Raghvendra SP, Alok RR (2019) Potential biocontrol and superlative plant growth promoting activity of indigenous *Bacillus mojavensis* PB-35 (R11) of soybean (*Glycine max*) rhizosphere. SN Appl Sci 1(10):1

Rady I, Siddiqui IA, Rady M, Mukhtar H (2017) Melittin, a major peptide component of bee venom, and its conjugates in cancer therapy. Cancer Lett 402:16–31

Rahoo AM, Mukhtar T, Gowen SR, Pembroke B (2011) Virulence of entomopathogenic bacteria *Xenorhabdus bovienii* and *Photorhabdus luminescens* against *Galleria mellonella* larvae. Pak J Zool 43:543–548

Reimer D (2018) Identification and characterization of selected secondary metabolite biosynthetic pathways from *Xenorhabdus nematophila*. Doctoral dissertation, Ph. D. thesis, Johann Wolfgang Groethe-Universität, Frankfurt

Reimer D, Cowles KN, Proschak A, Nollmann FI, Dowling AJ, Kaiser MR et al (2013) Rhabdopeptides as insect-specific virulence factorsfrom entomopathogenic bacteria. ChemBioChem 14:1991–1997

Reimer D, Nollmann FI, Schultz K, Kaiser M, Bode HB (2014) Xenortide biosynthesis by entomopathogenic *Xenorhabdus nematophila*. J Nat Prod 77:1976–1980

Richards GR, Goodrich-Blair H (2009) Masters of conquest and pillage: *Xenorhabdus nematophila* global regulators control transitions from virulence to nutrient acquisition. Cell Microbiol 11: 1025–1033

Richards GR, Goodrich-Blair H (2010) Examination of *Xenorhabdus nematophila* lipases in pathogenic and mutualistic host interactions reveals a role for xlpA in nematode progeny production. Appl Environ Microbiol 76:221–229

Riga E, Lacey LA, Guerra N, Headrick HL (2006) Control of the oriental fruit moth, *Grapholitamolesta*, using entomopathogenic nematodes in laboratory and fruit bin assays. J Nematol 38:168–171

Rodou A, Ankrah DO, Stathopoulos C (2010) Toxins and secretion systems of *Photorhabdusluminescens*. Toxins 2:1250–1264

Rosales C, Vonnie S (2017) Cellular and molecular mechanisms of insect immunity. In: Insect physiology and ecology. InTeach, Rijeka, pp 179–212

Sadekuzzaman M, Kim Y (2017) Specific inhibition of *Xenorhabdus hominickii*, an entomopathogenic bacterium, against different types of host insect phospholipase A2. J Invertebr Pathol 149:97–105

Sajnaga E, Kazimierczak W, Skowronek M, Lis M, Skrzypek T, Waśko A (2018) *Steinernemapoinari* (Nematoda: Steinernematidae): a new symbiotic host of entomopathogenic bacteria *Xenorhabdus bovienii*. Arch Microbiol 200:1307–1316

Schöner TA, Kresovic D, Bode HB (2015) Biosynthesis and function of bacterial dialkylresorcinol compounds. Appl Microbiol Biotechnol 99:8323–8328

Seo S, Lee S, Hong Y, Kim Y (2012) Phospholipase A2 inhibitors synthesized by two entomopathogenic bacteria, *Xenorhabdus nematophila* and *Photorhabdus temperata* subsp. temperata. Appl Environ Microbiol 78:3816–3823

Sergeant M, Baxter L, Jarrett P, Shaw E, Ousley M, Winstanley C, Morgan JA (2006) Identification, typing, and insecticidal activity of *Xenorhabdus* isolates from entomopathogenic nematodes in United Kingdom soil and characterization of the xpt toxin loci. Appl Environ Microbiol 72:5895–5907

Shapiro-Ilan DI, Gaugler R (2002) Production technology for entomopathogenic nematodes and their bacterial symbionts. J Ind Microbiol Biotechnol 28:137–146

Shapiro-Ilan DI, Gouge DH, Koppenhöfer AM (2002) 16 Factors affecting commercial success: case studies in cotton, turf and citrus. In: Entomopathogenic nematology. CAB International, Wallingford, pp 333–349

Shapiro-Ilan DI, Cottrell TE, Mizell RF III, Horton DL, Davis J (2009) A novel approach to biological control with entomopathogenic nematodes: prophylactic control of the peachtree borer, *Synanthedonexitiosa*. Biol Control 48:259–263

Shapiro-Ilan DI, Cottrell TE, Mizell RF III, Horton DL, Behle RW, Dunlap CA (2010) Efficacy of *Steinernemacarpocapsae* for control of the lesser peachtree borer, *Synanthedonpictipes*: improved aboveground suppression with a novel gel application. Biol Control 54:23–28

Shapiro-Ilan DI, Han R, Qiu X (2014) Production of entomopathogenic nematodes. In: Mass production of beneficial organisms. Academic Press, pp 321–355

Shi YM, Bode HB (2018) Chemical language and warfare of bacterial natural products in bacteria–nematode–insect interactions. Nat Prod Rep 35:309–335

Shi W, Li C, Li M, Zong X, Han D, Chen Y (2016) Antimicrobial peptide melittin against *Xanthomonas oryzae* pv. oryzae, the bacterial leaf blight pathogen in rice. Appl Microbiol Biotechnol 100:5059–5067

Sicard M, Le Brun N, Pages S, Godelle B, Boemare N, Moulia C (2003) Effect of native *Xenorhabdus* on the fitness of their *Steinernema* hosts: contrasting types of interaction. Parasitol Res 91:520–524

Sicard M, Brugirard-Ricaud K, Pagès S, Lanois A, Boemare NE, Brehélin M et al (2004) Stages of infection during the tripartite interaction between *Xenorhabdus nematophila*, its nematode vector, and insect hosts. Appl Environ Microbiol 70:6473–6480

Sicard M, Ramone H, Le Brun N, Pagès S, Moulia C (2005) Specialization of the entomopathogenic nematode *Steinernemascapterisci* with its mutualistic *Xenorhabdus* symbiont. Naturwissenschaften 92:472–476

Siegel JP, Lacey LA, Higbee BS, Noble P, Fritts R Jr (2006) Effect of application rates and abiotic factors on *Steinernemacarpocapsae* for control of overwintering navel orange worm (Lepidoptera: Pyralidae, *Amyeloistransitella*) in pistachios. Biol Control 36:324–330

Silva CP, Waterfield NR, Daborn PJ, Dean P, Chilver T, Au CP, Sharma S, Potter U, Reynolds SE, Ffrench-Constant RH (2002) Bacterial infection of a model insect: *Photorhabdus luminescens* and *Manduca sexta*. Cell Microbiol 4:329–339

da Silva WJ, Pilz-Júnior HL, Heermann R, da Silva OS (2020) The great potential of entomopathogenic bacteria *Xenorhabdus* and *Photorhabdus* for mosquito control: a review. Parasit Vectors 13:1–4

Singh J, Banerjee N (2008) Transcriptional analysis and functional characterization of a gene pair encoding iron-regulated xenocin and immunity proteins of *Xenorhabdus nematophila*. J Bacteriol 190:3877–3885

Singh RP, Singh RN, Manchanda G, Tiwari PK, Srivastava AK et al (2014a) Biofuels: a way ahead. Ind J Appl Res 4(1):3–6

Singh S, Reese JM, Casanova-Torres AM, Goodrich-Blair H, Forst S (2014b) Microbial population dynamics in the hemolymph of *Manduca sexta* infected with *Xenorhabdus nematophila* and the entomopathogenic nematode *Steinernemacarpocapsae*. Appl Environ Microbiol 80:4277–4285

Singh S, Orr D, Divinagracia E, McGraw J, Dorff K, Forst S (2015) Role of secondary metabolites in establishment of the mutualistic partnership between *Xenorhabdus nematophila* and the entomopathogenic nematode *Steinernemacarpocapsae*. Appl Environ Microbiol 81:754–764

Singh RP, Manchanda G, Yang Y, Singh D, Srivastava AK et al (2020a) Deciphering the factors for nodulation and symbiosis of Mesorhizobium associated with *Cicer arietinum* in Northwest India. Sustainability 12:1–17

Singh RP, Handa R, Manchanda G (2020b) Nanoparticles in sustainable agriculture: an emerging opportunity. J Control Release 329:1234–1248. https://doi.org/10.1016/j.jconrel.2020.10.051

Snyder H, Stock SP, Kim SK, Flores-Lara Y, Forst S (2007) New insights into the colonization and release processes of *Xenorhabdus nematophila* and the morphology and ultrastructure of the bacterial receptacle of its nematode host, *Steinernemacarpocapsae*. Appl Environ Microbiol 73: 5338–5346

Socarras KM, Theophilus PAS, Torres JP, Gupta K, Sapi E (2017) Antimicrobial activity of bee venom and melittin against *Borrelia burgdorferi*. Antibiotics (Basel) 6:31–50

Srisailam S, Arunkumar AI, Wang W, Yu C, Chen HM (2000) Conformational study of a custom antibacterial peptide cecropin B1: implications of the lytic activity. Biochim Biophys Acta Protein Struct Mol Enzymol 1479:275–285

Stein ML, Beck P, Kaiser M, Dudler R, Becker CF, Groll M (2012) One-shot NMR analysis of microbial secretions identifies highly potent proteasome inhibitor. PNAS 109:18367–18371

Steiner H, Hultmark D, Engström Å, Bennich H, Boman HG (1981) Sequence and specificity of two antibacterial proteins involved in insect immunity. Nature 292:246–248

Stock SP, Kusakabe A, Orozco RA (2017) Secondary metabolites produced by *Heterorhabditis* symbionts and their application in agriculture: what we know and what to do next. J Nematol 49: 373–383

Sundar L, Chang FN (1993) Antimicrobial activity and biosynthesis of indole antibiotics produced by *Xenorhabdus nematophilus*. Microbiology 139:3139–3148

Theodore CM, King JB, You J, Cichewicz RH (2012) Production of cytotoxic glidobactins/ luminmycins by *Photorhabdus asymbiotica* in liquid media and live crickets. J Nat Prod 75: 2007–2011

Tobias NJ, Mishra B, Gupta DK, Sharma R, Thines M, Stinear TP, Bode HB (2016) Genome comparisons provide insights into the role of secondary metabolites in the pathogenic phase of the *Photorhabdus* life cycle. BMC Genomics 17:537–548

Tobias NJ, Heinrich AK, Eresmann H, Wright PR, Neubacher N, Backofen R, Bode HB (2017) *Photorhabdus*-nematode symbiosis is dependent on hfq-mediated regulation of secondary metabolites. Environ Microbiol 19:119–129

Tobias NJ, Shi YM, Bode HB (2018) Refining the natural product repertoire in entomopathogenic bacteria. Trends Microbiol 26:833–840

Torres NI, Noll KS, Xu S, Li J, Huang Q, Sinko PJ, Wachsman MB, Chikindas ML (2013) Safety, formulation and *in vitro* antiviral activity of the antimicrobial peptide subtilosin against herpes simplex virus type 1. Probiot Antimicrob Prot 5:26–35

Ullah I, Khan AL, Ali L, Khan AR, Waqas M, Hussain J, Lee IJ, Shin JH (2015) Benzaldehyde as an insecticidal, antimicrobial, and antioxidant compound produced by *Photorhabdus temperata* M1021. J Microbiol 53:127–133

Vega FE, Kaya HK (2012) Insect pathology. Academic Press, Cambridge, MA

Vizioli J, Bulet P, Charlet M, Lowenberger C, Blass C, Müller HM, Dimopoulos G, Hoffmann J, Kafatos FC, Richman A (2000) Cloning and analysis of a cecropin gene from the malaria vector mosquito, *Anopheles gambiae*. Insect Mol Biol 9:75–84

Wang YH, Feng JT, Zhang Q, Zhang X (2008) Optimization of fermentation condition for antibiotic production by *Xenorhabdus nematophila* with response surface methodology. J Appl Microbiol 104:735–744

Waterfield N, George Kamita S, Hammock BD, Ffrench-Constant R (2005) The *Photorhabdus* Pir toxins are similar to a developmentally regulated insect protein but show no juvenile hormone esterase activity. FEMS Microbiol Lett 245:47–52

Waterfield NR, Ciche T, Clarke D (2009) *Photorhabdus* and a host of hosts. Annu Rev Microbiol 63:557–574

Webster JM, Chen GE, Hu K, Li J (2002) Bacterial metabolites. Entomopathog Nematol 2002:99–114

Wu X, Singh AK, Wu X, Lyu Y, Bhunia AK, Narsimhan G (2016) Characterization of antimicrobial activity against Listeria and cytotoxicity of native melittin and its mutant variants. Colloids Surf B: Biointerfaces 143:194–205

Xiao Y, Meng F, Qiu D, Yang X (2012) Two novel antimicrobial peptides purified from the symbiotic bacteria *Xenorhabdus budapestensis* NMC-10. Peptides 35:253–260

Yamanaka S, Hagiwara A, Nishimura Y, Tanabe H, Ishibashi N (1992) Biochemical and physiological characteristics of *Xenorhabdus* species, symbiotically associated with entomopathogenic nematodes including *Steinernemakushidai* and their pathogenicity against *Spodoptera litura* (Lepidoptera: Noctuidae). Arch Microbiol 158:387–393

Yasir M, Willcox MD, Dutta D (2018) Action of antimicrobial peptides against bacterial biofilms. Material 11:2468–2483

Zhang LJ, Gallo RL (2016) Antimicrobial peptides. Curr Biol 26:R14–R19

Zhao L, Kaiser M, Bode HB (2018) Rhabdopeptide/Xenortide-like peptides from *Xenorhabdusinnexi* with terminal amines showing potent antiprotozoal activity. Org Lett 20: 5116–5120

Zhou Q, Dowling A, Heide H, Wöhnert J, Brandt U, Baum J, Ffrench-Constant R, Bode HB (2012) Xentrivalpeptides A–Q: depsipeptide diversification in *Xenorhabdus*. J Nat Prod 75:1717–1722

Zhu S, Gao B (2013) Evolutionary origin of β-defensins. Dev Comp Immunol 39:79–84

Chapter 5
Microbial Abundance and Strategies of Adaptation in Various Extreme Environments

Pooja Misra, Dipti Singh, and Atul Kumar Srivastava

Abstract Microbes are versatile community of universe and able to survive at diverse conditions or niches such as deep sea vent, volcanic areas, Polar Regions, core of earth, etc. are the region where microbes can live and reproduce easily by their extremophilic adaptive nature. The adaptation and survival are basically governed by the morpho-phenotypic, biochemical, genetic, and enzymatic secretion strategies. Microbes survive in extreme condition possess a unique feature like thermostability, resistant against chemical denaturants and stability in extreme acidic or alkaline condition. Researchers, who work on the molecular biology isolated many important genes from extremophiles helpful in the industry like paper industry, dairy industry, and in food processing industry. In addition, extremophiles also produce biologically active enzymes like starch, cellulose, chitin, and protein degrading enzymes. The details of morphology to molecular survival strategies are included in this chapter with extremophile approaches to industrial applications.

Keywords Extremophiles · Adaptation · Strategy · Industrial application

5.1 Introduction

The extreme conditions generally refer to the unfavourable conditions which are basically lethal or dangerous for mesophilic organisms. Some of the microorganisms have capability to survive in extreme radiation, high pressure, extremely high and low temperature and even in any harsh environmental condition. Organisms which

P. Misra
Crop Protection Division, CSIR-Central Institute of Medicinal and Aromatic Plants, Lucknow, Uttar Pradesh, India

D. Singh
Department of Microbiology, Veer Bahadur Singh Purvanchal University, Jaunpur, India

A. K. Srivastava (✉)
Research and Development Department, Uttaranchal University, Dehradun, Uttarakhand, India

R. P. Singh et al. (eds.), *Microbes in Microbial Communities*,
https://doi.org/10.1007/978-981-16-5617-0_5

adopt such wide range of extreme environmental condition are called as extremophiles. Extremophilic organisms are primarily prokaryotic but few are eukaryotic also. The biology of extremophiles revealed the origin of life, arose from hot environment and support the theory that extremophiles are primordial organisms therefore they are the model organism of the ancient life forms. Instead of this extremophiles are also hot research topic in the field of astrobiology because of the majority of solar system is in frozen form and cold temperature loving microbes live there. The microbes use arsenic rather than phosphorus for their growth is one of another interest of astrobiologist. Thus, the deep study of extremophiles provides us vast knowledge to understand the life on the extreme environment and in future, discover the new habitat/solar bodies for us. Extremophilic organisms are used in the new industrial enzymes like chaperons (zigzag proteins) found on the cellular surface giving them resistance to denaturation enzymes, proteolytic enzymes, etc. Most importantly they produce tolerating enzymes of DNA polymerase that help in the diagnosis of severe pathogens (Irwin 2010). The types of extreme condition in which some organism can survive are extreme of pH, temperature, altitude, pressure, salinity, etc. (Singh et al. 2020a).

An extreme environment is the habitat of extremely high or low temperature, highly acidic or alkaline condition, deepest sea or top of mountain niche, under low or high pressure, on high or low radiation, highly salty or extremely shrill conditions (Horneck et al. 2010). Polar ice caps, dry spots in deserts and abysmal depths in the ocean are the remarkable extreme environments of the earth (Suyal et al. 2021). The state when the environmental particular clause is risen at the level of intolerant to living organisms, known as extreme condition. Extreme environment is the habitat not easy to survive for living beings. But in nature, scientific community found the group of prokaryotes.

This extreme environment is very hard to survive for any life forms. Nature is the treasure of microbes, including those microbes which can survive in that extreme environment where none of the plants and animals can dare to survive (Sayed et al. 2020). It is very hard to survive for other life forms live on earth and the microbes which survive in extreme condition are globally known as extremophiles. The species of extremophiles have either adopted the respective extreme environment over time or they have resided generation to generation. Most of the extremophiles belong to the domain archaea, but some are included in bacteria and multicellular organisms. Globally extremophilic organism survives in all the adverse places of earth and beyond the earth also. To find a life on other galaxies, beyond the earth, extremophiles work as a primary indicator for astrobiologist. The extremophiles are divided into various classes for our understanding that we see detail in the next paragraph about remarkable extremophiles.

5.1.1 Acidity and Alkalinity

The metabolism of microbes act best at their neutral cytoplasmic pH (Krulwich et al. 2011) and the alteration of pH have a significant effect on the microbial consortia at every level of growth. Normal range of cytoplasmic pH of acidophilic bacteria is 6.0 and the cytoplasmic pH of alkaliphilic bacteria is 7.2–8.7 (Krulwich et al. 2011).

Earth has possessed all types of ecological niches of extremely high and low pH values. Iron Mountain (Shasta County, CA, United States) is reported as extremely low pH environment (acidic condition: pH value is 3.6) and Gorka Lake of Chrazanow region, Poland found extremely high pH (alkaline condition: pH value is 13.3) areas (Czop et al. 2011). *Thermoplasmales*, *Acidithiobacillus ferrooxidans*, and *Leptospirillum ferrooxidans* are some micro flora reported in the Iron Mountain (Singh et al. 2011). *Picrophilus oshimae* and *P. torridus* are two hyperacidophilic archaea recoded as lowest pH value 0.06, which was isolated from solfataric hot spring of Noboribestsu, Japan. Chemiosmosis is reported in both bacteria and archea (Lane et al. 2010) however, various microbes use homeostasis through proton or other ions using various transporters as ion utilizing ATP synthase likely one of the first function developed in earliest cells (Lane and Martin 2012). Several authors have been reported the various alkalotolerant microbial species such as *Mesorhizobium ciceri*, *M. muleiense*, etc. (Singh et al. 2016; Zhang et al. 2017; Singh et al. 2019). Besides intracellular pH, microorganisms also manipulate the pH of immediate surrounding with the help of organic metabolites like lactic acid or acetic acid (Zhang et al. 2017).

5.1.2 Temperature

Temperature range of earth varied from −98.6 to 495 °C, with higher temperature possible in the volcanic areas and lower temperature in Polar Regions (Scambos et al. 2018). Microbes present in extreme temperature condition have significant effect on metabolic activity. In general, the microbial life can survive and metabolically active extended at −25 °C (*Deinococcus geothermalis*) to 130 °C (*Geogemma barossii*) (Kashefi 2003). The high saline and high pressure conditions could be possible by extreme temperature adaptation (Deming 2007). In geothermal environment, temperature gradient plays a key role in affecting the structure of microbiome (Sharpton 2014), than other ecological factors (e.g., soil), whereas, pH and salinity also act as major dominants (Rai et al. 2012).

5.1.3 Pressure

On Earth surface the pressure ranges recorded from 0.1 to 112 MPa and at the top of Subducting plate of Mariana Forearc observed higher pressures is around 900 MPa (Mottl et al. 2004). However the microbial life could supported around 340 MPa pressure on Subducting plate of Mariana Forearc (Plümper et al. 2017). *Thermococcus piezophilus* is a archaean recorded to survive at 125 MPa (Dalmasso et al. 2016). At higher temperature and lower atmospheric pressure, the generation time of piezophiles decreases (Bartlett et al. 2007). Some of the Piezophilic prokaryotes, fungi, and lichen that have potency to sporulate and form biofilm have capacity to survive several months to years under space condition (Onofri et al. 2018).

5.1.4 Radiation

Radiation like UV rays, gamma rays, X-rays can impact directly or indirectly on the biotic component including microbial cells. Microorganisms form the reactive oxygen by combined with the radiation and it can damage the nucleic acid, proteins, and lipid metabolism (Subhashini et al. 2017). Whomever radiation resistant bacteria also present in the environment, *Deinococcus radiodurans* is the first radiation resistant microbe and used as model organism for radiation tolerance (Krisko and Radman 2013).

5.2 How the Microbes Survive in Extreme Condition

Humans are survived in mild condition and hence limited environmental space available however they live in almost all land areas on earth and also searching for life availability on other planets like mars (Singh et al. 2012). On the contrary microorganisms live everywhere in the ecosystem but especially extremophiles, can survive in extreme environment like high and low temperature, pressure, salinity, etc. (details are shown in Table 5.1). The strategies behind their survivorship of different extremophiles for the different parameters are discussed here.

The strategy like rigidity help to survive thermophile organism in high temperature (mental, deep sea hydrothermal vents and hot springs) by a process of oligomerization, increased in disulphide bonds and hydrophobic core, surface charges and salt bridges (Koschinsky et al. 2008; Bischoff and Rosenbauer 1988). *Methanopyrus kandleri* is a hyperthermophile recorded to grow at 122 °C (Takai et al. 2008). In reverse psychrophiles acquire the flexibility for survive in low temperature environments (polar regions, deep sea, and alpine regions) by enrichment with nonpolar and glycine amino acid residues, decreased with proline and

Table 5.1 List of possessed extremophiles with examples

S. No.	Extremophiles	Survival condition	Examples	References
1.	Acidophile	Organism survive at highly acidic (pH 3 or lower) condition	*T. acidophilus*, *Acidithiobacillus ferrooxidans*, *Picrophilus torridus*	Mirete et al. (2017)
2.	Alkaliphile	Organism survive at highly alkaline (pH 9 or above) condition	*Halorhodospira halochloris*, *Natronomonas pharaonis*, *Plectonema nostocorum*, *hydrogenophaga* sp.	Singh (2012)
3.	Thermophile	Organism grow under high temperature 41 and 122 °C (106 and 252 °F)	*Bacillus stearothermophilus*, *Thermus aquaticus*, *Geogemma barossii*	Takai et al. (2008)
4.	Piezophile	Organism grow under high hydrostatic pressure	*Pyrococcus yayanosii*, *Shewanella benthica*	Zeng et al. (2009)
5.	Cryptoendolith	Organisms able to colonize the empty spaces or pores inside a rock with the connotation of being hidden	Cryptoendoliths—Cryptoendoliths, Cryptoendolithic lichen	Wierzchos et al. (2011)
6.	Halophile (salt-loving)	Organism that thrive in high salt concentrations.	*Chromohalobacter beijerinckii*, *Halorubrum sodomense*, *Haloferax volcanii*	Ollivier et al. (1994)
7.	Capnophile	Organism survive in high concentrations of CO_2	*Campylobacter jejuni*, *Aggregatibacter actinomycetemcomitans*	Sahuquillo-Arce et al. (2017)
8.	Anaerobe	commonly found in the gastrointestinal tract and do not live or grow when oxygen is present	*Staphylococcus* spp., *Escherichia coli*, *Listeria* spp.	Ueki et al. (2018)
9.	Psychrophiles or cryophiles	Grow and reproduce in low temperature (−20 to 10 °C)	*Psychromonas ingrahamii*	Clarke et al. (2013)
10.	Xerophile	Grow on low availability of water (Tolerant of dry conditions)	*Trichosporonoides nigrescens*	Chen and Jiang (2018)
11.	Hypolith	Live in underneath rocks (climatically in extreme deserts)	*Chroococcidiopsis* sp.	Chan et al. (2012)
12.	Sulphophile	Organism grow in high sulphur concentration	*Sulfurovum* sp.	Meier et al. (2017)
13.	Radioresistant	Microbes resist the radiation (UV rays, X-rays, Gamma rays)	*Halobacterium* sp., *Deinococcus radiodurans*	DeVeaux et al. (2007)
14.	Osmophile	Adopted to high osmotic pressure such as high glucose concentration	*Saccharomyces rouxii*, *Wallemia sebi*	Karaman and Sagdic (2019)

arginine amino acid residues, and increased the substrate binding site with hydrogen bonds. *Planococcus halocryophilus* is recorded to grow at low temperature (−15 °C) organism found in high arctic permafrost (Mykytczuk et al. 2013; Feller 2010) The lowest temperature recorded in atmosphere is −89 °C, where water body is in the form of ice (Feller 2010).

Deep sea has a high hydrostatic pressure (110 MPa in the challenger deep of the Mariana Trench) recorded. Microbes that survived in high pressure (deep sea and the centre of the Earth) follow the strategies like reduce water penetration into the core of protein and rigidness by making compact and dense hydrophobic core (reduce the large hydrophobic residues in core), occupying smaller hydrogen-bonding amino acid residues, and multimerization (Kusube et al. 2017; Kato et al. 1997). *Colwellia marinimaniae* bacteria isolated from deep sea and found that it can be grown at 140 MPa hydrostatic pressures, highest value among piezophiles (Kusube et al. 2017).

Earth's surface is covered by about 71% of water, and salt water covers around 96.5% of all water. Organisms that are characterized as halophiles found in salt water (salt lakes, saltern soils, and deep hypersaline sea basin), acquired the strategy of solubility in high salt concentration by increasing the acidic amino acid residues on the surface and decreasing the hydrophobic residues (Van Der Wielen et al. 2005; Hallsworth et al. 2007; Oren 1983). Halophiles are characterize on the basis of optimum NaCl concentration as slight, moderate, borderline extreme and extreme (Kushner 2020).

In the case of extreme pH, organism that live in highly acidic environment are acidophile (grow in acidic hot springs, acidic mines, and solfataric field), and that live in highly alkaline environment are alkaliphile (alkaline lakes, ground water, and gold mines). Both acidophiles and alkaliphiles use strategies for survival is to avoid insolubilization by aggregation. Less salt bridges, negative surface charges, and less hydrogen bond enhance the living capacity of acidophiles (Schleper et al. 1995a, b). Moreover, organism that belongs to alkaliphiles has increased hydrophobic residues in the inter-subunit contacts help to improve the stability in high alkaline pH (Czop et al. 2011; Takai et al. 2001).

5.3 Extremophile as a Model Organism

A model organism is necessary for knowing the molecular strategies of extremophilic organisms. Extremophiles must be cloned and transformed into a suitable model organism (*Escherichia coli*). The drawback is that it is limited to study only targeted genes at a time under laboratory condition and lack of specific model system as the condition determine for the model organism as mostly mesophylls. Fortunately researchers find the model organism for extremophile; the details are given in Table 5.2.

Table 5.2 Some representative model organism of extreme conditions

S. No.	Extremophiles	Related model organism	References
1.	Acidophile	*Leptospirillum ferriphilum*, *Sulfolobus solfataricus*	Christel et al. (2018), Cárdenas et al. (2016), Quehenberger et al. (2017), Figueiredo et al. (2017)
2.	Alkaliphile	*Natronomonas pharaonis*	Falb et al. (2005), Oren (2013)
3.	Piezophile	*Thermococcus barophilus*	Birien et al. (2018), Oger et al. (2016)
4.	Radiophile	*D. radiodurans*, *Halobacterium* sp. *NRC-1*	Krisko and Radman (2013), Zhou et al. (2017), Kennedy et al. (2001), Coker et al. (2009)
5.	Psychrophile	*Pseudoalteromonas*, *Halorubrum lacusprofundi*	Parrilli et al. (2019), Mocali et al. (2017), Liao et al. (2016)
6.	Thermophile	*Thermococcus kodakarensis*, *Sulfolobus solfataricus*, *Thermus thermophilus*	Aslam et al. (2017), Atomi and Reeve (2019), Quehenberger et al. (2017), Figueiredo et al. (2017), Alvarez et al. (2014)
7.	Halophile	*Bacillus halodurans*, *Halobacterium volcanii*, *Halobacterium* sp. *NRC-1*, *Wallemia ichthyophaga*, *Halorubrum lacusprofundi*, *Natronomonas pharaonis*	Van-Thuoc et al. (2013), Coker (2019), Liao et al. (2016), Falb et al. (2005), Oren (2013)

5.4 Biochemical and Molecular Strategies for Extremophiles

Extremophiles are a biotechnology interest because they produce extremozymes that are functional and allocate the extremophilic organism to grow in such different extreme conditions. Various genes and pathways are responsible for survival of extremophiles in varied environmental conditions. Beautifully the molecular strategies were explained by Orellana et al. (2018) and summarized in Table 5.3 and Fig. 5.1.

ATP synthase, Chaperones, and DNA repair proteins are play a key role in the adaptation of acidophilic microorganism (Singh et al. 2020b). Besides this the membrane of acidophiles are highly permeable to the protons and potassium antiporter start to release protons to the extracellular spaces. The long chains of lipid molecule with monomers of fatty acid and some are added with dicarboxylic fatty acids are the characteristic changes in the thermophiles. Enhancement of the pyruvate dehydrogenase complex, polyamines like spermidine and chaperones are also the important factors for the survival of thermophilic organisms. Primary and secondary chlorine transporters, potassium uptake into the cells and ATP synthase are the molecular key features of halophiles to survive in high salt condition. Whereas uptake of proline betaine, and ectoine maintains the osmotic balance in the low salt environment that established the turgor pressure of the cell. A short fatty acid chain with unsaturated compounds is the primary feature of psychrophiles. Cold

Table 5.3 Molecular adoptive strategies of important extremophiles (Orellana et al. 2018)

Extremophiles	Molecular adoptive strategies
Alkaliphiles	• For the proton accumulation, they used Na+ and H+ gradient by electrogenic antiporters.
	• They used sodium ion solute uptake system.
	• Cytochrome c-552 enhanced the terminal oxidation function by electron and protons accumulation.
UV resistance	• They reduced iron levels by manganese accumulation.
	• Produced antioxidants like glutathione.
	• Chaperones
	• Presence of DNA repair proteins
Xeric resistance	• They used evasion mechanism and bacteria sporulation.
	• Extracellular polymeric substances (EPS) increases by the process of spourlation.
	• DNA repair proteins active.
	• Accumulation of osmoprotectants like glycine, and trehalose.
Psychrophiles	• Lipid contains unsaturated fatty acids with cyclopropane containing fatty acid short chain.
	• Availability of CSP (Cold shock proteins).
	• Presence of chaperones
	• For restrict the growth of frost, AFPs (Anti-freeze proteins) are present on the protein surface.
	• To prevent protein aggregation in the cytosol, Psychrophilic cells used cryo-protectants like mannitol and other compatible compound.
	• To maintain the fluidity of cell membrane, star shaped carotenoids present between the lipid and also protect the cell from UV radiation.
Acidophiles	• Accumulation of protons on the extracellular spaces increases due to the presence of potassium-proton antiport system.
	• ATP synthase active in the membrane of acidophilic cell.
	• The cell membrane is highly impermeable to protons.
	• Chaperones present in the cytoplasm.
	• DNA-repair proteins found that help to repairing of cell damage.
Thermophiles	• Upregulated glycolysis proteins like PDC (Pyruvate Dehydrogenase Complex) are found.
	• Iso-branched chain fatty acids and long chain dicarboxylic fatty acids are present in the lipid of cell membrane.
	• Spermidine are found in the cell of thermophiles.
	• Chaperones occur.
Halophiles	In case of High Salt in situation:
	• Presence of primary and secondary chloride transporters.
	• By the ATP synthase and bacteriorhodopsin process the potassium uptake increases in the cell.
	In case of Low Salt situation:
	• De novo synthesis or uptake of osmoprotectants (proline betaine, ectoine) that maintain osmotic balance and maintain the turgor pressure under various salt concentrations.

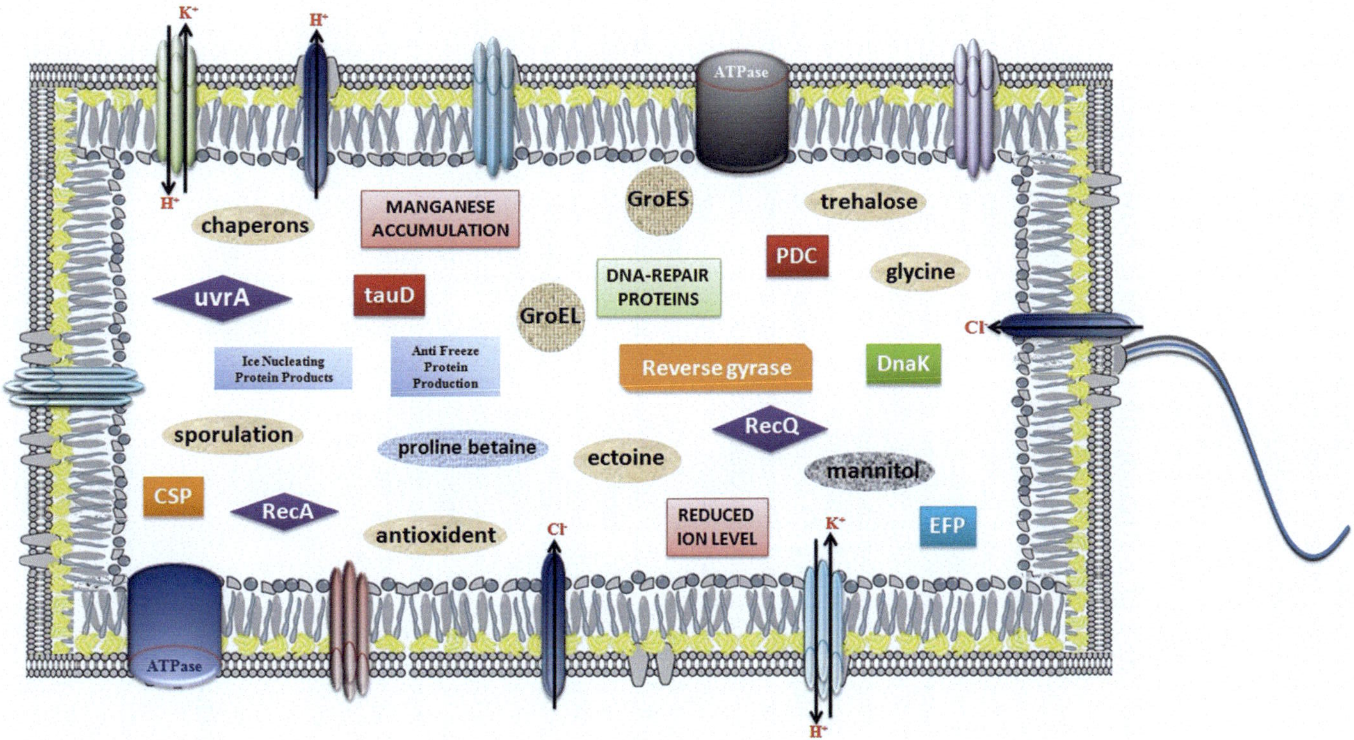

Fig. 5.1 Molecular strategies (responsible genes) of representative extremophilic genes (e.g. Hsp20, CDC48, Thermosome TF55, GroEL, GroES, DnaK, PDS for Thermophiles; AFP (Anti Freeze Protein), CSP (Cold Shock Protein) for Psychrophiles; uvrA, reqA, reqQ for UV resistant organisms)

sock and anti-freeze proteins inhibit the growth of ice on protein surface. Star shaped carotenoids maintained the fluidity of membrane in the psychrophilic organisms. Sporulation and accumulation of osmoprotectants made habitual to survive in xeric condition while antioxidants and DNA repair proteins help to survive in UV resistant microbes.

5.5 Significance of Extremophiles to Their Environment

The specific metabolic processes and the biological function of extremophiles are arbitrated by specific enzymes and proteins that can remain functional in extreme conditions. The enzymes and peptides which are obtained from these unusual microorganisms display unique features like thermostability, resistant against chemical denaturants (detergents, chaotropic agents, organic solvents) and stability in extreme pH (Kikani and Singh 2012). Therefore, these organisms can be used as model for constructing as well as designing the proteins and enzymes having new properties that are useful for industrial applications (Table 5.4). The elevated temperature plays a significant role in running biotechnological processes because it increases the solubility and bioavailability of organic compounds. Raised temperature is complimented with the decrease in viscosity and increase in organic compounds diffusion coefficient which consequently increases the reaction rate (Egorova and Antranikian 2005). Some special interests are the reaction having less soluble hydrophobic substrates (polyaromatic, aliphatic hydrocarbons and fat) and polymeric compounds (starch, cellulose, hemi-cellulose and proteins). The elevated temperatures also increase the bioavailability of some harsh biodegradable pollutants and allow the efficient bioremediation. Moreover, if we perform any biological process at elevated temperature like above 60 °C, the risk of contamination decreased and control under strict conditions. The use of number of thermophilic genes in mesophilic clones are increasing abruptly. So, the majority of proteins by mesophilic hosts can maintain their thermostability and the degree of enzyme purity is also satisfying for industrial application by using these thermophilic origin genes. Figure 5.2 revealed the detailed extremophiles application and their documented portfolio.

5.6 Conclusion

Dawn of Earth, many ecosystem evolve themselves by using various morphological to molecular strategies for survival. These ecosystems differ broadly from extreme ranges of pH, temperature, radiation, salinity, pressure and metal contaminated areas. Microorganisms evolve themselves to survive in extreme conditions like Deep Ocean, polar region, volcanic areas, and core of the Earth by adopting different strategies as compare to mesophilic microbes. Flexibility in genome probably helps

Table 5.4 Consequences of enzyme originated from extremophiles

Types of enzymes from extremophiles	Enzymes	Main sources with references	
Starch-degrading enzymes: biochemistry at the boiling point of water	Heat-stable amylases (a-Amylase) and glucoamylases (Glucoamylase)	*Pyrococcus* sp. *Thermococcus profundus*, *Desulfurococcus mucosus*, *Aspergillus niger*	Tachibana et al. (1996), Chung et al. (1995), Canganella et al. (1994), Antranikian (2001)
	Cyclodextringlycosyltransferases (CGTases)	*Thermoanaero bacter* sp., *Thermoanaerobacterium thermosulfurogenes*	Pedersen et al. (1995), Wind et al. (1995)
	Thermoactivepullulanases (type I and II)	*Fervidobacterium pennavorans*, *Desulfurococcus mucosus*	Koch et al. (1997), Canganella et al. (1994)
Degradation of cellulose, the most abundant polymer in nature	Thermoactivecellulases (Endocellulases, cellulase and hemicellulase)	*T. maritime*, *Caldocellum saccharolyticum*, *Anaerocellum thermophilum*	Bronnenmeier et al. (1995), Te'o et al. (1995), Zverlov et al. (1998)
	Xylan-degrading enzymes (Thermo-stable xylanases Endoglucanase, Exoglucanase, β-Glucosidase)	*Dictyoglomus thermophilum*, *T. maritime*, *Pyrodictium abyssi*	Gibbs et al. (1995), Winterhalter and Liebl (1995), Niehaus et al. (1999)
Chitin degradation	Thermostable chitin-hydrolysing enzymes (Endo and exochitinases)	*Streptomyces thermoviolaceus*, *B. licheniformis*, *Thermococcus chitonophagus*	Tsujibo et al. (1995), Takayanagi et al. (1991), Huber et al. (1995)
Protein degradation	Heat-stable proteolytic enzymes (Serine protease, Thiol protease, Acidic protease, Carboxypeptidase)	*Desulfurococcus mucosus*, *Pyrococcus* sp., *Sulfolobus acidocaldarius*, *Sulfolobus solfataricus*	Cowan et al. (1987), Fujiwara et al. (1996), Fusek et al. (1990), Fusi et al. (1991)
DNA-processing enzymes: DNA polymerases	DNA polymerase I, 9°N-7 DNA polymerase, *Pfu* polymerase	*Thermus aquaticus*, *Thermococcus* sp., *Pyrococcus furiosus*	Chien et al. (1976), Southworth et al. (1996), Lundberg et al. (1991)

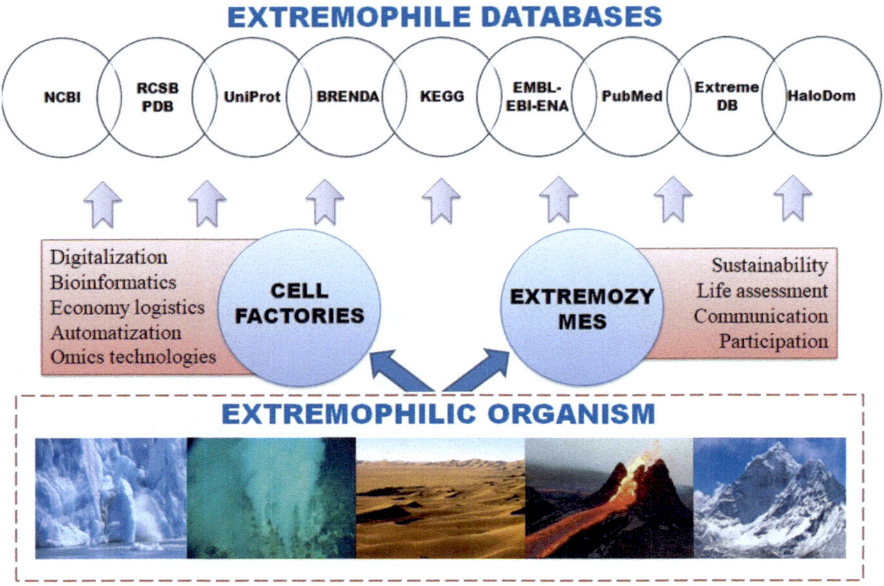

Fig. 5.2 Industry and databases information of extremophilic organisms

these microorganisms to survive in extreme conditions. As a result prokaryotes now divided in to thermophiles, psychrophiles, barophiles, alkalophile, acidophiles, halophiles and radiation and metal resistant, etc. The molecular biologist now discovered many important genes related to these extremophiles which are useful in many industries. The enzymes secreted by extremophiles are helpful in starch related industries, paper industries, food processing, dairy industry, etc. and also helpful in diagnosis of severe infectious diseases. This chapter is basically focused on the various survival strategies of extremophilic organisms as well as it also provides basic frame of extremophiles in industrial applications.

References

Alvarez L, Bricio C, Blesa A, Hidalgo A, Berenguer J (2014) Transferable denitrification capability of *Thermus thermophilus*. Appl Environ Microbiol 80(1):19–28

Antranikian G (2001) Extremophiles as a source of unique enzymes for biotechnological applications. In: Proceedings of the Microbiological Society of Korea Conference. The Microbiological Society of Korea, Seoul, pp 39–45

Aslam M, Horiuchi A, Simons JR, Jha S, Yamada M, Odani T, Atomi H (2017) Engineering of the hyperthermophilic archaeon *Thermococcus kodakarensis* for chitin-dependent hydrogen production. Appl Environ Microbiol 83(15):e00280–e00217

Atomi H, Reeve J (2019) Microbe profile: *Thermococcus kodakarensis*: the model hyperthermophilic archaeon. Microbiology 165(11):11–66

Bartlett DH, Eloe EA, Lauro FM (2007) Microbial adaptation to high pressure. In: Charles G, Nicolas G (eds) Physiology and biochemistry of extremophiles. American Society of Microbiology, Washington, DC, pp 333–348. https://doi.org/10.1128/9781555815813.ch25

Birien T, Thiel A, Henneke G, Flament D, Moalic Y, Jebbar M (2018) Development of an effective 6-methylpurine counterselection marker for genetic manipulation in *Thermococcus barophilus*. Genes 9(2):77

Bischoff JL, Rosenbauer RJ (1988) Liquid-vapor relations in the critical region of the system NaCl-H2O from 380 to 415 C: a refined determination of the critical point and two-phase boundary of seawater. Geochim Cosmochim Acta 52(8):2121–2126

Bronnenmeier K, Kern A, Liebl W, Staudenbauer WL (1995) Purification of Thermotogamaritima enzymes for the degradation of cellulosic materials. Appl Environ Microbiol 61(4):1399–1407

Canganella F, Andrade CM, Antranikian G (1994) Characterization of amylolytic and pullulytic enzymes from thermophilic archaea and from a new *Fervidobacterium* species. Appl Microbiol Biotechnol 42(2):239–245

Cárdenas JP, Quatrini-Nyqvist RC, Holmes DS (2016) Progress in acidophile genomics

Chan Y, Lacap DC, Lau MC, Ha KY, Warren-Rhodes KA, Cockell CS, Pointing SB (2012) Hypolithic microbial communities: between a rock and a hard place. Environ Microbiol 14 (9):2272–2282

Chen GQ, Jiang XR (2018) Next generation industrial biotechnology based on extremophilic bacteria. Curr Opin Biotechnol 50:94–100

Chien A, Edgar DB, Trela JM (1976) Deoxyribonucleic acid polymerase from the extreme thermophile *Thermus aquaticus*. J Bacteriol 127(3):1550–1557

Christel S, Herold M, Bellenberg S, El Hajjami M, Buetti-Dinh A, Pivkin IV, Dopson M (2018) Multi-omics reveals the lifestyle of the acidophilic, mineral-oxidizing model species *Leptospirillum ferriphilum* T. Appl Environ Microbiol 84(3):e02091–e02017

Chung YC, Kobayashi T, Kanai H, Akiba T, Kudo T (1995) Purification and properties of extracellular amylase from the hyperthermophilic archaeon *Thermococcus profundus* DT5432. Appl Environ Microbiol 61(4):1502–1506

Clarke A, Morris GJ, Fonseca F, Murray BJ, Acton E, Price HC (2013) A low temperature limit for life on Earth. PLoS One 8(6):e66207

Coker JA (2019) Recent advances in understanding extremophiles. F1000 Res 8:F1000 Faculty Rev-1917

Coker JA, DasSarma P, Capes M, Wallace T, McGarrity K, Gessler R, DasSarma S (2009) Multiple replication origins of Halobacterium sp strain NRC-1: properties of the conserved orc7-dependent oriC1. J Bacteriol 191(16):5253–5261

Cowan DA, Smolenski KA, Daniel RM, Morgan HW (1987) An extremely thermostable extracellular proteinase from a strain of the archaebacterium Desulfurococcus growing at 88 C. Biochem J 247(1):121–133

Czop M, Motyka J, Sracek O, Szuwarzyński M (2011) Geochemistry of the hyperalkaline Gorka pit lake (pH > 13) in the Chrzanow region, southern Poland. Water Air Soil Pollut 214(1):423–434

Dalmasso C, Oger P, Selva G, Courtine D, L'Haridon S, Garlaschelli A, Roussel E, Miyazaki J, Reveillaud J, Jebbar M, Takai K, Maignien L, Alain K (2016) Thermococcus piezophilus sp. nov. a novel hyperthermophilic and piezophilic archaeon with a broad pressure range for growth isolated from a deepest hydrothermal vent at the Mid-Cayman Rise. Syst Appl Microbiol 39 (7):440–444. https://doi.org/10.1016/j.syapm.2016.08.003

Deming JW (2007) Life in ice formations at very cold temperatures. In: Gerday C, Glansdorff N (eds) Physiology and biochemistry of extremophiles. ASM Press, Washington, DC, pp 133–145

DeVeaux LC, Müller JA, Smith J, Petrisko J, Wells DP, DasSarma S (2007) Extremely radiation-resistant mutants of a halophilicarchaeon with increased single-stranded DNA-binding protein (RPA) gene expression. Radiat Res 168(4):507–514

Egorova K, Antranikian G (2005) Industrial relevance of thermophilic Archaea. Curr Opin Microbiol 8(6):649–655

Falb M, Pfeiffer F, Palm P, Rodewald K, Hickmann V, Tittor J, Oesterhelt D (2005) Living with two extremes: conclusions from the genome sequence of *Natronomonas pharaonis*. Genome Res 15(10):1336–1343

Feller G (2010) Protein stability and enzyme activity at extreme biological temperatures. J Phys Condens Matter 22(32):323101

Figueiredo AS, Kouril T, Esser D, Haferkamp P, Wieloch P, Schomburg D, Schaber J (2017) Systems biology of the modified branched Entner-Doudoroff pathway in *Sulfolobuss olfataricus*. PLoS One 12(7):e0180331

Fujiwara S, Okuyama S, Imanaka T (1996) The world of archaea: genome analysis, evolution and thermostable enzymes. Gene 179(1):165–170

Fusek M, Lin XL, Tang J (1990) Enzymic properties of thermopsin. J Biol Chem 265 (3):1496–1501

Fusi P, Villa M, Burlini N, Tortora P, Guerritore A (1991) Intracellular proteases from the extremely thermophilic archaebacterium Sulfolobussolfataricus. Experientia 47(10):1057–1060

Gibbs MD, Reeves RA, Bergquist PL (1995) Cloning, sequencing, and expression of a xylanase gene from the extreme thermophile *Dictyoglomus thermophilum* Rt46B 1 and activity of the enzyme on fiber-bound substrate. Appl Environ Microbiol 61(12):4403–4408

Hallsworth JE, Yakimov MM, Golyshin PN, Gillion JL, D'Auria G, de Lima AF, McGenity TJ (2007) Limits of life in MgCl2-containing environments: chaotropicity defines the window. Environ Microbiol 9(3):801–813

Horneck G, Klaus DM, Mancinelli RL (2010) Space microbiology. Microbiol Mol Biol Rev 74 (1):121–156. https://doi.org/10.1128/MMBR.00016-09

Huber R, Stöhr J, Hohenhaus S, Rachel R, Burggraf S, Jannasch HW, Stetter KO (1995) *Thermococcus chitonophagus* sp nov, a novel, chitin-degrading, hyperthermophilic archaeum from a deep-sea hydrothermal vent environment. Arch Microbiol 164(4):255–264

Irwin JA (2010) Extremophiles and their application to veterinary medicine. Environ Technol 31 (8–9):857–869

Karaman K, Sagdic O (2019) Zygosaccharomyces bailii and Z rouxii induced ethanol formation in apple juice supplemented with different natural preservatives: a response surface methodology approach. J Microbiol Methods 163:105–659

Kashefi K (2003) Extending the upper temperature limit for life. Science 301(5635):934–934. https://doi.org/10.1126/science.1086823

Kato C, Li L, Tamaoka J, Horikoshi K (1997) Molecular analyses of the sediment of the 11000-m deep Mariana Trench. Extremophiles 1(3):117–123

Kennedy SP, Ng WV, Salzberg SL, Hood L, DasSarma S (2001) Understanding the adaptation of Halobacterium species NRC-1 to its extreme environment through computational analysis of its genome sequence. Genome Res 11(10):1641–1650

Kikani BA, Singh SP (2012) The stability and thermodynamic parameters of a very thermostable and calcium-independent α-amylase from a newly isolated bacterium, *Anoxybacillus beppuensis* TSSC-.1. Process Biochem 47(12):1791–1798

Koch R, Canganella F, Hippe H, Jahnke KD, Antranikian G (1997) Purification and properties of a thermostable pullulanase from a newly isolated thermophilic anaerobic bacterium, *Fervidobacterium pennavorans* Ven5. Appl Environ Microbiol 63(3):1088–1094

Koschinsky A, Garbe-Schönberg D, Sander S, Schmidt K, Gennerich HH, Strauss H (2008) Hydrothermal venting at pressure-temperature conditions above the critical point of seawater, 5 S on the Mid-Atlantic Ridge. Geology 36(8):615–618

Krisko A, Radman M (2013) Biology of extreme radiation resistance: the way of *Deinococcus radiodurans* Cold Spring. Harbor Perspect Biol 5(7):a012765

Krulwich TA, Sachs G, Padan E (2011) Molecular aspects of bacterial pH sensing and homeostasis. Nat Rev Microbiol 9(5):330–343. https://doi.org/10.1038/nrmicro2549

Kushner DJ (2020) Growth and nutrition of halophilic bacteria. In: The biology of halophilic bacteria. CRC Press, Boca Raton, FL, pp 87–103

Kusube M, Kyaw TS, Tanikawa K, Chastain RA Hardy KM, Cameron J, Bartlett DH (2017) *Colwelliamarinimaniae* sp nov, a hyperpiezophilic species isolated from an amphipod within the Challenger Deep, Mariana Trench. Int J Syst Evol Microbiol 67(4):824–831

Lane N, Martin WF (2012) The origin of membrane bioenergetics. Cell 151(7):1406–1416. https://doi.org/10.1016/j.cell.2012.11.050

Lane N, Allen JF, Martin W (2010) How did LUCA make a living? Chemiosmosis in the origin of life. BioEssays 32(4):271–280. https://doi.org/10.1002/bies.200900131

Liao Y, Williams TJ, Walsh JC, Ji M, Poljak A, Curmi PMG, Cavicchioli R (2016) Developing a genetic manipulation system for the Antarctic archaeon, *Halorubrum lacusprofundi*: investigating acetamidase gene function. Sci Rep 6(1):1–15

Lundberg KS, Shoemaker DD, Adams MW, Short JM, Sorge JA, Mathur EJ (1991) High-fidelity amplification using a thermostable DNA polymerase isolated from *Pyrococcus furiosus*. Gene 108(1):1–6

Meier DV, Pjevac P, Bach W, Hourdez S, Girguis PR, Vidoudez C, Meyerdierks A (2017) Niche partitioning of diverse sulfur-oxidizing bacteria at hydrothermal vents. ISME J 11 (7):1545–1558

Mirete S, Morgante V, González-Pastor JE (2017) Acidophiles: diversity and mechanisms of adaptation to acidic environments. In: Adaption of microbial life to environmental extremes. Springer, Cham, pp 227–251

Mocali S, Chiellini C, Fabiani A, Decuzzi S, de Pascale D, Parrilli E, Fani R (2017) Ecology of cold environments: new insights of bacterial metabolic adaptation through an integrated genomic-phenomic approach. Sci Rep 7(1):1–13

Mottl MJ, Wheat CG, Fryer P, Gharib J, Martin JB (2004) Chemistry of springs across the Mariana forearc shows progressive devolatilization of the subducting plate. Geochim Cosmochim Acta 68(23):4915–4933. https://doi.org/10.1016/j.gca.2004.05.037

Mykytczuk NC, Foote SJ, Omelon CR, Southam G, Greer CW, Whyte LG (2013) Bacterial growth at −15 C; molecular insights from the permafrost bacterium *Planococcus halocryophilus* Or1. ISME J 7(6):1211–1226

Niehaus F, Bertoldo C, Kähler M, Antranikian G (1999) Extremophiles as a source of novel enzymes for industrial application. Appl Microbiol Biotechnol 51(6):711–729

Oger P, Sokolova TG, Kozhevnikova DA, Taranov EA, Vannier P, Lee HS, Lebedinsky AV (2016) Complete genome sequence of the hyperthermophilic and piezophilicarchaeon *Thermococcus barophilus* Ch5, capable of growth at the expense of hydrogenogenesis from carbon monoxide and formate. Genome Announce 4(1):e01534

Ollivier B, Caumette P, Garcia JL, Mah RA (1994) Anaerobic bacteria from hypersaline environments. Microbiol Rev 58(1):27–38

Onofri A, Benincasa P, Mesgaran MB, Ritz C (2018) Hydrothermal-time-to-event models for seed germination. Eur J Agron 101:129–139. https://doi.org/10.1016/j.eja.2018.08.011

Orellana R, Macaya C, Bravo G, Dorochesi F, Cumsille A, Valencia R, Rojas C, Seeger M (2018) Living at the frontiers of life: extremophiles in chile and their potential for bioremediation. Front Microbiol 9. https://doi.org/10.3389/fmicb.2018.02309

Oren A (1983) *Halobacterium sodomense* sp nov, a Dead Sea halobacterium with an extremely high magnesium requirement. Int J Syst Evol Microbiol 33(2):381–386

Oren A (2013) Two centuries of microbiological research in the Wadi Natrun, Egypt: a model system for the study of the ecology, physiology, and taxonomy of haloalkaliphilic microorganisms. In: Polyextremophiles. Springer, Dordrecht, pp 101–119

Parrilli E, Tedesco P, Fondi M, Tutino ML, Giudice AL, de Pascale D, Fani R (2019) The art of adapting to extreme environments: the model system Pseudoalteromonas. Phys Life Rev 36:137

Pedersen S, Dijkhuizen L, Dijkstra BW, Jensen BF, Jorgensen ST (1995) A better enzyme for cyclodextrins. ChemTech 25(12):19–25

Plümper O, King HE, Geisler T, Liu Y, Pabst S, Savov IP, Rost D, Zack T (2017) Subduction zone forearc serpentinites as incubators for deep microbial life. Proc Natl Acad Sci 114(17):4324–4329. https://doi.org/10.1073/pnas.1612147114

Quehenberger J, Shen L, Albers SV, Siebers B, Spadiut O (2017) Sulfolobus–a potential key organism in future biotechnology. Front Microbiol 8:2474

Rai AR, Singh RP, Srivastava AK, Dubey RC (2012) Structure prediction and evolution of a haloacid dehalogenase of *Burkholderia mallei*. Bioinformatics 8(22):1111–1113

Sahuquillo-Arce JM, Chouman-Arcas R, Molina-Moreno JM, Hernández-Cabezas A, Frasquet-Artés J, López-Hontangas JL (2017) Capnophilic Enterobacteriaceae. Diagn Microbiol Infect Dis 87(4):318–319

Sayed AM, Hassan MH, Alhadrami HA, Hassan HM, Goodfellow M, Rateb ME (2020) Extreme environments: microbiology leading to specialized metabolites. J Appl Microbiol 128 (3):630–657

Scambos TA, Campbell GG, Pope A, Haran T, Muto A, Lazzara M, Reijmer CH, Broeke MR (2018) Ultralow surface temperatures in East Antarctica from satellite thermal infrared mapping: the coldest places on earth. Geophys Res Lett 45(12):6124–6133. https://doi.org/10.1029/2018GL078133

Schleper C, Piihler G, Kuhlmorgen B, Zillig W (1995a) Life at extremely low pH. Nature 375 (6534):741–742

Schleper C, Puehler G, Holz I, Gambacorta A, Janekovic D, Santarius UTE, Zillig W (1995b) *Picrophilus* gen nov, fam nov: a novel aerobic, heterotrophic, thermoacidophilic genus and family comprising archaea capable of growth around pH 0. J Bacteriol 177(24):7050–7059

Sharpton TJ (2014) An introduction to the analysis of shotgun metagenomic data. Front Plant Sci 5. https://doi.org/10.3389/fpls.2014.00209

Singh OV (2012) Extremophiles: sustainable resources and biotechnological implications. John Wiley & Sons, New York, NY

Singh RP, Singh RN, Srivastava AK, Kumar S, Dubey RC et al (2011) Structural analysis and 3D-modelling of fur protein from Bradyrhizobium japonicum. J Appl Sci Environ Sanit 6(3):357–366

Singh RP, Singh RN, Srivastava MK, Srivastava AK, Kumar S, Dubey RC, Sharma AK (2012) Structure prediction and analysis of MxaF from obligate, facultative and restricted facultative methylobacterium. Bioinformation 8(21):1042–1046. https://doi.org/10.6026/97320630081042

Singh RP, Manchanda G, Singh RN, Srivastava AK, Dubey RC (2016) Selection of alkalotolerant and symbiotically efficient chickpea nodulating rhizobia from North-West Indo Gangetic Plains. J Basic Microbiol 56:14–25. https://doi.org/10.1002/jobm.201500267

Singh RP, Manchanda G, Yang Y, Singh D, Srivastava AK, Dubey RC, Zhang C (2019) Deciphering the factors for nodulation and symbiosis of Mesorhizobium associated with Cicer arietinum in Northwest India. Sustainability 11(24):7216. https://doi.org/10.3390/su11247216

Singh R, Manchanda G, Maurya I, Wei Y (eds) (2020a) Microbial versatility in varied environments. Springer, Singapore. https://doi.org/10.1007/978-981-15-3028-9

Singh RP, Buttar HK, Kaur R, Manchanda G (2020b) The multifaceted life of microbes: survival in varied environments. In: Singh R, Manchanda G, Maurya I, Wei Y (eds) Microbial versatility in varied environments. Springer, Singapore. https://doi.org/10.1007/978-981-15-3028-9-1

Southworth MW, Kong H, Kucera RB, Ware J, Jannasch HW, Perler FB (1996) Cloning of thermostable DNA polymerases from hyperthermophilic marine Archaea with emphasis on Thermococcus sp 9 degrees N-7 and mutations affecting 3'-5'exonuclease activity. Proc Natl Acad Sci 93(11):5281–5285

Subhashini DV, Singh RP, Manchanda G (2017) OMICS approaches: tools to unravel microbial systems. Directorate of Knowledge Management in Agriculture, Indian Council of Agricultural Research, New Delhi. ISBN: 9788171641703. https://books.google.co.in/books?id=vSaLtAEACAAJ

Suyal DC, Soni R, Yadav AN, Goel R (2021) Cold adapted microorganisms: survival mechanisms and applications. In: Microbiomes of extreme environments. CRC Press, Boca Raton, FL, pp 177–191

Tachibana Y, Leclere MM, Fujiwara S, Takag M, Imanaka T (1996) Cloning and expression of the α-Amylase gene from the hyperthermophilic archaeon Pyrococcus sp. KOD 1, and characterization of the enzyme. J Ferment Bioeng 82:224–232

Takai K, Moser DP, Onstott TC, Spoelstra N, Pfiffner SM, Dohnalkova A, Fredrickson JK (2001) *Alkaliphilus transvaalensis* gen nov, sp nov, an extremely alkaliphilic bacterium isolated from a deep South African gold mine. Int J Syst Evol Microbiol 51(4):1245–1256

Takai K, Nakamura K, Toki T, Tsunogai U, Miyazaki M, Miyazaki J et al (2008) Cell proliferation at 122 C and isotopically heavy CH4 production by a hyperthermophilic methanogen under high-pressure cultivation. Proc Natl Acad Sci 105(31):10949–10954

Takayanagi T, Ajisaka K, Takiguchi Y, Shimahara K (1991) Isolation and characterization of thermostablechitinases from Bacillus licheniformis X-7u. Biochim Biophys Acta Prot Struct Mol Enzymol 1078(3):404–410

Te'o VSJ, Saul DJ, Bergquist PL (1995) cel A, another gene coding for a multidomain cellulase from the extreme thermophile *Caldocellum saccharolyticum*. Appl Microbiol Biotechnol 43 (2):291–296

Tsujibo H, Endo H, Miyamoto K, Inamori Y (1995) Expression in Escherichia coli of a gene encoding a thermostablechitinase from *Streptomyces thermoviolaceus* OPC-520. Biosci Biotechnol Biochem 59(1):145–146

Ueki A, Kaku N, Ueki K (2018) Role of anaerobic bacteria in biological soil disinfestation for elimination of soil-borne plant pathogens in agriculture. Appl Microbiol Biotechnol 102 (15):6309–6318

Van Der Wielen PW, Bolhuis H, Borin S, Daffonchio D, Corselli C, Giuliano L, Party BS (2005) The enigma of prokaryotic life in deep hypersaline anoxic basins. Science 307(5706):121–123

Van-Thuoc D, Hashim SO, Hatti-Kaul R, Mamo G (2013) Ectoine-mediated protection of enzyme from the effect of pH and temperature stress: a study using *Bacillus haloduransxylanase* as a model. Appl Microbiol Biotechnol 97(14):6271–6278

Wierzchos J, Cámara B, de Los RA, Davila AF, Sánchez Almazo IM, Artieda O, Ascaso C (2011) Microbial colonization of Ca-sulfate crusts in the hyperarid core of the Atacama Desert: implications for the search for life on Mars. Geobiology 9(1):44–60

Wind RD, Liebl W, Buitelaar RM, Penninga D, Spreinat A, Dijkhuizen L, Bahl H (1995) Cyclodextrin formation by the thermostable alpha-amylase of *Thermoanaerobacterium thermosulfurigenes* EM1 and reclassification of the enzyme as a cyclodextringlycosyltransferase. Appl Environ Microbiol 61(4):1257–1265

Winterhalter C, Liebl W (1995) Two extremely thermostable xylanases of the hyperthermophilic bacterium *Thermotogamaritima* MSB8. Appl Environ Microbiol 61(5):1810–1815

Zeng X, Birrien JL, Fouquet Y, Cherkashkov G, Jebbar M, Querellou J, Oger P, Cambon-Bonavita MA, Xiao X, Preur D (2009) *Pyrococcus* CH1, an obligate piezophilic hyperthermophile: extending the upper pressure-temperature limits for life. Int Soc Microb Ecol J 3(7):873–876

Zhang J, Yang X, Guo C, de Lajudie P, Singh RP, Wang E, Chen W (2017) *Mesorhizobium muleiense* and *Mesorhizobium* gsp. nov. are symbionts of Cicer arietinum L. in alkaline soils of Gansu, Northwest China. Plant Soil 410(1–2):103–112

Zhou Y, Shen P, Lan Q, Deng C, Zhang Y, Li Y, Xu P (2017) High-coverage proteomics reveals methionine auxotrophy in *Deinococcus radiodurans*. Proteomics 17(13–14):1700072

Zverlov V, Mahr S, Riedel K, Bronnenmeier K (1998) Properties and gene structure of a bifunctional cellulolytic enzyme (CelA) from the extreme thermophile 'Anaerocellum thermophilum' with separate glycosyl hydrolase family 9 and 48 catalytic domains. Microbiology 144 (2):457–465

Chapter 6
Bacterial Community Composition Dynamics in Rice Rhizosphere: A Metagenomic Approaches

Abha Manapure, Raghvendra Pratap Singh, and Alok R. Rai

Abstract The rhizosphere area of plant root surface shows bountiful diversity of microorganisms. Microbial community within rhizosphere inhabiting the rice field ecosystem have been studied previously. It is not possible to isolate the whole microbial genome by traditional culture dependent method. Metagenomic covers entire genome of all microbial community irrespective of any habitat without in vitro culturing. Present review has been aimed to summarize the past practices and recent issues of metagenomic analysis of paddy field bacterial communities within rhizosphere from different geographic locations. So, this chapter deals with the recent tools, platform, pipelines and software of metagenomics used with other techniques (e.g., 16S rRNA gene sequencing with V3-V4 hypervariable region, Pyrosequencing, Metaproteomic, etc.) for the study of bacterial composition from different regions such as rhizosphere, phyllosphere, bulk soil, wetland region of soil, irrigated soil, flooded and non-flooded soil, high prevalence of salt soil and high incidence of rice blast fungus contaminated soil. The findings from this review helps to enhance the crop production, improve soil quality by more use of biofertilizers and also helps in disease management with biocontrol agent.

Keywords Metagenomic · Pyrosequencing · Metaproteomic · Hypervariable region · Biocontrol agent

6.1 Introduction

Rice is widely consumed staple food for 50% of population of world (up to three billion people) especially in Asia and Africa. Interestingly, rice plants represent a habitat for a varied microbial population that colonizes the rhizosphere, a restricted

A. Manapure · A. R. Rai (✉)
Department of Microbiology, S K Porwal College, Nagpur, Maharashtra, India

R. P. Singh
Department of Research and Development, Azoth Biotech Pvt. Ltd., Noida, Uttar Pradesh, India

R. P. Singh et al. (eds.), *Microbes in Microbial Communities*,
https://doi.org/10.1007/978-981-16-5617-0_6

zone of the plant roots' surface (Kowalchuk et al. 2010). The root growth of the majority of plants in soil has altered the spatial structure (Angers and Caron 1998). According to Curtis et al. (2002) a soil can have up to 4×10^6 different types of microbial taxa, and 1 g of soil can have more than one million distinct microbial genomes which ultimately shows an enormous microbial diversity remains within the soil especially in rhizosphere, predicted by Gans (2005). Since, majority of microorganisms cannot be cultured by culture dependent or conventional method, is intrigued by unraveling soil microbial community structure as well as functionality remains as an attractive challenge for enhancing the plant health and crop production (Yang et al. 2019). About 20–50% of the plant photosynthate is transported below the ground level and it is totally depending upon the different plant species (Kuzyakov and Domanski 2000) and about 18% of plant photosynthate is discharged into the soil environment on average (Jones et al. 2009). The favorable impacts of the rhizosphere microbial population on rice plants, including the production of plant growth-promoting bacteria (PGPR) have been thoroughly investigated (Subhashini and Singh 2014; Majeed et al. 2015), phosphate solubilization (Elias et al. 2016), nitrogen fixation process, mycorrhizal fungi, also acts as biocontrol agents for management of various plants diseases (Massart et al. 2015) with a high level of stress tolerance (Tsurumaru et al. 2015).

Rice differs from most crops in that it is typically cultivated in flooded soil, which results in the formation of oxic and anoxic zones within the rice rhizosphere area of soil, which select for specific physiological groups of microbial community with either aerobic, anaerobic, or facultative metabolism (Brune et al. 2000). The structure of the microbial population in the rhizosphere of the rice field environment has previously been characterized. The majority of research has concentrated on isolating, identifying, and characterization of rice rhizospheric bacteria from various locales and types (Vacheron et al. 2013). The bacteria in the rhizosphere had been studied widely (Zhang et al. 2016; Prajakta et al. 2019; Yang et al. 2020a, b; Maheshwari et al. 2021). They also influence the rhizosphere microbiota's chemical composition and offer crucial microbial growth substrates through rhizodeposition (Lynch and Whipps 1990). The decomposition of organic matter in soil is also largely attributed to the microbial population (Kuzyakov 2002; Yang et al. 2020a). Recently developed technologists, provide relatively quick and prompt sequencing of metagenomic DNA samples at very moderate cost in short time (Subhashini et al. 2017; Yang et al. 2020b), metagenomic DNA sequencing, however completely sequenced whole genome sequencing, depends on the DNA extracted (Gautam et al. 2019).

Without in vitro culturing, prior individual identification, or gene amplification, a metagenomics study covers the entire genome of any microbial community inhabiting any habitat such as soil and water (Abulencia et al. 2006; Kunin et al. 2008). Metagenomic analysis in terms of the functions that they drive and regulate, analysis involves isolating DNA from an environmental sample, cloning the DNA into a suitable vector, transforming the clones into a host bacterium, and screening the resulting transformants (Zhang et al. 2019). Recent technological development has gradually increased our knowledge about the global ecological distribution of microbiota across the space and time and have furnished evidence for the contribution to ecosystem function (Chu et al. 2020). The use of metagenome sequencing techniques, such as Next Generation Sequencing technologies, has yielded enormous amounts of data, resulting in remarkable developments. To obtain detailed information on the diversity and ecology of microbial forms, the method involved isolating metagenomics DNA directly from an environmental niche (e.g., soil and water), fragmentation, generation of a sequence clone library, taxonomy and gene family community profiling, and high-throughput sequencing (Spence et al. 2014). The overall Metagenomics steps is illustrated in Fig. 6.1. The progress of metagenomics is totally dependent on high-throughput techniques for processing DNA from various environments and analyzing their sequence after running on high-end sequencers. Furthermore, examining millions or trillions of reads and putting them together to create a full genome is a difficult operation (Aguiar-Pulido et al. 2016a, b). Metagenomic analysis data provide the functional properties of a complex below-ground soil microbial community, such as intra and inter interactions, and so assist in the understanding their evolutionary aspect of microbial ecosystems as genetic and metabolic networks (Filippo et al. 2012; Ponomarova and Patil 2015).

This chapter article explains the current understanding of comparative metagenomic analysis of microbial diversity of paddy rhizospheric compartments and makes comparison of rhizospheric bacterial community structure among the different locations. A 16S rRNA gene profiling and shotgun metagenome analysis were used by Metagenomics. PCR will be used to amplify the V3-V4 region of 16S rRNA genes, which will then be sequenced on the Illumina Platform. Metagenomic library will be made and analyzed by different software. After then, a taxonomic analysis of a representative sequence from each OUT would be carried out to determine species distribution. The results will be represented in two-dimensional PCOA plots. The findings will be extremely useful since they may aid in the process of increasing rice output, improving crop quality, and reducing environmental impact owing to the usage of chemical fertilizers.

In this study, we focused on a variety of high-throughput sequencing investigations, collecting taxonomic data on bacterial communities at the genus level in the paddy rhizosphere and comparing them at the phylum level between rice plants from various places (Cox et al. 2010). Furthermore, this study explores metagenomic techniques to rhizospheric microbiomes and reports on the bacterial community composition in paddy rhizosphere (Mendes et al. 2013).

6.2 Approaches for Communities Structure Dynamics

Rhizospheric soil microbial communities play variety of roles in the function of soil by including enhancing organic and inorganic nutrient availability and nutrient cycling by boosting organic matter breakdown (Singh et al. 2019). The rhizospheric

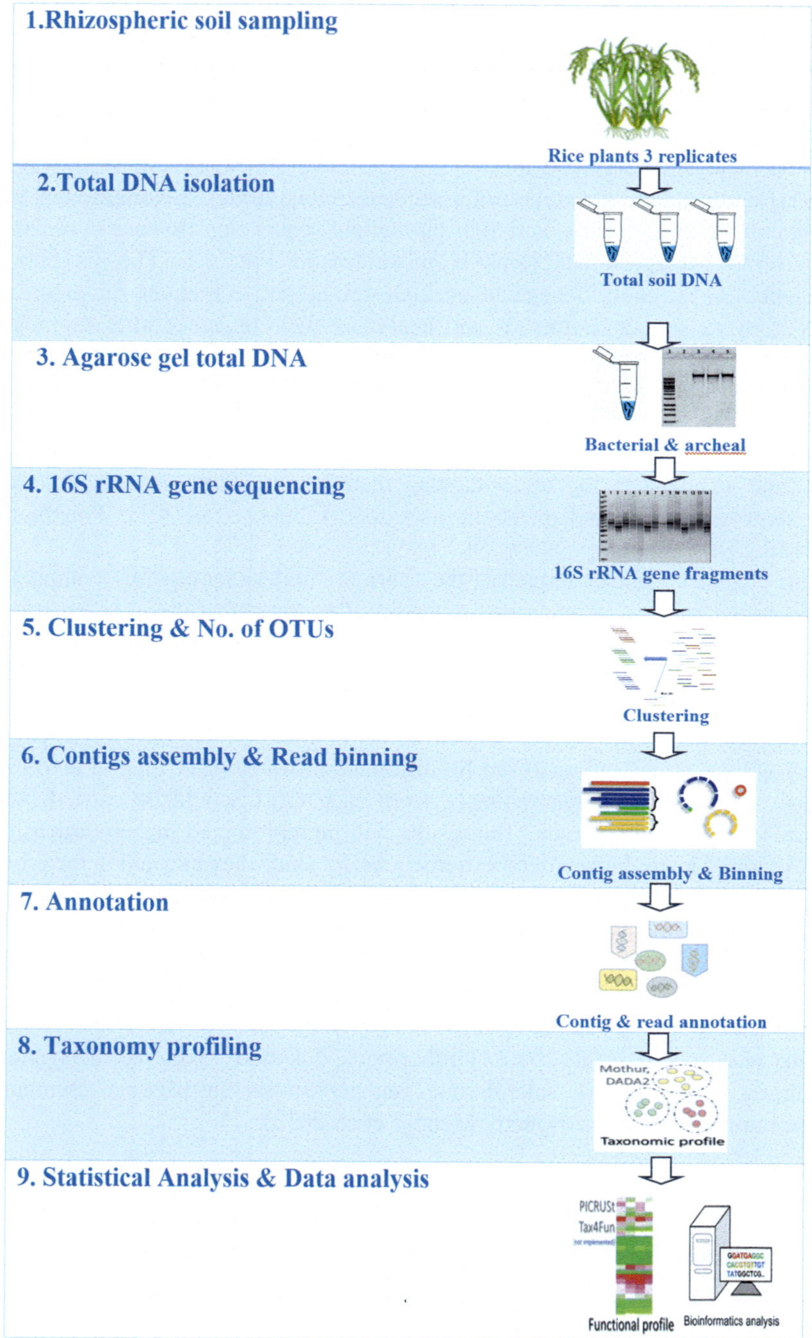

Fig. 6.1 Stepwise illustration of metagenomics

soil bacterial population is typically dominated by Proteobacteria, Actinobacteria, Acidobacteria, and Chloroflexi (Hussain et al. 2012).

In one of the studies in this research field, Arjun (2011), 16S rRNA sequencing retrieved from database found total 12 representative clones from the paddy field rhizosphere soil in Kuttanand, Kerala. About 600 bp were viewed and compiled and aligned using BioEdit version 5.0.6 software (Hall 2001) and generated phylogenetic tree by neighbor joining method with 1000 resampling bootstrap analysis by using Mega v.4 software (Tamura et al. 2007). The dominant taxa in the library were found to be Proteobacteria (7/12) followed by Firmicutes (2/12), Bacteriodetes (2/12) and Acidobacteria (1/12) (Table 6.1). These four phylotypes are also thought to describe the bacterial community structure in rice rhizospheric soil in previous investigations, and Proteobacteria are the largest and most metabolically diverse group of soil bacteria (Lu et al. 2006).

Knief et al. (2012) obtained 749,569 and 1,340,274 Rhizospheric and Phyllospheric soil sequences from paddy fields at the International Rice Research Institute in Los Banos, Philippines after a year. In the rhizospheric soil samples were found most abundantly Alphaproteobacteria and Deltaproteobacteria. Further more significant taxa such as Firmicutes, Actinobacteria, Gammaproteobacteria and Deinococcus—Thermus. Most abundant phyla include Archaea in the rhizosphere than phyllosphere region was detected. In this research article, scientists also studied Metaproteogenome and they were found majority of proteins within Alphaproteobacteria (33%) in these samples, proteins assigned particularly *Azospirillum*, *Bradyrhizobium*, *Rhodopseudomonas*, *Methylobacterium*, *Magnetospirillum*, and *Methylosinus* (Table 6.1). Based on metagenome readings and clone library analyses, the Betaproteobacteria (Acidovorax, Dechloromonas, and Herbaspirillum) and Deltaproteobacteria (Anaeromyxobacter, Desulfovibrio, and Geobacter) genera dominated the bacterial population. Furthermore, Sinclair et al. (2015), were focused microbial community structure in rice producing areas of Guadalquivir marshes (Seville). In the months of July (tillering or vegetative stage) and September (between blooming or ripening and maturity stage), rice rhizospheric soil was examined (Marschner et al. 2001). Total 240 cfu were obtained. The soil samples were collected from four different regions in rice yielding areas of Guadalquivir marshes (Seville). These areas were: Puebla, Colinas, Calonge and Rincon. The soil in these locations has two major issues that have impacted rice production: increased salt levels in irrigation water and rice plants infected with the rice blast fungus *Magnaporthe oryzae*, 25 different bacterial genera were identified based on 16S rRNA gene sequencing analysis, although only eight were found at both sample times, July and September. From July to September, the Paenibacillus, Bacillus, and Pantoea communities grew in dominance, whereas the Enterobacter, Pseudomonas, and Exiguobacterium communities decreased. In July, there was a 21.34% increase in Exiguobacterium and a 20.21% increase in Enterobacter. Conversely Bacillus (37.33%) was more abundant in September. According to 16S rRNA sequencing of total DNA from four areas found that Proteobacteria, Acidobacteria and Anaerolineae were found to be more significantly in all areas. Proteobacteria (Betaproteobacteria) was most abundantly detected group followed by Bacteriodetes

Table 6.1 Different Bacterial taxa identified in the different geographic locations of rice rhizospheric soil composition

S. No.	Geographic coordinates	Approach	Findings related to rhizospheric bacterial composition	References
1.	Kuttanand, Kerala, India.	16S rRNA Sequencing and Pyrosequencing	Bacterial Community in rice rhizosphere dominantly observed taxa were Proteobacteria, followed by Firmicutes, then Bacteroidetes and Acidobacteria.	Arjun (2011)
2.	Los Banos, Philippines	16S rRNA gene sequencing and Metaproteomic profiling.	Microbial community composition in rice rhizosphere includes Archaea (2.6%), Actinobacteria (8.5%), Chloroflexi (4.6%), Alphaproteobacteria (14%), Betaproteobacteria (16.6%), and Deltaproteobacteria (10.6%).	Knief et al. (2012)
3.	Guadalquivir marshes (Seville), Spain.	16S rRNA gene Sequencing	Most frequently present group was Proteobacteria Betaproteobacteria followed by Archaea, Bacteroidetes, Chloroflexi, Acidobacteria, Thermococci, Sphingobacteria, Vermicomicrobia, Bacillus, Enterobacteria, Exiguobacterium.	Lucas et al. (2013)
4.	South Korea, Philippine, Italy and China	16S rRNA, pmoA, and mcrA amplifications	16S rRNA gene sequencing, pmoA and mcrA amplification analysis observed that rice field methanogens mainly comprise Methanocella, Methanobacterium, and dominantly Methanosaeta all over the cultivation.	Hyo Jung Lee et al. (2014)
5.	Bogor, West Java and Indonesia.	16S rRNA gene sequencing and nif gene amplification	16S rRNA sequencing observed 5 genera of Actinomycetes including Geodermatophilus, Actinoplanes, Actinokineospora, Streptomyces, and Kocuria while nif gene amplification showed that strain member of species Rhizobium and Anaeromyxobacter.	Rusmana et al. (2015)
6.	Vercelli, Italy.	16S rRNA gene Pyrotag sequencing.	More abundance of Archaea and Acidobacteria in	Breidenbach et al. (2016)

(continued)

Table 6.1 (continued)

S. No.	Geographic coordinates	Approach	Findings related to rhizospheric bacterial composition	References
			rhizosphere observed. The rhizosphere also consists of higher relative abundances of Alphaproteobacteria, Betaproteobacteria, Gammaproteobacteria, Deltaproteobacteria, Cyanobacteria, Chloroflexi, Firmicutes, and Verrucomicrobia.	
7.	Venezuela.	16S rDNA taxonomic profiling.	Gammaproteobacteria was determined to be the most dominant phyla of Proteobacteria, followed by Betaproteobacteria and Alphaproteobacteria, Acidobacteria, Nitrospirae, Cyanobacteria, Firmicutes, Vermicomicrobia, Bacteriodetes, Caulobacter, and so on.	Venturi et al. (2018)
8.	Kerala, India	16S rRNA gene hypervariable V3-V4 region	In this sample, most detected phyla were Proteobacteria (26 ± 14%), Firmicutes (21 ± 9%), Actinobacteria (17 ± 6%), and Acidobacteria (14 ± 10%).	Imchen et al. (2018)
9.	Maritsa river and Zlato Pole wetland, Bulgaria.	V3-V5 hypervariable region of 16S rRNA amplicon sequencing.	Abundantly found phyla includes Proteobacteria (68%), Gammaproteobacteria (45%), Acinetobacter (54%), Alphaproteobacteria (21.4%), Actinobacteria (18.5%), Firmicutes (9.4%), and Bacteriodetes (8.3%).	Ivan et al. (2019)
10.	Faisalabad, Pakistan.	16S rRNA gene amplification.	Reports have been shown that dominant groups were Proteobacteria, Acidobacteria, Actinobacteria, Bacteriodetes, Chloroflexi, Firmicutes, Nitrospirae, Gaiella, Marmoricola, Clostridium.	Maria et al. (2020)

and Chloroflexi (Table 6.1). Thermococci archaea were identified in locations with high Magnaporthe oryzae frequency, while Sphingnobacteria archaea were discovered in areas with high salt occurrence. Conversely, Verrucomicrobiae class was only detected in control area.

Scientists performed rice field experiments at research farm located in Sacheon, South Korea (Lee et al. 2014). The rice field was ploughed and harrowed, and water was flooded up to 5 cm above the soil surface (Witt et al. 2000). Following that, 21-day-old Korean rice seeds (*Oryza sativa*, Japonica type) were planted. Every 30 days soil samples were collected in triplicate. For detecting the 16S rRNA gene copies Archaea and Bacteria and targeting the *pmo*A and *mcr*A gene (Breidenbach and Conrad 2015). 16S rRNA gene sequencing analysis obtained 80% of bacterial reads of four taxa including Proteobacteria, Acidobacteria, Chloroflexi and Actinobacteria during whole cultivation. 16S rRNA gene sequencing, *pmo*A and *mcr*A amplification analysis represents that the rice paddy methanogens mainly comprise Methanocella, Methanobacterium and dominantly Methanosaeta all over the cultivation (Table 6.1) (Vaksmaa et al. 2017; Wang et al. 2018).

Rusmana et al. (2015) collected rice rhizospheric soil samples from three different types of agroecosystems (irrigated rice, marshy tidal, and dry) in Indonesia while performing 16S rRNA gene and *nif*H gene amplification (Reichardt et al. 1997). 16S rRNA gene analysis found abundance of five genera of Actinomycetes mainly comprises *Actinokineospora*, *Actinoplans*, *Geodermatophilus*, *Kocuria* and *Streptomyces* (Nimnoi et al. 2010). Most abundantly found species within *Streptomyces* in almost all genera were *Streptomyces alboniger* and *Streptomyces acidiscabies* (Taj and Rajkumar 2016). The amplification of the *Nif* gene revealed a biological role that was closely related to Rhizobium and *Anaeromyxobacter* strains (Martina et al. 2008; Pereira et al. 2013).

6.3 Metagenomics Software as Bioinformatic Tools

Metagenomics is the study of genes in relation to their environment In addition, at the Rice Research Institute in Vercelli, Italy, a rhizospheric soil sample was taken from paddy fields. Rice plants were sampled at four different stages: Stage 1 (early vegetative or tillering), Stage 2 (late vegetative), Stage 3 (reproductive or flowering), and Stage 4 (maturity). Using the UPARSE workflow, 8685 OTUs with 97% identity were found from 16S rRNA Pyrotag sequence analysis (Edgar 2013). The Silva taxonomy and method were used to classify relative OTU sequences in MOTHUR version 1.31.2 (Schloss et al. 2009). Absolute abundance of Archaea was detected to be higher in rhizospheric soil than bulk soil sample. For Archaea, Methanosarcina and Methanosaeta were found more abundant in rhizospheric soil of Vercelli. Abundantly present genera such as Acidobacteria, Alphaproteobacteria, Deltaproteobacteria, Betaproteobacteria, Cyanobacteria, Chloroflexi, Firmicutes. Potential iron reducer (e.g., *Geobacter* and *Anaeromyxobacter*) (Conrad and Frenzel 2002; Hori et al. 2010). The VEGAN package version 2.2.1 was used to investigate OTU relative abundances (Oksanen et al. 2013). Fermenters (e.g., Clostridia and

Opitutus) and endophytic plant growth promoting bacteria (e.g., Herbaspirillum species) were reported to be more prevalent in rhizospheric soil (Andreesen and Schaupp 1973; Chin et al. 2001). Base pairs were viewed and compiled and aligned using BioEdit version 5.0.6 software (Hall 2001) and generated phylogenetic tree by neighbor joining method with 1000 resampling bootstrap analysis by using Mega v.4 software (Tamura et al. 2007). Multivariate analysis revealed considerable differences between the sites when comparing the taxonomic patterns of the bacterial communities. Ivan et al. (2019) studied V3-V5 hypervariable region of 16S rRNA amplicon sequencing using Miseq Illumina platform (Ebersberg, Germany). The gene expression of *Pmo*A and *mcr*A was studied using quantitative reverse transcriptase real-time PCR (qRT-PCR) (Lee et al. 2014). 16S rRNA gene sequencing, *pmo*A and *mcr*A amplification analysis perceived that rice paddy methanogens mainly consist of Methanocella, Methanobacterium and dominantly Methanosaeta all over the cultivation (Table 6.1) (Vaksmaa et al. 2017; Wang et al. 2018). The tools for deciphering the metagenome have been listed in Table 6.2.

6.4 Proteomics Analysis of Bacterial Community

Genomic, proteomic, metabolomic, metagenomic, and transcriptomic studies are all included in the term "omics." It refers to the joint characterization and measurement of biological molecule pools that translate into an organism's structure, function, and dynamics. Proteomics has enabled the identification of ever-increasing number of protein (Anderson and Anderson 1998; Blackstock and Weir 1990; Anwar et al. 2019). Recent research findings indicate that rhizosphere soil metagenomic analysis can provide a sketch of a protein domain's functional areas, which can be used for protein optimization and functional characterization (Jin et al. 2016). InterPro is a software used for access the information about Protein domains, protein activity, active site within the protein, protein families and function (Singh et al. 2016). Genes encoding dinitrogen reductase (*nifH*) and dinitrogenase (*nif*D and *nif*K) were often found in the phyllosphere and rhizosphere, according to metagenomic analysis in the current study (Zeng et al. 2005). In phyllosphere, the most abundant *nifH* sequence types were found to be Azorhizobium and Rhodopseudomonas while in rhizosphere, the *nifH* sequences was detected across diverse taxa such as Rhizobium, Methylococcus, Dechloromonas, Anaeromyxobacter, Syntrophobacter, and some methanogenic archaea (Knief et al. 2012; Singh et al. 2016). The metaproteomic analysis reveals that genus Methylobacterium were detected most dominant in phyllosphere community.

Table 6.2 Bioinformatics tools for metagenomic data analysis

S. No.	Software	Function of software	References
1.	MetaQUAST	For quality assessment of metagenomic assemblies.	Mikheenko et al. (2016)
2.	Mothur	Software for analysis of 16S rRNA gene sequencing	Schloss et al. (2009)
3.	MetaVelvet	Metagenomic de novo assembler	Namiki et al. (2012), Zerbino and Birney (2008)
4.	MG-RAST	Access to a number of tools for metagenomic analysis via a web-based platform.	Glass et al. (2010)
5.	IDBA-UD	For the building of contigs using a progressive cycle of rising k-mer values	Peng et al. (2012)
6.	Megahit	Useful in metagenomic analysis and uses similar approach to IDBA-UD	Li et al. (2015)
7.	UPSARSE	Pipeline for quality and length filtering of sequencing reads and OUT generation	Edgar (2013)
8.	MetAMOS	Ability to test multiple assembly tools and used for contigs length, contiguity, and error rates	Treangen et al. (2013)
9.	VEGAN	Software for diversity analysis and community ecology functions	Oksanen et al. (2013)
10.	InterPro	Software for access the information about protein domains, protein activity active site of protein, protein families and function	Mitchell et al. (2015)
11.	MegaGene Annotator	For high contig length and large number of predicted gene	Noguchi et al. (2008)
12.	RayMeta	Scalable software tool and assemblies are constructed on the basis of de Bruijin graphs	Boisvert et al. (2012), Pell et al. (2012)
13.	QUIIME	Quantitative insight into microbial ecology	Caporaso et al. (2010)
		Pipeline used for microbiome analysis from raw DNA sequencing data generated by Illumina platform	
14.	CONCOCT	Used to count the number of clusters and reconstruct pathogenic genomes (Shiga-toxin producing strain of *E. coli* outbreak in 2011)	Alneberg et al. (2014)
15.	CARMA	Used for Metagenomic analysis	Gerlach et al. (2009)
16.	Prokka	Pipeline used for annotation of bacterial genomes	Seemann 2014
17.	MEGAN	Software used for analysis of large metagenomic datasets	Huson and Weber (2013)
18.	Glimmer-MG	Software for gene prediction and provide accurate gene error-prone sequences than other method	Delcher et al. (2007)
19.	PICRUST	Used to connects taxonomic classifying metaprofiling results	Langille et al. (2013)

(continued)

Table 6.2 (continued)

S. No.	Software	Function of software	References
20.	MetaWatt	For metagenomic assembly, contig clustering or binning, and bin inspection for taxonomic signatures (through BLAST) and sequence coverage.	Strous et al. (2012)
21.	BioEdit	Software of biological sequence alignment editor	Hall (2001)
22.	FragGene Scan	Used to predicts fragments of gene from short reads	Rho et al. (2010)
23.	PIPITS	Used for processing of ITS amplicons	Gweon et al. (2015)
24.	EDGE	Software comprising QC, annotation, Assembly, binning, taxonomic profiling, and phylogenetic tree construction	Li et al. (2014)
25.	USEARCH	Open-source software	Edgar and Flyvbjerg (2015)
26.	VSEARCH	Open-source software	Rognes et al. (2016)
27.	EBI Metagenomics	Software used for data trimming and duplicates removal	Hunter et al. (2014)
28.	qRT-PCR	Real-time quantitative reverse transcriptase PCR	Lee et al. (2014)

6.5 Bacterial Community Structure at Different Level

There are several bacterial communities which present at different locations on geological areas of soil like some are associated with root endophytes, in phyllosphere, endorhizosphere, bulk soil, flooded and non-flooded soil, irrigated soil (Singh et al. 2020). Some bacterial communities are survived in high prevalence of *Magnaporthe oryzae* (Rice blast Fungus) and some in high incidence of salt.

6.5.1 Bacterial Community Composition Associated with Root Endophytes

Previously, research was conducted to investigate the microbial community structure of Indian rice root endophytes (Sengupta et al. 2017). Vittorio et al. used 16S rRNA taxonomy profiling of the rhizosphere and endorhizosphere of two high-yielding rice cultivars, Pionero 2010 FL and DANAC 6D 20A, which were cultivated intensively in Venezuela. Three Pionero 2010FL rhizosphere soil samples and three DANAC SD 20A rhizosphere soil samples were taken from Association of Certified Seed Producers of Western Plains paddy fields after 88 days of planting. After analyzing the complete rhizospheric and endorhizospheric bacterial community structure, they retrieved 326,496 bacterial readings. Proteobacteria accounted for 70–87% of all OTUs in the bacterial microbiota. Gammaproteobacteria was the most numerous

Proteobacteria class, followed by Alphaproteobacteria and Betaproteobacteria. Deltaproteobacteria and epsilonproteobacteria, on the other hand, were not found in the endorhizosphere. The colony of Acidobacteria and Nitrospirae was exclusively found in the rhizosphere. The phylum Cyanobacteria was also abundant in rhizospheric soil. Bacteriodetes and Verrucomicrobia abundant in Pionero 2010 FL. Caulobacter genus was significant and massively abundant in both rhizosphere and endorhizosphere soil sample.

Number of the researches have performed 16S rRNA gene amplification of hypervariable V3-V4 region and was amplified using primers set Pro341F and Pro805R (Takahashi et al. 2014; Merkel et al. 2019; Cichocki et al. 2020). They were collected rhizosphere and bulk soil sample from seven different areas of India. They obtained 28 phyla from all groups of bacteria. Among them the most dominant phyla were Proteobacteria (25.7 ± 14%) followed by Fermicutes (21 ± 8.7%) then Actinobacteria (16.7 ± 6%) and Acidobacteria (13 ± 10%). Candidatus koribacter (8 ± 19%) was most abundant genus in rhizosphere soil while Ktedonbacter (13%) most frequently detected in bulk soil sample. Furthermore, 18 methanogen genera were detected in all samples of rhizospheric and bulk soil (Lee et al. 2015). Most abundant genera of methanogen were detected Methanosaeta, followed by Methanobacterium and Methaocella (Rahalkar et al. 2016). Archaeal genera including type I and type II methanotrophs were significantly detected throughout the cultivation (Singh et al. 2016).

6.5.2 *Flooded and Non-flooded Located Bacterial Community*

Multivariate analysis revealed considerable differences between the sites when comparing the taxonomic patterns of the bacterial communities. Ivan et al. studied V3-V5 hypervariable region of 16S rRNA amplicon sequencing using Miseq Illumina platform (Ebersberg, Germany). At Zlato Pole, soil samples collected from flooded and non-flooded rice paddies, as well as sediments and non-flooded areas. Rice paddies are being located in wetlands along the Bulgarian side of the Maritza River, such as the Zlato Pole wetland and the Tsalapitsa paddy fields. After filtering of bacterial reads and OUT picking process, 181,328 sequences were obtained from flooded samples and 158,260 samples were obtained from non-flooded samples. Total 117 bacterial classes were identified among them 67 were detected in all soil samples. Proteobacteria (34%) in Plovdiv rice paddy sediments to (68%) in Zlato Pole sediments of all bacterial sequences. Alphaproteobacteria (21%) is the most common, followed by Gammaproteobacteria (13%), Betaproteobacteria (6.8%), and Deltaproteobacteria (4%). Moreover, abundant phyla were Actinobacteria (8–26%) and Acidobacteria (2–17%) detected the third most abundant phylum while Firmicutes (9%) and Bacteriodetes (8%) detected over Acidobacteria in Zlato pole sediments.

6.5.3 *Community Structure in Rhizosphere and Phyllosphere*

In Faisalabad, Pakistan, a comparison of 16S rRNA gene amplification studies of bacterial phyla in the rhizosphere and phyllosphere revealed that the rhizosphere had more diversity than the phyllosphere. According to reports, a total of 9383 16S rRNA sequences were retrieved from rhizospheric soil while 54,714 sequences were retrieved from Basmati rice phyllospheric soil (Yasmin et al. 2020). Eighteen different phyla detected from rhizosphere while seven phyla were from phyllosphere soil sample. Seven phyla were found in both compartments. Proteobacteria were most abundant phyla from both the compartments i.e., rhizosphere (37%) while phyllosphere (80%) followed by Firmicutes (10%), Bacteriodetes (9%), Chloroflexi (4%), Actinobacteria (1%) in phyllospheric soil sample. According to 16S rRNA gene amplification analysis was detected 208 different genera from rhizosphere while 24 genera from phyllosphere soil samples. In the bacterial community's rhizosphere and phyllosphere, 15 genera were determined to be common. Bacillariophyla (22%) was the most common genus in the phyllosphere, followed by Sphingomonas (9%), and Bradyrhizobium (7%). The most frequent genus in the rhizospheric soil sample was Thaurea (4%).

16S rRNA sequencing retrieved from database found total 12 representative clones from the paddy field rhizosphere soil in Kuttanand, Kerala (Arjun 2011). The dominant taxa in the library were found to be Proteobacteria (7/12) followed by Firmicutes (2/12), Bacteriodetes (2/12), and Acidobacteria (1/12). About 70–90% of total OTUs, Proteobacteria was dominated the bacterial microbiota. Gammaproteobacteria was the most important Proteobacteria phylum, followed by Alphaproteobacteria and Betaproteobacteria. In the endorhizosphere, deltaproteobacteria and epsilonproteobacteria were not found. Only the colony of Acidobacteria and Nitrospirae was found in the rhizosphere. Along with Cyanobacteria phylum was enriched in rhizospheric soil. Bacteriodetes and Verrucomicrobia abundant in Pionero 2010 FL. Caulobacter genus was significant and exclusively abundant rhizosphere as well as endorhizosphere (Sengupta et al. 2017).

6.5.4 *Bacterial Composition in Areas with High* **Magnaporthe oryzae** *Prevalence and High Salt Incidence*

Proteobacteria, Acidobacteria, and Anaerolineae were detected in all four areas, according to 16S rRNA sequencing of total DNA from the four regions. Proteobacteria (Betaproteobacteria) was most abundantly detected group followed by Bacteriodetes and Chloroflexi. Thermococci class archaea were identified in locations with high Magnaporthe oryzae incidence, while Sphingnobacteria class archaea were identified in areas of high salt incidence. The Verrucomicrobiae class, on the other hand, was only found in the control region (Lucas et al. 2013). The rhizosphere has a higher absolute abundance of Archaea than the bulk soil sample. For Archaea, Methanosarcina and Methanosaeta were found more abundant in rhizospheric soil of Vercelli (Breidenbacht et al. 2016). Abundantly present genera such as Acidobacteria, Alphaproteobacteria, Betaproteobacteria, Cyanobacteria, Chloroflexi, Deltaproteobacteria, Firmicutes. Potential iron reducer (e.g., *Geobacter* and *Anaeromyxobacter*) (Conrad and Frenzel 2002; Hori et al. 2010). Fermenters (e.g., Clostridia and Opitutus) and endophytic plant growth promoting bacteria (e.g., *Herbaspirillum* species) are also more abundant in the rhizospheric soil (Andreesen and Schaupp 1973; Chin et al. 2001). Furthermore, 18 methanogen genera were detected in all samples of rhizospheric and bulk soil (Lee et al. 2015). Most abundant genera of methanogen were detected Methanosaeta, followed by Methanobacterium and Methaocella (Rahalkar et al. 2016). Archaeal genera belong to type I and type II methanotrophs were found in entire cultivation (Singh et al. 2016). Total 117 bacterial classes were identified among them 67 were detected in all soil samples. Proteobacteria (34.2%) in Plovdiv rice paddy sediments to (68%) in Zlato Pole sediments of all bacterial sequences. Alphaproteobacteria (21%) is the most common, followed by Gammaproteobacteria (13%), Betaproteobacteria (7%), and Deltaproteobacteria (4%). Moreover, abundant phyla were Actinobacteria (8–26%) and Acidobacteria (2–17%) detected the third most abundant phylum while Firmicutes (9%) and Bacteriodetes (8%) detected over Acidobacteria in Zlato pole sediments (Ivan et al. 2019).

6.6 Future Perspective

We're working hard to figure out which bacterial genera are invading the rice rhizosphere. From this review article, we conclude that among all bacterial community in different samples from different locations most abundant phyla were detected Proteobacteria in rhizosphere soil samples followed by Acidobacteria then Actinobacteria, followed by Choroflexi and Firmicutes. Methylobacterium was detected as most dominant genus from Methylotrophs. Archaea were predominantly found in rhizosphere bulk soil, flooded soil, and wetland soil samples. Methanogenic archaea are also found in some rhizospheric soil samples. Streptomyces were detected from agroecosystem (irrigated rice and swampy rice) of rice plants. Furthermore, analyzing the structure of microbial communities is required in order to investigate the individual functions of bacteria. This understanding and insights aid in the development of methods for greater crop production, improved soil quality, and disease-causing microorganism protection in order to preserve natural resources and, ultimately, to produce more sustainable agricultural production.

We may choose these succeeding strains for formulation of a suitable inoculant as a biocontrol agent for administration in the rhizosphere of rice and disease management of rice plants due to decreased efficacy of natural nutrients available in soil. Biocontrol presumes special connotation being an environmental-friendly and cost-efficient strategy which can be used for effective rice disease management. Numerous microbial species are acts as a biocontrol agent against many plant pathogens. As a result, it is an inevitable step to gather as much microbial diversity as possible in order to provide a higher level of protection while retaining rice yields.

References

Abulencia CR et al (2006) Environmental whole-genome amplification to access microbial populations in contaminated sediments. Appl Environ Microbiol 72:3291

Aguiar-Pulido V, Huang W, Suarez-Ulloa V, Cickovski T, Mathee K, Narasimhan G (2016a) Metagenomics, metatranscriptomics, and metabolomics approaches for microbiome analysis. Evol Bioinformatics Online 12(Suppl 1):5

Aguiar-Pulido V et al (2016b) Metagenomics, metatranscriptomics, and metabolomics approaches for microbiome analysis: supplementary issue: bioinformatics methods and applications for big metagenomics data. Evol Bioinforma 12:EBO–S36436

Alneberg J, Bjarnason BS, de Bruijn I, Schirmer M et al (2014) Binning metagenomic contigs by coverage and composition. Nat Methods 11:1144–1146. https://doi.org/10.1038/nmeth.3103

Anderson NL, Anderson NG (1998) Proteome and proteomics: new technologies, new concepts and new words. Electrophoresis 19(11):1853–1861. https://doi.org/10.1002/elps.1150191103

Andreesen JR, Schaupp A (1973) Fermentation of glucose, fructose, and xylose by *Clostridium thermoaceticum*: effect of metals on growth yield, enzymes, and the synthesis of acetate from CO_2. J Bacteriol 114:743–751

Angers D, Caron J (1998) Plant induced changes in soil structure: processes and feedbacks. Biogeochemistry 42:55–72. https://doi.org/10.1023/A:1005944025343

Anwar MN, Li ZF, Gong Y, Singh RP, Li YZ (2019) Omics studies revealed the factors involved in the formation of colony boundary in *Myxococcus xanthus*. Cell 8(6):530

Arjun JK (2011) Metagenomic analysis of bacterial diversity in the rice rhizosphere soil microbiome. Biotechnol Bioinf Bioeng 1(3):361–367

Blackstock WP, Weir MP (1990) Proteomics: quantitative and physical mapping of cellular proteins. Trends Biotechnol 17(3):121–127. https://doi.org/10.1016/S0167-7799(98)01245-1

Boisvert S, Raymond F, Godzaridis E, Laviolette F, Corbeil J (2012) Ray Meta: scalable *de novo* metagenome assembly and profiling. Genome Biol 13:R122. https://doi.org/10.1186/gb-2012-13-12-r122

Breidenbach B, Conrad R (2015) Seasonal dynamics of bacterial and archaeal methanogenic communities in flooded rice fields and effect of drainage. Front Microbiol 5:752. https://doi.org/10.3389/fmicb.2014.00752

Breidenbach B, Pump J, Dumont MG (2016) Microbial community structure in the rhizosphere of rice plants. Front Microbiol 6. https://doi.org/10.3389/fmicb.2015.01537

Brune A, Frenzel P, Cypionka H (2000) Lifeattheoxic–anoxicinterface: microbial activities and adaptations. FEMS Microbiol Rev 24:691–710. https://doi.org/10.1016/S0168-6445(00)00054-1

Caporaso JG, Kuczynski J, Stombaugh J, Bittinger K, Bushman FD, Costello EK, Fierer N, Peña AG, Goodrich JK, Gordon JI, Huttley GA, Kelley ST, Knights D, Koenig JE, Ley RE, Lozupone CA, McDonald D, Muegge BD, Pirrung M, Reeder J, Sevinsky JR, Turnbaugh PJ, Walters WA, Widmann J, Yatsunenko T, Zaneveld J, Knight R (2010) QIIME allows analysis of high-throughput community sequencing data. Nat Methods 7(5):335–336. https://doi.org/10.1038/nmeth.f.303

Chin KJ, Liesack W, Janssen PH (2001) *Opitutus terraegen* nov., sp. nov., to accommodate novel strains of the division "Verrucomicrobia" isolated from rice paddy soil. Int J Syst Evol Microbiol 51:1965–1968. doi: https://doi.org/10.1099/00207713-51-6-1965

Chu H, Gao G-F, Ma Y, Fan K, Delgado-Baquerizo M, Shade A (2020) Soil microbial biogeography in a changing world: recent advances and future perspectives. mSystems 5(2). https://doi.org/10.1128/mSystems.00803-19

Cichocki N, Hübschmann T, Schattenberg F, Kerckhof FM, Overmann J, Müller S (2020) Bacterial mock communities as standards for reproducible cytometric microbiome analysis. Nat Protoc 15 (9):2788–2812

Conrad R, Frenzel P (2002) Flooded soils. In: Britton G (ed) Encylopedia of environmental microbiology. John Wiley & Sons, New York, NY, pp 1316–1333

Cox MP, Peterson DA, Biggs PJ (2010) SolexaQA: at-a-glance quality assessment of Illumina second-generation sequencing data. BMC Bioinformatics 11:485. https://doi.org/10.1186/1471-2105-11-485

Curtis TP, Sloan WT, Scannell JW (2002) Estimating prokaryotic diversity and its limits. Proc Natl Acad Sci 99(16):10494–10499. https://doi.org/10.1073/pnas.142680199

Delcher AL, Bratke KA, Powers EC, Salzberg SL (2007) Identifying bacterial genes and endosymbiont DNA with Glimmer. Bioinformatics 23:673–679. https://doi.org/10.1093/bioinformatics/btm009

Edgar RC (2013) UPARSE: highly accurate OUT sequences from microbial ampliconreads. Nat Methods 10:996–1000. https://doi.org/10.1038/nmeth.2604

Edgar RC, Flyvbjerg H (2015) Error filtering, pair assembly and error correction for next-generation sequencing reads. Bioinformatics 31(21):3476–3482. https://doi.org/10.1093/bioinformatics/btv401

Elias F, Woyessa D, Muleta D (2016) Phosphate solubilization potential of rhizosphere fungi isolated from plants in Jimma zone, Southwest Ethiopia. Int J Microbiol 2016:5472601

Filippo CD, Ramazzotti M, Fontana P, Cavalieri D (2012) Bioinformatic approaches for functional annotation and pathway inference in metagenomics data. Brief Bioinform 13:696–710

Gans J (2005) Computational improvements reveal great bacterial diversity and high metal toxicity in soil. Science 309(5739):1387–1390. https://doi.org/10.1126/science.1112665

Gautam SS, Rajendra KC, Leong KWC, Aogáin MM, O'Toole RF (2019) A step-by-step beginner. J Biol Methods 6(1):e110. https://doi.org/10.14440/jbm.2019.276

Gerlach W, Jünemann S, Tille F, Goesmann A, Stoye J (2009) WebCARMA: a web application for the functional and taxonomic classification of unassembled metagenomic reads. BMC Bioinformatics 10(1). https://doi.org/10.1186/1471-2105-10-430

Glass EM, Wilkening J, Wilke A, Antonopoulos D, Meyer F (2010) Using the metagenomics RAST server (MG-RAST) for analyzing shotgun metagenomes. Cold Spring Harb Protoc 5:1–11. https://doi.org/10.1101/pdb.prot5368

Gweon HS, Oliver A, Taylor J, Booth T, Gibbs M, Read DS, Griffiths RI, Schonrogge K (2015) PIPITS: an automated pipeline for analyses of fungal internal transcribed spacer sequences from the Illumina sequencing platform. Methods Ecol Evol 6(8):973–980. https://doi.org/10.1111/2041-210X.12399

Hall T (2001) BioEdit version 5.0.6. Deaprtment of Microbiology, North Carolina State University

Hori T, Müller AI, Garashi Y, Conrad R, Friedrich MW (2010) Identification of iron-reducing microorganisms in anoxic rice paddy soil by 13C-acetate probing. ISME J 4:267–278. https://doi.org/10.1038/ismej.2009.100

Hunter S, Corbett M, Denise H, Fraser M, Gonzalez-Beltran A, Hunter C, Jones P, Leinonen R, McAnulla C, Maguire E, Maslen J (2014) EBI metagenomics—a new resource for the analysis and archiving of metagenomic data. Nucleic Acids Res 42(D1):D600–D606

Huson DH, Weber N (2013) Microbial community analysis using MEGAN. Methods Enzymol 531:465–485. https://doi.org/10.1016/B978-0-12-407863-5.00021-6

Hussain Q, Liu Y, Zang A, Li L, Zhang X, Jin Z (2012) Microbial community dynamics and function associated with rhizosphere over periods of rice growth. Plant Soil Environ 58:55–61

Imchen M, Kumavath R, Barh D, Azevedo V, Ghosh P, Viana M, Wattam AR (2018) Author correction: searching for signatures across microbial communities: metagenomic analysis of soil samples from mangrove and other ecosystems. Sci Rep 8(1). https://doi.org/10.1038/s41598-017-18550-0

Ivan I, Mariana M, Sonya K, Velizar G, Marinela T, Angelina I, Galina Y, Vesselin B (2019) Metagenomic analysis of the microbial community structure in protected wetlands in the Maritza River Basin. Biotechnol Biotechnol Equip 33(1):1721–1732. https://doi.org/10.1080/13102818.2019.1697364

Jin Z, Di Rienzi SC, Janzon A, Werner JJ, Angenent LT, Dangl JL, Foweler DM, Ley RE (2016) Novel rhizosphere soil alleles for enzyme 1 – aminoacyclopropane-1-carboxylate deaminase queried for function with an in vivo competition assay. Appl Environ Microbiol 82:1050–1059

Jones DL, Nguyen C, Finlay RD (2009) Carbon flow in the rhizosphere: carbon trading at the soil–root interface. Plan Soil 321:5–33. https://doi.org/10.1007/s11104-009-9925-0

Knief C, Delmotte N, Chaffron S, Stark M, Innerebner G, Wassmann R, von Mering C, Vorholt JA (2012) Metaproteogenomic analysis of microbial communities in the phyllosphere and rhizosphere of rice. ISME J 6(7):1378–1390. https://doi.org/10.1038/ismej.2011.192

Kowalchuk GA, Yergeau E, Leveau JHJ, Sessitsch A, Bailey M (2010) Plant-associated microbial communities. In: Liu W-T, Jansson JK (eds) Environmental molecular microbiology. Caister Academic Press, Poole, pp 131–148

Kunin V, Copeland A, Lapidus A, Mavromatis K, Hugenholtz P (2008) A bioinformatician's guide to metagenomics. Microbiol Mol Biol Rev 72:557–578

Kuzyakov Y (2002) Review: factors affecting rhizosphere priming effects. J Plant Nutr Soil Sci 165:382–396

Kuzyakov Y, Domanski G (2000) Carbon input by plants into the soil. Review. J Plant Nutr Soil Sci 163(4):421–431. https://doi.org/10.1002/1522-2624(200008)163:4<421::AID-JPLN421>3.0.CO;2-R

Langille MG, Zaneveld J, Caporaso JG, McDonald D, Knights D, Reyes JA, Clemente JC, Burkepile DE, Thurber RLV, Knight R, Beiko RG (2013) Predictive functional profiling of microbial communities using 16S rRNA marker gene sequences. Nat Biotechnol 31(9):814–821

Lee HJ, Kim SY, Kim PJ, Madsen EL, Jeon CO (2014) Methane emission and dynamics of methanotrophic and methanogenic communities in a flooded rice field ecosystem. FEMS Microbiol Ecol 88:195–212. https://doi.org/10.1111/1574-6941.12282

Lee HJ, Jeong SE, Kim PJ, Madsen EL, Jeon CO (2015) High resolution depth distribution of Bacteria, Archaea, methanotrophs, and methanogens in the bulk and rhizosphere soils of a flooded rice paddy. Front Microbiol 6:639. https://doi.org/10.3389/fmicb.2015.00639

Li D, Liu CM, Luo R, Sadakane K, Lam TW (2015) MEGAHIT: an ultra-fast single-node solution for large and complex metagenomics assembly via succinct de Bruijn graph. Bioinformatics 31:1674–1676. https://doi.org/10.1093/bioinformatics/btv033

Lu Y, Rosencrantz D, Liesack W et al (2006) Structure and activity of bacterial community inhabiting rice roots and the rhizosphere. Environ Microbiol 8:1351–1360

Lucas JA, García-Villaraco A, Ramos B, García-Cristobal J, Algar E, Gutierrez-Mañero J (2013) J Appl Microbiol 115(1):218–235. https://doi.org/10.1111/jam.12225

Lynch JM, Whipps JM (1990) Substrate flow in the rhizosphere. Plant S 129:1–10. https://doi.org/10.1007/BF00011685

Maheshwari NK, Singh RP, Manchanda G, Dubey RC, Maheshwari DK (2021) Sunn Hemp (*Srotalaria juncea*) nodulating bacteria capable for high antagonistic potential and plant growth promotion attributes. J Microbiol Biotechnol Food Sci 10(3):385–389. https://doi.org/10.15414/jmbfs.2020.10.3.385-389

Majeed A, Abbasi KM, Hameed S, Imran A, Rahim N (2015) Isolation and characterization of plant growth promoting rhizobacteria from wheat rhizosphere and their effect on plant growth promotion. Front Microbiol 6:198

Maria R, Sumera Y, Sughra H, Ahmad Z, Babur M, Sajjad M (2020) Metagenomic analysis of bacterial community associated with rhizosphere and phyllosphere of basmati rice. bioRxiv:2020.04.09.034009. https://doi.org/10.1101/2020.04.09.034009

Marschner P, Yang CH, Lieberei R, Crowley DE (2001) Soil and plant specific effects on bacterial community composition in the rhizosphere. Soil Biol Biochem 33:1437–1445

Martina K, Jan K, Tamas F, Ladislav C, Marek O, Genevieve LG, Yvan ML, Marketa SM (2008) Development of a 16S rRNA genebased prototype microarray for the detection of selected actinomycetes genera. Antonie Van Leeuwenhoek 94:439e53

Massart S, Martinez-Medina M, Jijakli MH (2015) Biological control in the microbiome era: challenges and opportunities. Biol Control 89(98):108

Mendes R, Garbeva P, Raaijmakers JM (2013) The rhizosphere microbiome: significance of plant beneficial, plant pathogenic, and human pathogenic microorganisms. FEMS Microbiol Rev 37:634–663. https://doi.org/10.1111/1574-6976.12028

Merkel AY, Tarnovetskii IY, Podosokorskaya OA, Toshchakov SV (2019) Analysis of 16S rRNA primer systems for profiling of thermophilic microbial communities. Microbiology 88 (6):671–680

Mikheenko A, Saveliev V, Gurevich A (2016) MetaQUAST: evaluation of metagenome assemblies. Bioinformatics 32:1088–1090. https://doi.org/10.1093/bioinformatics/btv697

Namiki T, Hachiya T, Tanaka H, Sakakibara Y (2012) MetaVelvet: an extension of Velvet assembler to *de novo* metagenome assembly from short sequence reads. Nucleic Acids Res 40:e155. https://doi.org/10.1093/nar/gks678

Nimnoi P, Pongsilp N, Lumyong S (2010) Genetic diversity and community of endophytic actinomycetes within the roots of Aquilaria crassna Pierre ex Lex assessed by actinomycetes-specific PCR and PCR-DGGE of rRNA gene. Biochem Syst Ecol 38:595e601

Noguchi H, Taniguchi T, Itoh T (2008) MetaGeneAnnotator: detecting species-specific patterns of ribosomal binding for precise gene prediction in anonymous prokaryotic and phage genomes. DNA Res 15:387–396. https://doi.org/10.1093/dnares/dsn027

Oksanen J, Blanchet FG, Kindt R, Legendre P, Minchin PR, O'Hara RB et al (2013) Vegan: community ecology package. R Package Version 2.0-10. http://cran.r-project

Pell J, Hintze A, Canino-Koning R, Howe A, Tiedje JM, Brown CT (2012) Scaling metagenome sequence assembly with probabilistic de Bruijn graphs. Proc Natl Acad Sci U S A 109:13272–13277. https://doi.org/10.1073/pnas.1121464109

Peng Y, Leung HC, Yiu SM, Chin FY (2012) IDBA-UD: a *de novo* assembler for single-cell and metagenomic sequencing data with highly uneven depth. Bioinformatics 28:1420–1428. https://doi.org/10.1093/bioinformatics/bts174

Pereira SMC, Schloter-Hai B, Schloter M, Elsas JDV, Salles JF (2013) Temporal dynamics of abundance and composition of nitrogen fixing communities across agricultular soils. PLoS One 8(9):e74500

Ponomarova O, Patil KR (2015) Metabolic interactions in microbial communities: untangling the Gordian knot. Curr Opin Microbiol 27:37–44

Prajakta BM, Suvarna PP, Raghvendra SP, Alok RR (2019) Potential biocontrol and superlative plant growth promoting activity of indigenous *Bacillus mojavensis* PB-35 (R11) of soybean (*Glycine max*) rhizosphere. SN Appl Sci 1(10):1

Rahalkar MC, Pandit PS, Dhakephalkar PK, Pore S, Arora P, Kapse N (2016) Genome characteristics of a novel type I methanotroph (Sn10-6) isolated from a flooded Indian rice field. Microb Ecol 71:519–523. https://doi.org/10.1007/s00248-015-0699-z

Reichardt W, Mascarina G, Padre B, Doll J (1997) Microbial communities of continuously cropped, irrigated rice fields. Appl Environ Microbiol 63:233–238

Rho M, Tang H, Ye Y (2010) FragGeneScan: predicting genes in short and error-prone reads. Nucleic Acids Res 38:e191. https://doi.org/10.1093/nar/gkq747

Rognes T, Flouri T, Nichols B, Quince C, Mahé F (2016) VSEARCH: a versatile open source tool for metagenomics. PeerJ 4:e2584. https://doi.org/10.7717/peerj.2584

Rusmana MI, Lestari Y (2015) Metagenomic of actinomycetes based on 16S rRNA and nifH genes in soil and roots of four Indonesian rice cultivars using PCR-DGGE. HAYATI J Biosci 22 (3):113–121. https://doi.org/10.1016/j.hjb.2015.10.001

Schloss PD, Westcott SL, Ryabin T, Hall JR, Hartmann M, Hollister EB et al (2009) Introducing mothur: open source, platform independent, community supported software for describing and comparing microbial communities. Appl Environ Microbiol 75:7537–7541. https://doi.org/10.1128/AEM.01541-09

Seemann T (2014) Prokka: rapid prokaryotic genome annotation. Bioinformatics 30:2068–2069. https://doi.org/10.1093/bioinformatics/btu153

Sengupta S, Ganguli S, Singh PK (2017) Metagenome analysis of the root endophytic 467 microbial community of Indian rice (*O. sativa* L.). Genom Data 12:41–43

Sinclair L, Osman OA, Bertilsson S, Eiler A, Orlando L (2015) Microbial community composition and diversity via 16S rRNA gene amplicons: evaluating the illumina platform. PLoS One 10(2): e0116955. https://doi.org/10.1371/journal.pone.0116955

Singh RP, Manchanda G, Singh RN, Srivastava AK, Dubey RC (2016) Selection of alkalotolerant and symbiotically efficient chickpea nodulating rhizobia from North-West Indo Gangetic Plains. J Basic Microbiol 56:14–25. https://doi.org/10.1002/jobm.201500267

Singh RN, Singh RP, Sharma A, Saxena AK (2016) Modeling of PrnD protein from Pseudomonas fluorescens RajNB11 and its comparative structural analysis with PrnD proteins expressed in Burkholderia and Serratia. Turk J Biol 40:623–633. https://doi.org/10.3906/biy-1501-4

Singh RP, Manchanda G, Maurya IK, Maheshwari NK, Tiwari PK, Rai AR (2019) Streptomyces from rotten wheat straw endowed the high plant growth potential traits and agro-active compounds. Biocatalysis Agric Biotechnol 17:507–513. https://doi.org/10.1016/j.bcab.2019.01.014

Singh R, Manchanda G, Maurya I, Wei Y (eds) (2020) Microbial versatility in varied environments. Springer, Singapore. https://doi.org/10.1007/978-981-15-3028-9

Spence AK, Boddu J, Wang D, James B, Swaminathan K, Moose SP, Long SP (2014) Transcriptional responses indicate maintenance of photosynthetic proteins as key to the exceptional chilling tolerance of C4 photosynthesis in Miscanthus × giganteus. J Exp Bot 65(13):3737–3747. https://doi.org/10.1093/jxb/eru209

Strous M, Kraft B, Bisdorf R, Tegetmeyer HE (2012) The binning of metagenomic contigs for microbial physiology of mixed cultures. Front Microbiol 3:410. https://doi.org/10.3389/fmicb.2012.00410

Subhashini DV, Singh RP (2014) Isolation of endophytic actinomycetes from roots and leaves of tobacco (*Nicotiana tabacum* L.). Annal Plant Pro Sci 22(2):458–459

Subhashini DV, Singh RP, Manchanda G (2017) OMICS approaches: tools to unravel microbial systems. Directorate of Knowledge Management in Agriculture, Indian Council of Agricultural Research, New Delhi. ISBN: 9788171641703. https://books.google.co.in/books?id=vSaLtAEACAAJ

Taj ZZ, Rajkumar M (2016) Perspectives of plant growth-promoting actinomycetes in heavy metal phytoremediation. In: Plant growth promoting actinobacteria. Springer, Singapore, pp 213–231

Takahashi S, Tomita J, Nishioka K, Hisada T, Nishijima M (2014) Development of a prokaryotic universal primer for simultaneous analysis of bacteria and archaea using next generation sequencing. PLoS One 9:e105592. https://doi.org/10.1371/journal.pone.0105592

Tamura K, Dudley J, Nei M, Kumar S (2007) Mega 4, a molecular evolutionary genetic analysis MEGA Software version 4.0. Mol Biol Evol 24:596–1599

Treangen TJ, Koren S, Sommer DD, Liu B, Astrovskaya I, Ondov B et al (2013) MetAMOS: a modular and open source metagenomic assembly and analysis pipeline. Genome Biol 14:R2. https://doi.org/10.1186/gb-2013-14-1-r2

Tsurumaru H, Okubo T, Okazaki K, Hashimoto M, Kakizaki K, Hanzawa E, Hiroyuki T, Noriyuki A, Fukuyo T, Yasuyo S, Seishi I, Minamisawa K (2015) Metagenomic analysis of the bacterial community associated with the taproot of sugar beet. Microbes Environ 30 (1):63–69

Vacheron J, Desbrosses G, Bouffard ML, Touraine B et al (2013) Plant growth-promoting rhizobacteria and root system functioning. Front Plant Sci 4:356. https://doi.org/10.3389/fpls.2013.00356

Vaksmaa A, van Alen TA, Ettwig KF, Lupotto E, Valè G, Jetten MS et al (2017) Stratification of diversity and activity of methanogenic and methanotrophic microorganisms in a nitrogen-fertilized Italian paddy soil. Front Microbiol 8:2127. https://doi.org/10.3389/fmicb.2017.02127

Venturi V, Subramoni S, Sabag-Daigle A, Ahmer BMM (2018) Methods to study solo/orphan quorum-sensing receptors. Methods Mol Biol 1673:145–159. https://doi.org/10.1007/978-1-4939-7309-5_12

Wang H, Zeng Y, Guo C, Bao Y, Lu G, Reinfelder JR et al (2018) Bacterial, archaeal, and fungal community responses to acid mine drainageladen pollution in a rice paddy soil ecosystem. Sci Total Environ 616:107–116. https://doi.org/10.1016/j.scitotenv.2017.10.224

Witt C, Biker U, Galicia CC, Ottow JCG (2000) Dynamics of soil microbial biomass and nitrogen availability in a flooded rice soil amended with different C and N sources. Biol Fertil Soil 30:520–527

Yang YJ, Lin W, Singh RP, Xu Q, Chen Z, Yuan Y, Zou P, Li Y, Zhang C (2019) Genomic, transcriptomic and enzymatic insight into lignocellulolytic system of a plant pathogen *Dickeya* sp. WS52 to digest sweet pepper and tomato stalk. Biomolecule 9(12):753

Yang Y, Singh RP, Song D, Chen Q, Zheng X, Zhang C, Zhang M, Li Y (2020a) Synergistic effect of *Pseudomonas putida* II-2 and *Achromobacter* sp. QC36 for the effective biodegradation of the herbicide quinclorac. Ecotoxicol Environ Saf 188:109826

Yang Y, Liu L, Singh RP, Meng C, Ma S, Jing C, Li Y, Zhang C (2020b) Nodule and root zone microbiota of salt-tolerant wild soybean in coastal sand and saline-alkali soil. Front Microbiol 11:2178. https://doi.org/10.3389/fmicb.2020.523142

Yasmin S, Hakim S, Zaheer A, Mirza B, Mirza MS (2020) Metagenomic analysis of bacterial community associated with rhizosphere and phyllosphere of basmati rice. BioRxiv:1. https://doi.org/10.1101/2020.04.09.034009

Zeng LS, Liao M, Chen CL, Huang CY (2005) Variation of soil microbial biomass and enzyme activities at different growth stages of rice (*Oryza sativa*). Rice Sci 12:283–288

Zerbino DR, Birney E (2008) Velvet: algorithms for *de novo* short read assembly using de Bruijn graphs. Genome Res 18:821–829. https://doi.org/10.1101/gr.074492.107

Zhang JJ, Jing XY, de Lajudie P, Ma C, He PX, Singh RP, Chen WF, Wang ET (2016) Association of white clover (*Trifolium repens* L.) with rhizobia of sv. *trifolii* belonging to three genomic species in alkaline soils in North and East China. Plant Soil 407(1):417–427

Zhang J, Wang ET, Singh RP, Guo C, Shang Y, Chen J, Liu C (2019) Grape berry surface bacterial microbiome: impact from the varieties and clones in the same vineyard from central China. J Appl Microbiol 126(1):204–214

Chapter 7
Diversity and Application of Heavy-Metal Resistant Microbes

Armine Margaryan

Abstract Metal-rich natural and artificial habitats are extreme environments for the evolution of unique microbial communities, which have adapted to deal with the toxic levels of the metals. Diverse microbial groups belonging to Archaea and Bacteria domain possessing different metal-resistance strategies have been found in different metal-contaminated environments using cultivation and molecular approaches. Various metal-resistant bacteria belonging to *Bacillus*, *Arthrobacter*, *Pseudomonas*, *Ralstonia*, *Stenotrophomonas*, *Desulfovibrio*, and other genera were demonstrated a high capacity to the bisorbtion of the different heavy metals. Bacteria and archaea belonging to the genera *Acidithiobacillus*, *Leptospirillum*, *Sulfobacillus*, and *Ferroplasma* are mostly associated with metal minerals and are involved in the bioleaching processes. Thus, the microbial resistance to toxic heavy metals has fundamental importance in the bioremediation of metal-contaminated natural habitats and bioleaching of valuable metals from complex minerals.

Keywords Heavy metals · Metal resistant bacteria and archaea · Microbial diversity · Bioremediation · Bioleaching

7.1 Environmental Contamination by Heavy Metals

Natural biotopes with unusually high levels of heavy metals are widely found in our ecosystem. For instance, the copper concentration in the soils of the Valparaiso region (Chile), where the copper mine is located, reaches from 379 to 784 mg/kg, and in the sediments of the sea off the coast of Chile, the copper content reaches 1530 mg/kg (Besaury et al. 2013a, b; Altimira et al. 2012). The sediment of Lake Torch (Michigan, USA), as a result of crushed mine ore disposition over 100 years, contain high concentrations of copper: on the top 10 cm of sediment layer is around

A. Margaryan (✉)
Department of Biochemistry, Microbiology, and Biotechnology, Faculty of Biology, Yerevan State University, Yerevan, Armenia
e-mail: arminemargaryan@ysu.am

R. P. Singh et al. (eds.), *Microbes in Microbial Communities*,
https://doi.org/10.1007/978-981-16-5617-0_7

2000 ppm, and in deep areas, it can reach up to 5500 ppm, followed by decrease up to 1500 ppm. At different sites of the river high levels of nickel, chromium, and cobalt were found also. Iron concentrations vary from 20.981 to 88.717 ppm (the average concentration was 36.772–36.749 ppm) (Konstantinidis et al. 2003). The concentration of arsenic in Genola Warm, Saratoga Springs, and Diamond Fork Hot (Utah, USA) reaches 0.046 mg/L, 0.063 mg/L, and 0.15 mg/L, correspondingly (Rey et al. 2009). In the geothermal soils of Monterotondo Marittimo (Tuscany, Italy) the high concentration of arsenic (4–8 mg/kg) and antimony (0.4–0.9 mg/kg) was registered (Cuebas et al. 2011a, b). The mass-spectrometry analysis showed a high concentration of Fe, Cu, and Zn (6.55 mg/L, 0.498 mg/L and 5.87 mg/L, correspondingly) in the water samples of the river flowing next to Akhtala tailing dump (Akhtala town, Lori province, Armenia) (Margaryan et al. 2019). The Kajaran ore (Armenia) contains 0.2–1.5% copper and 0.03–0.15% molybdenum. The ore also contains Au 0.082 ppm, Ag 3.09 ppm, Re 1.01 ppm, Se 6.47 ppm, Te 4.45 ppm, Ge 1.74 ppm, Bi 10.7 ppm (Vehouni 2001).

The human impact on the environment resulted in the accumulation of heavy metals in some water bodies, soil, and even the atmosphere. The reason for heavy metal contamination of the water bodies can be the metallurgical industry and discharges of industrial waste into the environment.

The geogenic (volcano eruptions) and anthropogenic (electroplating, mining and smelting industries, petroleum refining, glass manufacturing, emissions from coal, caustic soda, peat producing industries, wood-burning, chemical synthesis, herbicides, fertilizers, solid waste, municipal sewage etc.) sources impacted on water and soil pollution with Cd, Cu, Cr, Pb, Zn, Ni, As, and other heavy metals. Airborne sources for heavy metal pollutions are chimneys emissions such as dust from storage facilities or waste accumulations (Lenart-Boroń and Boroń 2014).

The heavy metals, being discarded to the environment, get accumulated and may damage the soil ecosystem, including the microbial diversity (Wakelin et al. 2012; Mohammed et al. 2011; Almeida et al. 2009; Cervantes et al. 2001; Singh et al. 2020a). The environmental stress caused by toxic metals usually affects the native bacterial community, in which observing the domination of metal-tolerant or metallophilic microorganisms (Kozdroj and van Elsas 2001; Kuske et al. 1997; Teng et al. 2017). The biodiversity of metal-resistant microbial communities in different territories is different, due to the degree of contamination by metals and geographical parameters (Selenska-Pobell et al. 2001; Kozdroj and van Elsas 2001; Korehi et al. 2014; Pennanen et al. 1996).

7.2 Biodiversity of Metal-Resistant Bacteria and Archaea

Different bacterial species have adapted and evolved mechanisms to handle venomous concentrations of the metals, such as heavy metal efflux from the cells, metal absorption on the bacterial surface, periplasmic or intracellular accumulation, extracellular precipitation, and detoxification mechanisms (Nies 2007; Cervantes and Campos-García 2007; Riaz ul and Shakoori 2000).

For the first time, the impact of heavy metals on the soil microbiome was studied by Lipman and Burgess in 1914 (Lipman and Burgess 1914). Numerous studies have shown that in response to increasing concentrations of toxic metals, the microbial population is characterized by a change in biodiversity. Rajapaksha et al. by comparing the response of bacteria and fungi to the Zn and Cu concentrations in the soils sugested, that the bacterial population is more susceptible to the toxic levels of heavy metals than the fungal population (Rajapaksha et al. 2004). Many species of fungi may be effective biosorbent of Cu, Pb, Cd, Zn, and other heavy metals through the accumulation of them in their sporocarps. Nevertheless, these elements may inhibit their growth and reproduction (Baldrian 2003).

The accumulation of the toxic metals in the living cells may inactivate different enzymes, through the damage of their structure or conformation. Enzyme-associated metals can be displaced by toxic metals with analogous structure, consequently the function of the enzymes will be inhibited (Bruins et al. 2000; Singh et al. 2017). Furthermore, heavy metals modify the conformational structures of not only proteins, but also nucleic acids. Toxic metals, by accumulating in the cells, may also form complexes or chelates with structural proteins or essential metabolites, finally, disrupt integrity of the microbial cell membrane or whole cell (Bong et al. 2010; Sobolev and Begonia 2008).

The impact of various metal ions on different microbial populations depends on the metal concentration in the soil. For example, the existing levels of copper, zinc mercury in the soil affect microorganisms which are responsible for the nitrification and protein mineralization (Van Rossum et al. 2016; Lenart-Boroń and Boroń 2014; Kamal et al. 2010). Increased concentrations of lead in soil inhibit microbial hydrolysis of cellulose, as a result negatively affect processes of decomposition of organic matter. Lead resistant microorganisms able to accumulate metal up to 40% and may be used for bioremediation of polluted soil (Lenart-Boroń and Boroń 2014). Some bacteria can tolarate heavy metals with antiseptic activity, such as silver. For example, *Thiobacillus ferrooxidans* strains are silver-resistant and able to accumulate silver ions in high concentrations (Lenart-Boroń and Boroń 2014).

Metal-resistant microorganisms were found among various physiological groups, for example, in chemolithoautotrophic bacteria (Table 7.1). In 1921 S.A. Waxman and J.S. Joffie isolated from the mine drainage waters the autotrophic acidophilic bacterium *Thiobacillus ferrooxidans* (now *Acidithiobacillus ferrooxidans*), oxidizing sulfur and a number of its reduced compounds to sulfuric acid. However, in 1947 by Colmer A.R. and Hinkle M. proved the biological oxidation of ferrous oxide with the bacteria *T. ferrooxidans*. During the oxidation of bivalent iron and the reduction of sulfur, bacteria form iron oxide and hydrogen sulfide, which bind to toxic metal ions, forming insoluble precipitates, thereby ensuring metal tolerance (Korehi et al. 2014; Besaury et al. 2013a, b; Kondratyeva et al. 1999).

In addition to *T. ferrooxidans*, oxidizing bivalent iron and sulfur, *Leptospirillum ferrooxidans*, *Sulfobacillus thermosulfidooxidans*, *T. thiooxidans*, *T. acidophilus* oxidizing copper sulfide minerals, and uranium are also widely known (Korehi

Table 7.1 Biodiversity and heavy metal resistance of different bacteria and archaea

Microbe	Isolation sours	Heavy metals	MTC[a] (mM)	Reference
Aerobic bacteria				
Acinetobacter sp.	Jaduguda uranium mine (India)	Zn(II), Cd (II), Cu(II), Cr(II)	2, 1.5, 1, 0.5	Islam and Sar (2011)
Acidithiobacillus ferrooxidans	Copper mine site (Mynydd Parys); Cae Coch sulfur mine, North Wales, UK; Copper mine, Chili Ping Xiang coal mine, China; Spoil drainage of copper mine, Norway, cobalt/copper mine, ID, USA; Iron Mountain, CA, USA; Wheal Jane, England; Sextus and Killingdal mine dump, Norway; Rio Tinto, Spain; Waste pile from "Haberland Halde" uranium mine, Johanngeorgenstadt, Saxony, Germany; Sarcheshmeh copper mine, Kerman, Iran; Tandzut polymetallic ore, Armenia; Drmbon copper ore, Nagorno-Karabakh	Ni(II), Zn (II), Cu(II), Cd(II), As (III)	1000, 1071, 800, 500, 84	Hallberg et al. (2006), Hallberg et al. (2010), Dopson et al. (2003), Martínez-Bussenius et al. (2017), Chen et al. (2019), Blake Ii and Barrie Johnson (2000), Tzvetkova et al. (2002), Vardanyan and Vardanyan (2018)
Acidiphilium spp.	Storwartz mine, Norway; Sextus and Killingdal mine dump, Norway; Rio Tinto, Spain	Ni(II), Cd (II), Zn(II), Cu(II)	500, 500, 100, 12.5	Johnson et al. (2001), San Martin-Uriz et al. (2014), San Martin-Uriz et al. (2011), Chakravarty and Banerjee (2008)
Acidiphilium cryptum	Acidic coal mine situated in the Lusatian mining area, Germany	Fe(III), Cr (VI)	65, 0.05	Küsel et al. (1999), Cummings et al. (2007)
Leptospirillum ferrooxidans	Wheal Jane, England; Sextus and Killingdal mine dump, Norway; Cae Coch sulfur mine, North Wales, UK; Rio Tinto, Spain; Iron Mountain, California; Waste pile from "Haberland Halde"	Ni(II)	30–40	Blake Ii and Barrie Johnson (2000), Tian et al. (2007), Tzvetkova et al. (2002), Vardanyan and Vardanyan (2018)

(continued)

Table 7.1 (continued)

Microbe	Isolation sours	Heavy metals	MTC[a] (mM)	Reference
	uranium mine, Johanngeorgenstadt, Saxony, Germany; Mine in Jiangxi, China; Akhtala copper ore, Alaverdi copper ore, Theghut copper-molybdenum ore, Armenia			
At. thiooxidans	Wheal Jane, England; Rio Tinto, Spain; Ping Xiang coal mine, China	Ni(II)	600 (with addition of 5 mM Fe (III))	Chen et al. (2019), Blake Ii and Barrie Johnson (2000), Martínez-Bussenius et al. (2017), Dopson et al. (2003)
At. caldus	Iron Mountain, California	Zn(II), Cu (II)	200, 24	Mangold et al. (2013), Watkin et al. (2009), Blake Ii and Barrie Johnson (2000)
Acidocella sp.	Wheal Jane, England; Killingdal mine dump, Norway; Parys mine; Surda Copper Mine, Bihar, India	Zn(II), Cd (II), Ni(II), Cu(II)	1000, 1000, 200, 20	Blake Ii and Barrie Johnson (2000), Banerjee et al. (1996), Ghosh et al. (1997)
A. aminolytica			600, 400, 175, 40	
A. facilis		Ni(II), Zn (II)	150, 3	
"*Ferrimicrobium*" spp., *Ferrimicrobium acidiphilum*	Cae Coch sulfur mine, North Wales, UK	Fe(II), Fe (III)	200, 200	Johnson and Hallberg (2003), Johnson et al. (2009)
	Wheal Jane, England; Iron Mountain, California	Cu(II), Zn (II)	150, 50	
Ferrithrix thermotolerans	Beryl Spring/Gibbon river, Yellowstone National Park, USA	Fe(II), Cu (II), Zn(II), Fe(III)	200, 200, 200, 100	
Thiobacillus ferrooxidans	Uranium Denison mines, Uranium Rio Algom mines, Elliot Lake, Falconbridge nickel mine tailing areas, Ontario; Noranda mines, Quebec, Canada	Cu(II), Ni (II), UO_2(II), Th (II)	160, 160, 4, 4	Leduc et al. (1997)
Thiomonas cuprina	Solfatara fields, Iceland; Uranium mine waste heap, Germany	(Ni), Zn (II), Cu(II),	170, 150, 7.9, 1.3, 0.09	Schippers (2007), Schippers et al. (1995)

(continued)

Table 7.1 (continued)

Microbe	Isolation sours	Heavy metals	MTC[a] (mM)	Reference
		As(III), Cd (II),		
T. arsenitoxydans	Disused gold mine site, France, mining sites in Norway	As(III)	6	Battaglia-Brunet et al. (2006), Arsène-Ploetze et al. (2010)
A. ferrivorans	Copper mine spoil drainage, Norway; Cobalt/copper mine in Cobalt, ID, USA	Zn(II), Cu (II), Mo(II)	300, <50, <0.1	Blake Ii and Barrie Johnson (2000), Johnson et al. (2001), Hallberg et al. (2010)
Alkalibacterium sp.	Chañaral coastline near copper mining industries in the north of Chile	Cu(II)	1.5	Besaury et al. (2013a, b)
Acinetobacter lwoffi			1.5	
Pseudomonas sp.			4.7	
B. firmus			4.7	
Bacillus safensis			1.5	
Bacillus arsenicus			1.5	
B. pumilus			6.3	
Virgibacillus pantothenticus			1.5	
Sphingomonas sp.			1.5	
Arthrobacter protophormiae			1.5	
Alcaligenes sp.	Jaduguda uranium mine, India	Ni(II), Zn (II), Cd(II), Cu(II), Cr (VI)	2, 2, 1.5, 1, 0.5	Islam and Sar (2011)
Alcaligenes faecalis	Sewage wastewater at Taif province, Saudi Arabia	Ag(II), Sn (II), Cd(II)	1.2, 1.2, 0.9	Abo-Amer et al. (2015)
Arthrobacter sp.	Cu pollution agricultural soils from Valparaiso region, central Chile	Ni(II), Cu (II), Co(II), Cd(II), Zn (II), Hg(II), Cr(VI)	8.5, 3.9, 2.5, <0.4, 0.8, 0.1, 4.3	Altimira et al. (2012)
Sphingomonas sp.			0.9, 3.1, 0.8, <0.4, <0.8, 0.1, 4.3	
Stenotrophomonas maltophilia			17, 4.7, 2.5, <0.4, 8.5, 0.4, <0.4	
B. circulans	Tigris River, contaminated by Ergani copper mine wastewater, Maden-Elazig, Turkey	Mn(II), Zn (II), Ni(II), Cu(II), Co (II), Cd(II)	24, 22, 10, 2.5, 2, 2	Yilmaz (2003)

(continued)

Table 7.1 (continued)

Microbe	Isolation sours	Heavy metals	MTC[a] (mM)	Reference
B. subtilis	Atacama Desert region closely located to copper and lead-zinc mine tailings, Chile;	Ni(II), Fe (III), Cu (II), Co(II), Cd(II), Zn (II)	4, 2, 2, 1, 0.5, 0.5	Moreno et al. (2012)
B. licheniformis			4, 3, 1, 4, 0.5, 0.5	
B. subtilis	Sotk Gold Mine, Armenia	Ni(II), Cu (II), Zn(II), Cd(II)	4.5, 3.5, 1, 0.5	Margaryan et al. (2013)
Bacillus sp.	Artsvanik tailing dump, Armenia	Cr(VI)	250	Abrahamyan and Margaryan (2019)
Brevibacillus brevis	Cd- and Zn-polluted soils, Spain	Zn(II), Cd (II)	4, 0.1	Vivas et al. (2005)
Burkholderia dabaoshanensis	Dabaoshan Mining Area Soil, China	Cd(II), Pb (II)	22, 6	Zhu et al. (2012)
Paenibacillus sp.	Mt. Lofty, South Australia	Zn(II), Cd (II), Cu(II)	7, 1.77, 0.011	Rathnayake et al. (2009)
Pseudomonas sp.	Jaduguda uranium mine, India; Chañaral coastline near copper mining industries in the north of Chile; Artsvanik tailing dump, Armenia	Cu(II), Cr (II), Ni(II), Zn(II)	4.7, 0.2, 0.5, <0.5	Besaury et al. (2013a, b), Islam and Sar (2011), Ayvazyan and Margaryan (2018)
Stenotrophomonas sp.	Jaduguda uranium mine, India; Cu pollution agricultural soils from Valparaiso region, central Chile	Ni(II), Zn (II), Cu(II), Co(II), Cr (II), Cd(II)	17, 8.5, 3.9, 2.5, 0.4, 2	Islam and Sar (2011), Altimira et al. (2012)
Enterobacter sp.	Jaduguda uranium mine, India	Zn(II), Ni (II), Cu(II), Cr(VI), Cd (II)	2, 2, 0.5, 0.4	Islam and Sar (2011)
Exiguobacterium sp.			20.5, 1, 0.4, 0.1	
Microbacterium sp.			<0.5, 0.5, 0.2, 0.4, 0.1	
Yanghaparkia sp.			21, 2, 0.2, 0.1	
Cupriavidus metallidurans			5, 2, 1, 0.4, 3.5	
Ralstonia metallidurans (formerly known as *Alcaligenes eutrophus* and *R. eutropha*)	Zinc factory basin, Lie'g; Zinc ores storage, Overpelt, zinc, copper, and antimony purification plants, Beerse, mining areas, Zaire (Likasi and Shizuru), and zinc-	Co(II), Ni (II), Cd(II), Zn(II), Pb (II)	20, 2.5, 2.5, 12, 0.3	Mergeay et al. (2003), Mergeay et al. (1985), Diels and Mergeay (1990)

(continued)

Table 7.1 (continued)

Microbe	Isolation sours	Heavy metals	MTC[a] (mM)	Reference
	desertified area, Limburg, Belgium			
Rhodococcus sp.	Sewage sludge compost tea, Santa Fe, Granada, Spain	Cu(II), Zn (II), Pb(II), Cd(II)	16, 16, 16, 10	Vela-Cano et al. (2014)
Geobacillus kaustophilus	Geothermal field located in the area surrounding Monterotondo, Tuscany, Italy	As(III), Sb (II)	15, 5	Cuebas et al. (2011a, b)
Anaerobic/Aerotolerant bacteria				
Desulfovibrio senezii	Chañaral coastline near copper mining industries in the north of Chile	Cu(II)	1.5–3	Besaury et al. (2013a, b)
D. capillatus			3	
D. palmitatis			>15	
Archaeae				
Ferroplasma acidarmanus	Gold-containing arsenopyrite/pyrite ore concentrate from Bakyrtchik (Kazakhstan); Iron Mountain, CA, USA; Tinto River, Spain	As(V), As (III)	133	Baker-Austin et al. (2007), Golyshina et al. (2000), González-Toril et al. (2003)
Sulfolobus metallicus	Solfataric field in Krafla area, Iceland	Cu(II), Cd (II)	200, 3	Huber and Stetter (1991), Orell et al. (2013)
S. solftaricus	Sulfurous thermal springs in Yellowstone National Park	Zn(II), Cd (II), Cu(II), Ni(II)	10,	Miller et al. (1992), Brock et al. (1972)
S. acidocaldarius			10, 1, 0.1	
Metallosphaera sedula	Acidic drain from the hot water pond, Pisciarelli Solfatara, Italy	Zn(II), Cu (II), As (III), Cd (II), Co(II), Sb(II), UO_2(II)	150, 16, 1.3, 0.9, 0.85, 0.8, 0.4	Huber et al. (1989)

[a]*MTC* metal tolerance concentration

et al. 2014; Kojima and Fukui 2011). Some representatives of the genera *Sulfolobus* and *Acidianus* are also capable of oxidizing S^0, Fe(II), and sulfide minerals. Among these microorganisms, mesophilic and thermotolerant forms, extreme acidophilus, and acidothermophilia are described. For all these microorganisms, the oxidation processes of inorganic substrates are an energy source (Vardanyan and Vardanyan 2018; Singh et al. 2020b).

Alyssum murale, growing in serpentine soils of the desert northwest of California, have a notable ability to hyper accumulate Ni from insoluble Ni containing soils.

From the plant rhizosphere were isolated and characterized 46 metal tolerant bacterial cultures, belonging to the genus *Arthrobacter*, *Alcaligenes*, *Acidovorax*, *Bacillus*, *Burkholderia*, *Caulobacter*, *Nocardioides*, *Sinorhizobium*, *Stenotrophomonas*, *Sphingomonas*, *Paenibacillus*, *Pseudomonas*, *Phyllobacterium*, *Rhizobium*, *Variovorax*. The study of the resistance of cultures to various heavy metals revealed their high tolerance to ions Zn(II), Pb(II), Cu(II), Ni(II), and Co(II). *Arthrobacter rhombi*, *Clavibacter xyli*, *Microbacterium arabinogalactanolyticum*, *Rhizobium mongolense*, and *Varicovorax paradoxus* strains were also resistant to As(II), Hg (I), Cd(II), and Cr(VI) (Abou-Shanab et al. 2006).

Ozdemir and others (Ozdemir et al. 2012) showed high tolerance of thermophilic chemoorganoheterotrophic bacteria *G. toebii* subsp. *decanicum* and *G. thermoleovarans* subsp. *stromboliensis* to Ni(II), Cu(II), Zn(II), Cd(II), and Mn (II), as well as their accumulating ability of these metals. Cuebas and others (2011a, b) have isolated *G. kaustophilus* tolerant to high concentrations of arsenic (minimal inhibitory concentration 80 mM) from the geothermal soils of Monterotondo (Tuscany, Italy).

Metal-resistant bacteria are most common in the ores of various metals. For example, the strains of *Ralstonia pickettii* DX-T3-01 and *Sphingomonas* sp. DX-T3-03, resistant to high concentrations of Cd(II), Zn(II), and Cu(II), were isolated from the tailings of the Asia-Dexing copper mine (China) (Xie et al. 2010).

Besauri and his colleagues (Besaury et al. 2013a, b) used both cultivation and cultivation-independent methods to study the diversity of microorganisms in the samples of contaminated marine sediments on the coast of Chile. They showed domination of the genus *Bacillus* (up to 86.6%), while the remaining 23.4% represented the genera *Acinetobacter*, *Arthrobacter*, *Sphingomonas*, *Virgibacillus*, and *Pseudomonas*. From anaerobic copper-resistant microorganisms, they isolated *Desulfovibrio* and *Desulforomonas*. 16S rDNA clone libraries showed the presence of 15 operative taxonomic units (OTE), which included both cultivated and uncultivated species.

Acidophilic archean *Ferroplasma acidarmanus* Fer1, isolated from the iron mountain (CA, USA), demonstrate high tolerance to arsenic (Gihring et al. 2003; Baker-Austin et al. 2007).

High metal resistance has been described among halophilic archaea *Halobacterium* sp., *H. salinarum*, *Haloquadratum walsbyi*, *Haloarcula marismortui*, *Haloferax volcanii*, *Haloterrigena turkmenica*, and *Halorubrum lacusprofundi* (Orell et al. 2013; Srivastava and Kowshik 2013).

Acidophilic bacteria *Acidithiobacillus*, *Sulfobacillus* and *Leptospirillum*, belonging to the iron-oxidizing chemolithotrophic bacterial group, were isolated and studied from samples of the Alaverdi and Kapan complex polymetallic ores in Armenia (Vardanyan and Vardanyan 2018). The study of the bacterial diversity in the sludge samples from acid Akhtala tailing, suggested that the obligate autotrophic *Thiobacillus* and *Sulfuritalea* bacterial species are the main primary producers in the studied tailing (Margaryan et al. 2019).

Several metagenomic-based studies revealed that Proteobacteria is abounded in the various ecosystems, including extreme environments. The studies are suggesting that oligotrophic bacteria from the Proteobacteria phylum have developed different strategies to thrive under different stress conditions (Wakelin et al. 2012). For instance bacteria from the phylum Bacteroidetes (26%), Proteobacteria (23%), Actinobacteria (23%), and Firmicutes (16%) has founded in the surface samples of the Copper mine tailing located next to Sahuarita Arizona after 3 weeks of biosolid amendment (Pepper et al. 2012). 454 pyrosequencing analyses of the samples from the Akhtala copper mine tailing (Lori province, Armenia), which pH 2.6, characterized with the abundance of Proteobacteria (49%) and Bacteroidetes (43%) (Margaryan et al. 2019). In the samples from highly acidic Malanjkhand copper project tailing (<2) located in Balaghat, India the abundance of Proteobacteria (48%) and Actinobacteria (18%) have been founded (Gupta and Diwan 2017). The same results have been reported by Mardanov and colleagues (Mardanov et al. 2017). They studied the microbial diversity in the samples from Komsomolskaya gold mine tailing (Kemerovo region, Russia) using the shotgun library (GS FLX). The abundance of the bacterial phylum Proteobacteria (41%) and Saccharibacteria (23%) have been reported in acid mine drainage at Bjørndalen in Norway (pH 2.8–3.5) (García-Moyano et al. 2015).

7.3 Application of Heavy Metal Resistant and Metalophilic Microbes

Continuously accumulation of large amounts of industrial and agricultural waste is resulting in increased concentrations of the heavy metal in the environment, which can cause serious ecological problems and become a major human health hazard. The ability of microorganisms to adsorb heavy metals or change the forms of their presence in the environment attracts wide attention of researchers in connection with the possibility of biotechnological use of heavy metal resistant bacteria or archaea for wastewater treatment, bioremediation of contaminated environments, as well as in biogeotechnology of metals (Volesky 1994; White and Gadd 2000; Gadd 2005).

Various clean-up techniques, based on microbial cells or their enzymes, have been suggested and practiced for the clean-up of heavy metals from polluted areas (Okoduwa et al. 2017; Siddiquee et al. 2015). Bioremediation using microorganisms is receiving much attention due to their good performance and employed to transform toxic heavy metals into a less adverse form (Akcil et al. 2015; Watanabe 2001). This technique is cost-effective and environmentally friendly for revitalization of the polluted environment (Turpeinen et al. 2004; Ma et al. 2016; Yang et al. 2020). In Table 7.2 showed a number of microbes that may be used for removing metal ions from liquids. Nonetheless, bioremediation of heavy metals has number of limitations, such as the microbial production of toxic metabolites and heavy metal non-biodegradability.

Bacterial biosorption of heavy metals mainly connected with the bacteria cell walls and surface structures. In Gram-positive bacteria the phosphodiester bonds

Table 7.2 Microbial accumulation of the heavy metals from the solutions (modified from Igiri et al. 2018; Ghosh et al. 2015)

Bacterial species	Metal ions	Initial metal concentration (mg/L)	Removal (%)
Acinetobacter sp.	Cr	16	87
Sporosarcina saromensis		50	82.5
Bacillus cereus		1500	81
B. cereus (immobilized)		1500	96
B. circulans		1100	71.4
B. subtilis		570–2	99.6
B. subtilis (immobilized)		570–2	99.6
Desulfovibrio desulfuricans (immobilize on zeolite)		200	56.1
		100	99.8
		50	99.6
Staphylococcus sp.		4.108	45
Micrococcus sp.		100	90
Acinetobacter sp.		15	81
		16	78
		30	67
Pseudochrobactrum saccharolyticum		130	95
Streptomyces sp.		6.42	72
P. aeruginosa (immobilized)		570–4	99.3
P. aeruginosa		570–2	99.6
Stenotrophomonas sp.		16.59	81.27
Rhodococcus opacus		20	70
Acinetobacter sp. + *Arthrobacter* sp.		16	78
P. aeruginosa + *B. subtilis*		570–2	99/5
Pseudomonas aeruginosa	Hg	150	29.83
Vibrio parahaemolyticus		10	80
Klebsiella pneumonia		100	28.65
Cellulosimicrobium sp.	Pb	50	99.33
		100	96.98
		200	84.62
		300	62.28
B. iodinium		100–1.8	87
Rhodococcus opacus		50	95
D. desulfuricans (immobilize on zeolite)	Cu	50	97.4
		100	98.2
		200	78.7
Stenotrophomonas maltophilia		50	90
Rhodococcus opacus		20	52
Enterobacter cloacae		100	20
D. desulfuricans (immobilize on zeolite)		100	98.2
A. faecalis		100–19.2	70
P. aeruginosa		100–17.4	75

(continued)

Table 7.2 (continued)

Bacterial species	Metal ions	Initial metal concentration (mg/L)	Removal (%)
Enterobacter cloaceae		100	65
D. desulfuricans (immobilize on zeolite)	Ni	50	90.3
		100	90.1
		200	90.1
Micrococcus sp.		50	55
Acinetobacter sp.		51	68.94
E. cloacae	Co	100	8
Pseudomonas sp.	Zn	1	49.8

connecting teichoic acid monomers give a negative charge, which is responsible for the biosorption of divalent, positive charged metal ions. Gram-negative bacteria unlike Gram-positive ones have a thin layer of peptidoglycan and lack of teichoic acids. However, the phospholipids and lipopolysaccharides in the outer layer of Gram-negative bacteria determine the comprehensive negative charge promoting metal binding (Kanamarlapudi et al. 2018). Thus, the potential active sorption structures in the bacterial cell wall, allow using bacteria as an excellent biosorbents for the removal of toxic metal ions from industrial waste. Some experimental examples for the sequestering of toxic metals from industrial wastewater using bacterial biomass are summarized in Table 7.3.

Metalophilic bacteria and archaea are essential for the bioleaching process of metals from ores, concentrates, rocks and solutions, thus they are widely used in biogeometallurgy. The participation of microbes in the process of metal extraction from the mine ores registered from ancient times. For illustrative purposes, the river Rio Tinto can be discussed. In the Seville province of Spain are located the silver and copper deposit, which was recovered from pre-Romans and Romans times. Today, this region is known as the Rio Tinto mine, which means red color. The river is highly acidic (pH slightly above 2), contains high concentrations of ferric ions. The mining activity in this region was renewed in 1750 and continues nowadays. Subsequently, it was founded that, when copper ores were soaked with this water, copper was leached rapidly. It was proposed that the process connected with anaerobic sulfate-reducing bacteria, which activity generated sulfides that affect with metals in the ore and form insoluble metal sulfides such as chalcopyrite, and pyrite. These products, sequentially, can be used by aerobic sulfur-oxidizing bacteria like *Acidithiobacillus*, *Thiobacillus*, and *Leptospirillum* as electron donors. Finally, sulfur-oxidizing bacteria can turn ores into soluble metal sulfates. Described process is famous as "bioleaching" (Vardanyan and Vardanyan 2018; Siezen and Wilson 2009; Berlemont and Gerday 2011).

The main microorganisms actively involved in bioleaching process are presented in Table 7.4.

Table 7.3 Bacterial species used for biosorption of heavy metals

Bacteria	Metal ion	Sorption efficiency	Key notes	Reference
B. cereus	Zn	66.6 mg/g	Langmuir and Freundlich isotherm model was used for the studies. Amino, hydroxyl, carboxyl, and carbonyl groups involved in biosorption process. Physic-chemical adsorption and ion exchange was registered.	Joo et al. (2010)
B. subtilis	Ni	98.54, 99.2%	Carboxyl, phosphate amino and hydroxyl groups involved in biosorption process.	Al-Gheethi et al. (2017)
B. jeotgali	Cd, Zn	57.9, 128.2 mg/g	Langmuir isotherm model was used for the studies. Ion exchange was registered.	Green-Ruiz et al. (2008)
B. pumilus	Pb	28.06 mg/g	Langmuir isotherm model was used for the studies.	Çolak et al. (2011)
B. thuringiensis	Ni	15.7%		Öztürk (2007)
B. thioparans	Cu, Pb	27.3, 210.1 mg/g		Rodríguez-Tirado et al. (2012)
Arthrobacter sp.	Cu	32.64 mg/g		Hasan and Srivastava (2009)
Rhizobium leguminosarum	Cd, Co	135.3, 167.5 mg/g		Abd-Alla et al. (2012)
B. coagulans	Cr	39.9 mg/g		Vijayaraghavan and Yun (2008)
Lactobacillus delbruckii subsp. *Bulgaricus*, *Streptococcus thermophiles*	Fe, Zn	100, 90%	Carboxyl and hydroxyl groups involved in biosorption process.	Sofu et al. (2015)
E. coli	Cd, Ni, Cr	10.3, 6.9, 4.9 mg/g	Redlich-Peterson isotherm model was used for the studies. Carboxyl and hydroxyl groups involved in biosorption process.	Quintelas et al. (2009)
			Ion exchange was registered.	

7.4 Conclusion

The bacterial and archaeal community structure in the different samples worldwide with high content of toxic metals shows strong spatial variations. Toxic concentrations of heavy metals in the environment have an extensive effect on different microbial communities, which limited their diversity. Most of the bacterial and

Table 7.4 Different bioleaching bacteria and archaea are associated with minerals found in nature

Bioleaching microbes	Mineral (chemical composition)	Reference
A. ferrooxidans, *A. thiooxidans*	Hematite (Fe_2O_3)	Huang et al. (2013)
A. ferrooxidans, *A. thiooxidans*, *L. ferrooxidans*	Pentlandite ($(Fe, Ni)_9S_8$)	Brierley and Brierley (2001)
A. ferrooxidans	Pyrolusite (MnO_2)	Acharya et al. (2003)
A. ferrooxidans, *A. thiooxidans*, *Acidianus brierleyi*, *A. ambivalens*, *L. ferrooxidans*, *Sulfolobus solfataricus*	Molybdenite (MoS_2)	Rastegar et al. (2014), Pistaccio et al. (1994), Romano et al. (2001)
A. ferrooxidans	Magnetite ($(Fe, V)_3O_4$)	Liu et al. (2013)
Paenibacillus polymyxa, *A. ferrooxidans*, *A. thiooxidans*, *L. ferrooxidans*, *Ferroplasma acidiphilum*	Gibbsite ($Al(OH)_3$)	Natarajan (2016)
A. caldus, *Metallosphaera sedula*, *Sulfobacillus* sp., *S. thermosulfidooxidans*, *Leptospirillum* sp., *A. tandzuti*, *A. brierleyi*, *A. infernus*, *Sulfolobus shibatae*, *S. acidocaldarius* and *S. metallicus*	Chalcopyrite ($CuFeS_2$)	Vardanyan and Vardanyan (2018), Panda et al. (2015), Stott et al. (2003)
A. caldus, *A. thiooxidans*, *A. albertensis*, *L. ferriphilum*, *S. thermotolerans*, *S. thermosulfidooxidans*	Chalcocite (Cu_2S)	Xingyu et al. (2010), Khachatryan et al. (2021)
Alicyclobacillus sp.	Cuprite (Cu_2O)	Chaerun et al. (2017)
S. thermosulfidooxidans, *Leptospirillum* sp.	Pyrite (FeS_2)	Vardanyan and Vardanyan (2018)
S. thermosulfidooxidans, *Thermoplasma acidophilum*	Azurite ($Cu_3(CO_3)_2(OH)_2$)	Ilyas et al. (2012)
At. caldus, *S. thermosulfidooxidans*	Arsenopyrite ($FeAsS$)	Vardanyan and Vardanyan (2018)
A. ferrooxidans, *A. thiooxidans*	Realgar (AsS)	Zhang et al. (2007)
A. ferrooxidans, *S. sibiricus*	Orpiment (As_2S_3)	Zhang et al. (2015)
A. ferrooxidans, *A. thiooxidans*, *Leptospirillum* sp.	Sphalerite ($(Zn, Fe)S$)	Rodríguez et al. (2003), Xia et al. (2008)
A. ferrooxidans	Galena (PbS)	Baba et al. (2011)
A. ferrooxidans, *A. acidophilus*, *A. thiooxidans*, *L. ferrooxidans*, *L. ferriphilum*	Pitchblende (UO_2)	Chen et al. (2016)
Bacillus megaterium	Ilmenite ($FeTiO_3$)	Jonglertjunya and Rubcumintara (2013)

archaeal groups founded in the different environments with high levels of heavy metals are, undeniably, adapted and thrive in such a toxic environment. Different adaptation mechanisms of metal-resistant bacteria make them a prospective tool for different biotechnological applications such as bioremediation of the polluted environment and bioleaching of valuable metals.

Funding This work was supported by grants from the Eurasia Programme of the Norwegian Center for International Cooperation in Education (CPEA-2011/10081, CPEA-LT-2016/10095) and partially supported by Research Grant from Science Committee of MESCS RA to AM (19YR-1F065), and the Armenian National Science and Education Fund based in New York, USA, to AM (ANSEF Microbio-2390 and Microbio-3869).

References

Abd-Alla MH, Morsy FM, El-Enany A-WE, Ohyama T (2012) Isolation and characterization of a heavy-metal-resistant isolate of *Rhizobium leguminosarum* bv. viciae potentially applicable for biosorption of Cd^{2+} and Co^{2+}. Int Biodeterior Biodegradation 67:48–55

Abo-Amer AE, El-Shanshoury AE-RR, Alzahrani OM (2015) Isolation and molecular characterization of heavy metal-resistant alcaligenes faecalis from sewage wastewater and synthesis of silver nanoparticles. Geomicrobiol J 32(9):836–845

Abou-Shanab RAI, Angle JS, Chaney RL (2006) Bacterial inoculants affecting nickel uptake by Alyssum murale from low, moderate and high Ni soils. Soil Biol Biochem 38(9):2882–2889

Abrahamyan N, Margaryan A (2019) Study and description of hexavalent chromate of stable alkaliphilic microorganisms from samples of Artsvanik tailing dump. Collection of scientific articles of YSU SSS. Nat Phys Math Sci 13(29):105–113

Acharya C, Kar RN, Sukla LB (2003) Studies on reaction mechanism of bioleaching of manganese ore. Miner Eng 16(10):1027–1030

Akcil A, Erust C, Ozdemiroglu S, Fonti V, Beolchini F (2015) A review of approaches and techniques used in aquatic contaminated sediments: metal removal and stabilization by chemical and biotechnological processes. J Clean Prod 86:24–36

Al-Gheethi A, Mohamed R, Noman E, Ismail N, Kadir OA (2017) Removal of heavy metal ions from aqueous solutions using *Bacillus subtilis* biomass pre-treated by supercritical carbon dioxide. Clean Soil Air Water 45:1700356

Almeida WI, Vieira RP, Cardoso AM, Silveira CB, Costa RG, Gonzalez AM, Paranhos R, Medeiros JA, Freitas FA, Albano RM, Martins OB (2009) Archaeal and bacterial communities of heavy metal contaminated acidic waters from zinc mine residues in Sepetiba Bay. Extremophiles 13(2):263–271

Altimira F, Yáñez C, Bravo G, González M, Rojas LA, Seeger M (2012) Characterization of copper-resistant bacteria and bacterial communities from copper-polluted agricultural soils of central Chile. BMC Microbiol 12:193–193

Arsène-Ploetze F, Koechler S, Marchal M, Coppée J-Y, Chandler M, Bonnefoy V, Brochier-Armanet C, Barakat M, Barbe V, Battaglia-Brunet F, Bruneel O, Bryan CG, Cleiss-Arnold J, Cruveiller S, Erhardt M, Heinrich-Salmeron A, Hommais F, Joulian C, Krin E, Lieutaud A, Lièvremont D, Michel C, Muller D, Ortet P, Proux C, Siguier P, Roche D, Rouy Z, Salvignol G, Slyemi D, Talla E, Weiss S, Weissenbach J, Médigue C, Bertin PN (2010) Structure, function, and evolution of the thiomonas spp. genome. PLoS Genet 6(2):e1000859

Ayvazyan I, Margaryan A (2018) New heavy metal resistant alkaliphilic bacteria isolated from Artsvanik tailing. Collection of scientific articles of YSU SSS. Nat Phys Math Sci 11 (24):216–223

Baba AA, Adekola FA, Atata RF, Ahmed RN, Panda S (2011) Bioleaching of Zn(II) and Pb (II) from Nigerian sphalerite and galena ores by mixed culture of acidophilic bacteria. Trans Nonferrous Metals Soc China 21(11):2535–2541

Baker-Austin C, Dopson M, Wexler M, Sawers RG, Stemmler A, Rosen BP, Bond PL (2007) Extreme arsenic resistance by the acidophilic archaeon 'Ferroplasma acidarmanus' Fer1. Extremophiles 11(3):425–434

Baldrian P (2003) Interactions of heavy metals with white-rot fungi. Enzym Microb Technol 32 (1):78–91

Banerjee PC, Ray MK, Koch C, Bhattacharyya S, Shivaji S, Stackebrandt E (1996) Molecular characterization of two acidophilic heterotrophic bacteria isolated from a copper mine of India. Syst Appl Microbiol 19(1):78–82

Battaglia-Brunet F, Joulian C, Garrido F, Dictor MC, Morin D, Coupland K, Barrie Johnson D, Hallberg KB, Baranger P (2006) Oxidation of arsenite by Thiomonas strains and characterization of Thiomonas arsenivorans sp. nov. Antonie Van Leeuwenhoek 89(1):99–108

Berlemont R, Gerday C (2011) 1.18 – Extremophiles. In: Moo-Young M (ed) Comprehensive biotechnology, 2nd edn. Academic Press, Burlington, WA, pp 229–242

Besaury L, Ghiglione J-F, Quillet L (2013a) Abundance, activity, and diversity of archaeal and bacterial communities in both uncontaminated and highly copper-contaminated marine sediments. Mar Biotechnol 16(2):230–242

Besaury L, Marty F, Buquet S, Mesnage V, Muyzer G, Quillet L (2013b) Culture-dependent and independent studies of microbial diversity in highly copper-contaminated Chilean marine sediments. Microb Ecol 65(2):311–324

Blake Ii R, Barrie Johnson D (2000) Phylogenetic and biochemical diversity among acidophilic bacteria that respire on iron. In: Environmental microbe-metal interactions. American Society for Microbiology, Washington, DC, pp 53–78

Bong CW, Malfatti F, Azam F, Obayashi Y, Suzuki S (2010) The effect of zinc exposure on the bacteria abundance and proteolytic activity in seawater. In: Hamamura N, Suzuki S, Mendo S, Barroso CM, Iwata H, Tanabe S (eds) Interdisciplinary studies on environmental chemistry, biological responses to contaminants. Terrapub, Tokyo, pp 57–63

Brierley JA, Brierley CL (2001) Present and future commercial applications of biohydrometallurgy. Hydrometallurgy 59(2):233–239

Brock TD, Brock KM, Belly RT, Weiss RL (1972) Sulfolobus: a new genus of sulfur-oxidizing bacteria living at low pH and high temperature. Arch Mikrobiol 84(1):54–68

Bruins MR, Kapil S, Oehme FW (2000) Microbial resistance to metals in the environment. Ecotoxicol Environ Saf 45(3):198–207

Cervantes C, Campos-García J (2007) Reduction and efflux of chromate by bacteria. In: Nies DH, Silver S (eds) Molecular microbiology of heavy metals. Springer Berlin Heidelberg, Berlin, pp 407–419

Cervantes C, Campos-García J, Devars S, Gutiérrez-Corona F, Loza-Tavera H, Torres-Guzmán JC, Moreno-Sánchez R (2001) Interactions of chromium with microorganisms and plants. FEMS Microbiol Rev 25(3):335–347

Chaerun SK, Putri FY, Mubarok MZ, Minwal WP, Ichlas ZT (2017) Bioleaching of supergene porphyry copper ores from Sungai Mak Gorontalo of Indonesia by an iron- and sulfur-oxidizing mixotrophic bacterium. Solid State Phenom 262:20–23

Chakravarty R, Banerjee PC (2008) Morphological changes in an acidophilic bacterium induced by heavy metals. Extremophiles 12(2):279–284

Chen G, Sun Z, Liu Y (2016) Continued multicolumns bioleaching for low grade uranium ore at a certain uranium deposit. J Nanomater 2016:1683416

Chen A, Hao X, Xiao Y, Zou K, Liu H, Liu X, Yin H, Qiu G, Liang Y (2019) Responses of *Acidithiobacillus thiooxidans* A01 to Individual and Joint Nickel (Ni^{2+}) and Ferric (Fe^{3+}). Minerals 9(2):1–12

Çolak F, Atar N, Yazıcıoğlu D, Olgun A (2011) Biosorption of lead from aqueous solutions by *Bacillus strains* possessing heavy-metal resistance. Chem Eng J 173(2):422–428

Cuebas M, Sannino D, Bini E (2011a) Isolation and characterization of arsenic resistant Geobacillus kaustophilus strain from geothermal soils. J Basic Microbiol 51(4):364–371

Cuebas M, Villafane A, McBride M, Yee N, Bini E (2011b) Arsenate reduction and expression of multiple chromosomal ars operons in *Geobacillus kaustophilus* A1. Microbiology (Reading) 157(Pt 7):2004–2011

Cummings DE, Fendorf S, Singh N, Sani RK, Peyton BM, Magnuson TS (2007) Reduction of Cr (VI) under acidic conditions by the facultative Fe(lll)-reducing bacterium *Acidiphilium cryptum*. Environ Sci Technol 41(1):146–152

Diels L, Mergeay M (1990) DNA probe-mediated detection of resistant bacteria from soils highly polluted by heavy metals. Appl Environ Microbiol 56(5):1485

Dopson M, Baker-Austin C, Koppineedi PR, Bond PL (2003) Growth in sulfidic mineral environments: metal resistance mechanisms in acidophilic micro-organisms. Microbiology 149 (Pt 8):1959–1970

Gadd G (2005) Microorganisms in toxic metal-polluted soils. In: Varma A, Buscot F (eds) Microorganisms in soils: roles in genesis and functions. Soil biology. Springer, Berlin, pp 325–356

García-Moyano A, Austnes AE, Lanzén A, González-Toril E, Aguilera Á, Øvreås L (2015) Novel and unexpected microbial diversity in acid mine drainage in Svalbard (78° N), revealed by culture-independent approaches. Microorganisms 3(4):667–694

Ghosh S, Mahapatra NR, Banerjee PC (1997) Metal resistance in Acidocella strains and plasmid-mediated transfer of this characteristic to *Acidiphilium multivorum* and *Escherichia coli*. Appl Environ Microbiol 63(11):4523–4527

Ghosh A, Ghosh Dastidar M, Sreekrishnan TR, Asce AM (2015) Recent advances in bioremediation of heavy metals and metal complex dyes. J Environ Eng ASCE 78(4):496–505

Gihring TM, Bond PL, Peters SC, Banfield JF (2003) Arsenic resistance in the archaeon "*Ferroplasma acidarmanus*": new insights into the structure and evolution of the ars genes. Extremophiles 7(2):123–130

Golyshina OV, Pivovarova TA, Karavaiko GI, Kondratéva TF, Moore ER, Abraham WR, Lünsdorf H, Timmis KN, Yakimov MM, Golyshin PN (2000) *Ferroplasma acidiphilum* gen. nov., sp. nov., an acidophilic, autotrophic, ferrous-iron-oxidizing, cell-wall-lacking, mesophilic member of the *Ferroplasmaceae* fam. nov., comprising a distinct lineage of the Archaea. Int J Syst Evol Microbiol 50(Pt 3):997–1006

González-Toril E, Llobet-Brossa E, Casamayor EO, Amann R, Amils R (2003) Microbial ecology of an extreme acidic environment, the Tinto River. Appl Environ Microbiol 69(8):4853

Green-Ruiz C, Rodriguez-Tirado V, Gomez-Gil B (2008) Cadmium and zinc removal from aqueous solutions by *Bacillus jeotgali*: pH, salinity and temperature effects. Bioresour Technol 99(9):3864–3870

Gupta P, Diwan B (2017) Bacterial Exopolysaccharide mediated heavy metal removal: a review on biosynthesis, mechanism and remediation strategies. Biotechnol Rep (Amst) 13:58–71

Hallberg KB, Coupland K, Kimura S, Johnson DB (2006) Macroscopic streamer growths in acidic, metal-rich mine waters in North Wales consist of novel and remarkably simple bacterial communities. Appl Environ Microbiol 72(3):2022

Hallberg KB, González-Toril E, Johnson DB (2010) *Acidithiobacillus ferrivorans*, sp. nov.; facultatively anaerobic, psychrotolerant iron-, and sulfur-oxidizing acidophiles isolated from metal mine-impacted environments. Extremophiles 14(1):9–19

Hasan SH, Srivastava P (2009) Batch and continuous biosorption of Cu($^{2+}$) by immobilized biomass of *Arthrobacter* sp. J Environ Manag 90(11):3313–3321

Huang T, Gong W, Bao G, Lei S (2013) Bioleaching of phosphorus from hematite with mixed bacteria. Adv Mater Res 823:613–617

Huber G, Stetter KO (1991) *Sulfolobus metallicus*, sp. nov., a novel strictly Chemolithoautotrophic Thermophilic Archaeal species of metal-mobilizers. Syst Appl Microbiol 14(4):372–378

Huber G, Spinnler C, Gambacorta A, Stetter KO (1989) *Metallosphaera sedula* gen, and sp. nov. Represents a new genus of aerobic, metal-mobilizing, Thermoacidophilic Archaebacteria. Syst Appl Microbiol 12(1):38–47

Igiri BE, Okoduwa SIR, Idoko GO, Akabuogu EP, Adeyi AO, Ejiogu IK (2018) Toxicity and bioremediation of heavy metals contaminated ecosystem from tannery wastewater: a review. J Toxicol 2018:2568038

Ilyas S, Chi R, Bhatti HN, Bhatti IA, Ghauri MA (2012) Column bioleaching of low-grade mining ore containing high level of smithsonite, talc, sphaerocobaltite and azurite. Bioprocess Biosyst Eng 35(3):433–440

Islam E, Sar P (2011) Culture-dependent and -independent molecular analysis of the bacterial community within uranium ore. J Basic Microbiol 51(4):372–384

Johnson DB, Hallberg KB (2003) The microbiology of acidic mine waters. Res Microbiol 154 (7):466–473

Johnson DB, Rolfe S, Hallberg KB, Iversen E (2001) Isolation and phylogenetic characterization of acidophilic microorganisms indigenous to acidic drainage waters at an abandoned Norwegian copper mine. Environ Microbiol 3(10):630–637

Johnson DB, Bacelar-Nicolau P, Okibe N, Thomas A, Hallberg KB (2009) *Ferrimicrobium acidiphilum* gen. nov., sp. nov. and Ferrithrix thermotolerans gen. nov., sp. nov.: heterotrophic, iron-oxidizing, extremely acidophilic actinobacteria. Int J Syst Evol Microbiol 59 (Pt 5):1082–1089

Jonglertjunya W, Rubcumintara T (2013) Titanium and iron dissolutions from ilmenite by acid leaching and microbiological oxidation techniques. Asia Pac J Chem Eng 8(3):323–330

Joo J-H, Hassan SHA, Oh S-E (2010) Comparative study of biosorption of Zn^{2+} by *Pseudomonas aeruginosa* and *Bacillus cereus*. Int Biodeterior Biodegradation 64(8):734–741

Kamal S, Prasad R, Varma A (2010) Soil microbial diversity in relation to heavy metals. In: Soil heavy metals. Soil biology, vol 19. Springer, New York, NY, pp 31–63

Kanamarlapudi SLRK, Chintalpudi VK, Muddada S (2018) Application of biosorption for removal of heavy metals from wastewater, Chapter 4. In: Derco J, Vrana B (eds) Biosorption. IntechOpen, London, pp 69–116

Khachatryan A, Vardanyan N, Vardanyan A, Zhang R, Castro L (2021) The effect of metal ions on the growth and ferrous iron oxidation by *Leptospirillum ferriphilum* CC isolated from Armenia mine sites. Metals 11(3):425

Kojima H, Fukui M (2011) *Sulfuritalea hydrogenivorans* gen. nov., sp. nov., a facultative autotroph isolated from a freshwater lake. Int J Syst Evol Microbiol 61(7):pp. 1651–1655

Kondratyeva TF, Pivovarova TA, Muntyan LN, Karavaiko GI (1999) Strain diversity of *Thiobacillus ferrooxidans* and its significance in biohydrometallurgy. In: Amils R, Ballester A (eds) Process metallurgy. Elsevier, Amsterdam, pp 89–96

Konstantinidis KT, Isaacs N, Fett J, Simpson S, Long DT, Marsh TL (2003) Microbial diversity and resistance to copper in metal-contaminated lake sediment. Microb Ecol 45(2):191–202

Korehi H, Blöthe M, Schippers A (2014) Microbial diversity at the moderate acidic stage in three different sulfidic mine tailings dumps generating acid mine drainage. Res Microbiol 165 (9):713–718

Kozdroj J, van Elsas JD (2001) Structural diversity of microorganisms in chemically perturbed soil assessed by molecular and cytochemical approaches. J Microbiol Methods 43(3):197–212

Küsel K, Dorsch T, Acker G, Stackebrandt E (1999) Microbial reduction of Fe(III) in acidic sediments: isolation of *Acidiphilium cryptum* JF-5 capable of coupling the reduction of Fe(III) to the oxidation of glucose. Appl Environ Microbiol 65(8):3633–3640

Kuske CR, Barns SM, Busch JD (1997) Diverse uncultivated bacterial groups from soils of the arid southwestern United States that are present in many geographic regions. Appl Environ Microbiol 63(9):3614–3621

Leduc LG, Ferroni GD, Trevors JT (1997) Resistance to heavy metals in different strains of *Thiobacillus ferrooxidans*. World J Microbiol Biotechnol 13(4):453–455

Lenart-Boroń A, Boroń P (2014) The effect of industrial heavy metal pollution on microbial abundance and diversity in soils, a review. Chapter 26. In: Hernandez-Soriano MC (ed) Environmental risk assessment of soil contamination, pp 759–783

Lipman CB, Burgess PS (1914) Effect of copper, zinc, iron and lead salts on ammonification and nitrification in soils. Agric Sci 1(6)

Liu X-R, Jiang S-C, Liu Y-J, Li H, Wang H-J (2013) Biodesulfurization of vanadium-bearing titanomagnetite concentrates and pH control of bioleaching solution. Int J Miner Metall Mater 20(10):925–930

Ma Y, Rajkumar M, Zhang C, Freitas H (2016) Beneficial role of bacterial endophytes in heavy metal phytoremediation. J Environ Manag 174:14–25

Mangold S, Potrykus J, Björn E, Lövgren L, Dopson M (2013) Extreme zinc tolerance in acidophilic microorganisms from the bacterial and archaeal domains. Extremophiles 17 (1):75–85

Mardanov AV, Beletsky AV, Ivasenko DA, Karnachuk OV, Ravin NV (2017) Metagenome sequence of a microbial community from the gold mine tailings in the Kuzbass Area, Russia. Genome Announce 5(49):e01355–e01317

Margaryan AA, Panosyan HH, Birkeland N-K, Trchounian AH (2013) Heavy metal accumulation and the expression of the copA and nikA genes in *Bacillus subtilis* AG4 isolated from the Sotk gold min in Armenia. Biolog J Armenia 3(65):51–57

Margaryan A, Panosyan H, Mamimin C, Trchounian A, Birkeland N-K (2019) Insights into the bacterial diversity of the acidic akhtala mine tailing in armenia using molecular approaches. Curr Microbiol 76(4):462–469

Martínez-Bussenius C, Navarro CA, Jerez CA (2017) Microbial copper resistance: importance in biohydrometallurgy. Microb Biotechnol 10(2):279–295

Mergeay M, Nies D, Schlegel HG, Gerits J, Charles P, Van Gijsegem F (1985) *Alcaligenes eutrophus* CH34 is a facultative chemolithotroph with plasmid-bound resistance to heavy metals. J Bacteriol 162(1):328

Mergeay M, Monchy S, Vallaeys T, Auquier V, Benotmane A, Bertin P, Taghavi S, Dunn J, van der Lelie D, Wattiez R (2003) *Ralstonia metallidurans*, a bacterium specifically adapted to toxic metals: towards a catalogue of metal-responsive genes. FEMS Microbiol Rev 27(2–3):385–410

Miller KW, Risanico SS, Bruno Risatti J (1992) Differential tolerance of Sulfolobus strains to transition metals. FEMS Microbiol Lett 93(1):69–73

Mohammed AS, Kapri A, Goel R (2011) Heavy metal pollution: source, impact, and remedies. In: Khan MS, Goel AZR, Musarrat J (eds) Biomanagement of metal-contaminated soils. Springer Science+Business Media, Dordrecht

Moreno ML, Piubeli F, Bonfá MR, García MT, Durrant LR, Mellado E (2012) Analysis and characterization of cultivable extremophilic hydrolytic bacterial community in heavy-metal-contaminated soils from the Atacama Desert and their biotechnological potentials. J Appl Microbiol 113(3):550–559

Natarajan KA (2016) Biomineralization and biobeneficiation of bauxite. Trans Indian Inst Metals 69(1):15–21

Nies D (2007) Bacterial transition metal homeostasis. In: Nies D, Silver S (eds) Molecular microbiology of heavy metals microbiology monographs. Springer, Berlin, pp 117–142

Okoduwa SIR, Igiri B, Udeh CB, Edenta C, Gauje B (2017) Tannery effluent treatment by yeast species isolates from watermelon. Toxics 5(1):6

Orell A, Remonsellez F, Arancibia R, Jerez CA (2013) Molecular characterization of copper and cadmium resistance determinants in the biomining thermoacidophilic archaeon Sulfolobus metallicus. Archaea 2013:289236

Ozdemir S, Kilinc E, Poli A, Nicolaus B, Guven K (2012) Cd, Cu, Ni, Mn and Zn resistance and bioaccumulation by thermophilic bacteria, Geobacillus toebii subsp. decanicus and Geobacillus thermoleovorans subsp. stromboliensis. World J Microbiol Biotechnol 28(1):155–163

Öztürk A (2007) Removal of nickel from aqueous solution by the bacterium *Bacillus thuringiensis*. J Hazard Mater 147(1):518–523

Panda S, Akcil A, Pradhan N, Deveci H (2015) Current scenario of chalcopyrite bioleaching: a review on the recent advances to its heap-leach technology. Bioresour Technol 196:694–706

Pennanen T, Frostegard A, Fritze H, Baath E (1996) Phospholipid fatty acid composition and heavy metal tolerance of soil microbial communities along two heavy metal-polluted gradients in coniferous forests. Appl Environ Microbiol 62(2):420

Pepper IL, Zerzghi HG, Bengson SA, Iker BC, Banerjee MJ, Brooks JP (2012) Bacterial populations within copper mine tailings: long-term effects of amendment with Class A biosolids. J Appl Microbiol 113(3):569–577

Pistaccio L, Curutchet G, Donati E, Tedesco P (1994) Analysis of molybdenite bioleaching by Thiobacillus ferrooxidans in the absence of iron (II). Biotechnol Lett 16(2):189–194

Quintelas C, Rocha Z, Silva B, Fonseca B, Figueiredo H, Tavares T (2009) Removal of Cd(II), Cr (VI), Fe(III) and Ni(II) from aqueous solutions by an *E. coli* biofilm supported on kaolin. Chem Eng J 149(1):319–324

Rajapaksha RMCP, Tobor-Kapłon MA, Bååth E (2004) Metal toxicity affects fungal and bacterial activities in soil differently. Appl Environ Microbiol 70(5):2966–2973

Rastegar SO, Mousavi SM, Rezaei M, Shojaosadati SA (2014) Statistical evaluation and optimization of effective parameters in bioleaching of metals from molybdenite concentrate using Acidianus brierleyi. J Ind Eng Chem 20(5):3096–3101

Rathnayake IVN, Megharaj M, Bolan N, Naidu R (2009) Tolerance of heavy metals by gram positive soil bacteria, World Academy of Science, Engineering and Technology. Int J Biol Biomol Agric Food Biotechnol Eng 3:270–274

Rey KA, Thompson SM, Curtis BY, White RC, Emerman SH (2009) Arsenic and other heavy metals in swimming pools fed by hot springs in Utah Valley and the Wasatch Range, hydrological study in the Rocky Mountains and Beyond, Session N 5. Geol Soc Am Abstr Programs 41(6):10

Riaz ul H, Shakoori AR (2000) Microorganisms resistant to heavy metals and toxic chemicals as indicators of environmental pollution and their use in bioremediation. Folia Biol (Krakow) 48 (3–4):143–147

Rodríguez Y, Ballester A, Blázquez ML, González F, Muñoz JA (2003) New information on the chalcopyrite bioleaching mechanism at low and high temperature. Hydrometallurgy 71 (1):47–56

Rodríguez-Tirado V, Green-Ruiz C, Gómez-Gil B (2012) Cu and Pb biosorption on *Bacillus thioparans* strain U3 in aqueous solution: kinetic and equilibrium studies. Chem Eng J 181–182:352–359

Romano P, Blázquez ML, Ballester A, González F, Alguacil FJ (2001) Selective copper–iron dissolution from a molybdenite concentrate using bacterial leaching. J Chem Technol Biotechnol 76(7):723–728

San Martin-Uriz P, Gomez MJ, Arcas A, Bargiela R, Amils R (2011) Draft Genome Sequence of the *Acidiphilium* sp. Strain PM (DSM 24941). J Bacteriol 193(19):5585

San Martin-Uriz P, Mirete S, Alcolea PJ, Gomez MJ, Amils R, Gonzalez-Pastor JE (2014) Nickel-resistance determinants in Acidiphilium sp. PM identified by genome-wide functional screening. PLoS One 9(4):e95041

Schippers A (2007) Microorganisms involved in bioleaching and nucleic acid-based molecular methods for their identification and quantification. In: Donati ER, Sand W (eds) Microbial processing of metal sulfides. Springer Netherlands, Dordrecht, pp 3–33

Schippers A, Hallmann R, Wentzien S, Sand W (1995) Microbial diversity in uranium mine waste heaps. Appl Environ Microbiol 61(8):2930–2935

Selenska-Pobell S, Kampf G, Flemming K, Radeva G, Satchanska G (2001) Bacterial diversity in soil samples from two uranium waste piles as determined by rep-APD, RISA and 16S rDNA retrieval. Antonie Van Leeuwenhoek 79(2):149–161

Siddiquee S, Rovina K, Azad SA, Naher L, Suryani S, Chaikaew P (2015) Heavy metal contaminants removal from wastewater using the potential filamentous fungi biomass: a review. J Microb Biochem Technol 7(6):384–393

Siezen RJ, Wilson G (2009) Bioleaching genomics. Microb Biotechnol 2(3):297–303

Singh RP, Manchanda G, Li ZF, Rai AR (2017) Insight of proteomics and genomics in environmental bioremediation. In: Bhakta JN (ed) Handbook of research on inventive bioremediation techniques. IGI Global, Hershey, PA. https://doi.org/10.4018/978-1-5225-2325-3

Singh R, Manchanda G, Maurya I, Wei Y (eds) (2020a) Microbial versatility in varied environments. Springer, Singapore. https://doi.org/10.1007/978-981-15-3028-9

Singh RP, Anwar MN, Singh D, Bahuguna V, Manchanda G, Yang Y (2020b) Deciphering the key factors for heavy metal resistance in gram-negative bacteria. In: Singh R, Manchanda G, Maurya I, Wei Y (eds) Microbial versatility in varied environments. Springer, Singapore. https://doi.org/10.1007/978-981-15-3028-9_7

Sobolev D, Begonia MF (2008) Effects of heavy metal contamination upon soil microbes: lead-induced changes in general and denitrifying microbial communities as evidenced by molecular markers. Int J Environ Res Public Health 5(5):450–456

Sofu A, Sayilgan E, Guney G (2015) Experimental design for removal of Fe(II) and Zn(II) ions by different lactic acid bacteria biomasses. Int J Environ Res 9(1):93–100

Srivastava P, Kowshik M (2013) Mechanisms of metal resistance and homeostasis in haloarchaea. Archaea 2013:732864

Stott MB, Sutton DC, Watling HR, Franzmann PD (2003) Comparative leaching of chalcopyrite by selected acidophilic bacteria and archaea. Geomicrobiol J 20(3):215–230

Teng W, Kuang J, Luo Z, Shu W (2017) Microbial diversity and community assembly across environmental gradients in acid mine drainage. Minerals 7(6):106

Tian J, Wu N, Li J, Liu Y, Guo J, Yao B, Fan Y (2007) Nickel-resistant determinant from *Leptospirillum ferriphilum*. Appl Environ Microbiol 73(7):2364

Turpeinen R, Kairesalo T, Häggblom MM (2004) Microbial community structure and activity in arsenic-, chromium- and copper-contaminated soils. FEMS Microbiol Ecol 47(1):39–50

Tzvetkova T, Selenska-Pobell S, Groudeva V (2002) Recovery and characterization of *Leptospirillum ferrooxidans/Leptospirillum ferriphilum* and *Acidithiobacillus ferrooxidans* natural isolates from uranium mining waste piles. Biotechnol Biotechnol Equip 16(1):111–117

Van Rossum T, Pylatuk MM, Osachoff HL, Griffiths EJ, Lo R, Quach M, Palmer R, Lower N, Brinkman FSL, Kennedy CJ (2016) Microbiome analysis across a natural copper gradient at a proposed northern Canadian mine site. Front Environ Sci 3:84

Vardanyan NS, Vardanyan AK (2018) Thermophilic chemolithotrophic bacteria in mining sites. In: Egamberdieva D, Birkeland N-K, Panosyan H, Li W-J (eds) Extremophiles in Eurasian ecosystems: ecology, diversity, and applications. Springer, Singapore, pp 187–218

Vehouni A (2001) The geology of the Armenian highlands and the entrails' riches, vol 435. Yerevan Univesity Press, Yerevan

Vela-Cano M, Castellano-Hinojosa A, Vivas AF, Toledo MVM (2014) Effect of heavy metals on the growth of bacteria isolated from sewage sludge compost tea. Adv Microbiol 4(10):644–655

Vijayaraghavan K, Yun Y-S (2008) Bacterial biosorbents and biosorption. Biotechnol Adv 26 (3):266–291

Vivas A, Barea JM, Azcón R (2005) *Brevibacillus brevis* isolated from cadmium- or zinc-contaminated soils improves in vitro spore germination and growth of glomus mosseae under high Cd or Zn concentrations. Microb Ecol 49(3):416–424

Volesky B (1994) Advances in biosorption of metals: selection of biomass types. FEMS Microbiol Rev 14(4):291–302

Wakelin SA, Anand RR, Reith F, Gregg AL, Noble RR, Goldfarb KC, Andersen GL, DeSantis TZ, Piceno YM, Brodie EL (2012) Bacterial communities associated with a mineral weathering profile at a sulphidic mine tailings dump in arid Western Australia. FEMS Microbiol Ecol 79 (2):298–311

Watanabe K (2001) Microorganisms relevant to bioremediation. Curr Opin Biotechnol 12 (3):237–241

Watkin EL, Keeling SE, Perrot FA, Shiers DW, Palmer ML, Watling HR (2009) Metals tolerance in moderately thermophilic isolates from a spent copper sulfide heap, closely related to *Acidithiobacillus caldus, Acidimicrobium ferrooxidans* and *Sulfobacillus thermosulfidooxidans*. J Ind Microbiol Biotechnol 36(3):461–465

White C, Gadd GM (2000) Copper accumulation by sulfate-reducing bacterial biofilms. FEMS Microbiol Lett 183(2):313–318

Xia L-X, Liu J-S, Xiao L, Zeng J, Li B-M, Geng M-M, Qiu G-Z (2008) Single and cooperative bioleaching of sphalerite by two kinds of bacteria - *Acidithiobacillus ferriooxidans* and *Acidithiobacillus thiooxidans*. Trans Nonferrous Metals Soc China 18(1):190–195

Xie X, Fu J, Wang H, Liu J (2010) Heavy metal resistance by two bacteria strains isolated from a copper mine tailing in China. Afr J Biotechnol 9(26):4056–4066

Xingyu L, Biao W, Bowei C, Jiankang W, Renman R, Guocheng Y, Dianzuo W (2010) Bioleaching of chalcocite started at different pH: response of the microbial community to environmental stress and leaching kinetics. Hydrometallurgy 103(1):1–6

Yang Y, Singh RP, Song D, Chen Q, Zheng X, Zhang C, Zhang M, Li Y (2020) Synergistic effect of *Pseudomonas putida* II-2 and *Achromobacter* sp. QC36 for the effective biodegradation of the herbicide quinclorac. Ecotoxicol Environ Saf 188:109826

Yilmaz EI (2003) Metal tolerance and biosorption capacity of *Bacillus circulans* strain EB1. Res Microbiol 154(6):409–415

Zhang J, Zhang X, Ni Y, Yang X, Li H (2007) Bioleaching of arsenic from medicinal realgar by pure and mixed cultures. Process Biochem 42(9):1265–1271

Zhang G, Chao X, Guo P, Cao J, Yang C (2015) Catalytic effect of Ag+ on arsenic bioleaching from orpiment (As2S3) in batch tests with *Acidithiobacillus ferrooxidans* and *Sulfobacillus sibiricus*. J Hazard Mater 283:117–122

Zhu H, Guo J, Chen M, Feng G, Yao Q (2012) *Burkholderia dabaoshanensis* sp. nov., a heavy-metal-tolerant bacteria isolated from Dabaoshan mining area soil in China. PLoS One 7(12): e50225

Chapter 8
Microbial Syntrophy-Mediated Fortification for Eco-enterprising

Mian Nabeel Anwar, Zhi Feng Li, and Raghvendra Pratap Singh

Abstract Anaerobic digestion (AD) is fundamentally a chronological complex, chemical, and biochemical process in the environment. It basically depends on the activity of extremely diverse microbial populations involving hydrolytic, acidogenic, and syntrophic acetogenic bacteria as well as methanogenic archaea. During this process, microbes convert organic waste materials into biogas (high methane content), a combustible source of energy, and utilize as an important environmental technology. The generation of biogas through improving and updating hardware model and engineering process level is well known, but very little information is available about the complex role of microbes with the process, especially, the role of microbes and enzymes and their interaction at different level during the hydrolysis and conversion of organic matter to methane. Hence, in this chapter, we focus on the role of specific genes and enzymes co-related with microbes in different pathways such as hydrogenogenic, acetogenesis, and methanogenesis and generating hydrogen, acetate, and methane, respectively, along with the role of syntrophic relationship in the production of biogas and conclude with a section pointing out some main questions that remain unanswered and can be point of interest for future research.

Keywords Environment · Anaerobes · Enzyme · Biogeochemical cycles · Biogas

M. N. Anwar (✉)
Environmental Engineering Department, University of Alberta, Edmonton, AB, Canada
e-mail: miannabe@ualberta.ca

Z. F. Li
State Key Laboratory of Microbial Technology, Shandong University China, Jinan, China

R. P. Singh (✉)
Department of Research and Development, Azoth Biotech Pvt. Ltd., Noida, Uttar Pradesh, India

R. P. Singh et al. (eds.), *Microbes in Microbial Communities*,
https://doi.org/10.1007/978-981-16-5617-0_8

8.1 Introduction

The process of anaerobic digestion (AD) is a complex process in which multi-flora anaerobic microorganisms metabolize organic matter. The use of human and animal manure for anaerobic digestion and comprehensive utilization can not only recover energy but also purify the environment and achieve better economic benefits, so it has been widely used (Toerien and Hattingh 1969; Shih 1987; Wilkie Ann 2005; Chen et al. 2008).

The AD process is divided into four fundamental sequential stages: hydrolysis, acidogenesis, acetogenesis, and methanogenesis that strongly rely on the collaboration between the various functional groups of bacteria and archaea (Segers 1998; Kim et al. 2006; Nguyen et al. 2007; Gao et al. 2012; Li et al. 2019). In the hydrolysis stage, insoluble organic matter such as macromolecular carbohydrates, proteins, and lipids are decomposed into monosaccharides, amino acids, and short-chain fats, which are soluble in water, and can be decomposed by hydrolytic enzymes and hydrolytic acid-producing bacteria (Detman et al. 2017; Meegoda et al. 2018; Richard et al. 2019). Immediately afterwards, acid-producing bacteria convert the degraded substrates into fatty acids and alcohols such as acetic acid, propionic acid, butyric acid, valeric acid, caproic acid, etc. Such as, in the process of hydrogen production and acetic acid production, homo-acetogenic bacteria and hydrogen-producing bacteria actively participated. The hydrolysate is converted into hydrogen, acetic acid, and carbon dioxide; then in the methanogenesis stage, the hydrogen and acetic acid in the methanogen use the device to convert into methane (Fig. 8.1).

The hydrolysis stage is the rate-limiting step of the entire anaerobic digestion process. Regarding the research of various hydrolytic enzyme activities in anaerobic digestion, a large amount of literature is limited to the determination of enzyme activity before and after biogas fermentation (Zhang et al. 2007; Kim et al. 2012), and there are few reports on the changes of enzyme activity in the process. Therefore, whether the hydrolysis is sufficient is the key to the thoroughness of anaerobic digestion, which will directly affect the amount of gas produced. Our objective is to provide information on various factors influencing anaerobic digestion and the role of specific microbes and enzymes that play a key role during the process (Rai et al. 2012; Singh et al. 2012). This will help people deepen their understanding of the three-stage theory of anaerobic digestion and at the same time provide a scientific basis for improving the utilization of anaerobic digestion of raw materials and effectively increasing the methane production rate.

After a three-stage reaction, the complex organic solid waste in the system is transformed into small molecules that are harmless to the environment. From the perspective of the sustainable development of material protection and energy utilization, anaerobic digestion is the most effective way to stabilize, reduce, and recycle the organic matter in waste and is harmless (Parawira et al. 2005; Zhang et al. 2007; Kim et al. 2012; Adekunle and Okolie 2015).

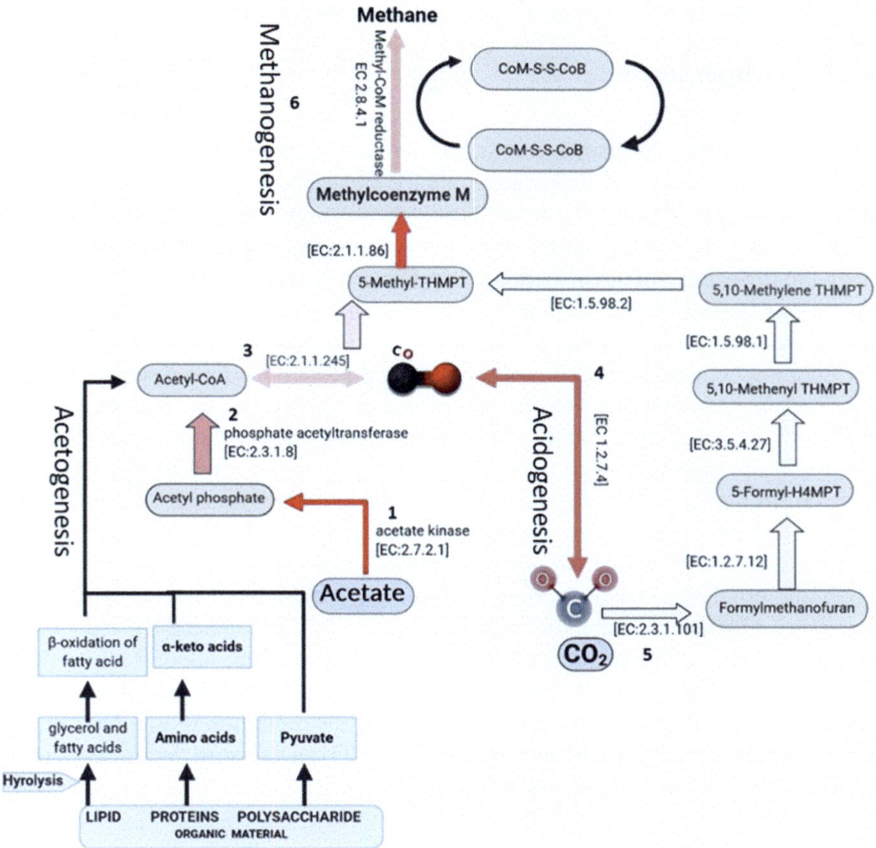

Fig. 8.1 Four successive stages involve in the process of anaerobic digestion (AD): hydrolysis, acidogenesis, acetogenesis, and methanogenesis. Methanogenic pathway of carbon dioxide (hydrogenotrophic pathway); acetate (acetoclastic pathway) revised from Park et al. (2008) and KEGG database. Hydrogenotrophic methanogens use H_2 for the reduction of CO_2 (or CO or formate). Methane is commonly produced using these two methods, acetate can also be converted to methane by a certain syntrophic association where syntrophic acetate-oxidizing bacteria (SAOB) play a central role. In addition, SAOB forms a syntrophic association with hydrogenotrophic methanogens and afterwards yields H_2 as an electron donor for hydrogenotrophic methanogens through the oxidation of acetate

Compared with anaerobic digestion alone, anaerobic co-digestion can significantly increase the methane production potential of biological substrates. Anaerobic co-digestion can dilute the potentially harmful substances in the biological matrix, adjust the humidity and pH in the system, and improve the buffering capacity for the mixed components. The co-fermentation of food waste and excess activated sludge can also increase the concentration of biodegradable substances in the system (Mata-Alvarez et al. 2014; Siddique and Wahid 2018; Esposito et al. 2012).

8.2 Influencing Factors of Anaerobic Digestion

8.2.1 Temperature

Temperature is an important indicator that affects anaerobic biodegradation. The types of microorganisms in sludge are complex, and each microorganism has its own adaptive temperature zone. The anaerobic digestion temperature is divided into three sections: low temperature, medium temperature, and high temperature. Therefore, anaerobic digestion can be divided into low temperature digestion (20 °C), medium temperature digestion (35 °C), and high temperature (50 °C) digestion (Iqbal et al. 2019; Lukitawesa et al. 2020).

Under low-temperature conditions, anaerobic digestion has low energy consumption, inactive microbial activities, low utilization of substrates in the system, and low final methane production, which cannot effectively remove viruses and harmful bacteria in the digestion matrix. In reality, low-temperature digestion is generally not used. In the anaerobic digestion process, moderate temperature and high temperature can generally achieve better digestion results. The medium temperature reaction is generally carried out under the condition of 30–40 °C, which has higher processing efficiency, good reaction stability, relatively mature technology, higher processing efficiency, and more practical use. The high-temperature reaction is generally carried out under the condition of 50–60 °C, with short reaction time, small space, and high load. Although high-temperature digestion contributes to the killing efficiency of pathogens, it is susceptible to the inhibition of high concentrations of ammonia nitrogen (Malý and Fadrus 1971; Donoso-Bravo et al. 2009; Appels et al. 2010; Iqbal et al. 2019).

8.2.2 pH Value

The pH value is also an important parameter that affects anaerobic biodegradation. Microorganisms are very sensitive to changes in pH when they grow (Singh et al. 2016). By adjusting the initial pH value of the reaction through the inner loop, microbial anaerobic digestion can achieve different effects. The acid production period of anaerobic digestion can gradually reduce the pH value of the reactor environment. With the change of pH value, some acid-producing bacteria need to adapt to the change of pH value within the corresponding range.

When the acid production of the system reaches a certain value, pH 5.5–5.6, the growth of acid-producing bacteria is the best. The initial reaction pH of anaerobic digestion is around 5.6–7, and the activity of methanogens is the best. If the pH is too high or too low, it will affect the efficiency of methanogenesis (Lindner et al. 2015; Ravi et al. 2018). The starting pH is around 10, and the yield of volatile organic acids is the highest. During the hydrolysis and acidification stage of anaerobic digestion, there must be a lot of hydrolytic enzyme activities. Each hydrolytic enzyme will exert its best catalytic effect only under the adaptive pH, and an inappropriate pH will cause the inactivation of hydrolytic enzymes (Zhang et al. 2005; Lindner et al. 2015).

8.2.3 Process

By anaerobic digestion, the reaction chamber can be divided into single-phase and two-phase anaerobic digestion. Single-phase anaerobic digestion of acid-producing bacteria and methanogenic bacteria in one container has direct reaction, simple operation, and low investment. During the reaction, acid-producing bacteria and methanogens interact with each other. Appropriate amount of organic acid provides carbon source for methanogens, but if excessive, it will weaken the gas production efficiency of methanogens. Two-phase digestion of acidogenic and methane is divided into two phases (Ganesh et al. 2014; Leite et al. 2016). Acid-producing bacteria and methanogens work in two containers, respectively, and organic acids and methanogens do not affect each other. The disadvantage is that the equipment is complex, the load in the reactor is high, and the methane production potential is large (Demirel and Orhan 2002; Park et al. 2008).

8.2.4 Moisture Content

In general, anaerobic digestion is divided into dry and wet digestion methods. Under dry digestion, the moisture content of the material is about 60–80%. The moisture content of wet digestion material is above 85%. Under the conditions of dry digestion, because there are more organic substrates available for processing, the gas production rate is relatively large, and the ability to treat organic waste is relatively large. However, the dry digestion equipment is relatively high, and the processing process is easily affected by toxic substances. The cost of wet digestion equipment is low, the gas production rate is low, the process is susceptible to the influence of ammonia nitrogen and salt concentration, and the pretreatment is complicated (Lay et al. 1997; Fujishima et al. 2000; Zhang et al. 2019).

8.2.5 Food to Microorganism (F/M) Ratio

In the process of anaerobic digestion, the nutrients needed for the growth and secretion of microorganisms are provided by the substrate. The most important thing in the nutrient ratio is the C/N ratio. The F/M ratio is essentially the C/N ratio. Microbial reproduction and metabolism are affected by the C/N ratio. For anaerobic digestion, a suitable C/N ratio can promote the rapid degradation of macromolecular organic matter in the matrix and generate sufficient methane-generating substrates. If the C content is too high, the nitrogen content in the digestive juice is too low, and the digestion solution cannot quickly buffer the pH changes caused by organic acids, which leads to the disorder of microbial metabolism, and the organic acids are easy to accumulate and excess (Nguyen et al. 2021).

If the C content is too low, ammonia nitrogen will continue to accumulate, and the pH value will rise, resulting in ammonia nitrogen inhibition. The C/N ratio in sludge is about 7.1/1; the C/N in food waste is about 50/1. In the process of anaerobic co-digestion, protein can give the system a better buffer capacity and a wider range of nutrients, and high-concentration carbon-containing waste can balance the C/N ratio and reduce ammonia inhibition in the system (Liu et al. 2011). Therefore, food waste can be used as a high-concentration carbohydrate to dilute the sludge. In the anaerobic digestion system, when the C/N ratio is about 20/1 to 30/1, the higher the acid production during anaerobic fermentation. During digestion, organic carbon compounds are continuously degraded and converted into CH_4 and CO_2. At the same time, a part of organic carbon and nitrogen synthesize nutrients needed by microorganisms. The excess ammonia nitrogen is dissolved in the buffer, and the C/N ratio in the system continues to decrease. Therefore, the quality of carbohydrates at the beginning of the anaerobic reaction is generally higher than that of protein (Liu et al. 2011, 2012; Saha et al. 2018).

8.2.6 Additives

The efficiency of anaerobic digestion is easily affected by system substances because microbial activities are affected by the surrounding environment. For example, an appropriate amount of sodium hydroxide can accelerate cell rupture and enhance utilization of microbial substrates, thereby promoting overall anaerobic digestion process. Excessive grease wraps the cells of microorganisms, inhibits the use of surrounding nutrients by the cells, thereby inhibiting the entire anaerobic digestion process (Singh et al. 2020). In the process of anaerobic digestion, some carbon sources can be added appropriately to speed up the process of anaerobic digestion efficiency. Researchers speed up the efficiency of anaerobic digestion by adding some substances such as enzymes and surfactants. Therefore, in the process of anaerobic digestion of sludge in sewage treatment plants, some additives can be added to accelerate the reaction of microorganisms (Romero-Güiza et al. 2016; Ye et al. 2018; Paritosh et al. 2020; Tang et al. 2020).

A large number of microorganisms participate in the anaerobic digestion process. Therefore, it is necessary to classify the microorganisms at each stage of the digestion process. The microorganisms involved in each stage are different. Microorganisms in sludge include bacteria, fungi, molds, and protozoa. Among them, bacteria are the main microorganisms involved in the process of anaerobic digestion, hydrolysis, and acidification of sludge, including hydrolytic acidifying bacteria and acetogenic bacteria, collectively referred to as fermentation acid-producing bacteria (Riviere et al. 2009; Ziganshin et al. 2011; Adekunle and Okolie 2015).

During the hydrolysis stage, the bacteria can be preliminarily divided into the following categories according to the different substrates of the bacteria: Carbohydrate-degrading bacteria: endospore-shaped, rod-shaped bacteria take carbohydrates as food and occupy a dominant position in the reproduction of the microbial community. Carbohydrates are degraded by Clostridium to form monosaccharides, which are then converted into acetone, butanol, hydrogen, and acetic acid. Protein-degrading bacteria: protein is decomposed into amino acids. Microbes use part of the protein to synthesize the nutrients they need, and the other part is converted into ammonia, hydrogen sulfide, and organic acids. Some nitrogen-containing compounds can also be decomposed by protein-degrading bacteria. Fat-degrading bacteria: The typical fat-degrading bacteria is Vibrio. Bacteria can degrade large fat molecules into short-chain fat molecules, and then produce carbon dioxide and methane. Cellulose-degrading bacteria: Cellulose is decomposed and then converted into carbon dioxide, hydrogen, acetic acid and ethanol. Hydrogen-producing acetogens in anaerobic digestion systems often have a symbiotic relationship with methanogens. The main function of methanogens is to convert the substrates in the hydrolysis acidification stage and acetogenesis into methane (Riviere et al. 2009; Ziganshin et al. 2011; Baek et al. 2018).

The main methanogens are: *Methanococcus*, *Sarcina* methanogens, *Methanothrix*, and Methanogens. Hydrogen-producing acetogens convert various higher fatty acids and alcohols into acetic acid and hydrogen to provide a suitable substrate for methanogens. Hydrogen-producing acetogenic bacteria use the existing organic matter in the digestion tank to degrade aromatic acids and other organic acids to produce acetic acid, hydrogen, and carbon dioxide. When the substrate is an even-numbered carbon atom, acetic acid and hydrogen are generated; when the odd-numbered carbon is degraded, it can generate carbon dioxide. When microorganisms degrade carbohydrates, pyruvate is generally its intermediate product. Under anaerobic conditions, pyruvate is used by some hydrogen-producing bacteria to produce acetic acid, carbon dioxide, and release hydrogen. The main hydrogen- and acetogen-producing bacteria in the anaerobic digestion process are: *Cynobacterium*, *Clostridium*, *Coriobacterium*, and *Pseudomonas*, most of which are obligate anaerobes and facultative anaerobes (Xu et al. 2014; Chen et al. 2018).

8.3 Syntrophic Butyrate Metabolism

β-Oxidation pathway plays the role to proceed the syntrophic butyrate metabolism reported by Wofford et al. (1986). In the first step of this pathway, the transfer of one of the CoA groups from acetyl-CoA molecule helps in the activation of butyrate to butyryl-CoA and other in the synthesis of ATP (Fig. 8.3). Butyrate is converted to butyryl-CoA to crotonyl-CoA and further converted through different enzymes into (S)-3-hydroxybutyryl-CoA to acetoacetyl-CoA; at every step, different enzymes

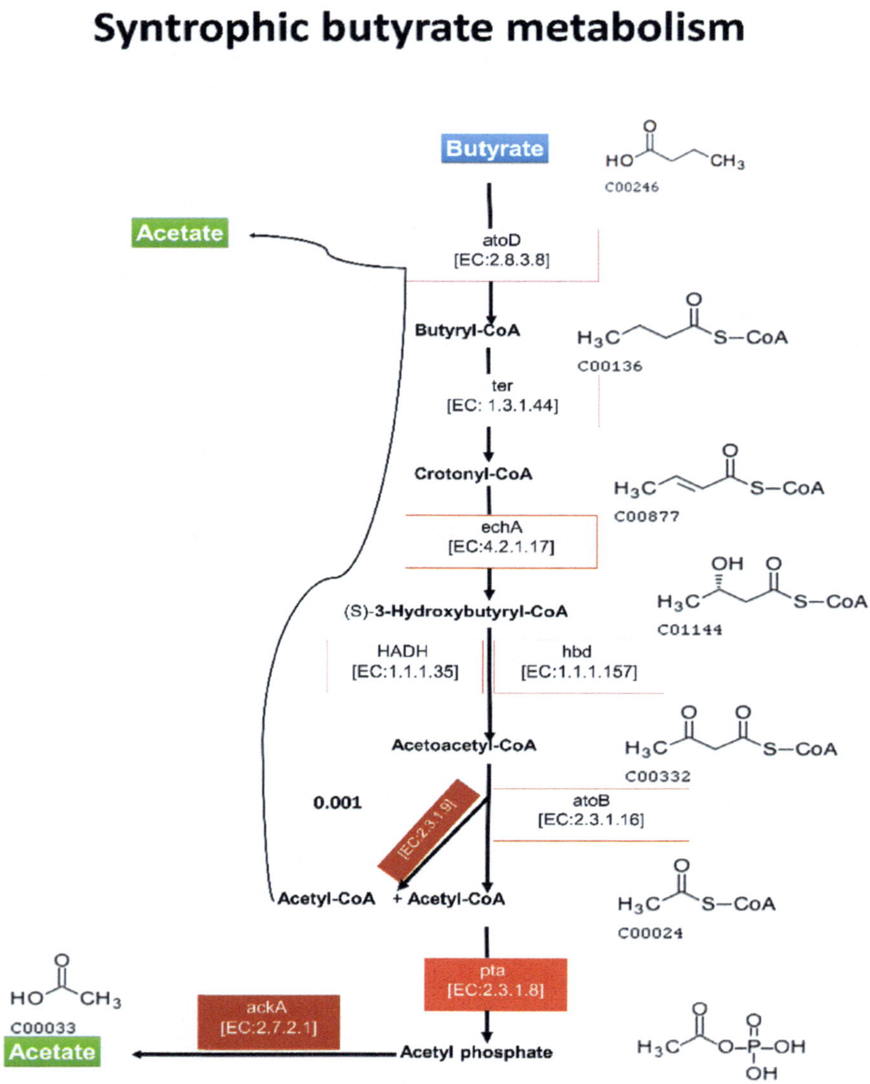

Fig. 8.2 The beta-oxidation pathway for butyrate metabolism revised from Wofford et al. (1986). The enzymes involved are: (atoD) CoA transferase; (ter) acyl-CoA dehydrogenase; (echA) enoyl-CoA hydratase; (HADH/hbd) L-(+)-3-hydroxybutyryl-CoA dehydrogenase; (atoB) 3-ketoacyl-CoA thiolase; (pta) phosphotransacetylase; (ackA) acetate kinase

paly their role in the degradation of the compounds through different microbes (Fig. 8.2). Phosphate acetyltransferase (PTA) [EC 2.3.1.8] and acetate kinase (AK) [EC 2.7.2.1] are the two main enzymes required in the standard mechanism of acetyl-CoA conversion to acetate (Wofford et al. 1986). Along with the production of acetate from acetyl-CoA in prokaryotes, combined with the formation of ATP during this process (Schäfer et al. 1993; Schäfer 2003).

$$\begin{aligned} \text{CoA} + P_i &\rightleftharpoons \text{acetyl phosphate} + \text{CoA(PTA)Aacetyl phosphate} + \text{ADP} \\ &\rightleftharpoons \text{acetate} + \text{ATP(AK)} \end{aligned}$$

Acetate is the final product in this β-oxidation pathway which is the key source during methanogenesis and leading towards methane production.

8.4 Propionate Metabolism

In anaerobic digestion of organic polymers propionate is a crucial mediator in total methanogenesis (~6%–35%) that degraded into acetate and H_2/CO_2 and finally into methane (Glissmann and Conrad 2000). In propionate metabolism, there are two pathways: the methylmalonyl-CoA pathway and the unique dismutation pathway (Fig. 8.3). The methylmalonyl-CoA pathway is reported frequently in many syntrophic propionate-oxidizing bacteria such as *Syntrophaceae* (McInerney et al. 2008; Chen et al. 2005). Syntrophic association is very essential for the oxidation of propionate in anaerobic digestion, mainly based on syntrophy between propionate-oxidizing bacteria and hydrogenotrophs (Li et al. 2012; Chen et al. 2005; Liu and Lu 2018). The methylmalonyl-CoA pathway for propionate metabolism was revised from Kosaka et al. (2006). This pathway is also known as the randomizing pathway, like to syntrophic butyrate metabolism, acetyl-CoA transfer CoA group for the initiation of propionate to propionate-CoA.

8.5 Immobilization of Enzymes

Enzymes refer to a class of chemical substances with catalytic function, which can efficiently treat wastewater and catalyze food (Yang et al. 2015). Enzyme activity is easily affected by temperature, pH, heavy metals, and activators (Table 8.1). Enzymes in a pure chemical state are unstable and easily inactivated. In general, it is necessary to fix the enzyme before using it. The high-efficiency catalysis of the enzyme is realized through the stabilization and protection of the carrier. However, the process of enzyme immobilization is generally accompanied by loss of enzyme activity. The current enzyme immobilization technologies mainly include: embedding method, adsorption method, peptide bond method, covalent bond method, and cross-linking method (Breure et al. 1986).

During the embedding process, the enzyme is encapsulated in the capsule (such as microcapsule embedding) and the lattice of polymer (such as gel embedding) and reacts with the substrate that has penetrated into the lattice. Adsorption methods

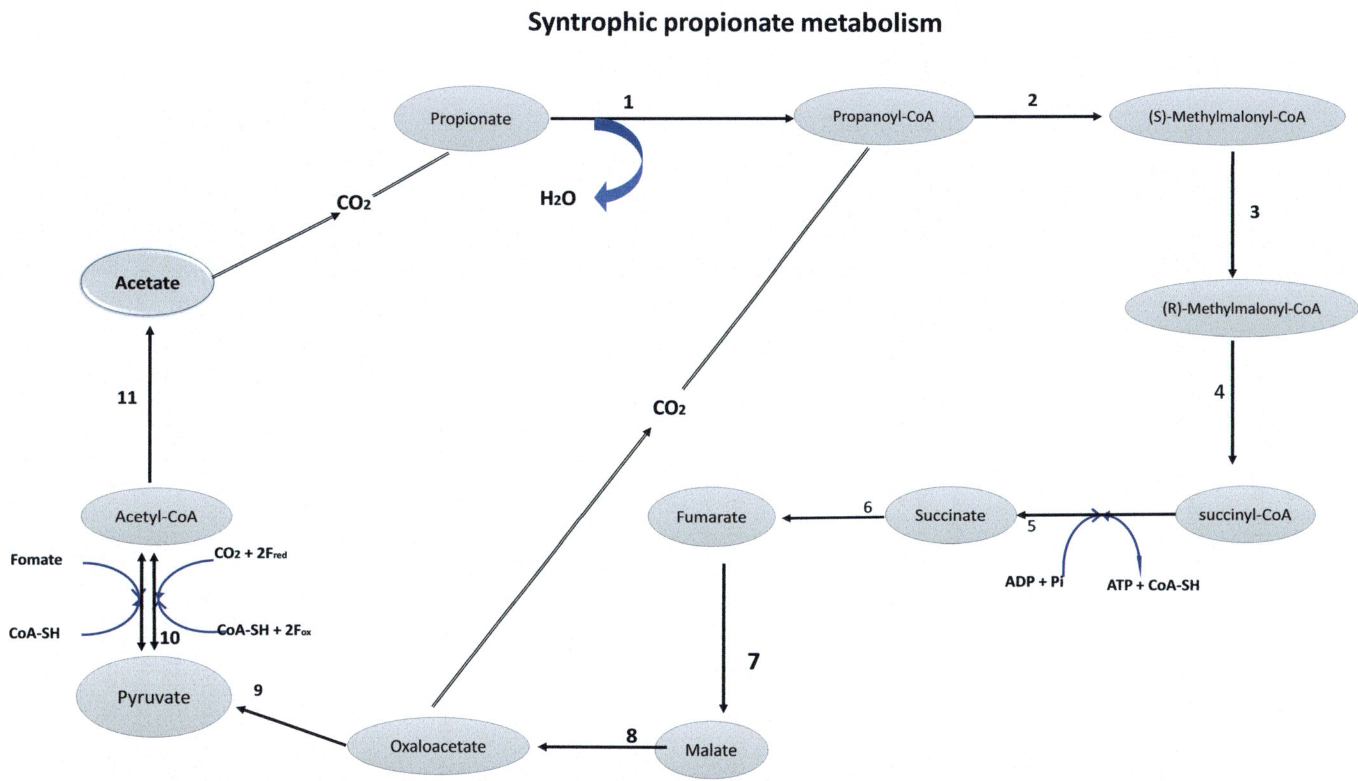

Fig. 8.3 The methylmalonyl-CoA pathway for syntrophic propionate metabolism was revised from Kosaka et al. (2006). The enzymes involved are as follows: (1) propionate-CoA transferase; (2) propionyl-CoA: oxaloacetate transcarboxylase; (3) methylmalonyl-CoA epimerase; (4) methylmalonyl-CoA mutase; (5) succinyl-CoA synthetase; (6) succinate dehydrogenase/fumarate reductase; (7) fumarate hydratase; (8) malate dehydrogenase; (9) pyruvate dehydrogenase; (10) pyruvate: formate lyase; (11) acetyl-CoA synthase; and (12) acetate kinase. Fd is ferredoxin

Table 8.1 Factors affecting enzyme activity during anaerobic digestion

Factors	Effect on enzyme activity
Temperature	Enzyme activity increases with increasing temperature. When the temperature increases to a certain level, the enzyme activity reaches its maximum value, and then the enzyme activity decreases as the temperature increases
pH	Any enzyme can only exhibit a highly active catalytic effect under a specific pH range. For example, the optimal catalytic pH of amylase is 6–8; the optimal catalytic pH of neutral protease is around 7
Substrate concentration	The concentration of the substrate is directly proportional to the reaction rate. Generally, the faster the reaction speed of the enzyme-catalyzed substrate, the greater the enzyme activity. When the amount of the enzyme is fixed and the substrate is within a certain concentration range, the activity of the enzyme expression is directly proportional to the concentration of the substrate
Enzyme concentration	The higher the enzyme concentration, the greater the reaction rate. The concentration of hydrolytic enzymes present in the sludge is generally very low. The higher the concentration of the substrate within a certain range, the larger the contact area between the enzyme and the substrate, and the higher the displayed activity. When industrial enzymes are used for anaerobic digestion of sludge, only a small volume of enzymes are needed to achieve a good hydrolysis effect
Heavy metal	Some heavy metals can destroy the disulfide bonds in the protein, denature the protein, and cause the enzyme to lose live. Microorganisms that secrete hydrolytic enzymes are affected by heavy metals during sludge hydrolysis. For example, low concentrations of Cu promote the secretion of hydrolytic enzymes of microorganisms in crop composting process, the secreted hydrolase inhibition in the high concentrations of Cu composting process
Inhibitors and activators	Inhibitors and activators, and through a specific binding site of the enzyme, inactivation of enzyme activity to achieve or inactive enzymes to improve the process activity

include physical adsorption and ion adsorption, which are connected by ionic bonds, van der Waals forces, and hydrogen bonds. Covalent methods include peptide bond method, diazonium method, alkyl and arylation method (Breure et al. 1986). The immobilized enzyme is easy to industrialize and batch production. In beer production, amylase in its natural state is not easy to completely hydrolyze starch. The immobilized amylase can sustainably catalyze the substrate starch in the fermentation broth and use it repeatedly (Xu and Lin 2007). There are abundant enzymes in biological cells, and the abundant enzyme solution in the cells can be extracted after being broken. The methods of separating intracellular enzymes mainly include centrifugal filtration, membrane separation technology, extraction filtration and precipitation technology, chromatography, and electrophoresis (Breure et al. 1986; Xu and Lin 2007).

8.6 Enzymes Derived from Excess Activated Sludge and Organic Waste

There are three forms of enzymes in sludge: enzymes on the cell surface, enzymes in sludge flocs, and enzymes in solution. In the anaerobic digestion process, because the concentration of enzymes dissolved in water is very low, the enzymes that hydrolyze nutrients in the system are mainly present in flocs (Table 8.2). At present, there are many pretreatment methods used to improve the hydrolysis efficiency in the anaerobic digestion process of sludge. No matter which pretreatment method is used, the reaction between the enzyme and the substrate always runs through the hydrolysis process. The pretreatment method can increase the concentration of enzymes in the water phase of the sludge. At present, the pretreatment methods for enhancing hydrolytic enzymes in sludge mainly include: ultrasonic enhanced pretreatment, radiation pretreatment technology, low temperature heat treatment, cation binder, different electron acceptors, and temperature pH methods. When the concentration of enzyme in the sludge is relatively high, the stability and purification of the enzyme can further improve the functionality and application range of the enzyme (Dang and Zhang 2004; Qin and Yu 2011). In the sewage treatment process, the hydrolytic enzymes involved in the hydrolysis process include protease, amylase, lipase, cellulase, and lytic enzymes.

The enzyme contained in the floc mainly refers to the enzyme contained in the extracellular polymer. The extracellular polymer (EPS) of the remaining activated sludge contains 17% 1-Leu-aminopeptidase, 5% α-glucosidase, 23% protease, and

Table 8.2 Types of hydrolases in AD

Hydrolase	Action principle
Protease	According to the optimal pH value, it can be divided into neutral protease, acid protease, and alkaline protease. It can be divided into endopeptidases and exopeptidases according to the different positions of the protein molecules. The enzymes used in industry are mainly endopeptidases (Dang and Zhang 2004; Xu and Lin 2007; Qin and Yu 2011)
Lysozyme	It mainly acts on the β-1,4-glycosidic bond between *N*-acetylglucosamine and *N*-acetylmuramic acid. The anaerobic digestion of sludge mainly acts on Gram-positive bacteria with thicker cell walls (Xu and Lin 2007; Qin and Yu 2011)
Amylase	Divided into α-amylase, β-amylase, γ-amylase, and isoamylase. α-Amylase and β-amylase mainly act on the α-1,4-chain of amylose and amylopectin. γ-Amylase acts on the α-1,4-chain glycosidic bond and the α-1,6-chain glycosidic bond to cause hemiacetal hydroxyl translocation to release β-glucose, and the final product is glucose. The main enzyme that catalyzes carbohydrates in sludge is α-amylase (Dang and Zhang 2004; Qin and Yu 2011; Bai and Zhang 2012)
Cellulase	Including β-glucosidase, endo β-glucosidase, exo β-glucosidase, and other enzymes. One mole of cellobiose can be hydrolyzed by β-glucosidase to 2 mol of glucose (Dang and Zhang 2004; Qin and Yu 2011; Bai and Zhang 2012)
Lipase	It mainly acts on triacylglycerol acyl groups to generate fatty acids, glycerol, and mono- or di-glycerides. The catalytic activity of lipase depends only on its protein structure (Esposito et al. 2012; Meng et al. 2017)

44% α-amylase [65]. The extracellular polymer (EPS) in the sludge flocs is mainly composed of humic acid, carbohydrates, protein, and lipids. It not only provides a stable living place for the hydrolytic enzymes but also provides enough for the microorganisms in the sludge (Singh et al. 2017). Therefore, destroying the flocs of sludge is an effective way to increase the reaction matrix and hydrolytic enzymes in the system. In addition, enzyme-catalyzed destruction of the cell's own structure is also an effective way to increase the reaction matrix.

Lysozyme plays an important role in destroying the cell wall of cells, especially Gram-positive bacteria. Gram-positive bacteria have thicker cell walls and higher peptidoglycan content. The combination of glycan chains in the lysozyme peptide bond sugar is the degradation of the peptidoglycan and the destruction of the cell wall, which leads to the imbalance of cell stability in the environment. In contrast, Gram-negative bacteria have thinner cell walls and thinner peptidoglycans. Compared with Gram-positive bacteria, the presence of bacteria in the environment does not rely much on the cell wall.

8.7 The Role of Hydrolase in the Anaerobic Digestion Process

During anaerobic digestion, the main enzymes put into the sludge are endogenous enzymes (secreted by microorganisms) and exogenous enzymes. The organic nutrients of sludge mainly include carbohydrates and protein. Yang and his workers put amylase, protease, mixed amylase, and protease into the sludge anaerobic fermentation system after putting in the hydrolase to compare the hydrolysis effect of the sludge. The results show that the mixed amylase and protease after being put into the sludge fermentation system, the hydrolysis effect is the best, and the ratio of amylase and protease is 1:3 (Esposito et al. 2012; Meng et al. 2017). In the system, carbohydrates are catalyzed by amylase to produce monosaccharides; under the action of proteases, peptide bonds are destroyed to produce polypeptides and monopeptides, which are further decomposed into amino acids. Monosaccharides and amino acids become water-soluble small molecules that pass through the cell wall, and then pass through the cell membrane to be used by the cytoplasm of the cell. The structure of an enzyme is a protein structure or an RNA structure, which itself is a nutrient substance that can be utilized by microorganisms. Therefore, the inactivated enzyme can be used directly by the cell.

The methods to accelerate the efficiency of anaerobic digestion generally start from three aspects: hydrolysis acidification, acetic acid production, and methane production. The hydrolysis stage can be accelerated by pretreatment. In the laboratory, alkali, heat, enzyme, and physical pre-methods have been used to accelerate the hydrolysis stage of the anaerobic digestion process of food waste or excess activated sludge (Cadoret et al. 2002; Chen et al. 2009). The purpose of the pretreatment method is to increase the nutrients available to the microorganisms in the solution.

Although some methods are more efficient for the rate of methane production, they can easily pollute the environment if they are improperly operated. Only the enzyme pretreatment method is the most environmentally friendly, because when the enzyme reacts with the substrate, the energy input to the environment is relatively low, the catalytic efficiency is high, and no other chemical substances are produced except for the products generated after the substrate is degraded. After the reaction, the structure and content of the enzyme will not change and will not cause secondary pollution to the environment. Therefore, the enzyme pretreatment method has been widely used in the treatment of waste by anaerobic digestion technology (Cadoret et al. 2002; Chen et al. 2009).

Appropriate addition of enzymes can accelerate the degradation of macromolecular substrates in the process of anaerobic digestion, making them small molecules that are dissolved in water. For example, when organic biomass such as wheatgrass, dairy cow manure, coconut milk effluent, solid cow manure, pasture silage, sewage sludge, and other organic biomass are used as substrates for anaerobic digestion, the addition of enzymes accelerates the decomposition of corresponding organic components, it is used by microorganisms more quickly (Chen et al. 2009; Singh et al. 2014). The rapid accumulation of hydrolyzed substrates provides more substrates for microorganisms working in the acetic acid and methanogenic stages of anaerobic fermentation, thereby increasing methane production as a whole and improving the efficiency of anaerobic digestion.

At present, there are two main sources of enzymes as additives into the reactor: direct addition of chemical forms of enzymes (exogenous or endogenous enzymes); and inoculation of certain microorganisms that secrete specific enzymes. Exogenous enzymes are generally refined through chemical reactions; endogenous enzymes are generally secreted by specific microorganisms under corresponding nutritional conditions; and inoculated with specific microorganisms, the inoculated microorganisms use the organic matter in the sludge to reproduce and metabolize. During the process of introducing exogenous enzymes, the source of endogenous enzymes and microorganisms is relatively wide. There are many microorganisms that can secrete enzymes in nature (Yu et al. 2007; Li et al. 2009, 2017; Yang et al. 2019).

Sludge, as an effective medium for wastewater treatment, includes abundant hydrolysis microorganisms. For example, Bacillus subtilis secretes the fastest at 30 °C, can secrete high concentrations of amylase, and can also secrete protease under specified conditions. *Aeromonas hydrophila* can secrete highly active proteases at 37 °C and pH 7. Researchers isolated a strain of *Brevibacterium KH3* from the sludge. Experiments have shown that *Brevibacterium KH3* can help degrade extracellular polymers under moderate temperature aerobic digestion and improve cell rupture efficiency, and the rapid propagation of *Brevibacterium KH3* inhibits the growth of other unrelated microorganisms, which is a hydrolysis process thar provides higher efficiency (Khan et al. 2016; Zhao et al. 2015; Wang et al. 2015). It can be seen that the sludge contains many microorganisms that promote hydrolysis. On the one hand, the abundant microorganisms in sludge can be used as a source of bacteria to secrete hydrolytic enzymes, and on the other hand, they can be used as inoculation microorganisms for the production of other by-products. The microorganisms in the sludge secrete products with higher purity after enrichment and separation. Therefore, it is more economical to add endogenous enzymes or inoculate microorganisms that secrete specific enzymes when promoting the anaerobic digestion process of organic matter. Although the hydrolase of extracellular polymer is relatively abundant, the amount is still very small. It was separated pure *Bacillus subtilis* and *Aeromonas hydrophila* from activated sludge, and then reproduced in solid medium. Then put *Bacillus subtilis* and *Aeromonas hydrophila* into liquid culture to multiply and secrete amylase and protease, respectively (Luo et al. 2012; Kiran et al. 2014; Wang et al. 2016). The results proved that the mixed liquid of *Bacillus subtilis* and *Aeromonas hydrophila* was put into the sludge anaerobic digestion system, and the digestion efficiency was significantly improved. The addition of the bacterial liquid accelerates the hydrolysis efficiency of the sludge and increases the production of short-chain organic acids (SCFAs). After 11 days of reaction, methane production increased by 23.1%. Waste is rich in organic nutrients that can be degraded by microorganisms and trace elements that are conducive to the growth and reproduction of microorganisms. Previous documents have proved that organic waste inoculated with specific microorganisms under fermentation conditions can generate specific high-activity enzymes such as protease, cellulase, amylase, lipase, and pectinase (Odnell et al. 2016; Bilal and Iqbal 2019; Luo et al. 2020).

8.8 Conclusion

This chapter summarizes some examples of organic waste used to produce enzymes, as well as the application of exogenous enzymes and some endogenous enzymes in the anaerobic digestion process, their types, applications, and influencing factors of hydrolytic enzymes, and an in-depth understanding of hydrolytic enzymes in the anaerobic digestion process promotes the mechanism of anaerobic digestion and hydrolysis stage. At present, there have been many studies on co-digestion of organic waste and surplus activated sludge, but there are few articles about the application of co-fermentation of organic waste and surplus activated sludge to produce enzymes to increase methane production in a two-phase anaerobic reactor. In this chapter, the co-fermentation experiment of food waste and surplus activated sludge mainly focuses on the following three aspects to demonstrate the production of enzymes by combined anaerobic fermentation of organic waste and activated sludge, and the addition of organic waste to the two-phase reaction system. The feasibility of methane generation potential, when garbage or sludge is digested separately, are stated below:

1. Studying different process parameters (temperature, pH, and mass ratio of organic waste to the remaining activated sludge), the activity of hydrolase during co-fermentation of organic waste and remaining activated sludge to determine the optimal hydrolysis and enzyme process parameters.

2. Using the excess activated sludge as the reaction substrate for anaerobic fermentation, put the enzyme solution from the anaerobic fermentation of the previous organic waste and the remaining activated sludge to study the effect of the enzyme solution on the anaerobic fermentation and hydrolysis stage of the remaining activated sludge.
3. When the waste sludge is fermented separately after the co-fermentation of organic waste and the remaining activated sludge to produce enzyme and the enzyme solution is added, it will be accompanied by anaerobic microbial activities, including the reproduction of enzyme-producing bacteria.

References

Adekunle KF, Okolie JA (2015) A review of biochemical process of anaerobic digestion. Adv Biosci Biotechnol 6(03):205

Appels L, Degrève J, Van der Bruggen B, Van Impe J, Dewil R (2010) Influence of low temperature thermal pre-treatment on sludge solubilisation, heavy metal release and anaerobic digestion. Bioresour Technol 101(15):5743–5748

Baek G, Kim J, Kim J, Lee C (2018) Role and potential of direct interspecies electron transfer in anaerobic digestion. Energies 11(1):107

Bai L, Zhang L (2012) Research progress of enzyme and protein separation and purification technology. Anhui Agric Sci 40(14):8018–8020

Bilal M, Iqbal HM (2019) Sustainable bioconversion of food waste into high-value products by immobilized enzymes to meet bio-economy challenges and opportunities–a review. Food Res Int 123:226–240

Breure AM, Mooijman KA, Van Andel JG (1986) Protein degradation in anaerobic digestion: influence of volatile fatty acids and carbohydrates on hydrolysis and acidogenic fermentation of gelatin. Appl Microbiol Biotechnol 24(5):426–431

Cadoret A, Conrad A, Block JC (2002) Availability of low and high molecular weight substrates to extracellular enzymes in whole and dispersed activated sludges. Enzym Microb Technol 31 (1):179–186

Chen S, Xiaoli L, Xiuzhu D (2005) *Syntrophobacter sulfatireducens* sp. nov., a novel syntrophic, propionate-oxidizing bacterium isolated from UASB reactors. Int J Syst Evol Microbiol 55 (3):1319–1324

Chen Y, Jay JC, Kurt S (2008) Creamer Inhibition of anaerobic digestion process: a review. Bioresour Technol 10:4044–4064

Chen Y, Jiang M, Ye Y et al (2009) Research progress of lysozyme. Chinese J Biol 26(2):64–66

Chen S, Ning L, Bin D, Wentao Z, Lingling D, Xiaohu D (2018) New insights into the enhanced performance of high solid anaerobic digestion with dewatered sludge by thermal hydrolysis: organic matter degradation and methanogenic pathways. J Hazard Mater 342:1–9

Dang H, Zhang B (2004) Preparation of immobilized enzyme and its application in food industry. Food Res Dev 25(3):68–72

Demirel B, Orhan Y (2002) Two-phase anaerobic digestion processes: a review. J Chem Technol Biotechnol 77(7):743–755

Detman A, Chojnacka A, Błaszczyk M, Kaźmierczak W, Piotrowski J, Sikora A (2017) Biohydrogen and biomethane (Biogas) production in the consecutive stages of anaerobic digestion of molasses. Pol J Environ Stud 26(3):1023–1029

Donoso-Bravo A, Retamal C, Carballa M, Ruiz-Filippi G, Chamy R (2009) Influence of temperature on the hydrolysis, acidogenesis and methanogenesis in mesophilic anaerobic digestion: parameter identification and modeling application. Water Sci Technol 60(1):9–17

Esposito G, Frunzo L, Giordano A, Liotta F, Panico A, Pirozzi F (2012) Anaerobic co-digestion of organic wastes. Rev Environ Sci Biotechnol 11(4):325–341

Fujishima S, Miyahara T, Noike T (2000) Effect of moisture content on anaerobic digestion of dewatered sludge: ammonia inhibition to carbohydrate removal and methane production. Water Sci Technol 41(3):119–127

Ganesh R, Torrijos M, Sousbie P, Lugardon A, Steyer JP, Delgenes JP (2014) Single-phase and two-phase anaerobic digestion of fruit and vegetable waste: comparison of start-up, reactor stability and process performance. Waste Manag 34(5):875–885

Gao R, Yuan X, Zhu W, Wang X, Chen S, Cheng X, Cui Z (2012) Methane yield through anaerobic digestion for various maize varieties in China. Bioresour Technol 118:611–614

Glissmann K, Conrad R (2000) Fermentation pattern of methanogenic degradation of rice straw in anoxic paddy soil. FEMS Microbiol Ecol 31(2):117–126

Iqbal ACC, Chang WL, Ju M (2019) Anaerobic digestion. Water Environ Res 91(10):1253–1271

Khan MA, Ngo HH, Guo WS, Liu Y, Nghiem LD, Hai FI, Deng LJ, Wang J, Wu Y (2016) Optimization of process parameters for production of volatile fatty acid, biohydrogen and methane from anaerobic digestion. Bioresour Technol 219:738–748

Kim JK, Oh BR, Chun YN, Kim SW (2006) Effects of temperature and hydraulic retention time on anaerobic digestion of food waste. J Biosci Bioeng 102(4):328–332

Kim HW, Nam JY, Kang ST, Kim DH, Jung KW, Shin HS (2012) Hydrolytic activities of extracellular enzymes in thermophilic and mesophilic anaerobic sequencing-batch reactors treating organic fractions of municipal solid wastes. Bioresour Technol 110:130–134

Kiran EU, Trzcinski AP, Ng WJ, Liu Y (2014) Enzyme production from food wastes using a biorefinery concept. Waste Biomass Val 5(6):903–917

Kosaka T, Uchiyama T, Ishii SI, Enoki M, Imachi H, Kamagata Y, Ohashi A, Harada H, Ikenaga H, Watanabe K (2006) Reconstruction and regulation of the central catabolic pathway in the thermophilic propionate-oxidizing syntroph *Pelotomaculum thermopropionicum*. J Bacteriol 188(1):202–210

Lay JJ, Li YY, Noike T (1997) Influences of pH and moisture content on the methane production in high-solids sludge digestion. Water Res 31(6):1518–1524

Leite WR, Gottardo M, Pavan P, Belli Filho P, Bolzonella D (2016) Performance and energy aspects of single and two phase thermophilic anaerobic digestion of waste activated sludge. Renew Energy 86:1324–1331

Li H, Lan Q, Li G et al (2009) Study on the degradation performance of amylase. Dril Fluid Compl Fluid 26(4):1–3

Li J, Ban Q, Zhang L, Jha AK (2012) Syntrophic propionate degradation in anaerobic digestion: a review. Int J Agric Biol 14:843

Li ZF, Li P, Jing YG, Singh RP, Li YZ (2017) Isolation and characterization of the epothilone gene cluster with flanks from high alkalotolerant strain *Sorangium cellulosum* (So0157-2). World J Microbiol Biotechnol 33(7):137. https://doi.org/10.1007/s11274-017-2301-y

Li Y, Chen Y, Wu J (2019) Enhancement of methane production in anaerobic digestion process: a review. Appl Energy 240:120–137

Lindner J, Zielonka S, Oechsner H, Lemmer A (2015) Effect of different pH-values on process parameters in two-phase anaerobic digestion of high-solid substrates. Environ Technol 36 (2):198–207

Liu P, Lu Y (2018) Concerted metabolic shifts give new insights into the syntrophic mechanism between propionate-fermenting *Pelotomaculum thermopropionicum* and hydrogenotrophic *Methanocella conradii*. Front Microbiol 9:1551

Liu J, Hu G, Li F et al (2011) Research progress of microbial flora in anaerobic digestion system. Ind Water Treat 31(10):10–14

Liu Y, Liu H, Cui L, Zhang K (2012) The ratio of food-to-microorganism (F/M) on membrane fouling of anaerobic membrane bioreactors treating low-strength wastewater. Desalination 297: 97–103

Lukitawesa PRJ, Millati R, Sárvári-Horváth I, Taherzadeh MJ (2020) Factors influencing volatile fatty acids production from food wastes via anaerobic digestion. Bioengineering 11(1):39–52
Luo K, Yang Q, Li XM, Yang GJ, Liu Y, Wang DB, Zheng W, Zeng GM (2012) Hydrolysis kinetics in anaerobic digestion of waste activated sludge enhanced by α-amylase. Biochem Eng J 15(62):17–21
Luo K, Xie X, Yang Q, Chen F, Zhong Y, Xie P, Wang G (2020) Multi-hydrolytic enzyme accumulation and microbial community structure of anaerobic co-digestion of food waste and waste-activated sludge. Environ Technol 41(4):478–487
Malý J, Fadrus H (1971) Influence of temperature on anaerobic digestion. J Water Pollut Control Fed 1:641–650
Mata-Alvarez J, Dosta J, Romero-Güiza MS, Fonoll X, Peces M, Astals S (2014) A critical review on anaerobic co-digestion achievements between 2010 and 2013. Renew Sustain Energy Rev 36:412–427
McInerney MJ, Struchtemeyer CG, Sieber J, Mouttaki H, Stams AJ, Schink B, Rohlin L, Gunsalus RP (2008) Physiology, ecology, phylogeny, and genomics of microorganisms capable of syntrophic metabolism. Ann N Y Acad Sci 1125(1):58–72
Meegoda JN, Li B, Patel K, Wang LB (2018) A review of the processes, parameters, and optimization of anaerobic digestion. Int J Environ Res Public Health 15(10):2224
Meng Y, Luan F, Yuan H, Chen X, Li X (2017) Enhancing anaerobic digestion performance of crude lipid in food waste by enzymatic pretreatment. Bioresour Technol 224:48–55
Nguyen PH, Kuruparan P, Visvanathan C (2007) Anaerobic digestion of municipal solid waste as a treatment prior to landfill. Bioresour Technol 98(2):380–387
Nguyen ML, Hung PC, Vo TP, Lay CH, Lin CY (2021) Effect of food to microorganisms (F/M) ratio on biohythane production via single-stage dark fermentation. Int J Hydog Energy 46 (20):11313–11324
Odnell A, Recktenwald M, Stensén K, Jonsson BH, Karlsson M (2016) Activity, life time and effect of hydrolytic enzymes for enhanced biogas production from sludge anaerobic digestion. Water Res 103:462–471
Parawira W, Murto M, Read JS, Mattiasson B (2005) Profile of hydrolases and biogas production during two-stage mesophilic anaerobic digestion of solid potato waste. Process Biochem 40 (9):2945–2952
Paritosh K, Yadav M, Chawade A, Sahoo D, Kesharwani N, Pareek N, Vivekanand V (2020) Additives as a support structure for specific biochemical activity boosts in anaerobic digestion: a review. Front Energy Res 11(8):88
Park Y, Hong F, Cheon J, Hidaka T, Tsuno H (2008) Comparison of thermophilic anaerobic digestion characteristics between single-phase and two-phase systems for kitchen garbage treatment. J Biosci Bioeng 105(1):48–54
Qin SL, Yu JS (2011) Research progress of enzyme immobilization techniques. Henan Chem Ind 5: 009
Singh RP, Singh RN, Srivastava MK, Srivastava AK, Kumar S, Dubey RC, Sharma AK (2012) Structure prediction and analysis of MxaF from obligate facultative and restricted facultative methylobacterium. Bioinformation 8(21):1042–1046. https://doi.org/10.6026/97320630081042
Rai AR, Singh RP, Srivastava AK, Dubey RC (2012) Structure prediction and evolution of a haloacid dehalogenase of Burkholderia mallei. Bioinformation 8(22):1111–1113. https://doi.org/10.6026/97320630081111
Ravi PP, Lindner J, Oechsner H, Lemmer A (2018) Effects of target pH-value on organic acids and methane production in two-stage anaerobic digestion of vegetable waste. Bioresour Technol 247:96–102
Richard EN, Hilonga A, Machunda RL, Njau KN (2019) A review on strategies to optimize metabolic stages of anaerobic digestion of municipal solid wastes towards enhanced resources recovery. Sustain Environ Res 29(1):1–3

Riviere D, Desvignes V, Pelletier E, Chaussonnerie S, Guermazi S, Weissenbach J, Li T, Camacho P, Sghir A (2009) Towards the definition of a core of microorganisms involved in anaerobic digestion of sludge. ISME J 3(6):700–714

Romero-Güiza MS, Vila J, Mata-Alvarez J, Chimenos JM, Astals S (2016) The role of additives on anaerobic digestion: a review. Renew Sustain Energy Rev 58:1486–1499

Saha B, Sathyan A, Mazumder P, Choudhury SP, Kalamdhad AS, Khwairakpam M, Mishra U (2018) Biochemical methane potential (BMP) test for Ageratum conyzoides to optimize ideal food to microorganism (F/M) ratio. J Environ Chem Eng 6(4):5135–5140

Schäfer M (2003) Die Verantwortung einer multidisziplinären Tourismuswissenschaft. Touris J 7 (3):377

Schäfer T, Selig M, Schönheit P (1993) Acetyl-CoA synthetase (ADP forming) in archaea, a novel enzyme involved in acetate formation and ATP synthesis. Arch Microbiol 159(1):72–83

Segers R (1998) Methane production and methane consumption: a review of processes underlying wetland methane fluxes. Biogeochemistry 41(1):23–51

Shih JC (1987) Ecological benefits of anaerobic digestion. Poult Sci 66(6):946–950

Siddique MN, Wahid ZA (2018) Achievements and perspectives of anaerobic co-digestion: a review. J Clean Prod 194:359–371

Singh RP, Singh RN, Manchanda G, Tiwari PK, Srivastava AK et al (2014) Biofuels: a way ahead. Ind J Appl Res 4(1):3–6

Singh RP, Manchanda G, Singh RN, Srivastava AK, Dubey RC (2016) Selection of alkalotolerant and symbiotically efficient chickpea nodulating rhizobia from North-West Indo Gangetic Plains. J Basic Microbiol 56:14–25

Singh RP, Manchanda G, Li ZF, Rai AR (2017) Insight of proteomics and genomics in environmental bioremediation. In: Bhakta JN (ed) Handbook of research on inventive bioremediation techniques. IGI Global, Hershey, PA. https://doi.org/10.4018/978-1-5225-2325-3

Singh R, Manchanda G, Maurya I, Wei Y (eds) (2020) Microbial versatility in varied environments. Springer, Singapore. https://doi.org/10.1007/978-981-15-3028-9

Tang Y, Dai X, Dong B, Guo Y, Dai L (2020) Humification in extracellular polymeric substances (EPS) dominates methane release and EPS reconstruction during the sludge stabilization of high-solid anaerobic digestion. Water Res 175:115686

Toerien DF, Hattingh WHJ (1969) Anaerobic digestion I. The microbiology of anaerobic digestion. Water Res 3–6:385–416

Wang D, Zhao J, Zeng G, Chen Y, Bond PL, Li X (2015) How does poly (hydroxyalkanoate) affect methane production from the anaerobic digestion of waste-activated sludge? Environ Sci Technol 49(20):12253–12262

Wang X, Li Z, Zhou X, Wang Q, Wu Y, Saino M, Bai X (2016) Study on the bio-methane yield and microbial community structure in enzyme enhanced anaerobic co-digestion of cow manure and corn straw. Bioresour Technol 219:150–157

Wilkie Ann C (2005) Anaerobic digestion: biology and benefits. Dairy Manure Manag Treat Handl Commun Relat 15:63–72

Wofford NQ, Beaty PS, McInerney MJ (1986) Preparation of cell-free extracts and the enzymes involved in fatty acid metabolism in Syntrophomonas wolfei. J Bacteriol 167(1):179–185

Xu J, Lin X (2007) The catalytic versatility of enzymes and their new developments in organic synthesis. Org Chem 27(12):1473–1478

Xu Z, Zhao M, Miao H, Huang Z, Gao S, Ruan W (2014) In situ volatile fatty acids influence biogas generation from kitchen wastes by anaerobic digestion. Bioresour Technol 163:186–192

Yang G, Zhang P, Zhang G, Wang Y, Yang A (2015) Degradation properties of protein and carbohydrate during sludge anaerobic digestion. Bioresour Technol 192:126–130

Yang YJ, Lin W, Singh RP, Xu Q, Chen Z, Yuan Y, Zou P, Li Y, Zhang C (2019) Genomic, transcriptomic and enzymatic insight into lignocellulolytic system of a plant pathogen *Dickeya* sp. WS52 to digest sweet pepper and tomato stalk. Biomolecule 9(12):753

Ye M, Liu J, Ma C, Li YY, Zou L, Qian G, Xu ZP (2018) Improving the stability and efficiency of anaerobic digestion of food waste using additives: a critical review. J Clean Prod 192:316–326

Yu XL, Wang L, Xu WM (2007) Progress in the studies of cellulose degradation by cellulase. J Ningbo Uni 1

Zhang B, Zhang LL, Zhang SC, Shi HZ, Cai WM (2005) The influence of pH on hydrolysis and acidogenesis of kitchen wastes in two-phase anaerobic digestion. Environ Technol 26 (3):329–340

Zhang B, He PJ, Lü F, Shao LM, Wang P (2007) Extracellular enzyme activities during regulated hydrolysis of high-solid organic wastes. Water Res 41(19):4468–4478

Zhang W, Xu Y, Dong B, Dai X (2019) Characterizing the sludge moisture distribution during anaerobic digestion process through various approaches. Sci Total Environ 675:184–191

Zhao J, Yang Q, Li X, Wang D, Luo K, Zhong Y, Xu Q, Zeng G (2015) Enhanced production of short-chain fatty acid from food waste stimulated by alkyl polyglycosides and its mechanism. Waste Manag 46:133–139

Ziganshin AM, Schmidt T, Scholwin F, Il'inskaya ON, Harms H, Kleinsteuber S (2011) Bacteria and archaea involved in anaerobic digestion of distillers grains with solubles. Appl Microbiol Biotechnol 89(6):2039–2052

Chapter 9
Lichen Microbiome: Diversity Biological Role and Biotechnological Application

R. R. Sargsyan, A. Tsurykau, and Hovik Panosyan

Abstract Lichens were traditionally considered as a remarkable assemblage of fungi with unicellular phototroph (algae or cyanobacteria) that have converged on similar symbiotic strategies. However, this view of lichens has recently been reconsidered by findings of miscellaneous associated microbes colonizing on or within the thallus causing no apparent effect. This hidden diversity includes filamentous fungi, lichen-inhabiting yeasts, as well as various prokaryotic bacteria. Despite these endothallic and exothallic organisms do not belong to constant lichen symbionts, they usually play important roles in lichen biology by participating in the lichen metabolism, regulating water relations, affecting thallus architecture, and being involved in the degradation processes. However, it is often difficult to understand microorganismal input and uptake and therefore determine their symbiotic outcome due to the complexity of lichen symbiosis. In this chapter, diversity and biological role of usually neglected or overlooked lichen microbial consortia are reviewed and their possible biotechnological application is discussed.

Keywords Lichen · Colonization · Symbiosis · Metabolism · Consortia

R. R. Sargsyan
Department of Biochemistry Microbiology and Biotechnology, Faculty of Biology, Yerevan State University, Yerevan, Armenia

A. Tsurykau
Department of Botany and Plant Physiology, Faculty of Biology, Francisk Skorina Gomel State University, Gomel, Belarus

Department of Ecology Botany, Nature Protection Institute of Natural Sciences, Samara National Research University, Samara, Russia

H. Panosyan (✉)
Department of Biochemistry, Microbiology and Biotechnology, Yerevan State University, Yerevan, Armenia
e-mail: hpanosyan@ysu.am

R. P. Singh et al. (eds.), *Microbes in Microbial Communities*,
https://doi.org/10.1007/978-981-16-5617-0_9

9.1 Introduction

Traditionally lichens are considered to be mutualistic associations encompassed of fungus as a mycobiont and either an algae or a cyanobacteria as a photobiont (Honegger 1991). About 20,000 species of lichens growing on rock stones (saxicolous lichens) and on the tree barks and branches (corticolous lichens or epiphytes) have been recorded worldwide (Ellis 2012). Lichens are found in various habitats from polar to equatorial regions. They are found in various geographical zones, from lowland layers to alpine levels. To withstand extreme conditions characterized by abnormal temperatures, periodic desiccation, high levels of UV radiation, and high concentrations of salts, lichens synthesize secondary metabolites (e.g., radioprotectants, cryoprotectants, compatible solutes) widely used in various biotechnologies (Suzuki et al. 2016; Subhashini et al. 2017; Sargsyan et al. 2021).

Since the recognition that a diverse microbial community is integral part to the traditionally recognized mycobiont and photobiont mutualistic association, lichens have progressively become a subject of research in ecological microbiology (Bates et al. 2011; Pankratov et al. 2017; Yang et al. 2020). It was revealed that lichens usually provide habitats for bacteria being different from those of nearby substrates. Moreover, lichens adapted to grow in different habitats usually host appropriate bacterial batch (Bates et al. 2011; Mushegian et al. 2011). These microbiomes often involve of non-photosynthetic diazotrophs, which usually provide benefits to the host lichen by their metabolic activities. There are many reports confirming that different groups of the lichen–bacterial associations are highly structured. Bacteria are not distributed only across the lichen thallus. More-consistent bacterial communities with different species, it is supposed to be in central parts of thalli (Kumar et al. 2014; Mushegian et al. 2011). In consequence, the long-established concept of mutualistic relationship between lichenized fungi and algae or cyano-bacteria is in need of revision and should also encompass bacterial component. It has been shown important role of lichenized bacterial community in the nutrient cycling of lichens (Grube and Berg 2009; Bates et al. 2011; Hawksworth and Grube 2020; Singh et al. 2020).

9.2 Microbial Diversity in Lichens

Lichen bacterial associations were first mentioned in the last century (Uphof 1925; Henckel and Yuzhakova 1936; Iskina 1938). Those studies were mainly based on traditional cultivation techniques and biochemical and morphological identification. The dominating genera were *Azotobacter*, *Pseudomonas* (Gammaproteobacteria), *Beijerinckia* (Alphaproteobacteria), *Bacillus* and *Clostridium* (Firmicutes) (Iskina 1938; Panosyan and Nikogosyan 1966).

Still only less than 1% of the microorganisms found in natural habitats have been cultivated and subsequently isolated so far. The development of molecular biology methodology largely promoted to expand our knowledge of environmental microbial diversity (Yang et al. 2019). The culture-independent studies uncovered a vast biodiversity of endothallic and exothallic bacteria (see Table 9.1) (González et al. 2005; Cardinale et al. 2006; Liba et al. 2006; Selbmann et al. 2010; Pankratov et al. 2017).

In the beginning to study lichen-associated bacteria, several culture-independent methods including denaturing gradient gel electrophoresis (DGGE), terminal restriction fragment length polymorphism (T-RFLP), and single-strand conformation polymorphism (SSCP) have been used (Liu et al. 1997; Muyzer and Smalla 1998; Schwieger and Tebbe 1998). Apart from genetic techniques, specific fluorescence in situ hybridization and confocal laser scanning microscopy (FISH-CLSM) methods were also used to uncover the lichen microbial content (Cardinale et al. 2008; Grube and Berg 2009; Muggia et al. 2013; Aschenbrenner 2015).

Currently, the molecular analyses (e.g., 16S rRNA gene clone library construction, pyrosequencing, metagenomics) are the most commonly used (Singh et al. 2016). Combination of classical microbiology with new molecular biology techniques has considerably enlarged our apprehension of the taxonomic and metabolic diversity of lichen microbiota.

CTAB method still remains to be one of the most available methods to extract the total DNA of lichens (Cardinale et al. 2006; Singh et al. 2016). Prior to the main extraction process, lichen thalli are usually washed by distilled sterile water, 5–9% H_2O_2, or ethanol. After that, it is required to clean the sterilized lichen thalli from the washing solution residues, usually distilled water is used in this step. After the "sterilization" of lichen thalli, the main DNA extraction process is initiated (Fig. 9.1).

Cardinale et al. (2012) revealed curious information regarding the lichen-inhabiting bacteria and age of the lichen thalli, its substrate, and growth conditions (e.g., solar irradiation, humidity). The youngest and therefore the most physiologically active lichen thalli were dominated by Alphaproteobacteria. In contrast, older parts of lichen thalli were associated with Actinobacteria. Density of Actinobacteria and Betaproteobacteria was also higher in shaded places while Alphaproteobacteria were highly predominant in the sunny sides regardless lichen age. The author noted that no members of Alphaproteobacteria were cultivated, which is possibly connected with special requirements for growth, e.g., the substances produced by the lichen thalli.

A core lichen microbiome was discovered by Sierra et al. (2020). Based on 16S rRNA gene amplicon sequence analysis, microbiome of representatives belonging to seven lichen genera (*Cora*, *Hypotrachyna*, *Usnea*, *Cladonia*, *Peltigera*, *Stereocaulon*, and *Sticta*) was screened. Phyla Proteobacteria, Acidobacteria, and Cyanobacteria were shown to be abundant in all studied lichens. The other microbiome members were also present but were varying from genera to genera.

Localization of bacteria in lichen thallus and possible dispersion of bacterial fraction were studied by Aschenbrenner et al. (2014). It was established that bacteria colonize also symbiotic propagules, which are intended for short-distance transmission of the lichen. *Cystobacterineae* (*Deltaproteobacteria*) prevailed in both the

Table 9.1 Some identified bacteria from different lichen species

Lichen species	Isolates			Identification method	Author
	Phylum	Order	Genus		
wl	Firmicutes	Caryophanales	*Bacillus Paenibacillus*	Molecular genetic identification	Cardinale et al. (2006)
	Proteobacteria	Burkholderiaceae	*Bukholderia*		
Cladonia coniocraea	Gammaproteobacteria	Xanthomonadaceae	*Luteibactor*	Molecular genetic identification	Cardinale et al. (2006)
Cladonia rangiferina	Alphaproteobacteria	–	–	FISH method	Pankratov et al. (2017)
	Actinobacteria				
	Acidobacteria				
	Firmicutes	Caryophanales	*Paenibacillus*	Molecular genetic identification	Cardinale et al. (2006)
	Proteobacteria	Burkholderiales	*Burkholderia*		
		Rhodospirillales	*Inquilinus*		
Hypogymnia physodes (on moss)	Firmicutes	Caryophanales	*Paenibacillus*	Molecular genetic identification	Cardinale et al. (2006)
Hypogymnia physodes (on bark)	Actinobacteria	Cellulomonadaceae	Agricultural soil bacterium isolate SI-14 (AJ252581)	Molecular genetic identification	Cardinale et al. (2006)
		Micromonosporales	*Micromonospora*		
		Streptomycetales	*Streptomyces*		
	Firmicutes	–	Uncultured bacterium clone X20 (DQ083105)		
Lecanora fuscobrunnea	Actinobacteria	Nakamurellales	*Saxeibacter*	Isolation by cultivation	Selbmann et al. (2010)
			Unclassified *Actinobacteria*		
Lobaria pulmonaria	Proteobacteria	Hyphomicrobiales	*Beijerinckia*	Molecular genetic identification	Erlacher et al. (2015)
			Bradyrhizobium		
			Brucellacea		
			Methylobacterium		
			Nitrobacter		

			Phyllobacterium		
			Rhodopseudomonas		
			Xanthobacter		
Lobaria retigera	Actinobacteria	Micrococcales	*Micrococcus*	Molecular genetic identification	Jiang et al. (2017)
			Subtercola		
		Mycobacteriales	*Rhodococcus*		
			Mycobacterium		
	Deinococcus-Thermus	Deinococcales	*Deinococcus*		
	Proteobacteria	Burkholderiales	*Burkholderia*		
			Variovorax		
		Hyphomicrobiales	*Beijerinckia*		
			Methylobacterium		
		Lysobacterales	*Luteibacter*		
		Rhodobacterales	*Paracoccus*		
		Sphingomonadales	*Porphyrobacter*		
			Sphingomonas		
Parmelia omphalodes	Actinobacteria	Frankiales	*Motilibacter*	Molecular genetic identification	Jiang et al. (2017)
		Geodermatophilales	*Modestobacter*		
		Kineosporiales	*Kineococcus*		
		Micrococcales	*Amnibacterium*		
			Curtobacterium		
			Microbacterium		
			Micrococcus		
		Mycobacteriales	*Corynebacterium*		
			Mycobacterium		
		Nakamurellales	*Nakamurella*		
		Propionibacteriales	*Friedmanniella*		

(continued)

Table 9.1 (continued)

Lichen species	Isolates			Identification method	Author
	Phylum	Order	Genus		
			Microlunatus		
			Nocardioides		
		Streptomycetales	*Streptomyces*		
	Firmicutes	Caryophanales	*Bacillus*		
		Eubacteriales	*Clostridium*		
	Proteobacteria	Burkholderiales	*Burkholderia*		
		Hyphomicrobiales	*Beijerinckia*		
			Methylobacterium		
		Pseudomonadales	*Acinetobacter*		
		Rhodospirillales	*Roseomonas*		
		Sphingomonadales	*Sandarakinorhabdus*		
			Sphingomonas		
Pseudevernia furfuracea	Firmicutes	Caryophanales	*Paenibacillus*	Molecular genetic identification	Cardinale et al. (2006)
Rusavskia elegans	Actinobacteria	Mycobacteriales	*Mycobacterium*	Isolation by cultivation	Selbmann et al. (2010)
	Firmicutes	Caryophanales	*Bacillus*		
			Paenibacillus		
	Proteobacteria	Pseudomonadales	*Pseudomonas*		
Solorina crocea	Actinobacteria	Nakamurellales	*Nakamurella*	Molecular genetic identification	Grube et al. (2015)
	Planctomycetes	Gemmatales	*Gemmata*		
		Isosphaerales	*Isosphaera*		
	Proteobacteria	Lysobacterales	*Dyella*		
		Myxococcales	*Byssovorax*		
		Sphingomonadales	*Novosphingobium*		
			Sphingomonas		

Umbilicaria cylindrica	Proteobacteria	Burkholderiales	*Burkholderia*	Isolation by cultivation	Grube and Berg (2009)
		Caryophanales	*Bacillus*		
		Pseudomonadales	*Acinetobacter*		
		Hyphomicrobiales	*Methylocystis*	Molecular genetic identification	
		Rhodospirillales	*Acidisoma*		
			Gluconoacetobacter		
		Sphingomonadales	*Sphingomonas*		
		–	–	FISH method	
Umbilicaria decussata	Actinobacteria	Micrococcales	*Knoellia*	Isolation by cultivation	Selbmann et al. (2010)
	Deinococcus-Thermus	Deinococcales	*Deinococcus*		
	Firmicutes	Caryophanales	*Paenibacillus*		
Umbilicaria esculenta	Actinobacteria	Mycobacteriales	*Rhodococcus*	Molecular genetic identification	Jiang et al. (2017)
		Micrococcales	*Arthrobacter*		
			Clavibacter		
			Microbacterium		
			Micrococcus		
		Micromonosporales	*Micromonospora*		
		Nakamurellales	*Nakamurella*		
	Deinococcus-Thermus	Deinococcales	*Deinococcus*		
	Firmicutes	Caryophanales	*Bacillus*		
	Proteobacteria	Burkholderiales	*Burkholderia*		
			Massilia		
		Enterobacterales	*Klebsiella*		
		Hyphomicrobiales	*Aureimonas*		
			Beijerinckia		
			Methylobacterium		

(continued)

Table 9.1 (continued)

Lichen species	Isolates			Identification method	Author
	Phylum	Order	Genus		
		Pseudomonadales	*Pseudomonas*		
		Sphingomonadales	*Novosphingobium*		
			Sphingomonas		
Umbilicaria pustulata	Actinobacteria	Actinomycetales	*Cellulomonas*	Molecular genetic identification	Greshake Tzovaras et al. (2020)
		Kineosporiales	*Kineococcus*		
			Kineosporia		
			Quadrisphaera		
			Thalassiella		
		Micrococcales	*Alpinimonas*		
	Acidobacteria	Acidobacteriales	*Acidipila*		
			Acidobacterium		
			Edaphobacter		
			Granulicella		
			Occallatibacter		
			Terriglobus		
		Bryobacterales	*Bryobacter*		
			Paludibaculum		
		Holophagales	*Geothrix*		
	Armatimonadetes	Armatimonadales	*Armatimonas*		
		Fimbriimonadales	*Fimbriimonas*		
	Bacteroidetes	Chitinophagales	*Ferruginibacter*		
			Flavisolibacter		
			Pseudoflavitalea		
			Puia		

			Terrimonas		
	Firmicutes	Eubacteriales	*Desulfotomaculum*		
			Heliobacterium		
	Proteobacteria	Hyphomicrobiales	*Beijerinckia*		
			Bradyrhizobium		
			Lichenibacter		
			Lichenihabitans		
			Methylocella		
			Methylocys		
			Rhodoblastus		
			Rhodomicrobium		
		Rhodospirillales	*Acidiphilium*		
			Acidisphaera		
			Acidomonas		
			Azospirillum		
			Endobacter		
			Gluconacetobacter		
			Gluconacetobacter		
			Granulibacter		
			Komagataeibacter		
			Rhodopila		
			Rhodovastum		
Usnea antarctica	Actinobacteria	Micrococcales	*Arthrobacter*	Isolation by cultivation	Selbmann et al. (2010)
	Proteobacteria	Pseudomonadales	*Pseudomonas*		
Not identified saxicolous lichen	Proteobacteria	Pseudomonadales	*Acinetobacter*	Molecular genetic identification	Hunanyan and Sargsyan (2019)

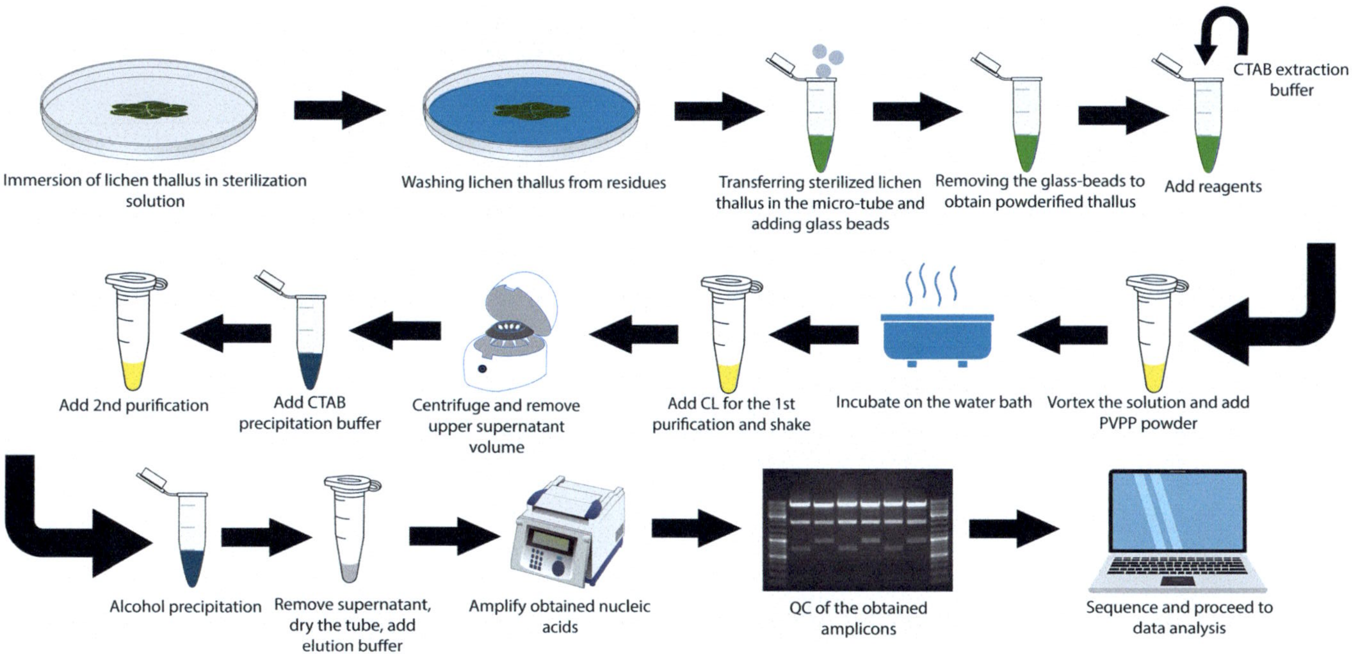

Fig. 9.1 Step-by-step guide of DNA isolation using CTAB method (Cubero and Crespo 2002)

lichen thalli and isidioid soredia, making up to 42% of all investigated microbes. In contrast, Alphaproteobacteria predominated within propagules. The presence of bacteria in vegetative propagules may indicate the need for the former for the functioning of lichen symbiosis. The microbial cargo is also confirmed by geographical shift: bacterial composition of lichen thalli sampled from the same region demonstrated closer similarity than those of distant populations (Aschenbrenner et al. 2014).

Except prokaryotes, yeasts seem to be another part of lichen symbiosis being mainly neglected until the recent discovery by Spribille et al. (2016). Correlation between yeast abundance and variations in lichen phenotype indicates their role as a potential symbiotic partner in the lichen mutualistic association (Palmqvist et al. 2017; Suryanarayanan and Thirunavukkarasu 2017; Zúñiga et al. 2017).

Up to now, a limited number of yeasts were detected in lichen thalli, mainly belonging to Cyphobasidiales and Microsporomycetaceae (Cystobasidiales), both Cystobasidiomycetes, Pucciniomycotina, Basidiomycota (Table 9.2).

Subsequent studies provided contradictory data. Obviously, the yeast-like fungi are not as ubiquitous as it was suggested. Despite Cernava et al. (2015) found these yeasts in 95% of the studied specimens of lichen genus *Cladonia* collected in various climatic conditions and habitats, other studies have not been as promising. Smith et al. (2020) detected Cystobasidiales yeasts in five of the 35 samples while Lendemer et al. (2019) confirmed Cyphobasidium or other Cystobasidiomycete yeasts only in nine of the 413 samples, and in nine of the 339 investigated lichen species. Furthermore, the study by Mark et al. (2020) contradicted previous findings of high mycobiont specificity of basidiomycete yeasts. This corresponds to the data presented in Table 9.2.

9.3 Microbiome Functions

The role of lichen-inhabiting bacterial communities remains largely elusive although its revealing may promise us high practical potential. The role of the bacterial composition in the symbiotic associations was stated by Schneider et al. (2011) who found correlations between microorganism composition and protein profile. The studies of the protein spectrum highlighted that *Bacteria* and *Archea* contribution was even more than input stated for the green algae. The main functional categories of *Bacteria* and *Archea* were the posttranslational modification, protein turnover and supply of chaperones (Fig. 9.2).

The antagonistic properties of bacterial community were investigated by Cernava et al. (2015). The isolated bacteria of *Lobaria pulmanoria* were most active against the lichenicolous fungus *Rhinocladoniella* sp., while the antibacterial activity was low.

Some progress has been made by applying the metaproteomic approaches which helped to reveal the involvement of lichen microbe communities in functions such as nutrient supply, resistance against stress factors, support of photosynthesis,

Table 9.2 Some basidiomycete yeasts detected in lichen thalli. Lichen nomenclature follows Wijayawardene et al. (2020)

Yeast taxa	Host lichen	Author
Buckleyzyma aurantiaca (Cystobasidiomycetes)	*Lecanora carpinea* s. lat., *Lecanora chlarotera*, *Lecanora pulicaris*, *Pseudevernia furfuracea*	Mark et al. (2020)
Cyphobasidium hypogymniicola (Cyphobasidiomycetes)	*Hypogymnia hultenii*, *Hypogymnia imshaugii*, *Hypogymnia incurvoides*, *Hypogymnia krogiae*, *Hypogymnia physodes*, *Hypogymnia vittata*, *Lecanora pulicaris*, *Parmelia sulcata*, *Pseudevernia furfuracea*	Diederich (1996, 2003, 2007), Holien (2005), Urbanavichene and Urbanavichus (2005), Hodkinson et al. (2009), Millanes et al. (2016), Mark et al. (2020)
Cyphobasidium usneicola (Cyphobasidiomycetes)	*Hypogymnia physodes*, *Hypogymnia tubulosa*, *Lecanora chlarotera*, *Lecanora pulicaris*, *Parmelia sulcata*, *Physcia adscendens/tenella*, *Pseudevernia furfuracea*, *Usnea articulata*, *Usnea brasiliensis*, *Usnea cornuta* s. lat., *Usnea galapagona*, *Usnea hirta*, *Usnea madeirensis*, *Usnea* cf. *praetervisa*, *Usnea silesiaca*, *Usnea subfloridana* s. lat., *Usnea subscabrosa*	Diederich (1996, 2003), Millanes et al. (2016), Mark et al. (2020)
Cyphobasidium spp. (Cyphobasidiomycetes)	*Bryoria nadvornikiana*, *Heterodermia leucomelos*, *Lecidea roseotincta*, *Opegrapha vulgata*, *Parmotrema hypotropum*, *Parmotrema subsumptum*, *Usnea cornuta*, *Usnea strigosa*, *Usnea subgracilis*	Lendemer et al. (2019)
Cyphobasidiales spp. (Cyphobasidiomycetes)	*Alectoria* sp., *Anzia* sp., *Asahinea* sp., *Brodoa* sp., *Bryoria* spp., *Bulbothrix* sp., *Cetraria* sp., *Cetrelia* sp., *Esslingeriana* sp., *Evernia* sp., *Flavopunctelia* sp., *Hypogymnia* sp., *Hypotrachyna* sp., *Imshaugia* sp., *Letharia* sp., *Melanelia* sp., *Menegazzia* sp., *Montanelia* sp., *Myelochroa* sp., *Nephromopsis* sp., *Nodobryoria* sp., *Omphalora* sp., *Oropogon* sp., *Parmelia* sp., *Parmelina* sp., *Parmotrema* sp., *Platismatia* sp., *Pseudevernia* sp., *Pseudoparmelia* sp., *Usnea* sp., *Xanthoparmelia* sp.	Spribille et al. (2016)
Hasegawazyma spp. (Cystobasidiomycetes)	*Hypogymnia tubulosa*, *Lecanora argentata*, *Lecanora carpinea* s. lat., *Lecanora chlarotera*,	Mark et al. (2020)

(continued)

Table 9.2 (continued)

Yeast taxa	Host lichen	Author
	Lecanora pulicaris, Parmelia sulcata, Pseudevernia furfuracea	
Lichenozyma pisutiana (Cystobasidiomycetes)	*Cladonia arbuscula, Cladonia cariosa, Cladonia chlorophaea* s. lat., *Cladonia cornuta, Cladonia deformis, Cladonia diversa, Cladonia floerkeana, Cladonia furcata, Cladonia gracilis, Cladonia merochlorophaea, Cladonia phyllophora, Cladonia pocillum, Cladonia polycarpoides, Cladonia pyxidata, Cladonia rangiferina, Cladonia rangiformis, Cladonia rei, Cladonia* cf. *subulata, Cladonia verticillata, Lecanora argentata, Lecanora carpinea* s. lat., *Lecanora chlarotera*	Cernava et al. (2015), Mark et al. (2020)
Microsporomycetaceae spp. (Cystobasidiomycetes)	*Cladonia cornuta, Cladonia foliacea, Cladonia furcata, Cladonia humilis, Cladonia* cf. *macroceras, Cladonia pocillum, Cladonia rangiformis, Cladonia rei, Cladonia subulata*	Cernava et al. (2015)
Microsporomyces cf. *pini* (Cystobasidiomycetes)	*Lecanora carpinea* s. lat., *Lecanora pulicaris*	Mark et al. (2020)
Microsporomyces spp. (Cystobasidiomycetes)	*Hypogymnia physodes, Hypogymnia tubulosa, Lecanora carpinea* s. lat., *Lecanora pulicaris, Parmelia sulcata, Physcia adscendens/tenella*	Mark et al. (2020)

production of hormone, detoxification of end metabolites, and lytic activity (Grube et al. 2015) (Fig. 9.3). A short summary is provided below.

Nutrient supply: High number of contigs were found to represent Ton and Tol transport systems, some of which involving in iron uptake. Phosphate metabolism was represented with a relatively small amount of contigs, including proteins involved in solubilization of phosphates.

Pathogen defense: Some contigs were corresponding to the multidrug resistance efflux pumps, as well as various antibiotic (luoroquinolone, vancomycin, methicillin, penicillin and cephalosporine) resistance. Relatively small amount of contigs represent the production of secondary metabolites which are well-known for their biological activity.

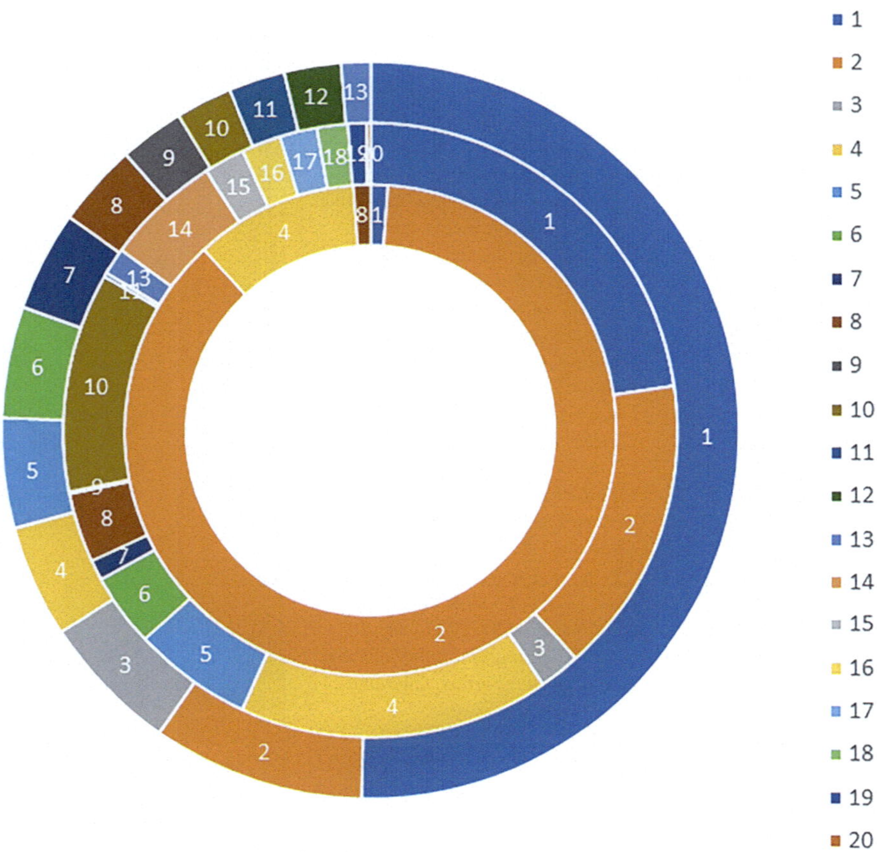

Fig. 9.2 Comparative description of orthologous group (COG/KOG) functions in bacteria (outer circle), fungus (middle circle), and algae (inner circle) associated to the lichen thalli. Compiled from data presented in Schneider et al. (2011). Note: 1—Posttranslational modification, protein turnover, chaperones; 2—Energy production and conversion; 3—Lipid metabolism; 4—Carbohydrate transport and metabolism; 5—Amino acid transport and metabolism; 6—General function prediction only; 7—Inorganic ion transport and metabolism; 8—Signal transduction mechanisms; 9—DNA replication, recombination, and repair; 10—Translation, ribosomal structure, and biogenesis; 11—Nucleotide transport and metabolism; 12—Transcription; 13—Secondary metabolites biosynthesis, transport, and catabolism; 14—Cytoskeleton; 15—Coenzyme metabolism; 16—Cell wall/membrane/envelope biogenesis; 17—Cell cycle control, cell division, chromosome partitioning; 18—Intracellular trafficking, secretion, and vesicular transport; 19—RNA processing and modification; 20—Chromatin structure and dynamics

Resistance against abiotic stress factors: Relatively high amount of contigs corresponds to the metal resistance. Moreover, about the same number of oxidative-stress protectants were observed.

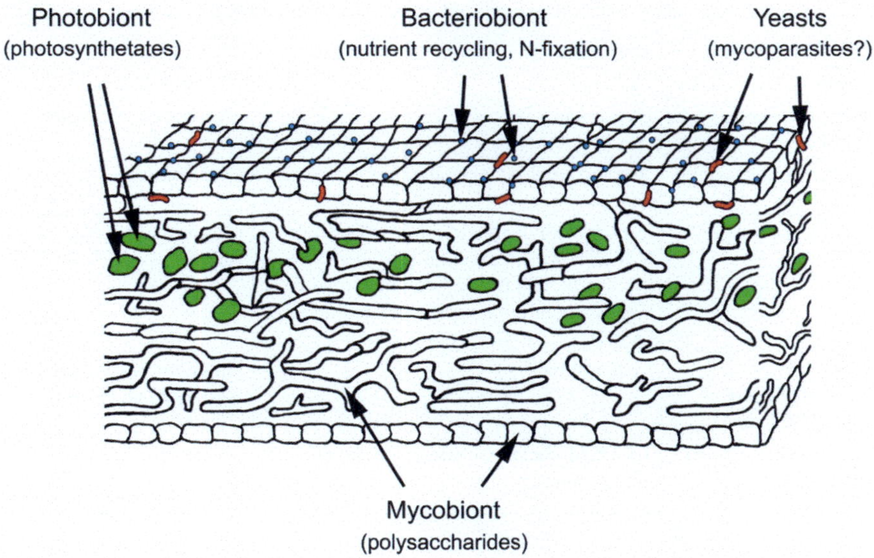

Fig. 9.3 Reconstruction of the lichen thallus based on data from Grube et al. (Grube and Berg 2009; Grube et al. 2015) and Spribille et al. (2016)

Photosynthesis support by vitamin B_{12} and hormone production. Numerous contigs were corresponding to biosynthesis of tetrapyrrole, coenzyme-B_{12}, thiamine, and biotin. Small amount of auxin biosynthesis corresponding contigs were found.

Detoxication of metabolites and lytic activity: High amount of indicated contigs were found to participate in xenobiotics metabolism and biodegradation. Some contigs corresponding to chitin and *N*-acetylglucosamine utilization were also detected, which can be responsible to degradation of older sections of the thallus for obtaining nutrients for the growing parts.

In contrast to bacteria, the role of basidiomycete yeasts in lichen symbiosis still remains unclear. Despite Spribille et al. (2016) shown the presence of yeast is correlated with amount of lichen secondary metabolites, Mark et al. (2020) confirmed the absence of any effect of medullary chemotype on the distribution of yeast community. The author supposed low probability of metabolites production directly by the yeasts.

It is also known that some Cyphobasidiomycetes cause gall formation and therefore can be associated with mycoparasites (Millanes et al. 2016) similar to different filamentous basidiomycete fungi (Diederich et al. 2018; Tuovinen et al. 2019). However, most of host lichen specimens lacked galls (Spribille et al. 2016; Cernava et al. (2015); Mark et al. 2020). More studies are required to make clear the interaction between basidiomycete yeasts, mycobiont, photobiont, and microbial consortia in lichens.

Table 9.3 Some secondary metabolites produced by lichen endobiont bacteria (Calcott et al. 2018)

Lichen endobiont	Metabolite	Activity
Streptomyces unicialis	Uncialamycin	Antibiotic, antitumor
Micromonospora cherisna	Dynemicin	Antitumor, antibiotic
Streptomyces sp. L-4-4	Aminocoumarin, coumabiocins A–F, novobiocin	Antibiotic
Streptomyces sp. L-9-10	2′-*O*-demethylherbicidin F (1), 9′-deoxy-8′,8′-dihydroxyherbicidin B, 9′-deoxy-8-′-oxoherbicidin B, 8′-epimer of herbicidin B, 9-(β-D-arabinofuranosyl) hypoxanthine (Ara-H)	Herbicidal agents, antibiotics
Streptomyces sp. isolated from *Stereocaulon* sp.	Lichostatinal	Inhibitor of the cysteine protease cathepsin K (CatK)
Unculturable *Streptomyces* sp.	1,1-Dichlorocyclopropane-containing angucycline	Antitumor, antibiotic
Streptomyces cyaneofuscantus	Aneodimycin, cyaneomycin, usnic acid	Antioxidant, antitumor, antibacterial, UV protectant
Nocardia ignorata	Brominated diketopiperazines	Antioxidant, anti-inflammatory

9.4 Biotechnological Potential/Relevance

Lichens have a huge role in traditional and evidence-based medicine practice. Recent discoveries of lichen-inhabiting bacteria and their biotechnological relevance can solve many issues connected with the lichen biotechnology. Obtaining biologically active substances from slow-growing lichen thalli is a difficult and time-consuming task, which can be solved with use of bacteria.

A short list of secondary metabolites produced by lichen endobionts and their application are presented in Table 9.3.

Definitely, this is not a really vast amount of characterized secondary metabolites from lichen endobionts. The main problem is the uncultivability of most bacteria, which makes the discovery of novel metabolites challenging task. In order to address this issue, novel methods such as transcriptomics may become handful.

9.5 Conclusions

The recent discoveries of diverse metacommunities of lichen thalli forced to redefine lichens as complex and dynamic ecosystems (Hawksworth and Grube 2020; Smith et al. 2020) and catalyzed a call to reconceptualize the symbiotic concept of lichen (Spribille et al. 2016; Hawksworth and Grube 2020). The traditional approach of lichens as bicomponent mutualistic association are outdated, and "ecosystem-like" concept is gaining increasing recognition.

However, physiological processes operating within the symbiosis as well as symbiont interactions remain to be a significant gap in our understanding the functioning of the association. Future studies are highly expected to uncover the diversity of the lichen-associated microbial communities and reveal their biotechnological potential.

Acknowledgments Funding The work was supported by the Science Committee of RA, in the frame of the research project No. 20AA-1F018.

References

Aschenbrenner IA (2015) Stability and change in the lichen microbiome. Doctoral thesis Graz University of Technology Graz

Aschenbrenner IA, Cardinale M, Berg G, Grube M (2014) Microbial cargo: do bacteria on symbiotic propagules reinforce the microbiome of lichens? Environ Microbiol 16 (12):3743–3752

Bates ST, Cropsey GW, Caporaso JG, Knight R, Fierer N (2011) Bacterial communities associated with the lichen symbiosis. Appl Environ Microbiol 77(4):1309–1314

Calcott MJ, Ackerley DF, Knight A, Keyzers RA, Owen JG (2018) Secondary metabolism in the lichen symbiosis. Chem Soc Rev 47(5):1730–1760

Cardinale M, Puglia AM, Grube M (2006) Molecular analysis of lichen-associated bacterial communities. FEMS Microbiol Ecol 57(3):484–495

Cardinale M, Vieira de Castro J Jr, Müller H, Berg G, Grube M (2008) In situ analysis of the bacterial community associated with the reindeer lichen *Cladonia arbuscula* reveals predominance of Alphaproteobacteria. FEMS Microbiol Ecol 66(1):63–71

Cardinale M, Steinová J, Rabensteiner J, Berg G, Grube M (2012) Age sun and substrate: triggers of bacterial communities in lichens. Environ Microbiol Rep 4(1):23–28

Cernava T, Müller H, Aschenbrenner IA, Grube M, Berg G (2015) Analyzing the antagonistic potential of the lichen microbiome against pathogens by bridging metagenomic with culture studies. Front Microbiol 6:620

Cubero OF, Crespo A (2002) Isolation of nucleic acids from lichens. In: Protocols in lichenology. Springer, Berlin, pp 381–391

Diederich P (1996) The lichenicolous heterobasidiomycetes. J. Cramer, Berlin

Diederich P (2003) New species and new records of American lichenicolous fungi. Herzogia 16: 41–90

Diederich PA (2007) New or interesting lichenicolous heterobasidiomycetes. Opuscula Philolich 4 (1):11

Diederich P, Lawrey JD, Ertz D (2018) The 2018 classification and checklist of lichenicolous fungi with 2000 non-lichenized obligately lichenicolous taxa. Bryology 121(3):340–425

Ellis CJ (2012) Lichen epiphyte diversity: a species community and trait-based review. Perspect Plant Ecol Evol Syst 14(2):131–152

Erlacher A, Cernava T, Cardinale M, Soh J, Sensen CW, Grube M, Berg G (2015) Rhizobiales as functional and endosymbiontic members in the lichen symbiosis of Lobaria pulmonaria L. Front Microbiol 6:53

González I, Ayuso-Sacido A, Anderson A, Genilloud O (2005) Actinomycetes isolated from lichens: evaluation of their diversity and detection of biosynthetic gene sequences. FEMS Microbiol Ecol 54(3):401–415

Greshake Tzovaras B, Segers FHID, Bicker A, Dal Grande F, Otte J, Anvar SY, Hankeln T, Schmitt I, Ebersberger I (2020) What Is in Umbilicaria pustulata? A metagenomic approach to reconstruct the Holo-Genome of a Lichen. Genome Biol Evol 12(4):309–324
Grube M, Berg G (2009) Microbial consortia of bacteria and fungi with focus on the lichen symbiosis. Fungal Biol Rev 23(3):72–85
Grube M, Cernava T, Soh J, Fuchs S, Aschenbrenner I, Lassek C, Wegner U, Becher D, Riedel K, Sensen CW, Berg G (2015) Exploring functional contexts of symbiotic sustain within lichen-associated bacteria by comparative omics. ISME J 9(2):412–424
Hawksworth DL, Grube M (2020) Lichens redefined as complex ecosystems. New Phytol 227 (5):1281
Henckel PA, Yuzhakova LA (1936) On the role of Azotobacter in the lichen symbiosis Bulletin of the Perm (Molotov). Biol Res Int 10:315
Hodkinson BP, Harris RC, Case MA (2009) A checklist of Virginia lichens. Evansia 26(2):64–88
Holien H (2005) Additions to the Norwegian flora of lichens and lichenicolous fungi III. Graphis Scripta 17(2):62–64
Honegger R (1991) Fungal evolution: symbiosis and morphogenesis. In: Margulis L, Fester R (eds) Symbiosis as a source of evolutionary innovation. MIT Press, Cambridge MA, pp 319–340
Hunanyan H, Sargsyan R (2019) Characterization of endophytic bacterial *Acinetobacter* Sp. a1 strain isolated from lithophytic lichen collection of scientific articles of YSU SSS 13. Nat Phys Math Sci:122–127
Iskina RY (1938) On nitrogen fixing bacteria in lichens. Isv Biol Inst Permsk 11:133–139
Jiang D, Wang H, Si H, Zhao L, Liu C, Zhang H (2017) Isolation and culture of lichen bacteriobionts. Lichenologist 49(2):175–181
Kumar J, Dhar P, Tayade AB, Gupta D, Chaurasia OP, Upreti DK, Arora R, Srivastava RB (2014) Antioxidant capacities phenolic profile and cytotoxic effects of saxicolous lichens from trans-Himalayan cold desert of Ladakh. PLoS One 9(6):e98696
Lendemer JC, Keepers KG, Tripp EA, Pogoda CS, McCain CM, Kane NC (2019) A taxonomically broad metagenomic survey of 339 species spanning 57 families suggests cystobasidiomycete yeasts are not ubiquitous across all lichens. Am J Bot 106(8):1090–1095
Liba C, Ferrara F, Manfio G, Fantinatti-Garboggini F, Albuquerque R, Pavan C, Ramos P, Moreira-Filho C, Barbosa H (2006) Nitrogen-fixing chemo-organotrophic bacteria isolated from cyanobacteria-deprived lichens and their ability to solubilize phosphate and to release amino acids and phytohormones. J Appl Microbiol 101:1076–1086
Liu WT, Marsh TL, Cheng H, Forney LJ (1997) Characterization of microbial diversity by determining terminal restriction fragment length polymorphisms of genes encoding 16S rRNA. Appl Environ Microbiol 63(11):4516–4522
Mark K, Laanisto L, Bueno CG, Niinemets Ü, Keller C, Scheidegger C (2020) Contrasting co-occurrence patterns of photobiont and cystobasidiomycete yeast associated with common epiphytic lichen species. New Phytol 227(5):1362–1375
Millanes AM, Diederich P, Wedin M (2016) *Cyphobasidium* gen nov a new lichen-inhabiting lineage in the Cystobasidiomycetes (*Pucciniomycotina Basidiomycota* Fungi). Fungal Biol 120 (11):1468–1477
Muggia L, Klug B, Berg G, Grube M (2013) Localization of bacteria in lichens from Alpine soil crusts by fluorescence in situ hybridization. Appl Soil Ecol 68:20–25
Mushegian AA, Peterson CN, Baker CC, Pringle A (2011) Bacterial diversity across individual lichens. Appl Environ Microbiol 77(12):4249–4252
Muyzer G, Smalla K (1998) Application of denaturing gradient gel electrophoresis (DGGE) and temperature gradient gel electrophoresis (TGGE) in microbial ecology. Antonie Van Leeuwenhoek 73(1):127–141
Palmqvist K, Franklin O, Näsholm T (2017) Symbiosis constraints: strong mycobiont control limits nutrient response in lichens. Ecol Evol 7(18):7420–7433
Pankratov TA, Kachalkin AV, Korchikov ES, Dobrovol'skaya TG (2017) Microbial communities of lichens. Microbiology 86(3):293–309

Panosyan AK, Nikogosyan VG (1966) The presence of Azotobacter in lichens Akad Nauk Armian SSR. Biol Zhurn Armen 19:3–11

Sargsyan R, Gasparyan A, Tadevosyan G, Panosyan H (2021) Antimicrobial and antioxidant potentials of noncytotoxic extracts of corticolous lichens sampled in Armenia. AMB Express 11(1). https://doi.org/10.1186/s13568-021-01271-z

Schneider T, Schmid E, de Castro JV, Cardinale M, Eberl L, Grube M, Berg G, Riedel K (2011) Structure and function of the symbiosis partners of the lung lichen (*Lobaria pulmonaria* L Hoffm) analyzed by metaproteomics. Proteomics 11(13):2752–2756

Schwieger F, Tebbe CC (1998) A new approach to utilize PCR–single-strand-conformation polymorphism for 16S rRNA gene-based microbial community analysis. Appl Environ Microbiol 64 (12):4870–4876

Selbmann L, Zucconi L, Ruisi S, Grube M, Cardinale M, Onofri S (2010) Culturable bacteria associated with Antarctic lichens: affiliation and psychrotolerance. Polar Biol 33(1):71–83

Sierra MA, Danko DC, Sandoval TA, Pishchany G, Moncada B, Kolter R, Mason CE, Zambrano MM (2020) The microbiomes of seven lichen genera reveal host specificity, a reduced core community and potential as source of antimicrobials. Front Microbiol 11:398

Singh RP, Manchanda G, Singh RN, Srivastava AK, Dubey RC (2016) Selection of alkalotolerant and symbiotically efficient chickpea nodulating rhizobia from North-West Indo Gangetic Plains. J Basic Microbiol 56:14–25. https://doi.org/10.1002/jobm.201500267

Singh R, Manchanda G, Maurya I, Wei Y (eds) (2020) Microbial versatility in varied environments. Springer, Singapore. https://doi.org/10.1007/978-981-15-3028-9

Smith HB, Dal Grande F, Muggia L, Keuler R, Divakar PK, Grewe F, Schmitt I, Lumbsch HT, Leavitt SD (2020) Metagenomic data reveal diverse fungal and algal communities associated with the lichen symbiosis. Symbiosis 82(1):133–147

Spribille T, Tuovinen V, Resl P, Vanderpool D, Wolinski H, Aime MC, Schneider K, Stabentheiner E, Toome-Heller M, Thor G, Mayrhofer H, Johannesson H, McCutcheon JP (2016) Basidiomycete yeasts in the cortex of ascomycete macrolichens. Science 353 (6298):488–492

Subhashini DV, Singh RP, Manchanda G (2017) OMICS approaches: tools to unravel microbial systems. Directorate of Knowledge Management in Agriculture, Indian Council of Agricultural Research. ISBN: 9788171641703. https://books.google.co.in/books?id=vSaLtAEACAAJ

Suryanarayanan TS, Thirunavukkarasu N (2017) Endolichenic fungi: the lesser-known fungal associates of lichens. Mycology 8(3):189–196

Suzuki MT, Parrot D, Berg G, Grube M, Tomasi S (2016) Lichens as natural sources of biotechnologically relevant bacteria. Appl Microbiol Biotechnol 100(2):583–595

Tuovinen V, Ekman S, Thor G, Vanderpool D, Spribille T, Johannesson H (2019) Two basidiomycete fungi in the cortex of wolf lichens. Curr Biol 29(3):476–483

Uphof JT (1925) Purple bacteria as symbionts of a lichen. Science 61(1568):67–67

Urbanavichene IN, Urbanavichus GP (2005) Southern lichen floristic elements in spruce forests of the White Sea coast of Murmansk Region. MA Bondartseva et al :240–244

Wijayawardene NNKD, Hyde LKT, Al-Ani L et al (2020) Outline of fungi and fungus-like taxa. Mycosphere 11(1):1060–1456

Yang YJ, Lin W, Singh RP, Xu Q, Chen Z, Yuan Y, Zou P, Li Y, Zhang C (2019) Genomic, transcriptomic and enzymatic insight into lignocellulolytic system of a plant pathogen *Dickeya* sp. WS52 to digest sweet pepper and tomato stalk. Biomolecule 9(12):753

Yang Y, Liu L, Singh RP, Meng C, Ma S, Jing C, Li Y, Zhang C (2020) Nodule and root zone microbiota of salt-tolerant wild soybean in coastal sand and saline-alkali soil. Front Microbiol 11:2178. https://doi.org/10.3389/fmicb.2020.523142

Zúñiga C, Leiva D, Carú M, Orlando J (2017) Substrates of Peltigera lichens as a potential source of cyanobionts. Microb Ecol 74(3):561–569

Chapter 10
Antagonists and Antibiosis: Game Changer of Agriculture and Health Sector

Atul Kumar Srivastava, Ashish Kumar Nayak, Arvind Saroj, and Pooja Misra

Abstract The microbial world always draws attention of research fraternity due to the phenomenon of host–parasite and microbe–microbe interactions. Competition/struggle among microbes in same niche is basically for space and nutrients. Therefore, every microbe has unique strategies to remain less affected in such microbe–microbe interaction (Antagonism). During the nineteenth century, the term antagonism was originated from the Greek word 'antagonizesthai' (struggle against). However, the term antibiosis was originated from the French word 'antibiose' that describes the antagonistic effects between microbes. Consequently, the findings of antagonistic activity leads to paradigm shift to constraint human, animal and plant parasites. The current chapter deals with the detailed discussion about various strategies and mechanisms involved in the phenomenon of antagonism and antibiosis.

Keywords Antagonism · Synergism · Biocontrol · Bacteria · Fungi

---

Atul Kumar Srivastava and Ashish Kumar Nayak contributed equally with all other contributors.

A. K. Srivastava
Research and Development Department, Uttaranchal University, Dehradun, Uttarakhand, India

A. K. Nayak
Microbial Genomics and Diagnostics Lab, Microbiology and Plant Pathology Division, Regional Plant Resource Centre, Bhubaneswar, Odisha, India

A. Saroj (✉)
Department of Horticulture, Molecular Plant Breeding, North East Agriculture University, Harbin, China

P. Misra
Crop Protection Division, CSIR-Central Institute of Medicinal Aromatic Plants, Lucknow, India

R. P. Singh et al. (eds.), *Microbes in Microbial Communities*,
https://doi.org/10.1007/978-981-16-5617-0_10

10.1 Introduction

Economy is predominantly based on agriculture; this is revealed by the FAO latest forecast on world cereal production which stands at almost 2790 million tons in 2020, which is the highest in the last 20 years (http://www.fao.org/3/ca9803en/ca9803en.pdf). Other than that, because of variation in Asian weather, farmers also used to grow cash crops like sugarcane, fruits, cotton, medicinal and aromatic plants, for improving their economic status. Thus, the agriculture farming regularly helps to increment the economy of our country. In the past decades, farmers were facing many challenges like high demand, reduction of farming area due to urbanization and infection-prone weather fluctuations. Pathogen attack is one of the major problems, which directly or indirectly affect the health, yield and productivity of plants (Yang et al. 2019). Some conventional pesticides are regularly used for the management of plant diseases. It is difficult to control fungal/bacterial/insect diseases because of their (pathogens) diversity by finding long-lasting resistance to the target pathogen in the host plants in a limited period of time and lack of effective chemical control (Jones et al. 2012). Very few bactericides or fungicides are available with low effectiveness for crop protection rather than a long list of chemical products applicable for the control of bacterial or fungal pathogens. Furthermore, thiram and tebuconazole are two fungicides used regularly to control diseases in vegetables by treatments of seeds, as well as methyl bromide that is used for soil treatments.

The use of different chemicals as pesticides for disease management creates many problems as the pathogenic resistance established against the applied bactericides/fungicides (Saha et al. 2012; Yang et al. 2020). The use of chemical pesticides is restricted or banned due to undesirable side effects on human health and environment, as well as evolution of new resistant pathogens (Saha et al. 2012; Singh et al. 2017). Chemical fungicides are banned worldwide due to undesirable outputs in long-term consumptions such as high toxicity. For long-term pathogen control, it is important to find an alternate environment-friendly measure. Nowadays, bio-pesticides have been considered as a safe and reasonable alternative way of chemical pesticides to achieve the goal of eco-friendly disease control as well as to improve plant health and productivity (Marcic et al. 2017) (shown in Fig. 10.1). These bio-pesticides are products of such natural materials like plants, bacteria, fungi and actinomycetes and have no adverse effect on environment. Microbes are regularly synthesized/secreted secondary metabolite in response to biotic/abiotic neighbours that may be used in the agriculture and health sectors. We need to know the details of antibiosis and antagonist responses in reference to agriculture and health sector.

Antagonism is the best to suit for recuperate from the plant and animal health-related issues. The microbes that are pathogenic are mainly found in the environment, and many of the soil microbes are found to be antagonistic nature. These antagonistic microbes attack on the neighbouring microbes by enzymatic degradation of cell wall and cellular content, and the protoplasmic material acts as a

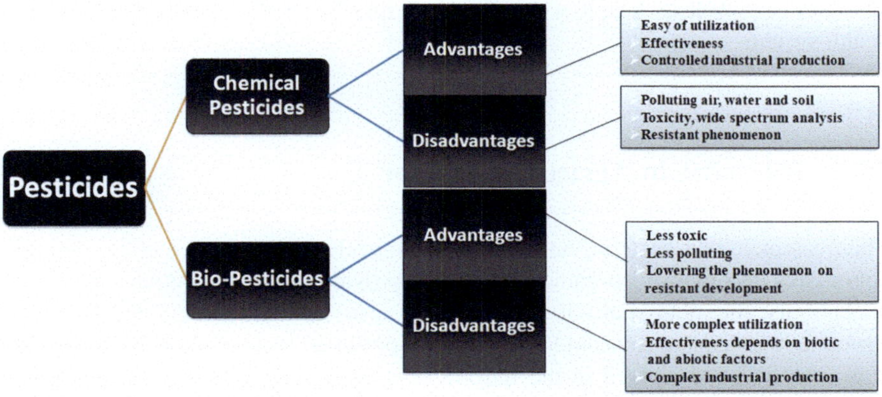

Fig. 10.1 Advantages and disadvantages of chemical and bio-pesticides

nutritious substance for the inhibitor organisms. For example, *Aspergillus* fungus affects the *Penicillium* and *Cladosporium*, whereas *Trichoderma* affects the actinomycetes. Similarly, bacteria like *Pseudomonas* spp. show the antagonistic behaviour on *Cladosporium* spp., and possibly many of practical importance is found after knowing that it produces antibiotics.

Direct inhibitory activity having opposite actions within the similar system between two organisms is referred as antagonism (Bhatti et al. 2017; Prajakta et al. 2019). This relationship is a highly prevalent phenomenon in the microbial world, where one microbial species suppresses or inhibits the growth and development of other microorganisms (Egorov 2004; Subhashini et al. 2017). Primarily, bacteria, actinomycetes and fungi possess the antagonistic relationship either with the same group or different group of microbes. In this relationship, the main mechanisms between bacteria are to maintain the colonization resistance and micro biocoenosis formation in human biotopes (Bukharin et al. 2006). The appearance of environmental conditions is the main factor for the microbial antagonism. In the last two decades, many studies have been focused on the antagonistic relationship mechanisms between bacteria and fungi that can be used as a biological control of plant pathogens in agricultural fields (Janisiewicz et al. 2000). The antagonist nature of different bacteria, fungi and actinomycetes primarily acts as biocontrol agents that reduce disease-causing pathogen numbers. Different antagonism mechanisms are conducted for the management of root and soil-borne pathogens such as parasitism, predation, antibiosis, competition for nutrient sources and induced resistance of the host plant (Killani et al. 2011; Maheshwari et al. 2021).

Directly or indirectly, the soil health may affect the human health because what we eat is grown in the soil. Conclusively, we can say that the microbial antagonism behaviour enhances the production of antibiosis by using different mechanism which surprisingly creates successive positive evolutionary changes in both agriculture crop and animal health. Thus, in this chapter we like to express details of the antagonistic and antibiosis characteristics of bacteria, fungi and actinomycetes in agriculture as well as health sector for providing better future in agriculture and health system.

10.2 Microbes in Agriculture Sector

Microbes including bacteria, fungi and actinomycetes have positive effect on agriculture like decomposing organic material and providing essential nutrient (nitrogen, phosphorus and potassium) to plants. They also have negative impact on the agricultural crops like resistant to crop control products, pathogenicity and diseases. However, as we discussed in the Introduction section, antagonism and antibiosis play an important role in enhancing the productivity of plants and health of animals. Here we describe about the antagonistic role of microbes in agriculture sector, one by one specifically.

10.3 Antagonistic Activity of Bacteria

Nowadays, many modern studies have focused on the use of microbial inoculants and their multiple modes of action as a biological control to antagonize the plant pathogens (Saha et al. 2012). In the bacterial world, antagonism is a highly prevalent phenomenon where one bacterium species suppresses or inhibits the growth and development of the other microorganisms (Pandey et al. 2020). Bacteria represent an important group of biocontrol agents through their antagonist activity to control plant parasite, nematodes and fungal pathogens. Biological control based on the use of antagonistic bacteria is potentially a promising alternative strategy currently employed in agricultural fields. Bacteria produce different antibiotics and synthesize bacteriocins which are effective against bacteria of same species. Bacteriocins are microbial compounds that have a bactericidal or bacteriostatic effect on other closely related species (Krzyzanowska et al. 2019). Bacteriocin-producing bacteria play crucial role to contain soil-borne plant pathogens such as *Fusarium* spp. In recent years, different strains of the genera *Bacillus*, *Pseudomonas* and *Enterobacter* have been studied, which are broadly used to treat diseases by suppressing their growth caused by phytopathogens. Bacteriocin production by Gram-negative bacteria has been investigated most extensively in *Bacillus* spp. (Han et al. 2016) and *Pseudomonas* spp. (Georgakopoulos et al. 2002), which we meticulously considered in the following section.

10.3.1 Genus Bacillus

Bacillus genus is a group of Gram-positive, aerobic, rod-shaped and endospore-forming bacteria. It is the most widespread bacteria among the microorganisms present in nature. It is also known as producer of antimicrobial substances such as antibiotics, peptides and bacteriocins. The bacteriocins are microbial compounds that have a bactericidal or bacteriostatic effect on other closely related species. These bioactive substances have major applications in various industrial areas. The genus is a well-known producer of antibiotics as secondary metabolites and supply of micro and macro-nutrient. *Bacillus* spp. play key roles as antagonists and protect the plants from phytopathogens. Out of all *Bacillus* species, *B. subtillis* is used as biological control (Lin et al. 2001). *B. subtilis* can survive in extreme conditions because of the production of endospores. It is non-toxigenic and non-pathogenic to humans, animals and plants. The antagonism activity of *B. subtilis* is conducted by the secretion of antifungal compounds against these soil-borne pathogens as well as suppression root colonization (Li et al. 2013).

10.3.1.1 Role of Genus Bacillus

- *Bacillus* spp. are known as important members of PGPB, which induce plants to tolerate biotic and/or abiotic stresses in a broad manner.
- The antimicrobial activity of the species that produces the antifungal compounds which main mode of action by the antagonistic bacteria.
- It is also used as biocontrol and biofertilizers agents.
- *Bacillus* spp. do not cause any morphological changes or visible damage to the host. Therefore, these bacterial species can be supportive for the survival of the host species against microbial competition and different environmental stresses.
- The species produce different catalytic enzymes (proteases, chitinases and glucanases) and peptide antibiotics (fengimycin, bacilizyn, bacilin, bacitracin, bacilomycin B) which are known as antibacterial and antifungal substances (Pal and McSpadden 2006; Stein 2005).

10.3.2 Genus Pseudomonas

Genus *Pseudomonas* is a group of common, Gram-negative, rod-shaped species and abundantly found in soil, water, plants and surface environments (Gross and Joyce 2009). *Pseudomonas* spp. are known to produce different antimicrobial compounds henceforth increases plant defence mechanisms (Ramamoorthy et al. 2002; Singh et al. 2016). In addition to their beneficial effect on the improvement of plant growth in the presence of pathogens, *Pseudomonas* spp. produce different bio-active metabolites (siderophores, polyketides, lipopeptides and volatile compounds) which help to directly compete with plant pathogens and also help to induce systemic plant resistance (Yadav et al. 2021).

10.3.2.1 Role of Genus Pseudomonas

- The siderophores produced by different *Pseudomonas* spp. directly help the plants in their growth and also indirectly improve the nutritional quality of crops, by enhancing the iron content in grain (Veerendra and Janakiram 2015).
- They also stimulate the systemic resistance in plants and cooperate in direct competition with both root- and soil-borne plant pathogens for bioavailable iron (Kraemer et al. 2017).
- The soluble compounds formed by *Pseudomonas* spp. are also adequate to suppress the improvement of different plant pathogens such as *P. infestans* and *R. solani* (Priyaja et al. 2014).
- It also helps in inducing resistance in tomato plants infected by *P. infestans* zoospores (Priyaja et al. 2014).
- The polyketides metabolites produced by the *Pseudomonas* spp. have direct antagonistic activity against damping-off diseases in different crops (Ahmadzadeh and Tehrani 2009).
- Different fungicides are prepared by the species to protect the food from toxic fungi such as *Botrytis* and *Penicillium*.

10.3.3 Genus Enterobacter

Enterobacter genus is a Gram-negative, anaerobic, motile and facultative bacilli, belonging to the family of *Enterobacteriaceae*. These microorganisms are saprophytic in nature and found in the environment, soil and sewage (Mezzatesta et al. 2012).

10.3.3.1 Role of Genus Enterobacter

- Acylated peptide is produced by *Enterobacter* species, which are active against filamentous fungi and yeasts.
- It also produces pyrrolnitrin, which is effective against *Aspergillus niger, Candida albicans* and phytopathogenic fungi.
- It also inhibits the sporulation and growth of *Fusarium oxysporum, Penicillium viridicatum, Trichoderma viride* and some phytopathogens (Kerr et al. 1999).

10.4 Antagonistic Mechanisms of Bacteria

Inhibition of phytopathogen by bacteria is generally performed by antagonistic mechanisms, which make it more capable to control diseases in postharvest fruit (Table 10.1). Different modes of anti-phytopathogenic activity of bacteria have been studied, such as space and nutrient, parasitism, volatile compounds, competition and biofilms. The main antagonistic mechanisms and different steps of mode of action exerted by bacteria contrary to phytopathogens are described below (Fig. 10.2).

In this technique, pathogenic fungi antagonistic interaction among microbes is used by changing the microbiological environment to promote reproduction of antagonistic bacteria and inhibiting the growth of pathogenic microorganisms (Table 10.2).

10.5 Fungal Antagonism

Fungi (singular: fungus) are defined as unicellular to multicellular, eukaryotic, heterotrophs and having a chitinous cell wall. They play an essential role in the nutrient cycle of the ecosystem. Antagonism (interference competition) is defined as an active metabolite of one species that exerts a negative effect on other species in an ecosystem. If the causal organism is a fungus, then it may be called as fungal

Table 10.1 Bacterial antagonist successfully deployed for biological control of postharvest diseases of fruits

S. No.	Bacterial antagonists	Diseases	Target pathogen(s)	Host fruit (s)	Reference
1	*Bacillus subtilis*	Anthracnose	*C. musae*	Banana	Khleekorn and Wongrueng (2014)
		Rot	*Alternaria alternata*	Melon	
2	*B. Atrophaeus*	Anthracnose	*Colletotrichum acutatum*, *C. gloeosporioides*	Pepper	Chen et al. (2016)
3	*B. Amyloliquefaciens*	Brown rot	*Monilinia* sp.	Apple	Chen et al. (2016)
4	*P. fluorescens*	Blue mould	*Penicillium expansum*	Apple	Wallace et al. (2018)
5	*P. Syringae*	Green mould	*P. digitatum*	Citrus	Panebianco et al. (2015)
6	*Pantoea agglomerans*	Anthracnose	*C. musae*	Banana	Khleekorn and Wongrueng (2014)
7	*Enterobacter cloacae*	Dry rot	*F. sambucinum*	Potato	Kim et al. (2016)
8	*B. Megaterium*	Damping off	*Aspergillus flavus*	Peanut	Kong et al. (2010)

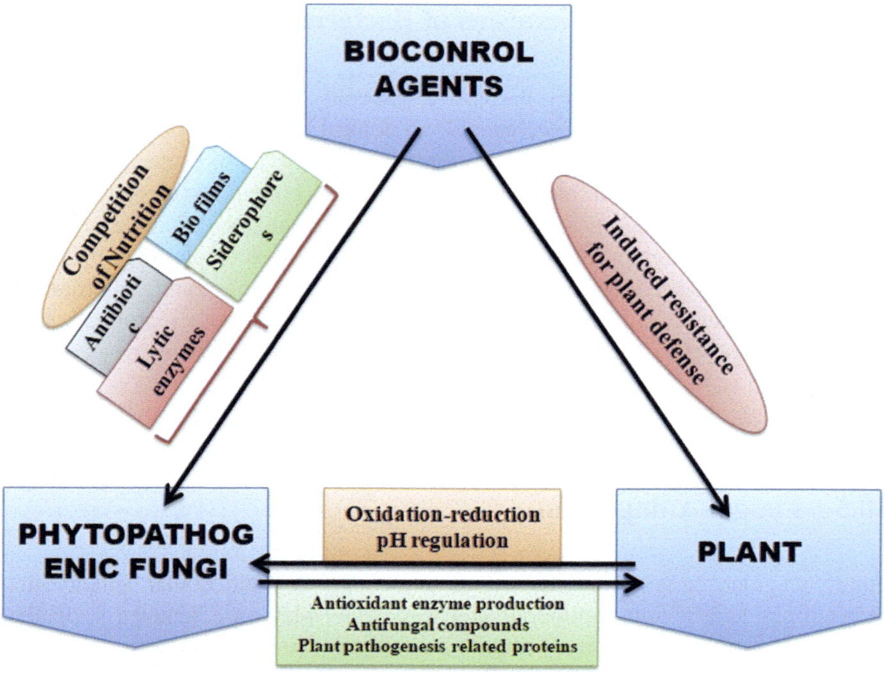

Fig. 10.2 Schematic presentation of microbes towards enhancing plant defence

antagonism. Fungi possess an essential role in the decomposition of organic matter and have shown both symbiotic and parasitic associations with other organism like bacteria, plants and animals. Fungal species are known to reproduce by two different stages: the anamorph i.e., asexual reproductive stage, and the teleomorph i.e., sexual reproductive stage (Kirk et al. 2002). The systematic study of fungi started in the seventeenth century after the discovery of the von Leeuwenhoek microscope. Still, the man who deserves to be the founder of mycological science is Pier Antonio Micheli for his book *Nova Plantarum Genera* (Alexopoulos 1962) about fungus. In the microbial world, fungal infections cause severe diseases in both plants and animals as well as in humans (Hyde et al. 2018).

In agriculture field, fungi are the major group of plant pathogen that possess the negative effect on the crop yield, and its productivity reduces the economy of the country and has shown major problems of farmers. Besides, this fungus produces different stimuli, antibiotics and extracellular enzymes that help in the degradation of organic material and provide direct nutrient to plants. Fungi are more popular in the scientific community for its antagonistic behaviour and act as biocontrol agents against many of the plant diseases with pesticidal and weedicidal property (Rangel et al. 2018). For the management of plant diseases, and increasement of proper plant growth as well to activate the defence mechanism, fungi are widely used. Thus, the industries developed mass production of that type of fungi for the commercially available biocontrol agents (Costa et al. 2013; Vega et al. 2009). Other than this,

Table 10.2 Modes of antagonistic actions in relation to development and use of bacterial species as biological control agents

S. No.	Mode of action	Method of screening	Pathogen specificity	Risk of resistance	Dependency on plant Physiology	Dependency on environmental conditions
1	Induced resistance	Complex bioassays on plants	Specific to broad	Low	High	Low
2	Competition	Simplified bioassays	Broad	Low	Low	High
3	Hyper-parasitism	Simplified bioassays	Pathogenic-specific interactions	Low	High	Low
4	Antimicrobial metabolites produced in situ	Simplified bioassays	Specific to broad	Low	Low	Moderate
5	Antimicrobial metabolites in product	In vitro assays	Broad	Moderate	Low	Low
6	Helper strains	Complex bioassays	Depends on MBCA	Low	Reduced	Reduced
7	Assembled consortia combining different modes of action	In silico design followed by complex bioassays	Broad	Low	Low	Low
8	Modulation of indigenous microbiota	Complex site-specific bioassays	Broad	Low	Low	Medium

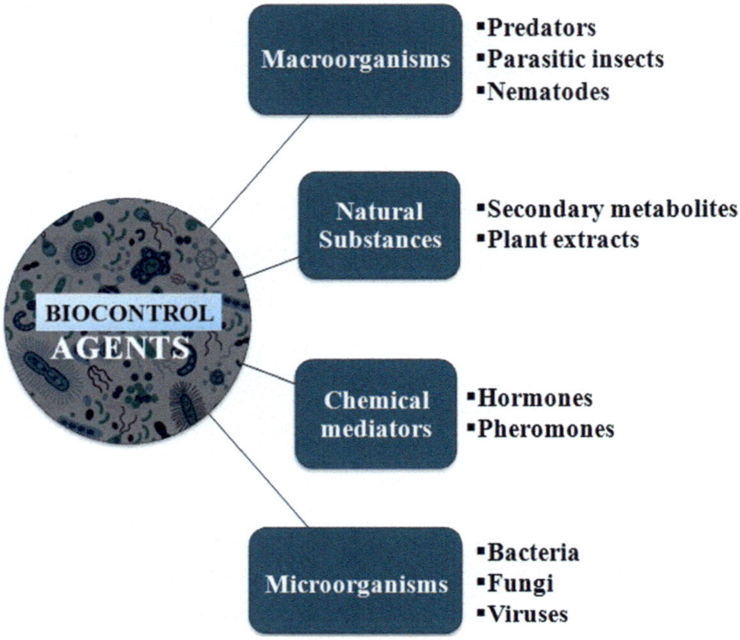

Fig. 10.3 Bio-control agents found in nature

these biocontrol agents also manage the terrestrial weeds, reduce aquatic weeds, plant-parasitic nematodes and insects (Siddiqui and Mahmood 1996; Alston et al. 2005; Li et al. 2010). Earlier, the agrochemical industries developed fungal-based natural products (Bills and Gloer 2016), but uncontrolled consumptions of these fungicides have led to the development of resistant species (Lucas et al. 2015). Natural substances, chemical mediators and micro and microorganisms are the major group of biocontrol agents seen in Fig. 10.3, including the secondary metabolites produced in the various growth stages of living organisms (Balog et al. 2017).

In the plant–soil ecosystem, the rhizospheric region is the dynamic association of plant and soil microbes commonly called as rhizospheric microbiota. The rhizospheric fungi hold their position tightly as compare to other microbiome, which is mutually benefitted by host–fungal interaction for ecosystem functioning and sustainability (Pattnaik and Busi 2019). Mycorrhiza is the best example of plant and fungal associations and works as plant-protecting agent.

10.6 Mycorrhiza: Plant-Protecting Agent

Most of the microbiomes get associated with plant through root especially in rhizospheric region. Mycorrhiza (Gr. mycos: fungus; rhiza: root) is a symbiotic association ship of fungi and plant roots (Manchanda et al. 2017; Singh et al.

2017). Around 80% angiosperms have mycorrhizal association (Santra and Banerjee 2020). Two types of mycorrhizae are present in nature: one is endomycorrhizae (AM), also known as arbuscular mycorrhizae (e.g. *Endogone* sp. and *Rhizophagus* sp.), and the other is ectomycorrhizae (EM) (e.g. *Amanita muscaria* and *Laccaria bicolor*). The mycorrhizal association is the oldest and natural privilege that protects plants by activating plant's defensive mechanism against phytopathogens, thereby working as a potential biocontrol agent (Güimil et al. 2005; Paszkowski 2006; Román et al. 2011). Other than this, mycorrhizal association modulate the rhizospheric microbiome, by either removing the phytopathogens or stimulating the microbial population that have antagonistic activity against phytopathogens (Santra and Banerjee 2020). The antagonistic fungus can produce many of the bioactive substances against plant and human pathogens reported widely seen in Table 10.3. and Fig. 10.4.

10.7 Actinomycetes Antagonism

Prevot (1961) placed strictly anaerobic bacteria like actino in group of actinomycetes and they have chemoautotrophic potential, both the qualities were not reported in fungal group. The chemoautotrophic property of actinomycetes made them grow easily in less nutrient availabilities like can be growing on water agar or other lean media. Actinomycetes generally show sensitivity against antibiotics (Lechevalier and Lechevalier 1967; Ansari et al. 2020) that work against Gram-positive bacteria; therefore, they support the fact that like bacteria and blue green algae they lack sterols. Actinomycetes are more prone to attack of phages (Marei and Elbaz 2013), but fungus is rarely seen to be attacked by viruses. The immunological evidence confirms that *Mycobacterium*, *Nocardia* and *Corynebacterium* groups were found to be closely related to actinomycetes group (Goodfellow and Williams 1983; Lechevalier and Lechevalier 1967). A group of Streptomyces have antagonistic activity against fungus such as *Alternaria* sp., *Pythium aphanidermatum, Colletotrichum higginsianum, Acremonium lactucum, Fusarium oxysporum* and *Rhizoctonia solani* (Lahdenpera et al. 1991; Hong et al. 2002; Singh et al. 2018).

10.8 Similarities between Actinomycetes and Bacterial and Fungal Groups

Actinomycetes are the group of spore-forming Gram-positive filamentous bacteria having branching of filaments, i.e. morphology similar to fungi. Actinomycetes have more GC content (57–75%) in their genome and belong to order Actiomycetales (Subhashini and Singh 2014; Bhatti et al. 2017). They have pathological similarities

Table 10.3 Bioactive antagonistic activity against plant and human pathogens

S. No	Metabolite	Source species	Comments	Reference
1	Strobilurins A–D and other strobilurins and oudemansins	*Strobilurus tenacellus*	First released in 1996, and 23–25% impact on the global fungicide sales of its products	Feng et al. (2020)
2	Gibberellic acid (GA)	*Fusarium fujikuroi*	GA was identified as a metabolic by-product of the fungi *Gibberalla fujikuroi* in 1926, affects rice plants. It is a plant growth hormone that helps in the high-value crops	Camara et al. (2018)
3	Sphaeropsidins A-F	*Diplodia cupressi*	Antimycotic, phytotoxic and insecticidal	Sparapano et al. (2004)
4	Afritoxinone A, B	*Diplodia africana*	Phytotoxic	Evidente et al. (2012)
5	Oxysporone (29)	*Diplodia africana*	Phytotoxic, antioomycetes and antifungal	Evidente et al. (2012); Andolfi et al. (2014)
6	(3*R*,4*S*)-4-Hydroxymellein, (3*R*,4*R*)-4-hydroxymellein, (3*R*,4*S*)-4-hydroxymellein	*Diplodia africana*	Phytotoxicity	Evidente et al. (2012)
7	Cytochalasins A	*Ascochyta heteromorpha*	Antifungal, antibacterial	Bottalico et al. (1990)
8	Viridepyronone (77)	*Trichoderma viride*	Antifungal	Evidente et al. (2003)
9	Fusaproliferin (78)	*Cleistothelebolus nipigonensis* and *Neogymnomyces virgineus*	Antifungal	Sarrocco (2016)
10	Terpestacin (79)	*Cleistothelebolus nipigonensis* and *Neogymnomyces virgineus*	Antifungal	Cimmino et al. (2016)
11	Gliotoxin (104)	*Neogymnomyces pseudofischeri*	Antibacterial	Liang et al. (2014)

(continued)

Table 10.3 (continued)

S. No	Metabolite	Source species	Comments	Reference
12	Penicillin G and V	*Penicillium rubens*, *Penicillium chrysogenum*	For the bacterial infections	Bennett and Chung (2001)
13	Cephalosporin C	*Acremonium chrysogenum*	For the bacterial infections	Hu and Zhu (2016)
14	Fusic acid	*Acremonium fudidioides*	For the bacterial infections	
15	Griseofulvin	*Penicillium griseofulvum*, *Penicillium aethiopicum*, *Penicillium coprophilum* and other *Penicillium* spp.	Treatment of systemic fungal infections	Blank et al. (1959)
16	Pneumocandin B_0	*Glarea lozoyensis*	Treatment of systemic fungal infections	Chen et al. (2013)
17	Echinocandin B	*Aspergillus pachycristatus* and other *aspergillus* spp.	Treatment of systemic fungal infections	Yue et al. (2015)
18	Enfumafungin	*Hormonema carpetanum*	Treatment of systemic fungal infections	Vicente et al. (2016)
19	Lovastatin (monacolin K)	*Aspergillus terreus*, *Monascus purpureus*, also occurs in basidiomata of *Pleurotus ostreatus*	Treatment of systemic fungal infections	Endo (2010)
20	Compactin (Mevastatin)	*Penicillium citrinum*, *Penicillium solitum* and other *Penicillium* spp.	Treatment of systemic fungal infections	Abe et al. (2002)
21	Cyclosporin A	*Tolypocladium inflatum*	Prevention of organ transplant and tissue graft rejection	Hess (1993)
22	Mycophenolic acid	*Penicillium brevicompactum* and other *Penicillium* spp.	Prevention of organ transplant and tissue graft rejection	Allison et al. (1993)
23	Myriocin (ISP-I)	*Isaria sinclairii*	Treatment of multiple sclerosis	Bills and Gloer (2017)
24	Ergotamine	*Claviceps purpurea*, *Claviceps fusiformis* and *Claviceps paspali*	Vasoconstrictor used as antimigraine agent, also combined with belladonna and phenobarbital for relief from menopausal hot flashes	Tfelt-Hansen et al. (2000)
25	Ergometrine (ergonovine)	*Claviceps purpurea*, *Claviceps fusiformis* and *Claviceps paspali*	Treatment of postpartum haemorrhage	Bills and Gloer (2017)

(continued)

Table 10.3 (continued)

S. No	Metabolite	Source species	Comments	Reference
26	Ergometrine (ergonovine)	*Claviceps purpurea*, *Claviceps fusiformis* and *Claviceps paspali*	Hallucinogen, treatment of psychosis and depression	Mower and Hancock (1975)
27	Ergocryptine	*Claviceps purpurea*, *Claviceps fusiformis* and *Claviceps paspali*	Treatment of reproductive disorders, e.g., galactorrhoea, prolactin dependent mammary carcinoma, amenorrhoea, etc.	Bills and Gloer (2017)
28	Mizoribine	*Penicillium brefeldianum*	Immunosuppressant used for renal transplants in Japan, Korea and China	Ishikawa (1999)
29	Kojic acid	*Aspergillus oryzae*, *aspergillus tamarii*, *aspergillus flavus*	Antioxidant in cosmetic products, used to lighten skin colour and treat abnormal hyperpigmentation	Rodrigues et al. (2014)
30	Fumagillin	*Aspergillus fumigatus*	Control of nosema disease in honey bees caused by *Nosema apis*	Huang et al. (2013)

with fungi, but its treatment can be done similar to bacterial disease. The similarities with bacterial and fungal kingdom were listed in Fig. 10.5.

Actinomycetes group becomes more scientifically fascinating because of its commercial point of view. This group can be freshly isolated from natural habitat like soil and then proceeded for classification on the basis of their morphological features. However, the strains that have been stored in laboratory for long time may lose their typical morphological feature (Lechevalierand Lechevalier1967), so it requires very sophisticated approach for correct generic designation. Ferdinand Cohn suggested that actinomycetales could be considered as sixth genera. The classification of different actino groups is listed in Table 10.4.

Soil is the common and main habitat of Actinomycetes, but they are also profoundly found in air, water and inside plant system. The thread like organism grows like hyphae that give it earthy smell of healthy soil (Rowbotham and Cross 1977). They are also very common in extreme environments (Bull 2011) especially in desert and cold environments.

10.8.1 Role of Actinomycetes

Some microbes have the ability to affect both functioning and formation of microbial community through symbiotic association. Actinomycetes stimulate the germination

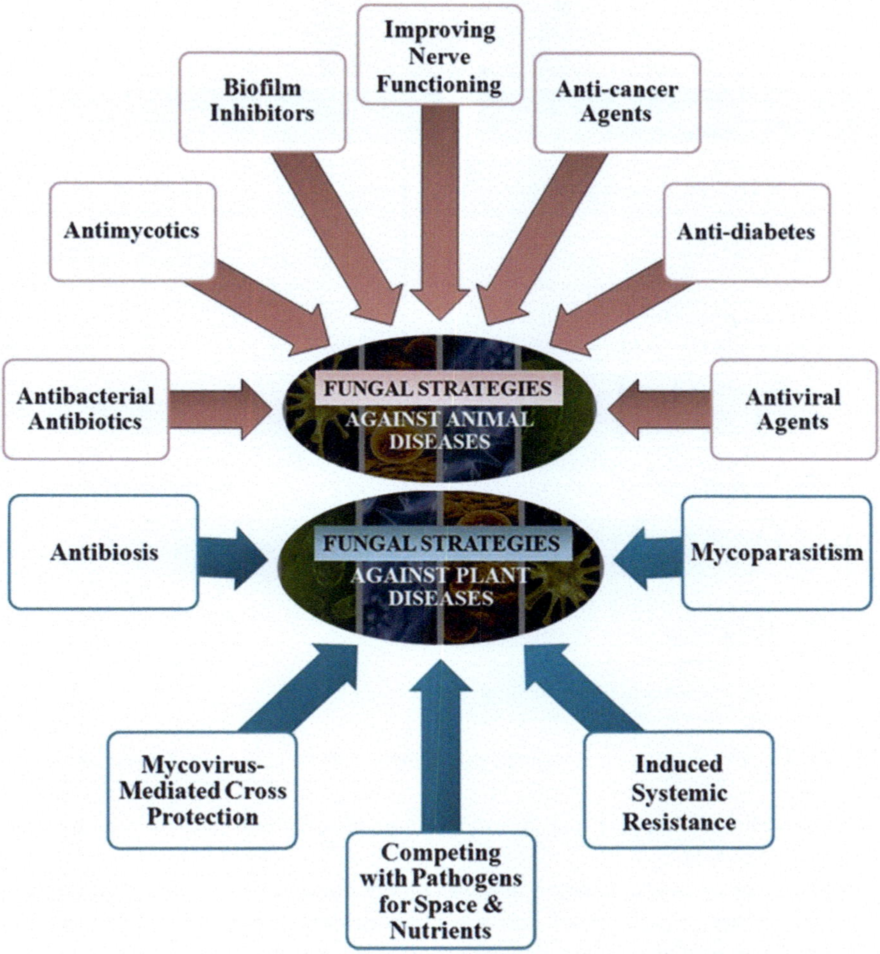

Fig. 10.4 Flowchart shows the fungal strategies involved in dealing against animal and plant diseases

of mycorrhizal spore, whereas the group *Thermobifida* inhibits the germination, but both of them promote the mycelial growth of *Glomus* sp. (Merzaeva and Shirokikh 2006). *Streptomyces orientalis* promotes the growth of *Gigaspora margarita* spores (Merzaeva and Shirokikh 2006).

Actinomycetes are worthy microorganism both economically and biotechnologically (Fig. 10.6). They are known for the production of various secondary metabolites like antibiotics, anticancer and immunosuppressant (Bhatti et al. 2017). Along with pharmaceutical benefits, they are also helpful in agriculture (Fig. 10.7) where they have the ability to inhibit the growth of various plant pathogens (Singh et al. 2019). As they can inhibit the growth of *Erwinia amylovora* (bacterial pathogen that

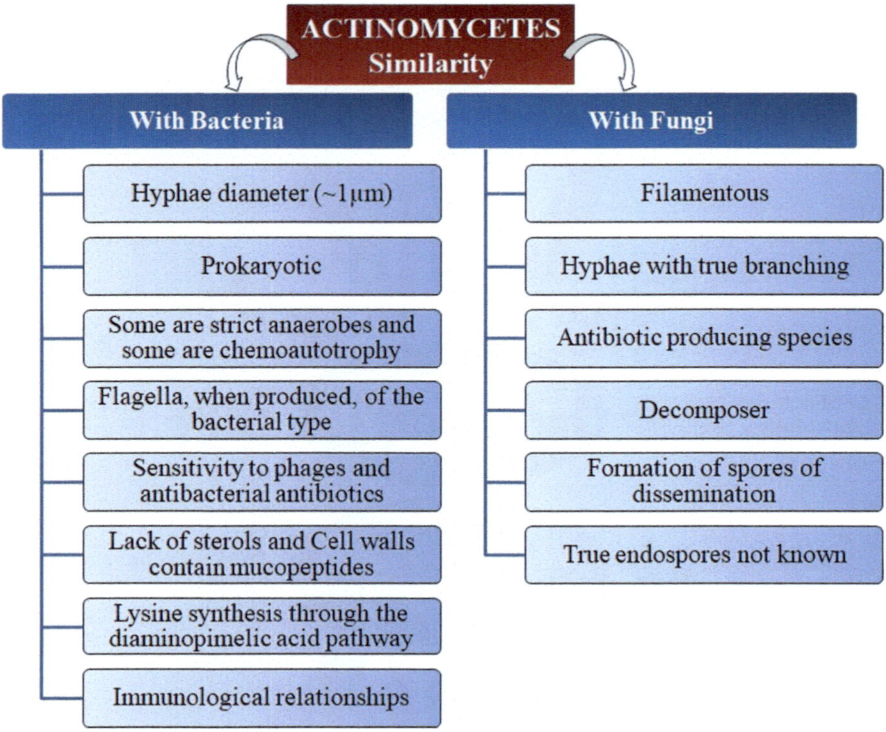

Fig. 10.5 Diagram showing the comparative similarity of actinomycetes with bacteria and fungi

cause fire blight in apple) and *Agrobacterium tumefaciens* (crown gall disease) (Oskay et al. 2004).

Actinomycetes can fix nitrogen in association with some plants (non-leguminous) and increase the availability of nitrogen for both the nearby plants and the host. Actinomycetes produce hydrolytic enzymes and are responsible for the decomposition and recycling of large range of organic matters. Therefore, they increase the availability of nutrient and minerals, along with this inhibit the phytopathogens and synthesize the plant growth regulators. They accomplish many functions like nitrogen fixation, phosphate solubilisation and siderophore production. Moreover, they are eco-friendly because they do not contaminate the environment; besides this, they are helpful in maintaining the biotic equilibrium.

Actinomycetes also degrade the pesticides like s-triazine, organochlorines, carbamates, triazinones, organophosphates, sulfonylurease, acetanilides etc. So, they are a good contestant for bioremediation of soil and recycling of complex organic carbon and polymers.

Table 10.4 Clinically important antibiotics produced from actinomycetes and their bio-control functions

S. No.	Antibiotic	Actinomycetes	Activity	Reference
1	Lomofungin	*Streptomyces lomondensis*	Antifungal	Das et al. (2008)
2	Sclerothricin	*Streptomyces scleogranulatus*	Antifungal	Kono et al. (1969)
3	Spoxamicin	*Streptosporangium oxazolinicum*	Antitrypanosomal	Inahashi et al. (2011)
4	Avermectin	*Streptomyces avermitilis*	Antiparasitic	Kitani et al. (2011)
5	Antimycin	*Streptomyces lucitanusus*	Antifungal	Han et al. (2012)
6	Rosamicin	*Micromonospora rosaria*	Antibacterial	Anzai et al. (2009)
7	Validamycin	*Streptomyces hygroscopicus*	Antifungal	Wu et al. (2012)
8	Azalomycin	*Streptomyces malaysiensis*	Antifungal	Cheng et al. (2010)
9	Roseoflavin	*Streptomyces davawensis*	Antibacterial	Matsui et al. (1979); Grill et al. (2008)
10	Rifamycin	*Micromonospora rifamycinica*	Antibacterial	Huang et al. (2008); Huang et al. (2009)
11	Salinomycin	*Streptomyces albus*	Antiparasite	Naidenova et al. (2001)

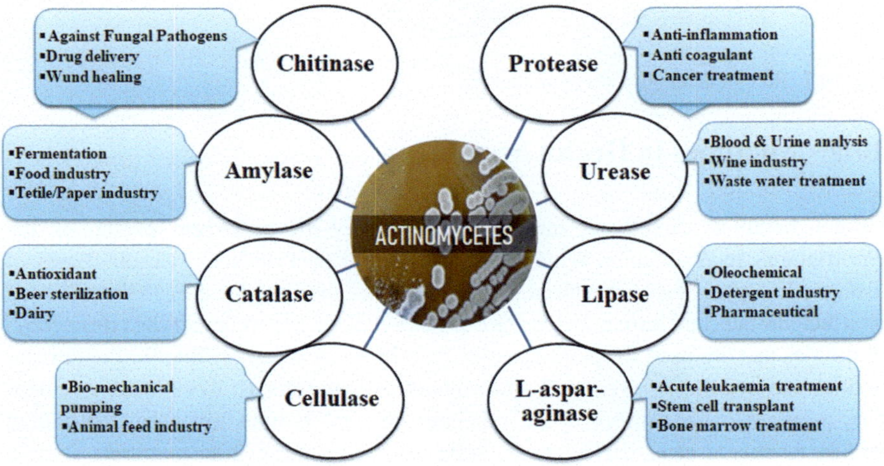

Fig. 10.6 Role of enzymes produced by actinomycetes

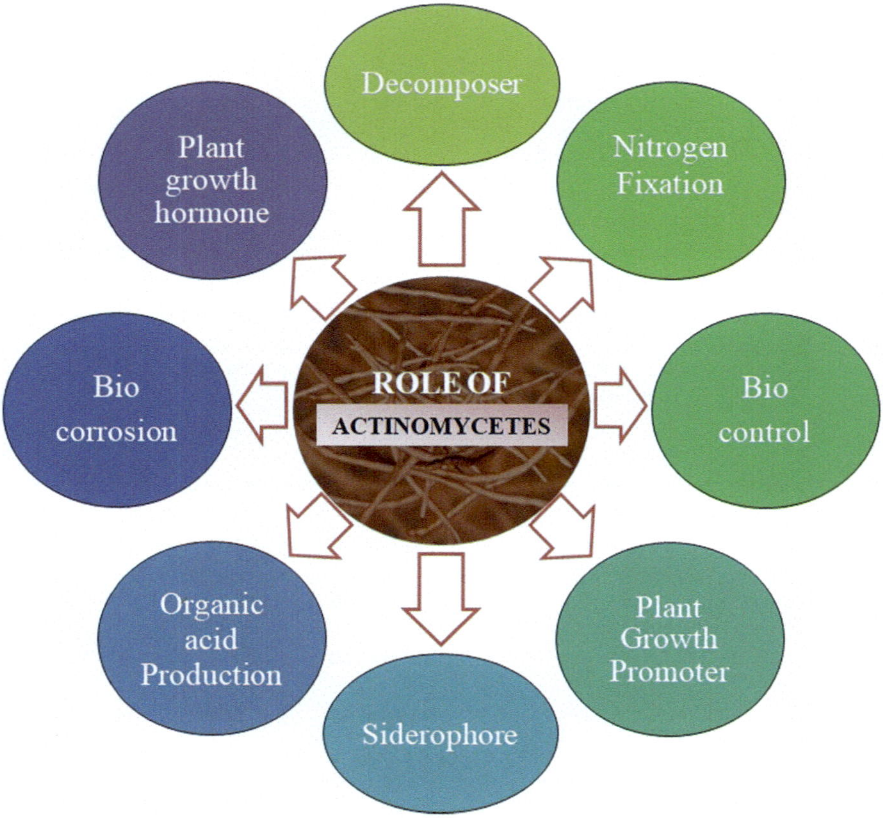

Fig. 10.7 Role of actinomycetes in agriculture

10.9 Microbes in Health Sector

Human health is directly or indirectly affected by the pathogens present in the environment in association with soil or air. In the past decades, various types of diseases like protozoans, bacterial, fungal and viral diseases have been documented to affect animals and human beings (Rodrigues et al. 2014). Whether the commercial pesticides that are used for plant diseases by farmers also responsible for the human/animal bad health. *Aspergillosis*, *Blastomycosis*, *Candidiasis*, *Coccidioidomycosis*, *Cryptococcosis*, *Dermatophytosis*, etc. are the vital human mycoses, distributed widely (Almeida et al. 2019). In contrast, some Basidiomycota members show a useful role as anti-infectives, immunosuppressants and other pharmaceutical properties (Badalyan et al. 2019).

In 1928, Alexander Fleming discovered that the first natural antibiotic "Penicillin" was isolated from the fungus *Penicillium notatum*, had antibacterial activity against Staphylococci (Flemming et al. 2000). Over the next 50 years, numerous antibiotics were discovered from the different group of fungi, bacteria and actinomycetes. The logic behind the production of antibiotics is the antagonistic mechanism of organism. The success of penicillin unlocks the development of microbial technology for the establishment of the pharmaceutical industry.

10.10 Conclusion

Since few decades synthetic pesticides seeks attention due to its harmful side effects on environment. Increase of uncontrolled use of synthetic pesticides resulted into water, air and soil pollution as well which resulted into serious consequences on human health. Therefore, use of biological controls get some attention from different research groups as alternative of synthetic pesticides for some extent. Nowadays, biological controls are remarkably included in integrated pest management (IPS) systems. Generally antagonistic organisms have ability to counter more than one pathogen at a time. Although, practical use of such biocontrol strategies at commercial level has been constrained by means of many factors. The major concerns which limit the use of biocontrol are cost effectiveness, efficacy and consistent performance on different environment.

Activity of the antagonist can be improved to a great extent by crop management strategies. Chiefly crop rotation plays an important role in disease management strategy globally and can positively affect antagonistic microorganisms in the soil. Because, when a crop is rotated yearly, the consortia of resident antagonists have a great likelihood to reduce the inoculum levels of pathogenic microorganism present on the surface of root and foliage while specific crop is absent.

References

Abe Y, Suzuki T, Ono C, Iwamoto K, Hosobuchi M, Yoshikawa H (2002) Molecular cloning and characterization of an ML-236B (compactin) biosynthetic gene cluster in *Penicillium citrinum*. Mol Gen Genomics 267:636–646

Ahmadzadeh M, Tehrani AS (2009) Evaluation of fluorescent pseudomonads for plant growth promotion, antifungal activity against *Rhizoctonia solani* on common bean, and biocontrol potential. Biol Control 48(2):101–107

Alexopoulos CJ (1962) Introductory mycology

Allison AC, Kowalski WJ, Muller CD, Eugui EM (1993) Mechanisms of action of mycophenolic acid. Ann N Y Acad Sci 696(1):63–87

Almeida M, Madeira T, Watanabe L, Meletti PC, Nixdorf SL (2019) Pesticide determination in water samples from a rural area by multi-target method applying liquid chromatography-tandem mass spectrometry. J Braz Chem Soc 30(8):1657–1666. https://doi.org/10.21577/0103-5053.20190066

Alston DG, Rangel DEN, Lacey LA, Golez HG, Jeong JK, Roberts DW (2005) Evaluation of novel fungal and nematode isolates for control of *Conotrachelus nenuphar* (Coleoptera: Curculionidae) larvae. Biol Control 35:163–171

Andolfi A, Maddau L, Basso S, Linaldeddu BT, Cimmino A, Scanu B, Deidda A, Tuzi A, Evidente A (2014) Diplopimarane, a 20-nor-ent-pimarane produced by the oak pathogen Diplodia quercivora. J Nat Prod 77:2352–2360

Ansari WA, Krishna R, Zeyad MT, Singh S, Yadav A (2020) Endophytic Actinomycetes-mediated modulation of defense and systemic resistance confers host plant fitness under biotic stress conditions. In: Singh R, Manchanda G, Maurya I, Wei Y (eds) Microbial versatility in varied environments. Springer, Singapore. https://doi.org/10.1007/978-981-15-3028-9_10

Anzai Y, Iizaka Y, Li W, Idemoto N, Tsukada S, Koike K (2009) Production of rosamicin derivatives in Micromonosporarosaria by introduction of Dmycinose biosynthetic gene with PhiC31-derived integration vector pSET152. J Ind Microbiol Biotechnol 36:1013e1021

Balog A, Hartel T, Loxdale HD, Wilson K (2017) Differences in the progress of the biopesticide revolution between the EU and other major crop-growing regions. Pest Manag Sci 73:2203–2208

Badalyan SM, Barkhudaryan A, Rapior S (2019) Recent progress in research on the pharmacological potential of mushrooms and prospects for their clinical application. In: Agrawal D, Dhanasekaran M (eds) Medicinal mushrooms. Springer, Singapore. https://doi.org/10.1007/978-981-13-6382-5_1

Bennett JW, Chung KT (2001) Alexander Fleming and the discovery of penicillin

Bhatti AA, Haq S, Bhat RA (2017) Actinomycetes benefaction role in soil and plant health. Microb pathogen 111:458–467

Bills GF, Gloer JB (2016) Biologically active secondary metabolites from the fungi. Microbiol Spectr 4:VE-0001-2014

Bills GF, Gloer JB (2017) Biologically active secondary metabolites from the fungi. The Fungal Kingdom:1087–1119

Blank H, Smith JG, Roth FJ, Zaias N (1959) Griseofulvin for the systemic treatment of dermatomycoses. J American Med Associat 171(16):2168–2173

Bottalico A, Capasso R, Evidente A, Randazzo G, Vurro M (1990) Cytochalasins: structure-activity relationships. Phytochemistry 29:93–96

Bukharin OV, Valyshev AV, Gilmutdinova FG (2006) [in Russian], Ekaterinburg

Bull AT (2011) Actinobacteria of the Extremobiosphere. In: Horikoshi K (ed) Extremophiles handbook. Springer, Tokyo. https://doi.org/10.1007/978-4-431-53898-1_58

Camara MC, Vandenberghe LP, Rodrigues C, de Oliveira J, Faulds C, Bertrand E, Soccol CR (2018) Current advances in gibberellic acid (GA 3) production, patented technologies and potential applications. Planta 248(5):1049–1062

Chen MC, Lai JN, Chen PC, Wang JD (2013) Concurrent use of conventional drugs with Chinese herbal products in Taiwan: a population-based study. J Tradit Complement Med 3:256–262

Chen X, Zhang Y, Fu X, Li Y, Wang Q (2016) Isolation and characterization of *bacillus amyloliquefaciens* PG12 for the biological control of apple ring rot. Postharvest Biol Technol 115:113–121

Cheng J, Yang SH, Palaniyandi SA, Han JS, Yoon TM, Kim TJ et al (2010) Azalomycin F complex is an antifungal substance produced by Streptomyces malaysiensis MJM1968 isolated from agricultural soil. J Korean Soc Appl Biol Chem 53:545e552

Cimmino A, Sarrocco S, Masi M, Diquattro S, Evidente M, Vannacci G, Evidente A (2016) Fusaproliferin, terpestacin and their derivatives display variable allelopathic activity against some Ascomycetous fungi. Chem Biodivers 13:1593–1600

Costa FG, Zucchi TD, de Melo IS (2013) Biological control of phytopathogenic fungi by endophytic actinomycetes isolated from maize (*Zea mays* L). Braz Arch Biol Technol 56(6):948–955

Das S, Lyla PS, Khan SA (2008) Distribution and generic composition of Culturable marine Actinomycetes from the sediments of Indian continental slope of bay of Bengal. Chin J Oceanol Limnol 26:166e177

Egorov NS (2004) Fundamentals of the theory of antibiotics: a manual [in Russian]. Moscow

Endo A (2010) A historical perspective on the discovery of statins. Proceedings of the Japan Academy, Series B 86(5):484–493

Evidente A, Cabras A, Maddau L, Serra S, Andolfi A, Motta A (2003) Viridepyronone, a new antifungal 6-substituted 2H-Pyran-2-one produced by *Trichoderma viride*. J Agric Food Chem 51:6957–6960

Evidente A, Masi M, Linaldeddu BT, Franceschini A, Scanu B, Cimmino A, Andolfi A, Motta A, Maddau L (2012) Afritoxinones a and B, dihydrofuropyran-2-ones produced by *Diplodia africana* the causal agent of branch dieback on *Juniperus phoenicea*. Phytochemistry 77:245–250

Feng Y, Zhan H, Huang Y, Bhatt P, Chen S (2020) An overview of strobilurin fungicide degradation: current status and future perspective. Front Microbiol 11:389

Flemming H, Wingender J, Mayer C, Körstgens V, Borchard W (2000) Cohesiveness in biofilm matrix polymers. In: Allison D, Gilbert P, Lappin-Scott H, Wilson M (eds) Community structure and co-operation in biofilms (Society for general microbiology symposia). Cambridge University Press, Cambridge, pp 87–106. https://doi.org/10.1017/CBO9780511754814.007

Georgakopoulos DG, Fiddaman P, Leifert C, Malathrakis NE (2002) Biological control of cucumber and sugar beet damping-off caused by *Pythium ultimum* with bacterial and fungal antagonists. J Appl Microbiol 92:1078–1086

Goodfellow & Williams (1983) Ecology of Actinomycetes. Ann Rev Microbial 37:189–216

Grill S, Busenbender S, Pfeiffer M, K€ohler U, Mack M (2008) The bifunctionallavokinase/flavin adenine dinucleotide synthetase from *Streptomyces davawensis* produces inactive flavin cofactors and is not involved in resistance to the antibiotic roseoflavin. J Bacteriol 190:1546e1553

Gross H, Joyce EL (2009) Genomics of secondary metabolite production by pseudomonas spp. Natural Products Report 26:1408–1446

Güimil S, Chang HS, Zhu T, Sesma A, Osbourn A, Roux C, Ioannidis V, Oakeley EJ, Docquier M, Descombes P, Briggs SP, Paszkowski U (2005) Comparative transcriptomics of rice reveals an ancient pattern of response to microbial colonization. Proc Natl Acad Sci 102:8066–8070

Han Z, Xu Y, McConnell O, Liu L, Li Y, Qi S et al (2012) Two antimycin a analogues from marine-derived actinomycete *Streptomyces lusitanus*. Mar Drugs 10:668e676

Han T, You C, Zhang L, Feng C, Zhang C, Wang J, Kong F (2016) Biocontrol potential of antagonist *Bacillus subtilis* Tpb55 against tobacco black shank. Biol Control 61:195–205

Hess AD (1993) Mechanisms of action of cyclosporine: considerations for the treatment of autoimmune diseases. Clin Immunol Immunopathol 68(2):220–228

Hong T, Cheng C, Huang J, Meng M (2002) Isolation and biochemical characterization of an endo-1, 3-b-glucanase from *Streptomyces sioyaensis* containing a C-terminal family 6 carbohydrate-binding module that binds to1, 3-bglucan. Microbiol 148:1151e1159

Hu Y, Zhu B (2016) Study on genetic engineering of *Acremonium chrysogenum*, the cephalosporin C producer. Syn Syst Biotechnol 1(3):143–149

Huang H, Lv J, Hu Y, Fang Z, Zhang K, Bao S (2008) *Micromonospora rifamycinica* sp. nov, a novel actinomycete from mangrove sediment. Int J Syst Evol Microbiol 58:17e20

Huang H, Wu X, Yi S, Zhou Z, Zhu J, Fang Z et al (2009) Rifamycin S and its geometric isomer produced by a newly found actinomycete, *Micromonospora rifamycinica*. Ant Van Leeuwenhoek 95:143e148

Huang WF, Solter LF, Yau PM, Imai BS (2013) Nosema ceranae escapes fumagillin control in honey bees. PLoS Pathog 9(3):e1003185

Hyde KD, Al-Hatmi AM, Andersen B, Boekhout T, Buzina W, Dawson TL, Eastwood DC, Jones EG, de Hoog S, Kang Y, Longcore JE (2018) The world's ten most feared fungi. Fungal Div 93: 161–194

Inahashi Y, Matsumoto A, Omura S, Takahashi Y (2011) Spoxazomicins A-C, novel antitrypanosomal alkaloids produced by an endophyticactinomycete, *Streptosporangium oxazolinicum* K07-0460T. J Antibiot 64:297e302

Ishikawa H (1999) Mizoribine and mycophenolate mofetil. Curr Med Chem 6:575–598

Janisiewicz WJ, Tworkoski TJ, Sharer C (2000) Characterizing the mechanism of biological control of postharvest diseases on fruits with a simple method to study competition for nutrients. Phytopathology 90(11):1196–2000

Jones JB, Vallad GE, Iriarte FB, Obradović A, Wernsing MH, Jackson LE, Momol MT (2012) Considerations for using bacteriophages for plant disease control. Bacteriophage 2(4):208–214. https://doi.org/10.4161/bact.23857

Kerr JR, Taylor GW, Rutman A, Høiby N, Cole PJ, Wilson R (1999) *Pseudomonas aeruginosa* pyocyanin and 1-hydroxyphenazine inhibit fungal growth. J Clin Pathol 52:385–387

Khleekorn S, Wongrueng S (2014) Evaluation of antagonistic bacteria inhibitory to *Colletotrichum musae* on banana. J Agric Technol 10:383–390

Killani AS, Abaidoo RC, Akintokun AK, Abiala MA (2011) Antagonistic effect of indigenous *Bacillus subtilis* on root-/soil-borne fungal pathogens of cowpea. Researcher 3:11–18

Kim YS, Balaraju K, Jeon Y (2016) Effects of rhizobacteria Paenibacillus polymyxa APEC136 and Bacillus subtilis APEC170 on biocontrol of postharvest pathogens of apple fruits. J Zhejiang Univ-Sci B 17:931–940

Kirk O, Borchert TV, Fuglsang CC (2002) Industrial enzyme applications. Curr Opin Biotechnol 13:345–351

Kitani S, Miyamoto KT, Takamatsu S, Herawati E, Iguchi H, Nishitomi K et al (2011) Avenolide, a Streptomyces hormone controlling antibiotic production in *Streptomyces avermitilis*. Proc Natl Acad Sci 108:16410e16415

Kong Q, Shan S, Liu Q, Wang Q, Yu F (2010) Biocontrol of *aspergillus flavus* on peanut kernels by use of a strain of marine *bacillus megaterium*. Int J Food Microbiol 139:31–35

Kono Y, Makino S, Takeuchi S, Yonehara H (1969) Sclerothricin, a new basic antibiotic. J Antibiot 22:583e589

Kraemer SA, Soucy JP, Kassen R (2017) Antagonistic interactions of soil pseudomonads are structured in time. FEMS Microbiol Ecol 93:1–9

Krzyzanowska DM, Maciag T, Siwinska J, Krychowiak M, Jafra S, Czajkowski R (2019) Compatible mixture of bacterial antagonists developed to protect potato tubers from Soft Rot caused by Pectobacterium spp. and Dickeya spp. Plant Dis 103(6):1374–1382

Lahdenpera ML, Simon E, Uoti J (1991) Mycostopda novel biofungicide based on Streptomyces bacteria. In: Bollen M, Gerlagh M, Ruissen M, Schippers B, Tempel A (eds) Biological control of plant disease on roots of conifer seedlings. Elsevier, Amsterdam, p 258e263

Lechevalier HA, Lechevalier MP (1967) Biology of actinomycetes. Annual Reviews in Microbiology 21(1):71–100

Li Z, Alves SB, Roberts DW, Fan M, Delalibera I, Tang J et al (2010) Biological control of insects in Brazil and China: history, current programs and reasons for their successes using entomopathogenic fungi. Biocontrol Sci Tech 20:117–136

Li S, Zhang N, Zhang Z, Luo J, Shen B, Zhang R, Shen Q (2013) Antagonist *Bacillus subtilis* HJ5 controls Verticillium wilt of cotton by root colonization and biofilm formation. Biol Fertil Soils 49:295–303

Liang WL, Le X, Li HJ, Yang XL, Chen JX, Xu J, Liu HL, Wang LY, Wang KT, Hu KC et al (2014) Exploring the chemodiversity and biological activities of the secondary metabolites from the marine fungus *Neosartorya pseudofischeri*. Mar Drugs 12:5657–5676

Lin D, Qu LJ, Gu H, Chen Z (2001) A 3.1-kb genomic fragment of Bacillus subtilis encodes the protein inhibiting growth of Xanthomonas oryzae pv. Oryzae.Journal of. Appl Microbiol 91 (6):1044–1050. https://doi.org/10.1046/j.1365-2672.2001.01475.x

Lucas JA, Hawkins NJ, Fraaje BA (2015) The evolution of fungicide resistance. Adv Appl Microbiol 90:29–92

Maheshwari NK, Singh RP, Manchanda G, Dubey RC, Maheshwari DK (2021) Sunn hemp (Crotalaria juncea) nodulating bacteria capable for high antagonistic potential and plant growth promotion attributes. J Microbiol Biotechnol Food Sci 10(3):385-389. doi:10.15414/jmbfs.2020.10.3.385-389

Manchanda G, Singh RP, Li ZF, Zhang JJ (2017) Mycorrhiza: creating good spaces for interactions. In: Mycorrhiza-function, diversity, state of the art. Springer, Cham

Marcic SM, Stepanovic M, Todorovic B, Duduk B, Stepanovic J, Rekanovic E, Ivana Potocnik I (2017) Biological control of green mould on *Agaricus bisporus* by a native Bacillus subtilis strain from mushroom compost. Eur J Plant Pathol 148:509–519

Marei EM, Elbaz RM (2013) Isolation and molecular characterization of three virulent Actinophages specific for *Streptomyces flavovirens*. J Virol Res 2(1):12
Matsui K, Juri N, Marukawa S, Kasa S (1979) Roseoflavin formation by Streptomyces davawensis. Agric Biol Chem 43:2189e2190
Merzaeva O, Shirokikh I (2006) Colonization of plant rhizosphere by Actinomycetes of different genera. Microbiol 75:226e230
Mezzatesta ML, Gona F, Stefani S (2012) *Enterobacter cloacae* complex: clinical impact and emerging antibiotic resistance. Future Microbiol 7(7):887–902
Mower RL, Hancock JG (1975) Sugar composition of ergot honeydew. Can J Bot 53:2813–2825
Naidenova MM, Michaĭilov YG, Ivanova I (2001) Oxadin effect on Actinomycetes strains, Antibiot. Khimioter 46:7e9
Oskay M, Same A, Azeri C (2004) Antibacterial activity of some Actinomycetes isolated from farming soils of Turkey. AJB 3:441e446
Pal KK, McSpadden GB (2006) Biological control of plant pathogens. The Plant Health Instructor 2:1117–1142. https://doi.org/10.1094/phi-a-2006-1117-02
Pandey S, Shukla N, Singh SS, Tripathi D, Tripathi T, Kant S (2020) Bacterial metabolic fitness during pathogenesis. In: Singh R, Manchanda G, Maurya I, Wei Y (eds) Microbial versatility in varied environments. Springer, Singapore. https://doi.org/10.1007/978-981-15-3028-9_12
Panebianco S, Vitale A, Polizzi G, Scala F, Cirvilleri G (2015) Enhanced control of postharvest citrus fruit decay by means of the combined use of compatible biocontrol agents. Biol Control 84:19–27
Paszkowski U (2006) Mutualism and parasitism: the yin and yang of plant symbioses. Curr Op Plant Biol 9:364–370
Pattnaik SS, Busi S (2019) Rhizospheric fungi: diversity and potential biotechnological applications. In: Yadav A, Mishra S, Singh S, Gupta A (eds) Recent advancement in white biotechnology through fungi. Fungal Biology. Springer, Cham. https://doi.org/10.1007/978-3-030-10480-1_2
Prajakta BM, Suvarna PP, Raghvendra SP, Alok RR (2019) Potential biocontrol and superlative plant growth promoting activity of indigenous *Bacillus mojavensis* PB-35 (R11) of soybean (*Glycine max*) rhizosphere. SN Appl Sci 1(10):1–1
Prevot AR (1961) Traite de Systematique Bacterienne, II. Dunod, Paris 771
Priyaja P, Jayesh P, Correya NS, Sreelakshmi B, Sudheer N, Philip R, Isaac Sarogeni I, Singh B (2014) Antagonistic effect of *Pseudomonas aeruginosa* isolates from various ecological niches on vibrio species pathogenic to crustaceans. Journal of Coastal Life Medicine 2(1):76–84
Ramamoorthy V, Raguchander T, Samiyappan R (2002) Induction of defense-related proteins in tomato roots treated with *Pseudomonas fluorescens* Pf1 and *fusarium oxysporum* f. sp. *lycopersici*. Plant Soil 239:55–68
Rangel DEN, Finlay RD, Hallsworth JE, Dadachova E, Gadd GM (2018) Fungal strategies for dealing with environment- and agriculture-induced stresses. Fungal Biol 122:602–612
Rodrigues APD, Farias LHS, Carvalho ASC, Santos AS, do Nascimento JLM, Silva EO (2014) A novel function for kojic acid, a secondary metabolite from aspergillus fungi, as antileishmanial agent. PLoS One 9(3):e91259
Román M, Fernández I, Wyatt T, Sahrawy M, Heil M, Pozo MJ (2011) Elicitation of foliar resistance mechanisms transiently impairs root association with arbuscular mycorrhizal fungi. J Ecol 99:36–45
Rowbotham TJ, Cross T (1977) *Rhodococcus coprophilus* sp. nov.: an aerobic nocardio form actinomycete belonging to the ' rhodochrous' complex. J gen microbio 100:123e138
Saha D, Purkayastha GD, Ghosh A, Isha M, Saha A (2012) Isolation and characterization of two new *Bacillus subtilis* strains from the rhizosphere of eggplantas potential biocontrol agents. J Plant Pathol 94(1):109–118. https://doi.org/10.4454/jpp.fa.2012.020
Santra HK, Banerjee D (2020) Natural products as fungicide and their role in crop protection. In natural bioactive products in sustainable agriculture (pp. 131-219). Springer, Singapore

Sarrocco S (2016) Dung-inhabiting fungi: a potential reservoir of novel secondary metabolites for the control of plant pathogens. Pest Manag Sci 72:643–652

Siddiqui ZA, Mahmood I (1996) Biological control of plant parasitic nematodes by fungi: a review. Bioresour Technol 58:229–239

Singh RN, Singh RP, Sharma A, Saxena AK (2016) Modeling of PrnD protein from *Pseudomonas fluorescens* RajNB11 and its comparative structural analysis with PrnD proteins expressed in Burkholderia and Serratia. Turkish J Biol 40(3):623–633

Singh RP, Manchanda G, Li ZF, Rai AR (2017) Insight of proteomics and genomics in environmental bioremediation. In: Bhakta JN (ed) Handbook of research on inventive bioremediation techniques. IGI Global, Hershey. https://doi.org/10.4018/978-1-5225-2325-3

Singh RP, Manchanda G, Anwar MN, Zhang JJ, Li YZ (2018) Mycorrhiza–helping plants to navigate environmental stresses. Microbes for Climate Resilient Agricultur 13:205–233. https://doi.org/10.1002/9781119276050.ch10

Singh RP, Manchanda G, Maurya IK, Maheshwari NK, Tiwari PK et al (2019) *Streptomyces* from rotten wheat straw endowed the high plant growth potential traits and agro-active compounds. Biocatal Agric Biotechnol 17:507–513. https://doi.org/10.1016/j.bcab.2019.01.014

Sparapano L, Bruno G, Fierro O, Evidente A (2004) Studies on structure-activity relationship of sphaeropsidins A–F, phytotoxins produced by *Sphaeropsis sapinea* f. sp. *cupressi*. Phytochemistry 65:189–198

Stein T (2005) *Bacillus subtilis* antibiotics: structures, syntheses and specific functions. Mol Microbiol 56:845–857. https://doi.org/10.1111/j.1365-2958.2005.04587.x

Subhashini DV, Singh RP (2014) Isolation of endophytic actinomycetes from roots and leaves of tobacco (*Nicotiana tabacum* L.). Annal Plant Pro Sci 22(2):458–459

Subhashini DV, Singh RP, Manchanda G (2017) OMICS approaches: tools to unravel microbial systems. Directorate of Knowledge Management in Agriculture, Indian Council of Agricultural Research. ISBN:9788171641703. https://books.google.co.in/books?id=vSaLtAEACAAJ

Tfelt-Hansen P, Saxena PR, Dahlöf C, Pascual J, Lainez M, Henry P, Diener HC, Schoenen J, Ferrari MD, Goadsby PJ (2000) Ergotamine in the acute treatment of migraine: a review and European consensus. Brain 123(1):9–18

Veerendra KM, Janakiram P (2015) Antagonistic activity exhibited by crude extracts of *Pseudomonas aeruginosa* (PIC-4) against *Aeromonas hydrophila* and *E. coli*. 4:36-40

Vega FE, Goettel MS, Blackwell M, Chandler D, Jackson MA, Keller S et al (2009) Fungal entomopathogens: new insights on their ecology. Fungal Ecol 2:149–159

Vicente F, Reyes F, Genilloud O (2016) Fungal secondary metabolites as source of antifungal compounds. Fungi Applications and Management Strategies:80–116

Wallace RL, Hirkala DL, Nelson LM (2018) Mechanisms of action of three isolates of *Pseudomonas fluorescens* active against postharvest grey mold decay of apple during commercial storage. Biol Control 117:13–20

Wu H, Qu S, Lu C, Zheng H, Zhou X, Bai L, Deng Z (2012) Genomic and transcriptomic insights into the thermo-regulated biosynthesis of valid amycin in *Streptomyces hygroscopicus* 5008. BMC Genomics 13:337

Yadav A, Singh RP, Singh AL, Singh M (2021) Identification of genes involved in phosphate solubilization and drought stress tolerance in chickpea symbiont Mesorhizobium ciceri Ca181. Arch Microbiol 203(3):1167–1174. https://doi.org/10.1007/s00203-020-02109-1

Yang YJ, Lin W, Singh RP, Xu Q, Chen Z, Yuan Y, Zou P, Li Y, Zhang C (2019) Genomic, transcriptomic and enzymatic insight into lignocellulolytic system of a plant pathogen *Dickeya* sp. WS52 to digest sweet pepper and tomato stalk. Biomolecule 9(12):753

Yang Y, Singh RP, Song D, Chen Q, Zheng X, Zhang C, Zhang M, Li Y (2020) Synergistic effect of *pseudomonas putida* II-2 and *Achromobacter* sp. QC36 for the effective biodegradation of the herbicide quinclorac. Ecotoxicol Environ Safety 188:109826

Yue Q, Chen L, Zhang X, Li K et al (2015) Evolution of chemical diversity in the echinocandin lipopeptide antifungal metabolites. Eukaryot Cell 14(7):76

Chapter 11
Role of Quorum Sensing in the Survival of Rhizospheric Microbes

Manohari Rathi, Geetanjali Manchanda, and Raghvendra Pratap Singh

Abstract Quorum sensing (QS) signaling is a cell-to-cell communication or coordination at microbial population level. However, the ecological role of QS in complex or multi-species communities, principally in the milieu of community assemblage, has neither been experimentally discovered nor theoretically revealed. QS comprises the production of secreted signals (diffusible), which can diverge across diverse types of microbes. Over the past decades, there has been a significant accretion of data of the molecular mechanisms, gene regulons, signal structures, and behavioral responses related with QS systems gained. More recent studies have focused on understanding quorum sensing in the context of bacterial sociality. Studies of the role of quorum sensing in cooperative and competitive microbial interactions have discovered, how QS coordinates interactions both within and between the species. Such studies of quorum sensing as a social behavior have relied on the development of "synthetic ecological" models that use nonclonal bacterial populations. Hence, the aim of this chapter is to understand how microbes might interact with one another in the plant root–associated soils using QS system.

Keywords Competition · Signals · Gene · Microbes · Rhizosphere

M. Rathi
Centre for Environmental Science and Technology, School of Environment and Earth Sciences, Central University of Punjab, Bathinda, Punjab, India

G. Manchanda
Department of Botany and Environmental Studies, DAV University, Jalandhar, Punjab, India

R. P. Singh (✉)
Department of Research and Development, Azoth Biotech Pvt. Ltd., Noida, Uttar Pradesh, India

R. P. Singh et al. (eds.), *Microbes in Microbial Communities*,
https://doi.org/10.1007/978-981-16-5617-0_11

11.1 Introduction

The environment around the plant root system is termed as "rhizosphere." It has a major influence on the plant growth and health, which are the basis for all life forms on the Earth. However, it is a biologically very active and complex system comprised of a series of interactions, influenced by abiotic and biotic factors. Microorganisms represent the most important constituent of the rhizosphere, and the composition of microbial population in the rhizosphere differs with change in the plant species due to different plant–microbe interactions (Singh et al. 2016a, b; Prajakta et al. 2019). These interactions involves both mutually beneficial or cooperative interactions such as that showed by plant growth–promoting bacteria (PGPR), nitrogen-fixing bacteria, and mycorrhizal symbiosis and competitive interactions such as in case of pathogenic microbes and antibiotic and biocontrol agent–releasing microbes (Zhang et al. 2016; Singh et al. 2018, 2019; Maheshwari et al. 2021).

It has been documented that the unicellular microbes such as bacteria can act not only as an individual cell, but under suitable conditions, when their numbers attain a significant level, they can alter their behavior in order to function as multi-cellular entities. This is because generally, bacteria do not live as a single cell, but they live in colonies or consortia to use the elaborative intercellular communication system that facilitates adaptation to the changing environmental conditions. The mechanism of microbial sensing and response via cell-to-cell communication through small signaling molecules has been revealed (Whitehead et al. 2001; Garg et al. 2014). Several molecule-mediated and cell density–dependent signaling pathways have now been demonstrated, some of these come under the scope of regulation, commonly called as quorum sensing (QS), the term first coined by Fuqua (Fuqua et al. 1994). Quorum sensing depends on the production of the low-mass signaling molecules called as autoinducers. The concentration of these "autoinducers" in the extracellular medium is associated with the producer organism population density. By sensing and acting according to these molecules, individual cell can sense the neighboring cell population to ensure whether there are sufficient bacteria i.e. quorum to commence to act in a multi-cellular approach. In general, the microbial-derived signaling molecules have been categorized as: i) short peptides and amino acid derivatives, normally utilized by Gram-positive bacteria and ii) fatty acid derivatives (AHLs) utilized by Gram-negative bacteria. The cellular processes that are regulated by quorum sensing in bacteria are varied and ranged from the development of genetic competence, i.e., to take up exogenous DNA as revealed in *Streptococcus pneumonia* and *Bacillus subtilis*, to biofilm formation and virulence in *Pseudomonas aeruginosa* and bioluminescence occurrence in *Vibrio fischeri* (Solomon et al. 1996; Sauer et al. 2002).

Several QS systems have been documented and categorized in the rhizosphere. These include plant growth–promoting rhizobacteria (PGPR) such as *Pseudomonas chlororaphis*, *Pseudomonas fluorescens*, *Burkholderia cepacia*, *Rhizobium elti*, *Rhizobium leguminosarum*, and *Sinorhizobium meliloti*; plant pathogens including

Agrobacterium rhizogenes, *Agrobacterium tumefaciens*, *Pseudomonas syringe*, and *Erwinia carotovora*; and saprophytes including *Chromobacter violaceum*, *Pseudomonas corrugate*, *Pseudomonas putida*, and *Nitrosomonas europaea* exhibit QS system to interact/communicate in the rhizosphere. Some recent studies also suggest that the stationary phase in bacteria, i.e., non-growth period of quiescent is also regulated by QS (Lazazzera 2000). Moreover, communication among bacteria to carry out various complex social behaviors such as cooperation has been documented from past half century.

11.2 The Rhizosphere: Niche for Microbes

Lorenz Hiltner used the term "rhizosphere" for the first time (Hiltner 1904). Rhizosphere is considered as the most dynamic environment inhabiting diverse microorganisms including bacteria (free living, root associated, and symbiotic), fungi, archaea, and viruses. Gram-negative bacteria such as *Pseudomonas* and *Agrobacterium* are predominant in the rhizosphere, whereas Gram-positive bacteria (*Bacillus* and *Clostridium*) are rare. The most common bacterial genera found in rhizosphere include *Arthrobacter*, *Azotobacter*, *Agrobacterium*, *Alcaligenes*, *Bacillus*, *Cellulomonas*, *Flavobacter*, *Mycobacterium*, *Micrococcus*, and *Pseudomonas*. *Pseudomonas* is the most common and widely studied genus, producing antibiotics and siderophores (Singh et al. 2016a, b; Singh et al. 2020). The larger part of microbiota in rhizosphere is constituted by bacteria *Actinomycetes* (Singh et al. 2018). *Actinomycetes* such as *Argania spinosa*, *Nocardia*, and *Streptomyces* sp. help in plant growth promotion and provide protection against fungal pathogens (Subhashini and Singh 2014). The fungi hardly get support from plant roots as bacteria. However, fungal genera commonly present in the rhizosphere are *Aspergillus*, *Fusarium*, *Penicillium*, and *Verticillium*. The majority of terrestrial plant roots establish mutualistic association with one or more fungal strains (Sylvia et al. 2005). These are reported to produce phytohormones and antibiotics that confers plant growth promotion and biocontrol activity as reported in *Trichoderma* and *Talaromyces flavus* (Fravel and Roberts 1991; Howell 1998). These rhizosphere microorganisms are responsible for nutrient availability and uptake by plants (Yang et al. 2018). They also accelerate soil remediation through improving plant growth and immunizing the plants to abiotic stress (Hrynkiewicz and Baum 2011; Yang et al. 2019). Root exudates such as organic acids, sugars, vitamins, enzymes, hormones, flavonoids, nucleotides, inorganic ions, etc. act as messengers that stimulate the interactions between plant roots and soil microorganisms. Consequently, due to very high microbial diversity, the rhizosphere is considered as the most dynamic system in the soil (Yang et al. 2020a, b). It is well documented that rhizosphere (Table 11.1) harbors higher microbial population as compared to the bulk soil (Bahadur et al. 2017; Kumar et al. 2017; Verma et al. 2017). Rhizosphere microbial population can also directly and indirectly shape the composition of plant species in natural ecosystems and vice-versa.

Table 11.1 Microbial population in rhizosphere and bulk soils

Microorganisms	Rhizosphere soil	Bulk soil	RS ratio
	Microorganisms g^{-1} dry soil		
Bacteria	1200×10^6	53×10^6	23
Actinomycetes	46×10^6	7×10^6	7
Fungi	12×10^5	1×10^5	12
Algae	5×10^3	27×10^3	0.2

The rhizosphere soil harbors efficient microbes that can be categorized as beneficial and harmful based on their influence on soil sustainability and plant growth and crop yield (Brimecombe et al. 2007; Singh et al. 2017). The rhizosphere microbial population is mainly recruited from the pool of microorganisms of bulk soil (Normander and Prosser 2000; Berg and Smalla 2009; Liu et al. 2018). Thus, the bulk soil is the key factor in shaping rhizosphere microbiome, and plant genotype is the driving force responsible for the recruitment of specific microorganisms from the bulk soil (Garbeva et al. 2008). Functions of rhizosphere microbiome are also influenced and commenced by the plant root exudations and quorum sensing. The microbial interactions and coordination in the rhizosphere, both intra-species and inter-species, occur through quorum sensing.

11.3 Quorum Sensing in Rhizosphere Microbes

Microbial communities are residing with plants and synergistically beneficial to them by the secretion of bioactive microbial compounds such as plant growth–promoting hormones, antagonistic compounds, etc. Within this association, plant roots are also releasing several organic compounds into the rhizosphere continually which is intensely beneficial to microorganisms and resisting to unwilling microbes (Subhashini et al. 2017). Though, a plethora of diverse community of microorganism is habituating with plant rhizosphere which have been survived socially by the behavior of synergistic, mutualistic, symbiotic, antagonistic, or tritagonistic. The above-described microbial community structure of rhizospheric region of plants is influenced by the resource competition, nutrient convenience, chemical interfering, and parasitism within them and between the host and plants. Among them, the cross talk between host and microbial communities or among the microbial assemblages is basically governed by the QS mechanism.

QS in the rhizosphere and is basically a type of density sensing that controls the physiological and chemical responses and behavior within bacteria. QS mechanism is very diverse as example the information carrier signal molecules which is different in different bacterial ($Gram^-$ or $Gram^+$) groups. However, each bacterial cell coordinates with the entire bacterial legacy by producing the self-inducing signal molecules. If the concentration of producing signal molecules grasps a specific threshold within the population density, the precise genes are expressed and start to regulate the adaptation of localized population. Overall, the QS programing regulates the several cellular mechanisms and processes, which largely comprise the regulation of bacterial biofilm formation, motility, tolerance, luminescence, virulence factors, disinfectants spore formation, toxin production, and resistance to drugs. The rhizosphere, a narrow zone of soil that surrounds and is influenced by plant roots is home to an overwhelming number of microorganisms and invertebrates and is considered to be one of the most dynamic interfaces on Earth. It is a "hot spot" of microbial activity, with increased microbial numbers, microbial interactions, and genetic exchange. Hence, the QS system accumulation in the local rhizospheric environment is very descriptive. The bacterial communities are sufficiently dense in this region and performs countless ecologically pertinent activities such as production of exopolysaccharide, extracellular enzyme, formation of biofilm, virulence traits expression, etc. QS is basically adapted by each bacteria residing in the rhizosphere, but the proteobacteria display the AHL-mediated QS. AHL-producing proteobacteria are highly abundant in the rhizosphere than natural bulky soil and constitute about 2/3 of the total rhizospheres microorganisms (Lagos et al. 2015). Metagenomic analysis of soil provides the evidence for QS in the plant rhizosphere (Williamson et al. 2005). Steidle et al. (2001) have proven it by comparative analysis of inoculated and uninoculated rhizosphere of sterile soil. QS study was clearly proven by the microcosm study of natural soil and compost soil but the QS role in soil processes is not investigated profoundly. Several authors have revealed that QS controls the extracellular enzyme activity during pathogenesis and was specifically described in pathogenic Gamma-proteobacteria, for example, *Aeromonas hydrophila*, *Enterobacteria* spp., *Erwinia carotovora, P. aeruginosa* PAO1, *P. fluorescens*, *Serratia* spp., and *Vibrio* spp. and beta-proteobacteria such as *Burkholderia* sp., *Chromobacterium violaceum*, etc. These reports displayed the QS controlled prevalence of pathogenesis by secretion of enzyme, and hence, it indicated the importance of QS in the soil nitrogen and other organic compounds cycling.

QS-based signal exchange in rhizosphere was reviewed by several authors and revealed its role in nitrogen-fixing behavior and coordination in legume-rhizobia symbiosis. During the nodulation process, the rhizobia increases in cell density toward the plant root chemotaxis and followed the procedure of exopolysaccharide production, symbiosome development, nodulation, nitrogen fixation, etc., and interestingly, the QS involves in all the mentioned process. Some basic structures of biomolecules that act as signals in rhizosphere are mentioned in Fig. 11.1 (Zhuang et al. 2013). Moreover, in the well-studied signal molecules, by bacterial quorum sensing signal compound AHLs are listed as symbiotic signal. Some of the typical rhizospheric systems are discussed in the next paragraph for deciphering the role of QS in rhizobia.

Among the rhizosphere and nitrogen-fixing rhizobia, *R. leguminosarum* bv. *viciae* has been treated as a model bacterium for the characterization of quorum sensing. Several review and research reports are disclosed the quorum-sensing systems such as *cin*, *rhi*, *rai,* and *tra*. Earlier, Cubo et al. (1992) have studied the *rhi* system and found that it is composed of *luxR* homolog (*rhiR and rhiI*) *and rhiABC* operon (*luxI* homolog), and all are found in symbiotic plasmid pRL1JI.

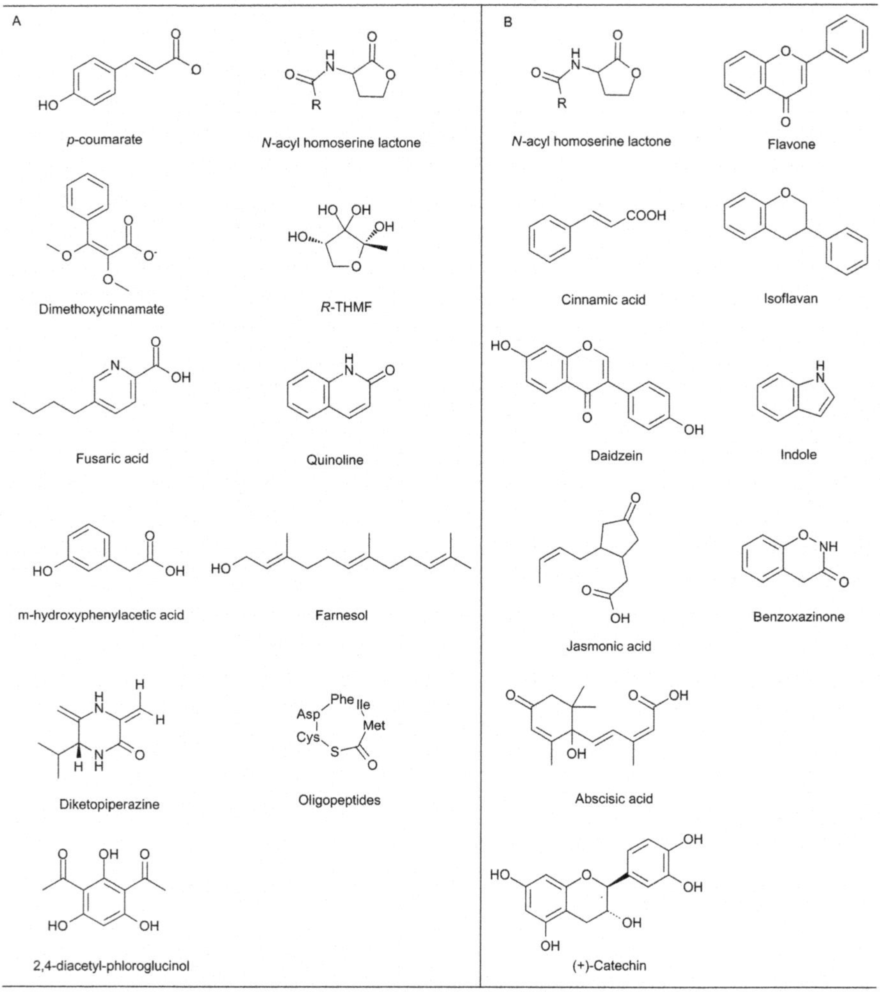

Fig. 11.1 Some basic structures of biomolecules act as signals in rhizosphere. (**a**) Molecules produced by microbes. (**b**) Examples of root exudates. [With the permission of MDPI (Zhuang et al. 2013)]

Also, they have displayed the mechanism of action and found that RhiR controlled the *rhiABC,* and hence, the flavonoids suppressed the expression of both *rhiR* and *rhiABC*.

Further, Rosemeyer et al. (1998) have studied the *R. etli* QS system which has provided the first evidence of linking the QS to symbiosis. Though, *R. etli* is less characterized compared to *R. leguminosarum as R. etli* is composed of up to seven AHLs, but only two QS systems (*raiRI* and *cinRI*) have been characterized. In *R. etli*, all the AHLs are synthesized by *raiRI* and *cinRI system, though the developed mutants of these two QS systems were failed to produce any detectable signals* (Daniels et al. 2002).

11.4 Quorum Sensing and Interactions in Rhizosphere Microbes

The rhizosphere microbial population utilizes quorum sensing for various and interesting interactions as symbiosis, competence, conjugation, motility, sporulation, antibiotic production, virulence, and biofilm formation (Miller and Bassler 2001). Co-operative interactions are biofilm formation, symbiosis, and plant growth promotion, etc.

Quorum sensing (QS) is involved in microbial cell aggregation and in rhizosphere colonization via formation of biofilm (Ng and Bassler 2009). Quorum sensing has been also demonstrated as an important factor responsible for symbiosis between rhizobia and legumes. Typically, rhizobia secretes diffusible AHL molecules in the surrounding, which can be used as signals to control the plant–microbe interaction. The AHL molecules recognized the several processes such as biofilm formation, secretion of extracellular polymeric substances, cell motility and gene expression for symbiosis, and nitrogen fixation starts (Yang et al. 2009). The QS regulatory system in rhizobia is based on LuxR-LuxI type. This is reliant on the threshold intensity of AHL molecules to induce the expression of target genes (Veliz-Vallejos et al. 2020). The rhizobia with mutations in their QS system has been reported with reduced ability to infect the root hairs or nodule formation. Moreover, several legumes have been reported to secrete substances that can interfere bacterial QS system (Table 11.2).

Plant growth–promoting bacteria (PGPB) interact with the plant roots via quorum sensing. The bacterial cells secrete QS molecules for colonization through biofilm formation and to exert effects on the plants to stimulate phenotypes including plant growth promotion and protection against stress and pathogens. *Kandelia obovata* secretes AHL in the rhizosphere to activate plant growth–promoting rhizobacteria (Ma et al. 2016). Similarly, *Burkholderia graminis* sp. imparts plant growth promotion and protection against salt stress in *Arabidopsis thaliana* through AHL quorum sensing (Barriuso et al. 2008). Recently, total 48 bacterial isolates with AHL and *N*-butyryl DL-homoserine lactone signals that were responsible for biofilm formation and plant growth promotion were isolated from rice root (Balasundararajan and Dananjeyan 2019).

QS generously uses by the several bacterial genera for the control of different types of secretion system. Fontaine et al. (2007) have studied the QS regulation for the bacteriocin production (Blp) in *Streptococcus thermophilus*, and the effectors of type VI secretion were reviewed thoroughly for *B. thailandensis* (Majerczyk et al. 2016). The secreted toxin management and control by QS was reviewed thoroughly and concluded that toxins are responsible for the competitive promotion with and within species of bacteria (Hibbing et al. 2010). Hence, it can be said as the QS will help to disclose the microbial community dynamics and the influencing species among them by mining of dominancy of secreted QS signals. Interestingly, this work was advantageous to the studies of the rhizosphere microbial communities associated with wheat (Mazzola et al. 1992). In the native soil, *P. fluorescens* 2–79 and

Table 11.2 Quorum sensing molecule and associated phenomena in rhizosphere microorganisms

Rhizosphere microorganism	Quorum sensing molecules	Functional attributes	References
Rhizobium etli CNPAF512	3-OH-slc-HSL	Symbiosis, plant growth, and nitrogen fixation	Daniels et al. (2002)
Rhizobium sp. NGR234	Short-chain AHL	Nodulation	He et al. (2003)
Mesorhizobium huakuii	C_8-HSL	Biofilm formation	Wang et al. (2004)
Bacillus thuringiensis	AHL-degrading enzyme (aiiA)	Antifungal activity	Park et al. (2008)
Pseudomonas putida	AHL	Effective communication between the cells	Gantner et al. (2006)
Burkholderia graminis	AHL	Protection against salt stress Plant growth promotion	Barriuso et al. (2008)
Pseudomonas chlororaphis 449	AHL: C4-AHL, C6-AHL and 30C6-AHL	Antibiotic production against phytopathogenic fungi	Veselova et al. (2008)
Pseudomonas putida	AHL	To coordinate in a population	Steidle et al. (2001)
Pseudomonas aureofaciens	AHL	AHL-mediated communication	Pierson et al. (1998)
Serratia liquefaciens MG1	AHL	Biocontrol activity against pathogens	Schuhegger et al. (2006)
Erwinia carotovora	3-oxo-C_6-HSL	Antibiotic activity	Sjöblom et al. (2006)
Serratia plymuthica HRO-C48	AHL	Biocontrol activity Disease suppression Plant growth hormone synthesis	Müller et al. (2009)
Acinetobacter	AHL	Attenuates the virulence factor produced in plant pathogens	Chan et al. (2011)
Pseudomonas CMR12a	*N*-Acylhomoserine lactone	Antagonistic activity against *Pythium myriotylum*	De Maeyer et al. (2011)
Serratia glossinae strain GS2	*N*-octanoyl-L-homoserine lactone *N*-hexanoyl-L-homoserine lactone	Plant growth promotion	Jung et al. (2017)
Serratia marcescens	AHL Bacterial degrading (AiiA) genes	Systemic resistance	Ryu et al. (2013)
Serratia glossinae	*N*-octanoyl-L-homoserine lactone and *N*-hexanoyl-L-homoserine lactone	Biofilm formation Plant growth promotion	Jung et al. (2017)

(continued)

Table 11.2 (continued)

Rhizosphere microorganism	Quorum sensing molecules	Functional attributes	References
Paenibacillus polymyxa (SQR-21)	AHL	Plant growth promotion	Yaoyao et al. (2017)
Pseudomonas syringae	AHL	Plant growth promotion Induced resistance to plant pathogens	Shrestha et al. (2020)
Pseudomonas	AHL	Antifungal activity against plant pathogen	Li et al. (2020)

P. aureofaciens 30–84 acts as saprophytes and biocontrol, respectively, and secreted the QS-regulated antibiotics phenazines to fight against the *Gaeumannomyces graminis* var. *tritici* and did the colonization of wheat plant.

In the rhizosphere, the QS is determined as a major factor for competition which is thought to be the controlled antibiotic production by the dominant competitive communities that might mitigate the metabolic production until the competitor population reached to threshold for killing. Another prospective for the competition might be the competitor ability to plinth a self-protective retort to antibiotic at subinhibitory level.

11.5 Conclusions

Previous research revealed the significant insight and advances in the bacterial QS system and displayed the mechanistic procedure within and between the species, genera, and with the associated hosts. Furthermore, the vital factor of gene regulatory networks of bacterial genome and its role in habitat adaptation have been described widely. Though, the QS signaling behavior and mechanism in rhizosphere still need highlights. Interestingly, the overall chapter has described the QS signaling in rhizospheric bacteria and its facilitation for residing, nodulation, or colonization with plant roots.

References

Bahadur I, Maurya BR, Meena VS, Saha M, Kumar A, Aeron A (2017) Mineral release dynamics of tricalcium phosphate and waste muscovite by mineral-solubilizing rhizobacteria isolated from indo-gangetic plain of India. Geomicrobiol J 34(5):454–466

Balasundararajan V, Dananjeyan B (2019) Occurrence of diversified N-acyl homoserine lactone mediated biofilm-forming bacteria in rice rhizoplane. J Basic Microbiol 59:1031–1039

Barriuso J, Solano BR, Fray RG, Cámara M, Hartmann A, Mañero FJG (2008) Transgenic tomato plants alter quorum sensing in plant growth-promoting rhizobacteria. Plant Biotechnol J 6:442–452

Berg G, Smalla K (2009) Plant species and soil type cooperatively shape the structure and function of microbial communities in the rhizosphere. FEMS Microbiol Ecol 68:1–13

Brimecombe MJ, De Leij FA, Lynch JM (2007) Rhizodeposition and microbial populations. The rhizosphere biochemistry and organic substances at the soil-plant interface. Taylor & Francis Group, New York

Chan KG, Atkinson S, Mathee K, Sam CK, Chhabra SR, Cámara M, Koh CL, Williams P (2011) Characterization of N-acylhomoserine lactone-degrading bacteria associated with the *Zingiber officinale* (ginger) rhizosphere: co-existence of quorum quenching and quorum sensing in Acinetobacter and Burkholderia. BMC Microbiol 11(1):1–4

Cubo MT, Economou A, Murphy G, Johnston AW, Downie JA (1992) Molecular characterization and regulation of the rhizosphere-expressed genes rhiABCR that can influence nodulation by *Rhizobium leguminosarum* biovar *viciae*. J Bacteriol 174(12):4026–4035

Daniels R, De Vos DE, Desair J, Raedschelders G, Luyten E, Rosemeyer V, Verreth C, Schoeters E, Vanderleyden J, Michiels J (2002) The cin quorum sensing locus of *Rhizobium etli* CNPAF512 affects growth and symbiotic nitrogen fixation. J Biol Chem 277(1):462–468

De Maeyer K, D'aes J, Hua GK, Perneel M, Vanhaecke L, Noppe H, Höfte M (2011) N-Acylhomoserine lactone quorum-sensing signalling in antagonistic phenazine-producing Pseudomonas isolates from the red cocoyam rhizosphere. Microbiology 157(2):459–472

Fontaine L, Boutry C, Guédon E, Guillot A, Ibrahim M, Grossiord B, Hols P (2007) Quorum-sensing regulation of the production of Blp bacteriocins in *Streptococcus thermophilus*. J Bacteriol 189(20):7195–7205

Fravel DR, Roberts DP (1991) In situ evidence for the role of glucose oxidase in the biocontrol of Verticillium wilt by *Talaromyces flavus*. Biocontrol Sci Tech 1(2):91–99

Fuqua WC, Winans SC, Greenberg EP (1994) Quorum sensing in bacteria: the LuxR-LuxI family of cell density-responsive transcriptional regulators. J Bacteriol 176(2):269–275

Gantner S, Schmid M, Dürr C, Schuhegger R, Steidle A, Hutzler P, Langebartels C, Eberl L, Hartmann A, Dazzo FB (2006) In situ quantitation of the spatial scale of calling distances and population density-independent N-acylhomoserine lactone-mediated communication by rhizobacteria colonized on plant roots. FEMS Microbiol Ecol 56(2):188–194

Garbeva P, Van Elsas JD, Van Veen JA (2008) Rhizosphere microbial community and its response to plant species and soil history. Plant S 302(1):19–32

Garg N, Manchanda G, Kumar A (2014) Bacterial quorum sensing: circuits and applications. Antonie van Leeu 105:289–305. https://doi.org/10.1007/s10482-013-0082-3

He X, Chang W, Pierce DL, Seib LO, Wagner J, Fuqua C (2003) Quorum sensing in *Rhizobium* sp. strain NGR234 regulates conjugal transfer (*tra*) gene expression and influences growth rate. J Bacteriol 185(3):809–822

Hibbing ME, Fuqua C, Parsek MR, Peterson SB (2010) Bacterial competition: surviving and thriving in the microbial jungle. Nat Rev Microbiol 8(1):15–25

Hiltner L (1904) Über neuere Erfahrungen und Probleme auf dem Gebiete der Bodenbakteriologie unter besonderer Berücksichtigung der Gründüngung und Brache. Arbeiten der Deutschen Landwirtschaftlichen Gesellschaft 98:59–78

Howell CR (1998) The role of antibiosis in biocontrol. In: Harman GE, Kubicek CP (eds) *Trichoderma* and *Gliocladium*, Vol. 2. Enzymes, biological control and commercial applications. Taylor & Francis, London, pp 173–184

Hrynkiewicz K, Baum C (2011) The potential of rhizosphere microorganisms to promote the plant growth in disturbed soils. In: Malik A, Grohmann E (eds) Environmental protection strategies for sustainable development. Springer, New York, pp 35–64

Jung BK, Khan AR, Hong SJ et al (2017) Quorum sensing activity of the plant growth promoting rhizobacterium *Serratia glossinae* GS2 isolated from the sesame (*Sesamum indicum* L.) rhizosphere. Ann Microbiol 67:623–632

Kumar A, Meena VS, Maurya BR, Raghuwanshi R, Bisht JK, Pattanayak A (2017) Towards the biological nitrogen fixation and nitrogen management in legume under sustainable agriculture. Appl Soil Ecol. https://doi.org/10.1016/j.apsoil.2017.05.013

Lagos L, Maruyama F, Nannipieri P, Mora ML, Ogram A, Jorquera MA (2015) Current overview on the study of bacteria in the rhizosphere by modern molecular techniques: a mini-review. J Soil Sci Plant Nutr (ahead):0–0. https://doi.org/10.4067/S0718-95162015005000042

Lazazzera BA (2000) Quorum sensing and starvation: signals for entry into stationary phase. Curr Opin Microbiol 3(2):177–182

Li Q, Wu Y, Wang J, Yang B, Chen J, Wu H, Zhang Z, Lu C, Lin W, Wu L (2020) Linking short-chain N-acyl homoserine lactone-mediated quorum sensing and replant disease: a case study of *Rehmannia glutinosa*. Front Plant Sci 17(11):787

Liu YH, Wang ET, Jiao YS, Tian CF, Wang L, Wang ZJ, Guan JJ, Singh RP, Chen WX, Chen WF (2018) Symbiotic characteristics of *Bradyrhizobium diazoefficiens* USDA 110 mutants associated with shrubby sophora (*Sophora flavescens*) and soybean (*Glycine max*). Microbiol Res 214:19–27

Ma ZP, Lao YM, Jin H, Lin GH, Cai ZH, Zhou J (2016) Diverse profiles of AI-1 type quorum sensing molecules in cultivable bacteria from the mangrove (Kandelia obovata) rhizosphere environment. Front Microbiol 7. https://doi.org/10.3389/fmicb.2016.01957

Maheshwari NK, Singh RP, Manchanda G, Dubey RC, Maheshwari DK (2021) Sunn Hemp (*Crotalaria juncea*) nodulating bacteria capable for high antagonistic potential and plant growth promotion attributes. J Microbiol Biotechnol Food Sci 10(3):385–389. https://doi.org/10.15414/jmbfs.2020.10.3.385-389

Majerczyk C, Schneider E, Greenberg EP (2016) Quorum sensing control of type VI secretion factors restricts the proliferation of quorum-sensing mutants. eLife 5:e14712. doi:https://doi.org/10.7554/eLife.14712

Mazzola M, Cook RJ, Thomashow LS, Weller DM, Pierso LS 3rd (1992) Contribution of phenazine antibiotic biosynthesis to the ecological competence of fluorescent pseudomonads in soil habitats. Appl Environ Microbiol 58(8):2616–2624

Miller MB, Bassler BL (2001) Quorum sensing in bacteria. Annu Rev Microbiol 55:165–199

Müller H, Westendorf C, Leitner E et al (2009) Quorum-sensing effects in the antagonistic rhizosphere bacterium *Serratia plymuthica* HRO-C48. FEMS Microbiol Ecol 67:468–478

Ng WL, Bassler BL (2009) Bacterial quorum-sensing network architectures. Annu Rev Genet 43:197–222

Normander B, Prosser JI (2000) Bacterial origin and community composition in the barley phytosphere as a function of habitat and presowing conditions. Appl Environ Microbiol 66:4372–4377

Park SJ, Park SY, Ryu CM, Park SH, Lee JK (2008) The role of AiiA, a quorum-quenching enzyme from *Bacillus thuringiensis*, on the rhizosphere competence. J Microbiol Biotechnol 18:1518–1521

Pierson EA, Wood DW, Cannon JA, Blachere FM, Pierson LS (1998) Interpopulation signaling via N-acyl-homoserine lactones among bacteria in the wheat rhizosphere. Mol Plant-Microbe Interact 11:1078–1084

Prajakta BM, Suvarna PP, Raghvendra SP, Alok RR (2019) Potential biocontrol and superlative plant growth promoting activity of indigenous *Bacillus mojavensis* PB-35 (R11) of soybean (*Glycine max*) rhizosphere. SN Appl Sci 1(10):1–1

Rosemeyer V, Michiels J, Verreth C, Vanderleyden J (1998) luxI-and luxR-homologous genes of *Rhizobium etli* CNPAF512 contribute to synthesis of autoinducer molecules and nodulation of *Phaseolus vulgaris*. J Bacteriol 180(4):815–821

Ryu CM, Choi HK, Lee CH, Murphy JF, Lee JK, Kloepper JW (2013) Modulation of quorum sensing in acylhomoserine lactone-producing or-degrading tobacco plants leads to alteration of induced systemic resistance elicited by the rhizobacterium *Serratia marcescens* 90–166. Plant Pathol J 29:182

Sauer K, Camper AK, Ehrlich GD, Costerton JW, Davies DG (2002) Pseudomonas aeruginosa displays multiple phenotypes during development as a biofilm. J Bacteriol 184:1140–1154

Schuhegger R, Ihring A, Gantner S et al (2006) Induction of systemic resistance in tomato by N-acyl-L-homoserine lactone-producing rhizosphere bacteria. Plant Cell Environ 29:909–918

Shrestha A, Grimm M, Ojiro I, Krumwiede J, Schikora A (2020) Impact of quorum sensing molecules on plant growth and immune system. Front Microbiol 11:1545. https://doi.org/10.3389/fmicb.2020.01545

Singh RN, Singh RP, Sharma A, Saxena AK (2016a) Modeling of PrnD protein from *Pseudomonas fluorescens* RajNB11 and its comparative structural analysis with PrnD proteins expressed in Burkholderia and Serratia. Turkish J Biol 40(3):623–633

Singh RP, Manchanda G, Singh RN, Srivastava AK, Dubey RC (2016b) Selection of alkalotolerant and symbiotically efficient chickpea nodulating rhizobia from North-West Indo Gangetic Plains. J Basic Microbiol 56:14–25. https://doi.org/10.1002/jobm.201500267

Singh RP, Manchanda G, Li ZF, Rai AR (2017) Insight of proteomics and genomics in environmental bioremediation. In: Bhakta JN (ed) Handbook of research on inventive bioremediation techniques. IGI Global, Hershey. https://doi.org/10.4018/978-1-5225-2325-3

Singh RP, Manchanda G, Anwar MN, Zhang JJ, Li YZ (2018) Mycorrhiza–helping plants to navigate environmental stresses. Microbes for Climate Resilient Agricultur 13:205–233. https://doi.org/10.1002/9781119276050.ch10

Singh RP, Manchanda G, Maurya IK, Maheshwari NK, Tiwari PK et al (2019) *Streptomyces* from rotten wheat straw endowed the high plant growth potential traits and agro-active compounds. Biocatal Agric Biotechnol 17:507–513. doi:https://doi.org/10.1016/j.bcab.2019.01.014

Singh RP, Manchanda G, Yang Y, Singh D, Srivastava AK et al (2020) Deciphering the factors for nodulation and Symbiosis of Mesorhizobium associated with *Cicer arietinum* in Northwest India. Sustainability 12:1–17

Sjöblom S, Brader G, Koch G, Palva ET (2006) Cooperation of two distinct ExpR regulators controls quorum sensing specificity and virulence in the plant pathogen *Erwinia carotovora*. Mol Microbiol 60:1474–1489

Solomon JM, Lazazzera BA, Grossman AD (1996) Purification and characterization of an extracellular peptide factor that affects two different developmental pathways in Bacillus subtilis. Genes Dev 10(16):2014–2024. https://doi.org/10.1101/gad.10.16.2014

Steidle A, Sigl K, Schuhegger R et al (2001) Visualization of N-acylhomoserine lactone-mediated cell-cell communication between bacteria colonizing the tomato rhizosphere. Appl Environ Microbiol 67:5761–5770

Subhashini DV, Singh RP (2014) Isolation of endophytic actinomycetes from roots and leaves of tobacco (*Nicotiana tabacum* L.). Annal Plant Pro Sci 22(2):458–459

Subhashini DV, Singh RP, Manchanda G (2017) OMICS approaches: tools to unravel microbial systems. Directorate of Knowledge Management in Agriculture, Indian Council of Agricultural Research. ISBN: 9788171641703. https://books.google.co.in/books?id=vSaLtAEACAAJ

Sylvia D, Fuhrmann J, Hartel P, Zuberer D (2005) Principles and applications of soil microbiology. Pearson Education Inc, New Jersey

Veliz-Vallejos DF, Kawasaki A, Mathesius U (2020) The presence of plant-associated bacteria alters responses to N-acyl homoserine lactone quorum sensing signals that modulate nodulation in *Medicago truncatula*. Plant 9:777

Verma R, Maurya BR, Meena VS, Dotaniya ML, Deewan P, Jajoria M (2017) Enhancing production potential of cabbage and improves soil fertility status of Indo-Gangetic plain through application of bio-organics and mineral fertilizer. Int J Curr Microbiol App Sci 6(3):301–309

Veselova MA, Klein SH, Bass IA et al (2008) Quorum sensing systems of regulation, synthesis of phenazine antibiotics, and antifungal activity in rhizospheric bacterium *pseudomonas chlororaphis* 449. Russ J Genet 44:1400

Wang H, Zhong Z, Cai T, Li S, Zhu J (2004) Heterologous overexpression of quorum-sensing regulators to study cell-density-dependent phenotypes in a symbiotic plant bacterium *Mesorhizobium huakuii*. Arch Microbiol 182:520–525

Whitehead NA, Barnard AM, Slater H, Simpson NJ, Salmond GP (2001) Quorum-sensing in Gram-negative bacteria. FEMS microbiol revi 25(4):365–404

Williamson LL, Borlee BR, Schloss PD, Guan C, Allen HK, Handelsman J (2005) Intracellular screen to identify metagenomic clones that induce or inhibit a quorum-sensing biosensor. Appl Environ Microbiol 71(10):6335–6344. https://doi.org/10.1128/AEM.71.10.6335-6344.2005

Yang M, Sun K, Zhou L, Yang R, Zhong Z, Zhu J (2009) Functional analysis of three AHL autoinducer synthase genes in *Mesorhizobium loti* reveals the important role of quorum sensing in symbiotic nodulation. Can J Microbiol 55:210–214

Yang SH, Chen WH, Wang ET, Chen WF, Yan J, Han XZ, Tian CF, Sui XH, Singh RP, Jiang GM, Chen WX (2018) Rhizobial biogeography and inoculation application to soybean in four regions across China. J Appl Microbiol 125(3):853–866

Yang YJ, Lin W, Singh RP, Xu Q, Chen Z, Yuan Y, Zou P, Li Y, Zhang C (2019) Genomic, transcriptomic and enzymatic insight into lignocellulolytic system of a plant pathogen Dickeya sp. WS52 to digest sweet pepper and tomato stalk. Biomolecules 9(12):753. https://doi.org/10.3390/biom9120753

Yang Y, Liu L, Singh RP, Meng C, Ma S, Jing C, Li Y, Zhang C (2020a) Nodule and root zone microbiota of salt-tolerant wild soybean in coastal sand and saline-alkali soil. Front Microbiol 11:2178. https://doi.org/10.3389/fmicb.2020.523142

Yang Y, Singh RP, Song D, Chen Q, Zheng X, Zhang C, Zhang M, Li Y (2020b) Synergistic effect of *Pseudomonas putida* II-2 and *Achromobacter* sp. QC36 for the effective biodegradation of the herbicide quinclorac. Ecotoxicol Environ Safety 188:109826

Yaoyao E, Yuan J, Yang F et al (2017) PGPR strain *Paenibacillus polymyxa* SQR-21 potentially benefits watermelon growth by re-shaping root protein expression. AMB Expr 7:104

Zhang JJ, Yang X, Chen G, de Lajudie P, Singh RP et al (2016) *Mesorhizobium muleiense* and *Mesorhizobium* gsp. nov. are symbionts of *Cicer arietinum* L. in alkaline soils of Gansu, Northwest China. Plant Soil 410:103–112

Zhuang X, Gao J, Ma A, Fu S, Zhuang G (2013) Bioactive molecules in soil ecosystems: masters of the underground. Inter J Mol Sci 14(5):8841–8868. https://doi.org/10.3390/ijms14058841

Chapter 12
Understanding the Link Between the Urinary Microbiome and Urinary Lithiasis Disease

Mangesh Suryavanshi, Mukesh Poddar, and Rekha Punchappady-Devasya

Abstract Urinary system has its own micro-environmental niche, and microbes with their genes and metabolic products are the components of the constitutes; hence, we referred them as urinary microbiome. Microbial residents of urinary system comprise the diversified types of microbes, their genes, genome, and metabolites and have the greatest impact on the urinary system performance. Unlike others, urinary lithiasis or stone formation in urinary system is the commonest reason to healthcare system burden. Such urolithiasis phenomena considered to be a part of lifestyle disorder is well perceived now. Whereas, prevalence in general population recorded as 10% and most of the times 20% reported from stone belt areas. Urinary stones are with different types of chemical nature, mostly derived from the metabolic product saturation inside the excretory systems. We are able to understand that the metabolic origin products in urinary system have the impact to derivatize the lithiasis activity, mostly along with supersaturation in solutes and by means of the micro-environment changes pertaining to the urinary niche.

However, events and mechanisms of lithiasis process are still undervest, and mostly, the urine micro-environmental changes could be the focused area. So, it is necessary to understand the triangle of urinary microbiome, urinary micro-environment, and lithiasis components, which may give us the clues to process and progression in hinderance of urinary system function. Here we aimed to discuss about the diversity in microbes and their products impedance on normal as well as in different types of lithiasis situation. Latest advent of "Omics" technologies like genomics of host and residents, and proteomics and metabolomics for the cases has doubled the knowledge about the micro-environment changes in the urinary system. Consensus suggests us different modes of mechanisms for the stone formation and disease progression for the entire urinary lithiasis event.

Lastly, correlating microbial phylotypes, certain metabolites from host and microbes, and changes in microenvironments depicts the role of urinary microbiome

---

M. Suryavanshi (✉) · R. Punchappady-Devasya
Yenepoya Research Centre, Yenepoya (Deemed to be University), Mangalore, Karnataka, India

M. Poddar
Department of Microbiology, Sikkim University, Gangtok, Sikkim, India

R. P. Singh et al. (eds.), *Microbes in Microbial Communities*,
https://doi.org/10.1007/978-981-16-5617-0_12

which have major impact in pathophysiology for urolithiasis condition. Hence, this review is to provide overview of current findings in urinary microbiome, describing the role of bacterial communities present in urine, and to discuss the possible role in lithiasis and the possible role of probiotics as preventive agents of urological disorder.

Keywords Urolithiasis · Kidney stones · Urinary microbiome · Rare phylotypes · Proteins and metabolites in kidney stone progress

12.1 Introduction

"Microbiome" term indicates commensal, pathogenic, and symbiotic microbial community covering a particular organ of the human body (Tang 2017; Subhashini et al. 2017). Microbiome consists of multiple communities of microorganisms. The healthy urinary system hosts different types of microorganisms such as viruses, bacteria, fungi, and protozoans. Microorganisms colonizing various sites of the urinary system are referred to as urinary microbiome or urobiome, which plays a crucial role in maintaining urinary health. There were remarkable differences in urobiome composition between healthy individuals and those with urologic diseases (Cho and Blaser 2012), and we learnt that microorganisms occupy 90% of the total cell of the body and that commensal microbes are found largely in our body (Savage 1977). Urobiome can change with seasons, life cycle, or environmental changes, which convert these commensals into opportunistic pathogens and cause disorders (Han 2015). It is difficult to define whether these organisms are beneficial for health or play a role in disease development (Peterson et al. 2009).

The aim of the Human Microbiome Project (HMP) is to develop an overall characterization of the human microbiome (Alfano et al. 2016). Initially, bladder and urine were considered sterile based on standard culture techniques. Due to certain technical problems, these methods could not find bacterial populations in urine and bladder. But, more advanced molecular biology technique has detected the presence of bacteria in urine and bladder of healthy people (Aagaard et al. 2014; Thomas-White et al. 2016; Tang 2017). Sequencing of highly variable fragments of 16S ribosomal RNA gene such as V1-V9 region has identified many types of microbial taxa present in the urinary system (Clemente et al. 2012). Metagenomics study is a well-defined genetic material study and is based on next-generation sequencing technology (Handelsman 2004; Wooley et al. 2010), which helps to identify and classify microorganisms to phylum, genus, or species, etc. It provides information regarding a quantitative and qualitative contribution of a particular microorganism to the urobiome (Wolfe et al. 2012). Each group of microbial clusters was different in their dominant species, composition, and diversity (Mueller et al. 2017). The low abundant microorganism in the urinary tract can also be characterized by using this metagenomics technology. Apart from the urinary system, microbiomes are also found in the gastrointestinal tract, skin, and respiratory tract

(Lazarevica et al. 2009). Those clearly define different microbial niche and its environmental condition in human body. Urobiome will help in uroepithelial integrity, homeostasis, maintaining adhesion junction, immunity, neurotransmission, urinary metabolism, and overall urinary health (Reid et al. 2003; Reid and Bruce 2006). Normal bacterial flora has a role in immunity to compete with foreign microbes and produce vitamin K for blood clotting (Wojciuk et al. 2019). Bacterial flora has a vital role due to interaction with host cells and the regulation of structural and functional characteristics of those cells. The intestinal tract, whose microbiome is responsible for large metabolic output, not only uses those end products for itself but also shares them with kidney and urinary tract by systemic translocation. Urobiome will also promote to produce neuroactive substances, hormones, and inflammatory reactions to maintain health of the urinary system (Tang 2017).

Urobiome is a result of microbiota reaching the urinary system from the gastrointestinal tract (Tang 2017). The existence of abundant microorganisms in the urobiome has already been discovered. Its abundance varies according to age, gender, etc. The total number of urinary microorganisms varies with different age groups. Microbiome diversity varies between males and females, and this may be due to differences in anatomical structure and hormonal balance of these genders (Whiteside et al. 2015).

12.2 Detection of Microbial Communities in Urobiome

12.2.1 Sample Collection Methods

For constructive study of urobiome, the sample collection method needs to be closely monitored and standardized. It is because error in sample collection could result in false interpretation. Thus, in order to avoid contamination and false interpretation of the result, the urine specimen should be collected aseptically. The accurate capturing of urinary microbiota is only possible if the sampling technique is well considered. In healthy condition, human urinary microbiome generally possesses low biomass of bacteria as compared to cases with UTI (urinary tract infection) or rUTI (recurrent urinary tract infection).

Common urine collection method includes suprapubic aspiration, intermittent transurethral catheterization, clean-catch midstream urine, and first-void urine. Each of these methods has its own significance. Of those, some are most widely used; they are as follows.

12.2.1.1 Suprapubic Aspiration

It is a sterile procedure of urine collection directly from bladder using needle (sterile). The needle is inserted into bladder through the suprapubic skin over the pubic bone. This method reduces the chances of vulvovaginal or urethral contamination. It is mostly recommended for the non-toilet-trained patients. Despite being most invasive method, it is considered as the gold standard technique for urine collection. As this method specifically profile bladder microbiome; thus, it is highly recommended for urobiome-related studies (May 2018). This method is also recommended in official clinical practice for UTI studies, but its use is very rare as it is a painful process. Urine collection through suprapubic aspiration requires experienced professional doctors as it could not be performed by nurses. One of the urine collection methods is suprapubic aspiration (Kouri et al. 2000), which is the correct method to hold non-toilet-trained children during collection.

12.2.1.2 Urethral Catheterization

It is one of the most commonly used sampling techniques practiced for many years. In comparison to SPA (suprapubic aspiration), it is a less invasive method that causes mild discomfort to the patient, thus it is most commonly preferred by patients over supra pubic aspiration (Newman et al. 2017). Unlike SPA technique that requires professional expert, urine collection through this method could be easily performed by nurses (Teo et al. 2016; May 2018). Although urine collection through suprapubic aspiration is recommended in official medical guideline, urethral catheterization is performed as an alternative when SPA is not possible. Urine obtained through catheterization provides great cultural sensitivity as well as specificity for the detection of microbial community.

12.2.1.3 Clean-Catch Void (CCV)

It is the least invasive method of urine collection. It is the most acceptable sampling procedure by patients. This technique does not rely on the nurses or clinical experts for urine collection. Thus, sample collection using this procedure is very easy. The major limitation of this procedure is contamination of urine by bacteria present on the skin of perineal area, vagina, or from fecal matter. However, if the periurethral area is cleaned using cleaning solution, there are chances of reduction in contamination. The role of collector during CCV is very important (Teo et al. 2016). Collectors are those parents whose children were not trained enough to collect urine sample for study. Collection of true midstream urine from a patient is very important for study. The use of CCV to study urogenital tract microbiota is more feasible as the chances of contamination are higher.

After detailed comparison of the listed methods, the best possible methods for the detection of bacterial communities present in bladder are suprapubic aspiration and transurethral catheterization (Wolfe et al. 2012). It is considered as the best method as it reduces the vulvovaginal contamination that could lead to misinterpretation of results. The study was conducted in the year 2014 by Wolfe et al. The data for this comparative study are listed in Tables 12.1 and 12.2, which show details of the

Table 12.1 Some common bacteria present in urobiome of healthy males

Age, years	Bacteria in urobiome of healthy males	Urine sample collection	Techniques used for identification of isolates	Reference
12–17	*Corynebacterium*, *Lactobacillus*, *Staphylococcus*, *Gardnerella*, *Streptococcus*, *Anaerococcus*, *Veillonella*, *Prevotella*, and *Escherichia*	First voided	16S rRNA sequencing, gene sequencing (GS)	Pearce et al. (2014)
18–49	*Lactobacillus*, *Corynebacterium*, *Escherichia*, *Streptococcus*, *Prevotella*, *Sneathia*, *Veillonella*, *Ureaplasma*, *Mycoplasma*, *Anaerococcus*, *Atopobium*, *Aerococcus*, *Actinobacteria*, *Staphylococcus*, *Gemella*, *Enterococcus*, *Finegoldia*, *Klebsiella*, *Gardnerella*, *Firmicutes*, and *Pseudomonas*	Midstream urine (MSU)	GS, 16S rRNA sequencing	Nelson et al. (2010), Dong et al. (2011), Fouts et al. (2012), Nelson et al. (2012), Wojciuk et al. (2019)
50–70	*Lactobacillus*, *Firmicutes*, *Klebsiella*, *Corynebacterium*, *Staphylococcus*, *Streptococcus*, *Aerococcus*, *Gardnerella*, *Prevotella*, *Escherichia*, *Enterococcus*, and *Bacteroidetes*	MSU	GS, 16S rRNA sequencing	Fouts et al. (2012)
Above 70	*Parvimonas*, *Pseudoramibacter*, *Saccharofermentans*, *Proteiniphilum*, *Jonquetella*, *Peptococcus*, *Aerococcus*, *Aminobacterium*, *Anaerococcus*, *Butyriciococcus*, *Campylobacter*, *Corynebacterium*, *Eubacterium*, *Fusobacterium*, *Gardnerella*, *Azospira*, *Catonella*, *Gemella*, *Gordonibacter*, *Mycoplasma*, *Peptococcus*, *Peptoniphilus*, *Pervotella*, *Sneathia*, and *Soehngenia*	MSU, a transurethral catheter (TUC)	16S rRNA sequencing, Expanded quantitative urine culture (EQUC)	Fouts et al. (2012), Wojciuk et al. (2019)

Table 12.2 Some common bacteria present in urobiome of healthy females

Age, years	Bacteria in urobiome of healthy females	Sample collection	Techniques for identification	Reference
20–50	*Lactobacillus*, *Klebsiella*, *Corynebacterium*, *Staphylococcus*, *Streptococcus*, *Aerococcus*, *Gardnerella*, *Prevotella*, *Escherichia*, *Enterococcus*, *Peptoniphilus*, *Dialister*, *Finegoldia*, *Rhodophila*, *Anaerococcus*, *Allisonella*, *Bifidobacterium*, *Enterobacteriaceae*, *Sneathia*, *Trueperella*, *Sutterella*, *Alloscardovia*, *Veillonella*, *Butyricicoccus*, *Azospira*, *Neisseria*, *Tepidimonas*, *Arcanobacterium*, and *Tessaracoccus*	MSU, TUC	Gene sequencing (GS), 16S rRNA sequencing	Dong et al. (2011), Lewis et al. (2013), Wojciuk et al. (2019)
51–70	*Lactobacillus*, *Klebsiella*, *Corynebacterium*, *Staphylococcus*, *Streptococcus*, *Aerococcus*, *Gardnerella*, *Prevotella*, *Escherichia*, *Enterococcus*, *Firmicutes*, *Peptoniphilus*, *Dialister*, *Finegoldia*, *Anaerococcus*, *Allisonella*, *Bifidobacterium*, *Sneathia*, *Trueperella*, *Alloscardovia and Veillonella*, *Brevibacterium*, *Catonella*, *Peptostreptococcus*, *Peptococcus*, *Sutterella*, and *Rhodopila*	TUC	16S rRNA sequencing, expanded quantitative urine culture (EQUC)	Dong et al. (2011), Fouts et al. (2012), Lewis et al. (2013), Whiteside et al. (2015)
Above 70	*Firmicutes*, *Parvimonas*, *Saccharofermentans*, *Proteiniphilum*, *Jonquetella*. *Actinomyces*, *Arthrobacter*, *Oligella*, *Rhodococcus*, *Gulosibacter*, and *Modestobacter*	TUC	GS, 16S rRNA sequencing	Fouts et al. (2012), Wojciuk et al. (2019)

technique used for collection. Also the bacterial profile resulted from both the techniques show similar results.

12.2.2 Technique Used for the Detection of Microbial Communities

Uropathogens present in urine sample can be detected by both culture-dependent and culture-independent methods. Culture-dependent methods involve standard or traditional urine culture in which urine sample is plated over 5% sheep blood agar or MacConkey agar. The plates were incubated at 35 °C for 24 h in aerobic condition. The main advantage of urine culture technique is the detection of viable bacterial population present in urine collected by transurethral catheterization (Price et al. 2016). Standard method of urine culture had many drawbacks, which make it less important (Wolfe et al. 2012; Lewis et al. 2013). For example, in aerobic atmosphere, the slow-growing bacteria die due to the presence of oxygen. Also the standard urine culture has limited detection capacity, i.e., $\geq 10^5$ culturable CFU/ml. Detail study of bacterial communities present in urine requires advance molecular technique like 16S rRNA gene sequencing. However, enhanced modified technique like expanded quantitative urine culture (EQUC) is widely used in clinical laboratories to detect uropathogens (Wolfe et al. 2012; Lewis et al. 2013; Pearce et al. 2014; Thomas-White et al. 2016).

12.2.2.1 Expanded Quantitative Urine Culture (EQUC)

EQUC is a modified form of standard urine culture in terms of volume of the sample, incubation condition, and cultural media used. It involves use of higher volume of urine plated over a variety of cultural media in combination with different incubation period. The plates are also incubated in different environmental condition like anaerobic, aerobic, or enhanced CO_2-rich atmosphere. Using a wide variety of cultural media with different environmental conditions, a diversity of bacterial communities can be detected employing this technique. This technique not only aims to detect uropathogens like *E. coli* but increases our understanding by detecting commensal flora of urogenital tract and bladder. Price et al. (2016) compared these two methods and concluded that EQUC can detect 67% more uropathogens as compared to standard urine culture. Also the standard urine culture missed 88% of non-*E. coli* uropathogens which are detected by EQUC. EQUC can detect many bacterial genera that remain undetected through standard method (Lewis et al. 2013). Instead of choosing random condition for culture, Price et al. formulated a modified and streamlined EQUC protocol, which involves the use of higher volume of urine of about 100 μl plated in different media. Most commonly used media includes BAP (blood agar plates), CNA (colistin-nalidixic acid agar plates), and MacConkey agar. The incubation period is of 48 h in 5% CO_2 atmosphere. The streamline protocol is of great clinical significance as it is widely used to detect uropathogens from patients having urinary tract infection. This method also proves that many bacteria that are identified through 16 s rRNA gene sequencing can be cultivated using this technique. As per a report by s, urine sample collected via transurethral catheter gives no growth in standard urine culture, whereas advance techniques like expanded quantitative urine culture (EQUC) lead to isolation of bacterial communities from 80% of the same sample (Lewis et al. 2013). EQUC protocol also detects all bacterial genera that show growth in standard urine culture method. Compared to standard urine culture, EQUC protocol detected all viable bacterial diversity with higher average CFU per ml (Price et al. 2016).

12.2.2.2 16S rRNA Gene Sequencing

It is a cultural-independent NGS-based metagenomic sequencing technique for the detection of microbial composition of specimen. Apart from this, another NGS-based technique which can be used is whole-genome shotgun metagenomic sequencing. 16S rRNA amplicon sequencing technique mainly involves extraction of community DNA from urine sample, thereby amplifying variable (V4) region of 16S rRNA gene of each DNA (Zhang et al. 2019). Amplification is done by using universal primers like 515F and 806R in two-step nested PCR. The amplicons after processing undergo sequencing. Sequencing data then matched with standard database which gives a complete profile of bacterial communities present in urine. On the other hand, it gives relative abundance of the bacterial population. The use of primers has an advantage that its high sensitivity can detect microbial community even with very low abundance. Thus, it can be used to study relative abundance and for taxonomic studies.

16S rRNA sequencing can detect complete bacterial diversity of sample, but EQUC protocol could not detect dead or ruptured bacteria. A rational theory suggests that every bacterial genera (including dead bacteria) identified through high-throughput sequencing were active in human body and played significant role. Thus,16S rRNA sequencing provides additional benefit over EQUC in understanding human urobiome.

Bacterial diversity of urobiome in different age group of healthy males and females were summarized in Tables 12.1 and 12.2, respectively. The urine samples were collected in different points of time during voiding. For complete detection of bacterial diversity, both (EQUC and 16S rRNA sequencing) the techniques have been used. However, on comparing EQUC and 16S rRNA gene sequencing results (Tables 12.1 and 12.2), it signifies that urobiome profile of both the techniques is not fully identical (Price et al. 2016). For example, bacterial genera *Atopobium* is detected only through 16S rRNA gene sequencing while the genus *Trueperella* is detected via EQUC method. This is due to the fact that for some bacterial genera, the EQUC growth condition might be unsuitable while for other bacterial communities that were under- represented in urine, it could be optimum.

12.3 Understanding Urobiome Profile in Health

As per multiple urobiome studies conducted so far, the most common and frequent bacterial genera reported are *Lactobacillus* and *Streptococcus*. A possible explanation is that both the genera belong to the group of lactic acid bacteria that are commensal flora of several tissues. Both the genera are associated with genitourinary tract and involved in protective roles against enteric pathogens (Shetty et al. 2017). A previous study also revealed that the frequency of bacterial genera like *Alloscardovia*, *Burkholderia*, *Jonquetella*, *Klebsiella*, *Saccharofermentans*, *Rhodanobacter*, and *Veillonella* is very less (Shetty et al. 2017). The overall composition of urobiome reported in various studies differs; it is because of the differences in sample collection method and age and sex of the patient. Due to high level of variation reported in studies, the core bacterial communities is still not accurately defined (Neugent et al. 2020).

It is not very surprising that urobiome composition differs among individual based on sex and age. The fact could be related to the extent of hygiene and voiding habits, as it differs among children, adults, and elders (Aragón et al. 2018). 16S rRNA gene sequencing of clean-catch mid-stream urine showed that bladder microbiome of females contains diverse bacterial population as compared to male bladder. *Actinomyces*, *Arthobacter*, and *Bacteroides* are some bacterial genera which are detected only in female urine by 16S rRNA sequencing. Another comparative study of healthy male and female showed that the number of *E. coli* cultured is higher in females as compared to males. 16S rRNA gene sequencing of urine sample collected from adolescent men showed that *Streptococcus*, *Gardnerella*, *Lactobacillus,* and *Veillonella* are present in predominant level. As per a study conducted by Fouts et al. (2012), *Corynebacterium* is abundantly present in male urine. This bacterial genus is commonly present in skin and distal urethra, thus the presence of this bacteria in urine could be related to contamination from skin when clean-catch voiding technique is used for urine collection. As per several other studies, *Enterococcus*, *Proteus*, *Klebsiella*, and *Aerococcus* were also present in male urine at high biomass (Shetty et al. 2017). The variation in the composition of urinary microbiome among male and female could be related to various factors. This involves difference in anatomical structure, production of different hormones, metabolites, etc. For example, creatinine excretion in male is higher as compared to female, whereas the production of citrate is higher in female. Also, the production of calcium and oxalate is less in female. Thus, the metabolites may promote or inhibit certain bacteria to survive in the niche. The overall microbial composition of various physiological systems and the ratio of *Firmicutes* to *Bacteroidetes* in gut microbiome alter with the age of the individual (Mariat et al. 2009). Similarly, the composition of urinary microbiome changes with age (Teo et al. 2016). However, the knowledge of age-related alteration in urinary microbiome is limited. Several studies conducted among adolescent girls before menarche, reproductively active women, and post-menopausal women for vaginal microbiome show variation (Osborne et al. 1979; Hickey et al. 2015). It has been seen that the genera

Lactobacillus is dominant in girls during puberty, whereas its biomass decreases after menopause. The composition of urinary microbiome changes when women bear pregnancy, whereas it remains unchanged or constant in normal condition. Several studies showed that the vaginal microflora is dominated by *Lactobacillus* in healthy reproductive women (Fouts et al. 2012). The vaginal microflora mainly consists of *Lactobacillus crispatus*, *Lactobacillus iners*, *Lactobacillus gasseri*, and *Lactobacillus jensenii*. As per evidence, the bacterial genera play significant role in maintaining the pH of vagina, protection against pathogen causing infection, STD (sexually transmitted disease), and UTI. As per the study conducted by Lewis et al., *Jonquetella*, *Proteiniphilum*, *Saccharofermentans,* and *Parvimonas* were detected only in older aged people, i.e., above 70 years (Lewis et al. 2013). The study was based on 16S rRNA sequencing of clean-catch voided mid-stream urine collected from male and females of wider age range. It also revealed the detection of other bacterial genera that are common irrespective of their age. The use of standard method for the cultivation of all four bacterial genera is difficult. This is due to the fact that all are anaerobic with fastidious nutrient requirement.

12.4 Role of Urobiome in Health and Disease

As per multiple studies conducted under Human Microbiome Project, researchers found a link between specific disease condition and variation in microbial diversity of different niche (gut, colon, etc.) in human body. Some studies reported variation in urinary microbiome with urological disorder like urinary incontinence (UI), bladder and prostate cancer, was neurogenic bladder dysfunction (NBD), interstitial cystitis (IC), sexually transmitted infections (STIs), and chronic prostatitis or chronic pelvic pain syndrome (CP or CPPS). The variation in urinary microbiome with different urological disorders is described in detail below.

12.4.1 Finding Link Between Urobiome with Urological Disorders

Urinary incontinence is a urological disorder involving involuntary leakage of urine. The condition is seen in both male and female, but its symptoms are more prevalent and frequent in women. It is more common in women due to dysfunction of bladder or pelvic floor muscle that commonly occurs during childbirth, whereas in male, it is a result of prostatic enlargement. It is a very common urological disorder that can be classified as urgency UI (UUI), stress UI (SUI), or mixed UI (MUI) (Thomas-White et al. 2018).

Pearce et al. (2014) and Karstens et al. (2016) conducted studies using 16S rRNA and EQUC of urine sample collected by transurethral catheterization from healthy volunteers and women with UUI. The study shows alteration in urinary microbiome in women with UUI. As compared to urinary microbiome of healthy women, the microbiome of women with UUI has higher load of *Gardnerella* and lower load of *Lactobacillus* (Aragón et al. 2018). Table 12.3 gives the details of different studies on how the samples were collected with a variety of techniques used for the detection of bacterial communities with relative biomass. Karstens et al. (2016) reported that the bacterial abundance between UUI and non-UUI patients varies by 14 OTUs (operational taxonomic units), and also the severity of UUI symptoms is greater in UUI patients with lower bacterial diversity (Aragón et al. 2018). Some studies also concluded the link between urinary microbiome and treatment of urgency urinary incontinence.

According to the International Urogynecological Association (IUGA) and the International Continence Society (ICS) standard definition of urgency urinary incontinence (UUI), it is leakage of urine with strong desire to void that is difficult to differ. UUI and SUI often coexist in some patients with a combination of symptoms and termed as mixed incontinence. As per study by Brubaker et al. (2015), urinary microbiome plays a role in UUI episodes, symptom severity, and posttreatment UTI risk. The study involves collection of catheterized urine sample from women with UUI. The UM contribution was based on bacterial DNA analysis of patients under ABC (Anti cholinergic versus botulinum toxin A comparison) trial using qPCR (Teo et al. 2016). Pearce et al. in the same trial used 16S rRNA gene sequencing to detect different bacterial communities present in women with UUI disorder who discovered different therapies like anticholinergic or on a botulinum toxin A (Pearce et al. 2015). The 16S rRNA sequence profiling of both the cohort studies shows no significant difference. However, it was also noted that women with post UTI treatment has lower *Lactobacillus* sequence as compared to women without UTI. At the end of the research, it was concluded that DNA of urinary bacteria was linked with treatment response, thus agreed the finding of Brubaker et al. (2015), regarding inference of bacterial DNA in urgency urinary incontinence episodes and posttreatment urinary tract infection (Nelson et al. 2012).

As per a comparative study, there is a significant variation in taxonomic composition, richness, and microbial diversity among females with interstitial cystitis and asymptomatic healthy women. The study was based on high-throughput sequencing of clean-catch mid-stream urine. High-throughput sequencing analysis of IC (interstitial cystitis) urine samples also showed increase in the overall biomass of *Lactobacillus genus* whereas decrease in overall microbial richness and ecological diversity (Pearce et al. 2014). Although some *Lactobacillus* genus were known to maintain acidic environment in vagina, thus preventing from other infection to occur, however few study showed that some species of *Lactobacillus* such as *Lactobacillus delbrueckii* and *L. gasseri* could be associated with urinary tract infection and urgency urinary incontinence, respectively. Another comparative study based on 16S rRNA sequencing of urine sample collected from male and female with neurogenic bladder dysfunction (NBD) and healthy bladder showed variation in urobiome. The overall richness of *Lactobacillus* and *Corynebacterium* genera were high among the control bladder urine, whereas *Klebsiella*,

Table 12.3 Urobiome detection in patients with UUI and SUI

Aim of study	Method of sample collection	Techniques used for identification of isolates	Conclusion	Main urotypes	Reference
Relationship of bacterial DNA to UUI and UTI risk ($n = 155$)	Cystoscopic injection	qPCR	38.5% bacterial DNA	NA	Brubaker et al. (2015)
Urinary microbiome of female with UUI (urgency urinary incontinence) and without UUI ($n = 60$ and 58)	Transurethral catheterization	16S rRNA gene sequencing and EQUC	Significant difference in bacterial frequency and abundance: >*Gardnerella* and <*Lactobacillus* in UUI	Non-UUI: *Lactobacillus* (60%), *Gardnerella* (12%), others (20%) UUI: *Lactobacillus* (43%), Gardnerella (26%), others (17%)	Thomas-White et al. (2018)
Urinary microbiome of female with stress urinary incontinence (SUI) ($n = 197$)	Clean catch voiding ($n = 174$) and transurethral catheterization ($n = 23$)	16S rRNA gene sequencing	Diversity of bacteria associated with UUI symptoms, hormonal status, and BMI increased; Urinary microbiome not associated with SUI symptoms	*Lactobacillus* (46–37%) Gardnerella (18–14%), diverse (12–5.1%)	May (2018)

Enterococcus, and *Escherichia* were predominant in NBD urine. Nelson and co-workers conducted a comparative study based on 16S rRNA gene sequencing of initial stream urine collected from patient (Male) with sexually transmitted infection and healthy males. As per the analysis of overall bacterial composition, they proposed possible impact of urogenital bacteria to increase STI risk. The urine sample collected from male patient with sexually transmitted infection was predominated by bacterial genera like *Sneathia*, *Gemella*, *Aerococcus*, *Anaerococcus*, *Prevotella*, and *Veillonella* which normally do not grow in standard urine culture conditions. In a study conducted among patients with chronic prostatitis, it was seen that the bacterial diversity and abundance of *Clostridia* class was relatively high as compared to control samples. The study was also based on the 16S rRNA sequencing of midstream urine collected from patients with CP (chronic prostatitis).

In summary, several studies conducted on urinary microbiome among healthy and patient with various disorders show a clear role of urobiome bacteria in different disorders and treatments. The findings also suggest some direct or indirect role of bacteria in different urinary disease like UTI, UUI, CP, STI, NBD, etc. Therefore, it is necessary to investigate the proper link between urobiome and different urological disorder to understand the fate of urobiome variation among individuals.

12.4.2 Finding Link Between Urobiome with Urolithiasis

A specific group of microorganisms may form clusters at a particular site of the urinary tract and perform their role in regulating urinary health. For example, *Lactobacilli* species present predominantly in the female vagina may colonize throughout the bladder. This colony is part of the urobiome which interacts with uroepithelial cells and maintains a good urinary environment. But, sometimes these colonies can be altered by sexual activity and may cause serious infectious diseases (Price et al. 2016). Urinary microbiota of healthy populations and those with clinical symptoms possess different microbial populations. Female lower urinary tract microbiota varies among each individual and their age group. The proportion of commensal *Lactobacillus* species is different among each healthy generation and those with a urinary disorder (Gajer et al. 2012). Bladder and vaginal urobiome of females are almost similar in their composition, as studied in a phylogenetic analysis of whole-genome of bacterial strains isolated from the bladder and vagina (Gajer et al. 2012).

Healthy urine is not completely sterile. As ecological expression, core microbiome where phylotypes diversity would not change and in-contrast transient microbiome, phylotypes diversity keeps on changing. As described in Fig. 12.1, in urolithiasis condition, literature suggests convergence would go down and the dysbiosis in the microbiota would be found to be increase. The urobiome plays an important role in health and homeostasis of the urinary tract by interacting with host cells, stabilizing the pH of the urinary tract, maintaining the integrity of host cells,

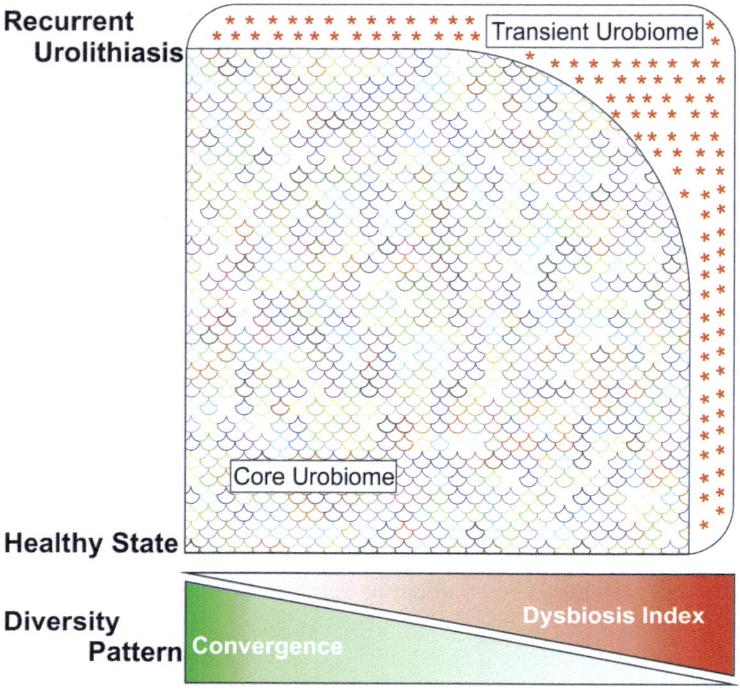

Fig. 12.1 Diagrammatic illustration of ecological implications for urobiome in urolithiasis condition. Urobiome inculcates with core and transient microbiome, and literature suggested that transient flora is more involved in lithiasis disease physio-pathologies. Microbial diversity patterns were found to be less convergence and high dysbiosis index in urolithiasis condition

promoting immunity and neurotransmission, etc. The microbial population in this human urobiome varies with age and sex, which are unique in each individual. Metagenomics analysis based on next-generation sequencing will help in characterizing bacterial type, even in low abundance.

Urobiome of healthy men and women consists of three predominant genera: *Lactobacillus*, *Corynebacterium,* and *Streptococcus* (Dong et al. 2011; Pearce et al. 2014). Urobiome of females is most stable in microbial composition during non-pregnant state, but during pregnancy, its composition will become unstable (Thomas-White et al. 2018). In case of an infant's urinary tract, acid-producing *Lactobacillus* was abundant, and it contributes to acidic conditions that protect against urinary tract infection (UTI) (Lee et al. 2009). Bacterial flora may produce enzymes and cofactors for metabolic reactions. For example, *Escherichia faecalis* species regulates *N*-acetyl glucosamine pathway, citrate, and aspartate metabolic pathway which suppresses glucose uptake; so urine contains more citrate than glucose (Shaykhutdinov et al. 2009; Vebø et al. 2010; Bouatra et al. 2013). A high level of glucose in the urine may promote the growth of microorganisms like *E. faecalis*, which could utilize sucrose and fructose by genomic upregulation (Tasevska et al. 2005; Guo and Li 2009). It also activates the conversion of aspartic

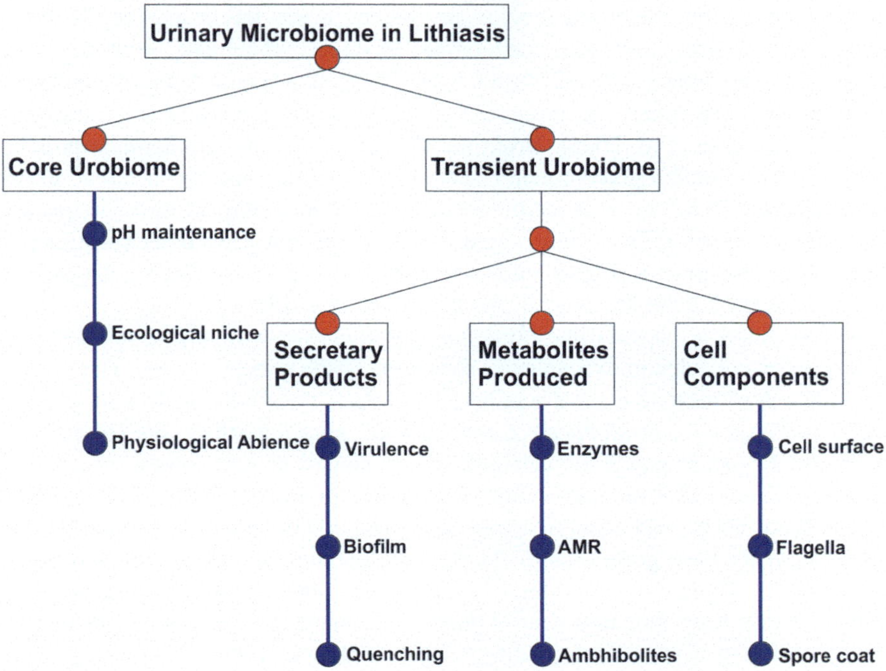

Fig. 12.2 Representation of urinary microbe segregation into core and transient urobiome-type associated with lithiasis condition. Here, we illustrate major activities performed by urobiome in lithiasis disease. We hypothesized that transient urobiome would impose the pathologies in urolithiasis events through different actions like secretary products, metabolites formation, and body components

acid and α-ketoglutaric acid into oxaloacetic acid and glutamic acid and thus helps in nitrogen metabolism (Guo and Li 2009). Species *E. coli* seemed to be osmoprotectant and is responsible for resistance urea toxicity in the urinary tract (Kunin et al. 1992).

Literature explained well till date, core microbes have major impact on maintaining the health status of urinary system especially urine physiology. As mentioned in Fig. 12.2, transient phylotypes would be found to be the culprit of lithiasis episodes with different mechanisms.

Microbiome with a diversified type of microbes, genes, genome, and metabolites has been explored on crystallization events. However, the chemical structure of crystals has been correlated to the kinds of bacterial species that may direct us to understand the specific roles and treatment choices in antibiotics use for recurrent episodes. Some literature on parametric analysis with microbial phylotypes and certain metabolites from host and microbes are discussed for a major impact in pathophysiologies for urolithiasis condition. Urine is usually antimicrobial due to low pH (about <6), hypertonicity, and a high concentration of urea which inhibits urinary pathogens (Kunin et al. 1992; Chambers and Lever 1996). Urine has glycine at a high level, and other amino acids are present in low levels such as D-serine, glutamine, proline, histidine, methionine, cysteine, glutamate, arginine, and branched-chain amino acids (Carlsson et al. 2001). D-serine is a highly abundant bacteriostatic constituent of urine, which inhibits the reproducible pathogenic growth (Tasevska et al. 2005). Urine mainly consists of an antimicrobial protein called Tamm–Horsfall glycoprotein (Jarvisalo et al. 1992), and low acidified urine containing nitrite will act as an antimicrobial agent by producing nitrite oxide species and help to prevent UTI (Kucheria et al. 2005). Urine consist of other constituents like glucose, creatine, and trace amount of fatty acids, sucrose, citrate, manganese, etc. (Guo and Li 2009; Minot et al. 2011; Foye et al. 2012).

Colonization of bacteria like *Oxalobacter formigenes* will help to reduce 70% risk of recurrent calcium oxalate stone formation in the urinary system. The rate of colonization of bacteria is greater in healthy subjects than individuals with urolithiasis. For example, *Oxalobacter formigenes* of gastrointestinal microbiota will decrease the development of calcium-oxalate stones in the urinary system by degrading dietary oxalate in the human body (Riesenfeld et al. 2004). Colonization of this genus is age-dependent; it is most dominant in children up to 8 years. By the age of 12 years, these genera start to decline and continue through adulthood (Pearce et al. 2014). Therefore, urolithiasis is most predominant in adults because of a comparatively lesser population of *O. formigenes* and hence resulting in increased concentration of oxalate in adult urine samples (Bik et al. 2010). Viruses are more abundant than bacteria, and they infect bacteria called bacteriophage. Gene transfer is mediated by bacteriophage, and it helps to create a reservoir of an antibiotic-resistant gene in many bacteria (Bik et al. 2010; Dethlefsen and Relman 2011). Diet will alter the bacterial community but not the virus because it is stable for a long time and it induces changes in the bacterial community (Dominguez-Bello et al. 2011).

As we refer transient flora, urobiome composition is in disequilibrium or unstable conditions, mainly due to dietary changes that may cause diversification of microorganisms in a specific area of a urinary system (Dominguez-Bello et al. 2011). Diversification is a recycling process of depletion and formation of new microbial composition, and it can form a new stable state (Cosloy and McFall 1973). Antibiotic exposure or infections also make a temporal variation in the diversity of urobiome composition (Kaufman et al. 2008). This temporary variation in a composition can cause urological infectious disease, and it has a chance to get recovered by probiotic intake such as *Lactobacillus* spp. (Kelly et al. 2011; Cho and Blaser 2012). In a life cycle, aging process is a cause of diversification of diversity and composition changes will alter with number and type of genera (Kelly et al. 2011).

Urinary stones are mostly derived from the metabolic product saturation inside the excretory systems. This metabolic origin crystalized products in the urinary system has derivatized the lithiasis activity and mostly involves supersaturation in solutes and microenvironment changes about the urinary niche. Sometimes, stone formation is exclusively associated with bacteria as in case with infectious stones and is considered to be a consequence of a UTI (Ciftçioglu et al. 1999). Presence of bacteria inside the nidus of kidney stones has already been demonstrated (Tavichakorntrakool et al. 2012), but the exact role of these bacteria in the nucleation and growth of stone is poorly understood. However, some explanations suggest that colonization of bacterial cells inside the nidus would exhibit either renal tubular acid tolerance (Heilberg and Schor 2006), degradation of amino acids (Kopple et al. 1978), and changing the pH of microenvironment milieu by ammonia production through urease activity (Torzewska et al. 2014). Another possible explanation towards the role of bacteria in the formation of stone is their ability to inhibit and/or alter the inhibitors of kidney stone formation (Flannigan et al. 2014). Role of some of these common bacteria in the lithiasis of CaOx stones has been dealt in mechanical and illustrative ways (Chutipongtanate et al. 2013). Multiple mechanisms suggest that intact bacterial cell may initiate the cascade of stone development by providing physical surface for attachment and nucleation (Flannigan et al. 2014). Calcium ion chelation property has been found to be associated with bacteria which mediate the nucleation and aggregation of calcium salts of oxalate and cascade the stone formation mechanism. In addition, excretion of carbonate apatite through the cell membranes of nanobacteria and exopolysaccharides from bacterial matrices in biofilms is also involved in stone genesis (Ciftçioglu et al. 1999). Some recent studies suggest cell surface components like cell wall, flagellar components like flagellin (Kanlaya et al. 2019), and spore coat components like calcium dipicolinate would have the direct role in urolithiasis events. Presence of bacteria in stones provides perception that in hyperoxaluric condition, bacteria are not only found in the kidney but may participate in the formation of kidney stones and other urological diseases (Whiteside et al. 2015). Also, some non-urease bacteria like *Kalamiella piersonii* would be responsible to struvite stone generation (Rekha et al. 2020). Thus, studies such as this will helpful to find out the exact bacterial composition of stone that is essential to predict the stone type as hypothesized earlier (Suryavanshi et al. 2018).

12.5 Probiotics for Maintaining Health of Urobiome

Meaning of probiotic is "life," with similar meanings in Latin (pro) and Greek (bios). Elie Metchnikoff introduced the concept of probiotics (Jiang et al. 2011). Probiotics are living microorganisms, which are mainly found in food as well as food supplements such as tablets, syrup, or any other additional forms in products (Donders 1999). Probiotics also form a part of the urobiome, which confer beneficial health effects on the host (Jiang et al. 2011). Probiotics play a role in many factors of urinary tract such as cell–cell adherence, prevention of cellular adherence of pathogenic microorganisms, helping in acid secretion and hydrogen peroxide secretion to control the growth of pathogens in the urinary tract (Preidis and Versalovic 2009). *Lactobacillus* and *Bifidobacterium* most frequently contain a source of probiotics, and action of these probiotics includes adherence and survival in a specific site of the urinary system and competes with urinary pathogen (Servin 2004).

Lactobacillus acts as a bladder targeting probiotic, which will function as a therapeutic agent for the system. However, these bacteria are most predominantly present in the female urinary system (Akgül and Karakan 2018). This species is associated with vaginal microflora, where it maintains an acidic condition, so that it acts as a strong barrier against infections (Lamont et al. 2011; Schwenger et al. 2015). Lactic acid-producing bacteria also reduce the risk of bladder cancer and kidney stones (Servin 2004; Preidis and Versalovic 2009; Lamont et al. 2011; Schwenger et al. 2015; Akgül and Karakan 2018). *Lactobacillus acidophilus* also prevents interstitial cystitis and reduces inflammation (Foye et al. 2012). It also creates an acidic environment in the urinary tract that prevents the growth of pathogenic bacteria (Li et al. 2016).

Lactobacilli can prevent adherence, colonization, and growth of urinary pathogenic bacteria. *Lactobacillus* spp. may produce biosurfactants in the cell surface, which are amphipathic and multifunctional molecules. It has antiadhesive properties against urinary pathogens (Aragón et al. 2018). Urinary health is mainly associated with particular microbial composition of urobiome, and sometimes, this composition can be altering because the effect of any changes in urinary environmental factor like pH will cause urinary disorder such as UTI and this alteration of composition may be used as biomarker in the identification of disease by clinicians, and in this diseased condition, biomarkers composition might help to type probiotic intake to reduce that particular disorder (Aragón et al. 2018). For example, a healthy population of *Lactobacillus* species strongly inhibits the effects of *Escherichia coli,* which is the main cause of UTI in women, and this *E. coli* is a biomarker of UTI (Beerepoot et al. 2013). In this condition, orally administered *Lactobacillus* species act as probiotics that prevent recurrent UTI in women.

Bifidobacterium and *Lactobacillus* stimulate the immune response, for example, *L. casei* is immune modulator in role, it helps to suppress recurrent bladder cancer by modulating immune system against tumor-specific antigen (Feyisetan et al. 2012), and these genera will modulate immunity to produce specific antibody IgA to prevent adhesion of pathogenic microorganism to host uroepithelial cell wall (Herich and Levkut 2002). These genera will disrupt the pathogenesis process because that may induce UTI by causing cytokinetic proinflammatory such as IL-6 (Wullt et al. 2003).

12.6 Conclusion and Future Perspective

A specific urobiome exists in each individual's urinary system, which has its specific bacterial characteristics and functional role. This urobiome is crucial for the maintenance of a healthy urinary system. Even the characteristics and functional role of highly abundant microorganisms in the urinary system are widely studied. Several studies provide evidences that the urinary tract of healthy individual possesses a unique microbiome that changes in different physiological and diseased condition. However, due to difference in sample collection method and lack of synchronization in research methodology, the bacterial genera mainly responsible could not be found in consensus. Still, much has to be learned about the low abundant microorganisms in urobiome and their biochemical requirements. Exact functional role and evolutionary aspects of these low abundant microorganisms are still unknown and that needs further study. Multiomics approach would be the best match to study the pathologies, and genomic applications would render the biomarker discoveries in urolithiasis research (Kachroo et al. 2021b). Role of metabolites and the whole or partial bacterial part in the lithiasis phenomenon would be the next phases of the urological research focus. Meta-analysis and systematic reviews by Kachroo et al. (2021a) (unpublished work) suggest more studies needed to address the age, comorbidities associated with urolithiasis beyond geographic regions. In addition, several studies provide the beneficial role of probiotics, prebiotics, and diet modification to regulate the urobiome in order to reduce the risk of urological disorders. Despite controversies for its use, many studies supported its beneficial role. Further studies are required in the field to gain sufficient knowledge which might have positive impact on society.

Acknowledgments The authors wish to thank and extend their sincere gratitude to all the faculty members, non-teaching staff of the Department of Microbiology, Sikkim University, and Yenepoya Research Centre, Mangalore for their continuous support and help throughout the study.

Funding This work has been funded by SERB-NPDF, DST, Govt. of India (Sanction No.-PDF/2018/000147/LS) to Dr. Mangesh Suryavanshi.

Conflict of Interest The authors declare that the research was conducted in the absence of any commercial or financial relationships that could be construed as a potential conflict of interest.

References

Aagaard K, Ma J, Antony KM et al (2014) The placenta harbors a unique microbiome. Sci Transl Med 6. https://doi.org/10.1126/scitranslmed.3008599

Akgül T, Karakan T (2018) The role of probiotics in women with recurrent urinary tract infections. Turkish J Urol 44:377–383. https://doi.org/10.5152/tud.2018.48742

Alfano M, Canducci F, Nebuloni M et al (2016) The interplay of extracellular matrix and microbiome in urothelial bladder cancer. Nat Rev Urol 13:77–90. https://doi.org/10.1038/nrurol.2015.292

Aragón IM, Herrera-Imbroda B, Queipo-Ortuño MI et al (2018) The urinary tract microbiome in health and disease. Eur Urol Focus 4:128–138. https://doi.org/10.1016/j.euf.2016.11.001

Beerepoot MAJ, Geerlings SE, Van Haarst EP et al (2013) Nonantibiotic prophylaxis for recurrent urinary tract infections: a systematic review and meta-analysis of randomized controlled trials. J Urol 190:1981–1989. https://doi.org/10.1016/j.juro.2013.04.142

Bik EM, Long CD, Armitage GC et al (2010) Bacterial diversity in the oral cavity of 10 healthy individuals. ISME J 4:962–974. https://doi.org/10.1038/ismej.2010.30

Bouatra S, Aziat F, Mandal R et al (2013) The human urine metabolome. PLoS One 8. https://doi.org/10.1371/journal.pone.0073076

Brubaker L, Charles NW, Holly RE et al (2015) Urinary bacteria in adult women with urgency urinary incontinence. Int Urogynecol 25:1179–1184. https://doi.org/10.1007/s00192-013-2325-2.Urinary

Carlsson S, Wiklund NP, Engstrand L et al (2001) Effects of pH, nitrite, and ascorbic acid on nonenzymatic nitric oxide generation and bacterial growth in urine. Nitric Oxide Biol Chem 5: 580–586. https://doi.org/10.1006/niox.2001.0371

Chambers ST, Lever M (1996) Betaines and urinary tract infections. Nephron 74:1–10. https://doi.org/10.1159/000189274

Cho I, Blaser MJ (2012) The human microbiome: at the interface of health and disease. Nat Rev Genet 13:260–270. https://doi.org/10.1038/nrg3182

Chutipongtanate S, Sutthimethakorn S, Chiangjong W et al (2013) Bacteria can promote calcium oxalate crystal growth and aggregation. J Biol Inorg Chem 18:299–308

Ciftçioglu N, Björklund M, Kuorikoski K et al (1999) Nanobacteria: an infectious cause for kidney stone formation. Kidney Int 56:1893–1898

Clemente JC, Ursell LK, Parfrey LW et al (2012) The impact of gut microbiota on human health: an integrative view. Cell 148:1258–1270. https://doi.org/10.1016/j.cell.2012.01.035

Cosloy SD, McFall E (1973) Metabolism of D-serine in Escherichia coli K-12: mechanism of growth inhibition. J Bacteriol 114:685–694. https://doi.org/10.1128/jb.114.2.685-694.1973

Dethlefsen L, Relman DA (2011) Incomplete recovery and individualized responses of the human distal gut microbiota to repeated antibiotic perturbation. Proc Natl Acad Sci U S A 108:4554–4561. https://doi.org/10.1073/pnas.1000087107

Dominguez-Bello MG, Blaser MJ, Ley RE et al (2011) Development of the human gastrointestinal microbiota and insights from high-throughput sequencing. Gastroenterology 140:1713–1719. https://doi.org/10.1053/j.gastro.2011.02.011

Donders GGG (1999) Microscopy of the bacterial flora on fresh vaginal smears. Infect Dis Obstet Gynecol 7:126–127. https://doi.org/10.1002/(SICI)1098-0997(1999)7:3<126:AID-IDOG2>3.0.CO;2-N

Dong Q, Nelson DE, Toh E et al (2011) The microbial communities in male first catch urine are highly similar to those in paired urethral swab specimens. PLoS One 6:1–5. https://doi.org/10.1371/journal.pone.0019709

Feyisetan O, Tracey C, Hellawell GO (2012) Probiotics, dendritic cells and bladder cancer. BJU Int 109:1594–1597. https://doi.org/10.1111/j.1464-410X.2011.10749.x

Flannigan R, Choy WH, Chew B et al (2014) Renal struvite stones--pathogenesis, microbiology, and management strategies. Nat Rev Urol 11:333–341

Fouts DE, Pieper R, Szpakowski S et al (2012) Integrated next-generation sequencing of 16S rDNA and metaproteomics differentiate the healthy urine microbiome from asymptomatic bacteriuria in neuropathic bladder associated with spinal cord injury. J Transl Med 10:1–17. https://doi.org/10.1186/1479-5876-10-174

Foye OT, Huang IF, Chiou CC et al (2012) Early administration of probiotic *Lactobacillus acidophilus* and/or prebiotic inulin attenuates pathogen-mediated intestinal inflammation and Smad 7 cell signaling. FEMS Immunol Med Microbiol 65:467–480. https://doi.org/10.1111/j.1574-695X.2012.00978.x

Gajer P, Brotman RM, Bai G et al (2012) Temporal dynamics of the human vaginal microbiota. Sci Transl Med 4. https://doi.org/10.1126/scitranslmed.3003605

Guo K, Li L (2009) Differential 12C-/13C-isotope dansylation labeling and fast liquid chromatography/mass spectrometry for absolute and relative quantification of the metabolome. Anal Chem 81:3919–3932. https://doi.org/10.1021/ac900166a

Han YW (2015) *Fusobacterium nucleatum*: a commensal truned pathogen. Curr Opin Microbiol:141–147. https://doi.org/10.1016/j.mib.2014.11.013.Fusobacterium

Handelsman J (2004) Metagenomics: application of genomics to uncultured microorganisms. Microbiol Mol Biol Rev 68:669–685. https://doi.org/10.1128/MBR.68.4.669

Heilberg IP, Schor N (2006) Renal stone disease: causes, evaluation and medical treatment. Arq Bras Endocrinol Metabol 50:823–831

Herich R, Levkut M (2002) Lactic acid bacteria, probiotics and immune system. Vet Med (Praha) 47:169–180. https://doi.org/10.17221/5821-VETMED

Hickey RJ, Zhou X, Settles ML et al (2015) Vaginal microbiota of adolescent girls prior to the onset of menarche resemble those of reproductive-age women. MBio 6:1–14. https://doi.org/10.1128/mBio.00097-15

Jarvisalo J, Olkinuora M, Kiilunen M et al (1992) Urinary and blood manganese in occupationally nonexposed populations and in manual metal arc welders of mild steel. Int Arch Occup Environ Health 63:495–501

Jiang J, Knight J, Easter LH et al (2011) Impact of dietary calcium and oxalate, and *Oxalobacter formigenes* colonization on urinary oxalate excretion. J Urol 186:135–139. https://doi.org/10.1016/j.juro.2011.03.006

Kachroo N, Lange D, Penniston K et al (2021a) Meta-analysis of clinical microbiome studies in urolithiasis reveal age, stone composition, and study location as the predominant factors in urolithiasis-associated microbiome composition. Mbio 12(4):e02007-21. https://doi.org/10.1128/mBio.02007-21

Kachroo N, Lange D, Stern J et al (2021b) Standardization of microbiome studies for urolithiasis: an international consensus agreement. Nat Rev Urol 18(5):303–311. https://doi.org/10.1038/s41585-021-00450-8

Kanlaya R, Naruepantawart O, Thangboonkerd V (2019) Flagellum is responsible for promoting effects of viable *Escherichia coli* on calcium oxalate crystallization, crystal growth, and crystal aggregation. Front Microbiol 10:2507. https://doi.org/10.3389/fmicb.2019.02507

Karstens L, Asquith M, Davin S et al (2016) Does the urinary microbiome play a role in urgency urinary incontinence and its severity? Front Cell Infect Microbiol 6:1–13. https://doi.org/10.3389/fcimb.2016.00078

Kaufman DW, Kelly JP, Curhan GC et al (2008) *Oxalobacter formigenes* may reduce the risk of calcium oxalate kidney stones. J Am Soc Nephrol 19:1197–1203. https://doi.org/10.1681/ASN.2007101058

Kelly JP, Curhan GC, Cave DR et al (2011) Factors related to colonization with O*xalobacter formigenes* in U.S. adults. J Endourol 25:673–679. https://doi.org/10.1089/end.2010.0462

Kopple JD, Jones M, Fukuda et al (1978) Amino acid and protein metabolism in renal failure. Am J Clin Nutr 31:1532–1540

Kouri TT, Gant VA, Fogazzi GB et al (2000) Towards European urinalysis guidelines: introduction of a project under European confederation of laboratory medicine. Clin Chim Acta 297:305–311. https://doi.org/10.1016/S0009-8981(00)00256-4

Kucheria R, Dasgupta P, Sacks SH et al (2005) Urinary tract infections: new insights into a common problem. Postgrad Med J 81:83–86. https://doi.org/10.1136/pgmj.2004.023036

Kunin CM, Hua TH, White LVA et al (1992) Growth of *Escherichia coli* in human urine: role of salt tolerance and accumulation of glycine betaine. J Infect Dis 166:1311–1315

Lamont RF, Sobel JD, Akins RA et al (2011) The vaginal microbiome: new information about genital tract flora using molecular based techniques. BJOG Int J ObstetGynaecol 118:533–549. https://doi.org/10.1111/j.1471-0528.2010.02840.x

Lazarevica V, Whitesona K, Huse S et al (2009) Metagenomic study of the oral microbiota by Illumina high- throughput sequencing. J Microbiol Methods 79:266–271. https://doi.org/10.1016/j.mimet.2009.09.012.Metagenomic

Lee JW, Shim YH, Lee SJ (2009) *Lactobacillus* colonization status in infants with urinary tract infection. Pediatr Nephrol 24:135–139. https://doi.org/10.1007/s00467-008-0974-z

Lewis DA, Brown R, Williams J et al (2013) The human urinary microbiome; bacterial DNA in voided urine of asymptomatic adults. Front Cell Infect Microbiol 4:1–14. https://doi.org/10.3389/fcimb.2013.00041

Li H, Zhang L, Chen L et al (2016) *Lactobacillus acidophilus* alleviates the inflammatory response to enterotoxigenic *Escherichia coli* K88 via inhibition of the NF-κB and p38 mitogen-activated protein kinase signaling pathways in piglets. BMC Microbiol 16:1–8. https://doi.org/10.1186/s12866-016-0862-9

Mariat D, Firmesse O, Levenez F et al (2009) The firmicutes/bacteroidetes ratio of the human microbiota changes with age. BMC Microbiol 9:1–6. https://doi.org/10.1186/1471-2180-9-123

May OW (2018) Urine collection methods in children: which is the best? Nurs Clin North Am 53: 137–143

Minot S, Sinha R, Chen J et al (2011) The human gut virome: inter-individual variation and dynamic response to diet. Genome Res 21:1616–1625. https://doi.org/10.1101/gr.122705.111

Mueller ER, Wolfe AJ, Brubaker L et al (2017) The female urinary microbiota. Curr Opin Urol 27: 282–286. https://doi.org/10.1097/MOU.0000000000000396

Nelson DE, van der Pol B, Dong Q et al (2010) Characteristic male urine microbiomes associate with asymptomatic sexually transmitted infection. PLoS One 5:1–7. https://doi.org/10.1371/journal.pone.0014116

Nelson DE, Dong Q, van der Pol B et al (2012) Bacterial communities of the coronal sulcus and distal urethra of adolescent males. PLoS One 7:1–9. https://doi.org/10.1371/journal.pone.0036298

Neugent ML, Hulyalkar NV, Nguyen VH et al (2020) Advances in understanding the human urinary microbiome and its potential role in urinary tract infection. MBio 11:1–15. https://doi.org/10.1128/mBio.00218-20

Newman DK, Rovner ES, Wein AJ (2017) Clinical application of urologic catheters, devices and products. Springer. https://doi.org/10.1007/978-3-319-14821-2

Osborne NG, Wright RC, Grubin L (1979) Genital bacteriology: a comparative study of premenopausal women with postmenopausal women. Am J Obstet Gynecol 135:195–198. https://doi.org/10.1016/0002-9378(79)90342-9

Pearce MM, Hilt EE, Rosenfeld AB et al (2014) The female urinary microbiome: a comparison of women with and without urgency urinary incontinence. MBio 5:1–12. https://doi.org/10.1128/mBio.01283-14

Pearce MM, Zilliox MJ, Rosenfeld AB et al (2015) The female urinary microbiome in urgency urinary incontinence. Am J Obstet Gynecol 213(347):e11. https://doi.org/10.1016/j.ajog.2015.07.009

Peterson J, Garges S, Giovanni M et al (2009) The NIH human microbiome project. Genome Res 19:2317–2323. https://doi.org/10.1101/gr.096651.109

Preidis GA, Versalovic J (2009) Targeting the human microbiome with antibiotics, probiotics, and prebiotics: gastroenterology enters the metagenomics era. Gastroenterology 136:2015–2031. https://doi.org/10.1053/j.gastro.2009.01.072

Price TK, Dune T, Hilt EE et al (2016) The clinical urine culture: enhanced techniques improve detection of clinically relevant microorganisms. J Clin Microbiol 54:1216–1222. https://doi.org/10.1128/JCM.00044-16

Reid G, Bruce AW (2006) Probiotics to prevent urinary tract infections: the rationale and evidence. World J Urol 24:28–32. https://doi.org/10.1007/s00345-005-0043-1

Reid G, Jass J, Sebulsky MT, McCormick JK (2003) Potential uses of probiotics in clinical practice. Clin Microbiol Rev 16:658–672. https://doi.org/10.1128/CMR.16.4.658-672.2003

Rekha P, Hameed A, Manzoor A et al (2020) First report of pathogenic bacterium *Kalamiella piersonii* isolated from urine of a kidney stone patient: draft genome and evidence for role in struvite crystallization. Pathogens 9(9):711

Riesenfeld CS, Goodman RM, Handelsman J (2004) Uncultured soil bacteria are a reservoir of new antibiotic resistance genes. Environ Microbiol 6:981–989. https://doi.org/10.1111/j.1462-2920.2004.00664.x

Savage DC (1977) Microbial ecology of the gastrointestinal tract. Annu Rev Microbiol 31:107–133. https://doi.org/10.1146/annurev.mi.31.100177.000543

Schwenger EM, Tejani AM, Loewen PS (2015) Probiotics for preventing urinary tract infections in adults and children. Cochrane Database Syst Rev. doi: https://doi.org/10.1002/14651858.CD008772.pub2

Servin AL (2004) Antagonistic activities of lactobacilli and bifidobacteria against microbial pathogens. FEMS Microbiol Rev 28:405–440. https://doi.org/10.1016/j.femsre.2004.01.003

Shaykhutdinov RA, MacInnis GD, Dowlatabadi R et al (2009) Quantitative analysis of metabolite concentrations in human urine samples using 13C{1H} NMR spectroscopy. Metabolomics 5: 307–317. https://doi.org/10.1007/s11306-009-0155-5

Shetty K, Paliyath G, Pometto A et al (2017) CRC Press, Taylor & Francis Group, New York

Subhashini DV, Singh RP, Manchanda G (2017) OMICS approaches: tools to unravel microbial systems. Directorate of Knowledge Management in Agriculture, Indian Council of Agricultural Research. ISBN: 9788171641703. https://books.google.co.in/books?id=vSaLtAEACAAJ

Suryavanshi MV, Bhute SS, Kajale SC et al (2018) Molecular imprints of bacteria in oxalate containing kidney stone: a preliminary report. J Clin Diagnostic Res 12:10–13. https://doi.org/10.7860/JCDR/2018/32561.11341

Tang J (2017) Microbiome in the urinary system—a review. AIMS Microbiol 3:143–154. https://doi.org/10.3934/microbiol.2017.2.143

Tasevska N, Runswick SA, McTaggart A et al (2005) Urinary sucrose and fructose as biomarkers for sugar consumption. Cancer Epidemiol Biomark Prev 14:1287–1294. https://doi.org/10.1158/1055-9965.EPI-04-0827

Tavichakorntrakool R, Prasongwattana V, Sungkeeree S et al (2012) Extensive characterizations of bacteria isolated from catheterized urine and stone matrices in patients with nephrolithiasis. Nephrol Dial Transplant 27:4125–4130. https://doi.org/10.1093/ndt/gfs057

Teo S, Cheek JA, Craig S (2016) Improving clean-catch contamination rates: a prospective interventional cohort study. EMA Emerg Med Australas 28:698–703. https://doi.org/10.1111/1742-6723.12697

Thomas-White KJ, Hilt EE, Fok C, Pearce MM et al (2016) Incontinence medication response relates to the female urinary microbiota. Int Urogynecol J 27:723–733. https://doi.org/10.1007/s00192-015-2847-x.Incontinence

Thomas-White K, Forster SC, Kumar N et al (2018) Culturing of female bladder bacteria reveals an interconnected urogenital microbiota. Nat Commun 9:1–7. https://doi.org/10.1038/s41467-018-03968-5

Torzewska A, Budzyńska A, Białczak-Kokot M et al (2014) In vitro studies of epithelium-associated crystallization caused by uropathogens during urinary calculi development. Microb Pathog 71:25–31. https://doi.org/10.1016/j.micpath.2014.04.007

Vebø HC, Solheim M, Snipen L et al (2010) Comparative genomic analysis of pathogenic and probiotic *Enterococcus faecalis* isolates, and their transcriptional responses to growth in human urine. PLoS One 5. https://doi.org/10.1371/journal.pone.0012489

Whiteside SA, Razvi H, Dave S et al (2015) The microbiome of the urinary tract - a role beyond infection. Nat Rev Urol 12:81–90. https://doi.org/10.1038/nrurol.2014.361

Wojciuk B, Salabura A, Ciechanowski K (2019) Urobiome: in sickness and in health. Microorganisms 7:548

Wolfe AJ, Toh E, Shibata N et al (2012) Evidence of uncultivated bacteria in the adult female bladder. J Clin Microbiol 50:1376–1383. https://doi.org/10.1128/JCM.05852-11

Wooley JC, Godzik A, Friedberg I (2010) A primer on metagenomics. PLoS Comput Biol 6. https://doi.org/10.1371/journal.pcbi.1000667

Wullt B, Bergsten G, Fischer H et al (2003) The host response to urinary tract infection. Infect Dis Clin N Am 17:279–301. https://doi.org/10.1016/S0891-5520(03)00028-X

Zhang J, Wang ET, Singh RP, Guo C, Shang Y, Chen J, Liu C (2019) Grape berry surface bacterial microbiome: impact from the varieties and clones in the same vineyard from Central China. J Appl Microbiol 126(1):204–214

Chapter 13
Understanding of Gut Microbial Ecology for New Therapeutics

Tulsi K. Joishy, Santanu Das, and Mojibur R. Khan

Abstract The human body harbors a plethora of microbes which are commonly termed as the microbiota. The dominance of microbiota is mainly observed in the intestine, with a count of 10^{14} cells in the colon. Gut microbiota is shaped by various factors, including diet, age, geography, and genetics. Gut microbial dysfunctions may lead to dysbiosis and affect host health. With the advent of genomic analysis techniques, our understanding of gut microbiota, their functionalities as determinants of health and diseases has rapidly increased over the last decade. Gut microbiota exhibits great potential toward revolutionizing disease etiology and medical treatments. A novel therapeutic approach such as probiotics, prebiotics, and fecal microbiome transplantation (FMT) approach may ameliorate specific disease symptoms. Probiotic therapy aims at altering the microbiota through exogenous administration of live microbes. In contrast, prebiotic therapy relies on compounds that are consumed with the intention of affecting the microbiota composition or function in a beneficial way. While probiotics and prebiotics treatment are unspecific therapeutic approaches, FMT is a specific method that involves the transfer of microbial community from a healthy donor to a diseased recipient in order to reduce the disease-associated microbiota. This chapter cumulates the role of the fecal microbe as the biomarker for the prognosis of diseases and ameliorating it with personalized microbiome-based therapy.

Keywords Fecal microbiota transplantation · Biomarkers · Live Biotherapeutics · Precision medicine

---

Tulsi K. Joishy and Santanu Das contributed equally with all other contributors.

T. K. Joishy · S. Das
Molecular Biology and Microbial Biotechnology Laboratory, Life Sciences Division, Institute of Advanced Study in Science and Technology (IASST), Guwahati, Assam, India

Department of Molecular Biology and Biotechnology, Cotton University, Guwahati, Assam, India

M. R. Khan (✉)
Molecular Biology and Microbial Biotechnology Laboratory, Life Sciences Division, Institute of Advanced Study in Science and Technology (IASST), Guwahati, Assam, India

R. P. Singh et al. (eds.), *Microbes in Microbial Communities*,
https://doi.org/10.1007/978-981-16-5617-0_13

13.1 Introduction

The human gut comprises trillions of microbes termed as microbiota (Lloyd-Price et al. 2016). It executes enormous functions vital for maintaining the metabolic and immune health of an individual (Nagpal et al. 2014). The diverse microbiota predominantly belongs to major phyla, *Actinobacteria, Bacteroides*, and *Firmicutes* (Tap et al. 2009). However, the abundance of these phyla varies among the populations. Various confounding factors are responsible for shaping the individual's microbiota, including dietary habits, lifestyle, age, geography, and ethnicity. The food habit of living beings show a critical part in microbial communities shape of gut and can be tempered by the consumption of long-term dietary practice (Ley et al. 2006; Wu et al. 2011) and short-term dietary habits (David et al. 2014). Gut microbial assessment could be temporally (David et al. 2014) or irreversibly altered (Sonnenburg and Bäckhed 2016). Western dietetic practice such as high-fat diet altered the metabolic activity of resident microbes which might lead to chronic illness. *Lactobacilli* in yogurt are associated with longevity in the Bulgarian peasant populations. *Prevotella* is common in non-westerners who uptake a vegetable-based diet, and its diverse strains are also associated with inflammatory conditions. Increase in *P. copri* improved glucose lenience in studied mice model and suggests that it might do so in humans too. Moreover, diverse strains of probiotic bacteria such as *Lactococcus, Lactobacillus, Leuconostoc, and Enterococcus* are prevalent in different dairy products. Ingestion of probiotics is reported for the noteworthy enhancements in harmonizing the intestinal penetrability and barrier function (Hiippala et al. 2018) and direct effects on atherosclerosis, on metabolic syndrome, colon cancer, and bowel diseases (Wang et al. 2011), and indirect effects on anxiety, anger, depression, and levels of stress hormones. The gut microbiota in the immune system protects against pathogens, harvest energy, and nutrients from the diet by fermenting indigestible starch and proteins. However, the disruption in gut microbiota can negatively affect the host's health too.

The diverse microbiota mainly resides in the distal part of GIT (gastrointestinal tract). These microbial communities contribute 70- to 100-fold extra catalogue of genes relative to the host. Interestingly, these reported genes encode numerous enzymes and proteins which play an immense role in regulating host physiology. In the last decade, advancements in techniques, including high-throughput next-generation sequencing, metabolomics, and transcriptomics, has deciphered that various microbiota is allied with health, fitness, and diseases (Gupta et al. 2017). The relation between gut microbiome and diseases as well as obesity, type 2 diabetes, hypertension, metabolic associated syndrome, diabetes, inflammatory bowel diseases, autism, non-alcoholic fatty liver diseases, and atopic allergies disorders (Bäckhed et al. 2004; Cani et al. 2008; Ley et al. 2006; Turnbaugh et al. 2006) has paved anew way for therapeutic targets. Gut microbiota influence drug metabolism, toxicity, and post-surgical recovery. Therefore, it is vital to explore the biomarkers based on microbiome for diagnostics and treatment purposes. Gut microbiota alterations are linked with etiology of several diseases but the underlying mechanisms are still indistinct. Also, population-specific variation in gut microbiome composition is a significant limitation to draw a baseline for microbiota-mediated therapeutic targets. Therefore, it is imperative to target a community-tailored approach to modulate the gut microbiome of dysbiotic conditions. Large-scale effort in defining core microbiome or healthy microbiome might be helpful in determining microbiome role in diseases across the geography.

13.2 Gastrointestinal Microbiota and its Regional Diversity

The gastrointestinal tract (GIT) is a multiorgan system, and each of the organs consists of a vast and diverse number of microbial entities specialized for various functions. The diversity of the organisms increases at the proximal region to intestinal distal end (Tropini et al. 2017). Several factors such as availability of oxygen, nutrients, pH, bile acids, mucus layers, and immunological factors determine the selection of microbial species in each site (Friedman et al. 2018). The GIT starts with the mouth, where the mastication of food materials takes place. The human oral cavity harbors complex and distinct microbes among which 700 belong to the bacteria (Dewhirst et al. 2010). The predominant member of the oral flora includes *Corynebacterium*, *Porphyromonas*, *Capnocytophaga*, *Streptococcus*, *Veillonella*, *Granulicatella*, *Fusobacterium*, *Leptotrichia*, *Haemophilus*, *Aggregatibacter*, *Neisseria*, etc. (Dewhirst et al. 2010). The commensals in the mouth prevent pathogenic microbes from forming dental biofilms which further endorse dental cavity. Additionally, some of the members such as *Granulicatella, Streptococcus*, and *Veillonella* have increased the production of anti-microbial peptides (AMPs), secrete inflammatory cytokines, which leads to the increase in mucosal thickness and epithelial barrier function (Shang et al. 2018). Transmission of oral microbiota to the distal GIT is a normal phenomenon and occurs in greater frequency than expected by ingestion alone. However, the transmission of a few species is correlated to the disease of the distal GIT (Atarashi et al. 2017). The next major organ in the GIT is the esophagus, which channels the food materials into the stomach. The microbial diversity of the esophagus is similar to the oral cavity. The dominant flora of the esophagus includes *Rothia*, *Prevotella*, *Streptococcus*, *Veillonella*, *Actinobacillus*, *Haemophilus*, *Novosphingobium*, and *Sphingomonas* (May and Abrams 2018). The composition of the esophageal microbiome depends on the diet of an individual. It was observed that regular intake of low-fat diet and dietary fiber promotes enrichment of Firmicutes, while reduces the abundance of Gram-negative bacteria coupled with Proteobacteria; on the other hand, a decreased fiber content in the diet is associated with enrichment of *Neisseria*, *Eikenella*, and *Prevotella* (Nobel et al. 2018). However, in stomach, the microbial count decreases to ~10^1–10^3 CFU/ml, and very few genera including *Veillonella*, *Prevotella*, *Rothia*, *Streptococcus*, and *Haemophilus* thrived. It might be due to gastric pH alteration, thickness of mucosa, and peristaltic movement which limits the growth of reside microbes (O'Hara and Shanahan 2006) (Nardone and Compare 2015).

The distal region of the GIT contains of the small intestine and the colon where the microbial load and diversity are significantly high. The small intestine contains diverse types of cells such as Paneth, tuft, absorptive, enteroendocrine, goblet, and other cell types. Each cell type is differentially distributed throughout the small intestine which determines microbial function, abundance, and diversity. The small intestine is a massive organ and further segmented into duodenum, jejunum, and ileum. The microbial load of duodenum ranges from 10^1 to 10^3 CFU/ml, jejunum harbors 10^4–10^7 CFU/mL, and ileum harbors 10^3–10^8 CFU/mL (Martinez-Guryn et al. 2019; O'Hara and Shanahan 2006). The duodenal microbiome mainly consists of Actinobacteria and Proteobacteria, and their genes are enriched with the metabolism of various nutrient factors (carbohydrates, lipids, protein, etc.) (Angelakis et al. 2015). The jejunum is the major spot for absorption of nutrients. However, the information regarding the microbiome is limited owing to its accessibility. Colonoscopy is the only method to study the jejunal microbiome. Several factors such as diet, bile acids concentration, transit time of food products, and oxygen level influence jejunal microbiota. Previous study reported that Firmicutes are the most dominant phyla in jejunum followed by Proteobacteria, Actinobacteria, and Bacteroidetes (El Aidy et al. 2013; Martinez-Guryn et al. 2019; O'Hara and Shanahan 2006). Similarly, *Clostridium*, *Bacteroidetes*, *Enterococcus*, *Enterobacteria*, *Veillonella*, and *Lactobacillus* are dominant in ileum (Martinez-Guryn et al. 2019; O'Hara and Shanahan 2006). Interestingly, the ileum resided in microbial communities plays the crucial part in the reuptake of bile juices and re-entry into enterohepatic circulation. Moreover, few genera also aid in the absorption of various micronutrients.

Colon is the distal part of the GIT which consists of significantly high and diverse microbial load in comparison to small intestine. The colon has distinctive functional regions such as cecum and proximal of which cecum and ascending colon which serve as major fermentation site, while in the distal colon electrolytes are extracted (O'Hara and Shanahan 2006). The colon harbors ~10^{10}–10^{12} CFU/ml of bacteria among which Firmicutes and Bacteroidetes are dominant (O'Hara and Shanahan 2006). Anaerobic bacterial genera including *Clostridia*, *Roseburia*, and *Eubacteria* are known nondigestible carbohydrate and fiber fermenters which generate short-chain fatty acids (SCFA), and they have played a very important role in maintaining healthy intestine (Koh et al. 2016) (Sommer and Bäckhed 2016). While, *Lactobacillus*, *Bacteroides*, *Clostridium*, and *Bifidobacteria* harbor bile salt hydrolase gene (BSH) which is accountable for deconjugation of taurine and glycine conjugated primary bile juice (Gérard 2014).

13.3 Human Gut Microbiota Functions

The useful prospects of the gut microbial communities are accessed by generating as well as integrating functional reads out of multi-omics (metaproteomic, metatranscriptomic, and metabolomics) data. The gut microbiota are attributed to three specialized functions, viz., crucial and specific metabolic, protective, and trophic functions.

13.3.1 Metabolism

A fundamental function of the colonic microflora is the metabolism of endogenous mucus and exogenous nondigestible complex dietary residue (Roberfroid et al. 1995). A diverse microbes and catalogue of their genes, enzymes, and biochemical pathways were identified by metagenomic studies (Roberfroid et al. 1995; Yadav et al. 2018). Metabolites produced during the carbohydrate fermentation are prime energy source for the colon. The indigestible polysaccharides (such as cellulose, resistant starch, hemicellulose, pectin, and gums), unabsorbed sugars, and oligosaccharides are metabolized to monosaccharides and then finally converted to SCFA, particularly acetate, butyrate, and propionate (Cummings et al. 1987; Cummings et al. 1996). Additionally, gut microbiomes also synthesize vitamins and absorb minerals including calcium, iron, and magnesium. Production of SCFA aids in the absorption of irons in caecum. All fatty acid has a significant role in physiology of host. Butyrate is the prime energy foundation for colonocytes, and it is entirely consumed by colonic epithelium (Cummings et al. 1987). Acetate and propionate are metabolized by muscle and liver, respectively, and act as substrates for gluconeogenesis and lipogenesis. SCFA such as butyrate and propionate modulate appetite and body weight by inducing the expression of leptin in adipose tissue (Villanueva-Millán et al. 2015).

13.3.2 Trophic

Gut microbiome exerts a trophic consequence on the intestinal epithelial cell differentiation activities (Guarner and Malagelada 2003). Acetate, butyrate, and propionate are the three major SCFAs which stimulate proliferation of epithelial cell and its differentiation. Though, butyrate inhibits the proliferation of epithelial cell at neoplastic origin and stimulates its differentiation in vitro. Previous reports suggest that butyrate also promotes cell reversion from neoplastic to non-neoplastic phenotypes, reduces apoptosis of normal enterocytes, and accelerates intestinal mucosa maturation and repair activity after an injury.

13.3.3 Protective

Resident bacteria are crucial in preventing the invasion and colonization of exogenous microbes. They also act as a continuous barrier to the opportunistic pathogens whose abundance is low in gut. In normal conditions, the resident species maintain equilibrium and stability in the ecosystem. However, this ecological balance gets disrupted with the use of antibiotics; thereby, the proliferation of pathogens such as *Clostridium difficle* takes place (Van der Waaij et al. 1996).

13.4 Gut Microbiome Dysbiosis

The human gastrointestinal tract (GIT) is a long tube and arranged as a multi-organ system. Each GIT organ regulates complex digestive and metabolic process. GIT plays a crucial role in digesting and absorbing nutrients from the food which contains viable microorganisms, antigens, and bacterial-derived molecules. The intestinal mucosa exerts an important role in excluding food-borne microbes and macromolecules from systemic circulations while absorbing important nutrients and minerals. Diets rich in protein and sulfate produce toxic metabolites, including phenols, ammonia, hydrogen sulfide, and indoles, which if absorbed by host leads to bowel toxemia. In 400 BC, Hippocrates explained that death lies in the bowel, and indigestion is the root of all evil.

Eventually, the bowel toxemia theories evolved into the intestinal dysbiosis hypothesis, which was originally coined by Metchnikoff. Dysbiosis is characterized as altered gut microflora and their metabolic activities that has detrimental effect in host health. Often dysbiotic gut is related to a rise in pathobionts which may induce various gastrointestinal complications (Carding et al. 2015) and two categories are used to classified as (1) taxonomic and (2) functional.

13.4.1 Taxonomic

The altered composition of microbial species coupled with reduced diversity and richness is referred to as taxonomic dysbiosis. Loss of keystone taxa results in increase of pathogens which affects host health. This condition is often implicated with an increased abundance of pathogens, reduction of alpha diversity, and keystone taxa (Das et al. 2019; Frank et al. 2007). As it is exemplified that the abundance of Bacteroides and Firmicutes decreases in case of intestinal bowel disease while Enterobacteriaceae is increased. Therefore, in such individuals, the butyrate production by Firmicutes is reduced while sulfate reductions increase by Enterobacteriaceae which primes to increased permeability of epithelial cells and causes GIT inflammation (Koh et al. 2016). Similarly, gut bacteria-derived metabolites act as xenobiotics to the hepatocytes that initiate inflammation in the liver, which leads to non-alcoholic steatohepatitis (NASH) and non-alcoholic fatty liver disease (NAFLD) (Jiang et al. 2015; Yu et al. 2016). In case of colorectal cancer (CRC), the abundance of an indigenous gut bacterium, i.e., *Fusobacterium nucleatum,* increases (Arthur et al. 2012; Das et al. 2019). While butyrate synthesis is abridged in case of type 2 diabetes (Qin et al. 2012).

13.4.2 Functional

In case of functional dysbiosis, there is no change in signature taxa, but there is alteration in the abundance of microbial-derived repertoire of metabolites. For example, concentrations of SCFAs and glutamine are highly different in individuals suffering from celiac diseases relative to normal individual (Das et al. 2019). However, sulfur reduction increases with decreases in lactate productions in individuals suffering from constipated-irritable bowel syndrome (C-IBS) (Chassard et al. 2012). Plasma level of trimethylamine-N-oxide (TMAO), a key metabolite produced by gut commensals, is positively allied with cardiovascular disease (CVD) (Wang et al. 2011).

Understanding the mechanism of dysbiosis is key to develop strategies to manage gastrointestinal conditions. Introduction of next-generation probiotics coupled with prebiotics might support the growth of a subset of beneficial microbiota and reduce pathogens (Carding et al. 2015; Das et al. 2019).

13.5 Restoration of Dysbiotic Gut

Therapeutic modulations of gut microbiota are in practice for millennia, for example, implications of traditional herbs medications or by transplanting fecal microbiota (De Groot et al. 2017). Evidence suggests a potential linkage between etiology of diseases with gut microbiome which can also be restored by bacteria-based therapy such as FMT, probiotics, and prebiotics.

13.5.1 Fecal Microbiota Transplant (FMT)

FMT is the unique approach of improving undesired microbiome states of patients suffering from chronic diseases. In this approach, first healthy donor is selected from whom stool is collected and transferred into the recipient GIT. In the fourth century, Ge Hong, a clinician, described the importance of human fecal samples to treat patients suffering from food poisoning and severe diarrhea (Zhang et al. 2012). Later in the sixteenth century, different forms of fecal suspension including fermented, fresh, and dry were employed to treat diarrhea, vomiting, fever, and constipation (Zhang et al. 2012). In the twenty-first century, FMTs are reported to be exceptionally effective in the treatment of diseases such as IBD, recurrent *Clostridium difficile* infection (RCDI), and irritable bowel syndrome (IBS) (Aroniadis and Brandt 2013;

Van Nood et al. 2013). There are several techniques of fecal transplantation, including enema, nasogastric tube, and colonoscopy. However, colonoscopy is the most effective method in which the colon and ileum are inoculated with desired microbiota in comparison to an enema, which targets till splenic flexure. Donor stools are mostly used within 8 hours of passage and colonoscopically administered. The stool is suspended in non-bacteriostatic saline, and further, to remove larger granular matter, and the fecal suspension is filtered through the gauze pads. Several studies report that in RCDI patients, the alpha diversity increases, and the microbial community shifts towards donor microbiome composition post-FMT (Fuentes et al. 2014; Seekatz et al. 2014). Additionally, the donor strains is reported to coexist for prolonged periods in recipient's post-FMT (Li et al. 2016; Moss et al. 2017). However, the underlying mechanism of the microbiome's response is yet to be elucidated. FMT provides a distinctive platform to learning the microbial colonization and pliability.

13.5.2 Probiotics

Probiotics are well-defined as the live beneficial microbes. Probiotics deliberate health benefits by reducing pathogens in the gut (Hill et al. 2014). In contrast to FMT, probiotics therapy is besieged modulations of the gut microbiota (Schmidt et al. 2018), therefore, influence human health. Probiotics employed for human medications must have a "generally regarded as safe" position with a confirmed lower risk of etiology of diseases (O'Toole et al. 2017). Microorganisms that are able to tolerate pepsin, pancreatin, higher concentration of conjugated as well as deconjugated bile acids coupled with antimicrobial resistance can be claimed as probiotic organisms. Several strains of *Lactobacilli* as well as *Bifidobacteria* are widely characterized and are reported to reduce gastrointestinal infections (Nagpal et al. 2012). Probiotics harbor several potential benefits (Fig. 13.1& 13.2). However, the effect of probiotics on microbiome levels might be minuscule (Kristensen et al. 2016).

13.5.3 Prebiotics

Prebiotic is a nondigestible substrate which is metabolized selectively by the host ecosystem deliberating health benefits. It is a targeted mode of microbiome manipulations that confers specific beneficial changes in compositions and activity of gastrointestinal microbiota. Fiber-rich diet increases the abundances of *Prevotella* and *Bacteroides* ratio and improves glucose metabolism (Kovatcheva-Datchary et al. 2015). Additionally, it revealed that consumption of fiber-rich diet by obese children shifted their functional gut microbial community composition and led to weight-loss (Zhang et al. 2015). The prebiotics in the colon are fermented by commensal

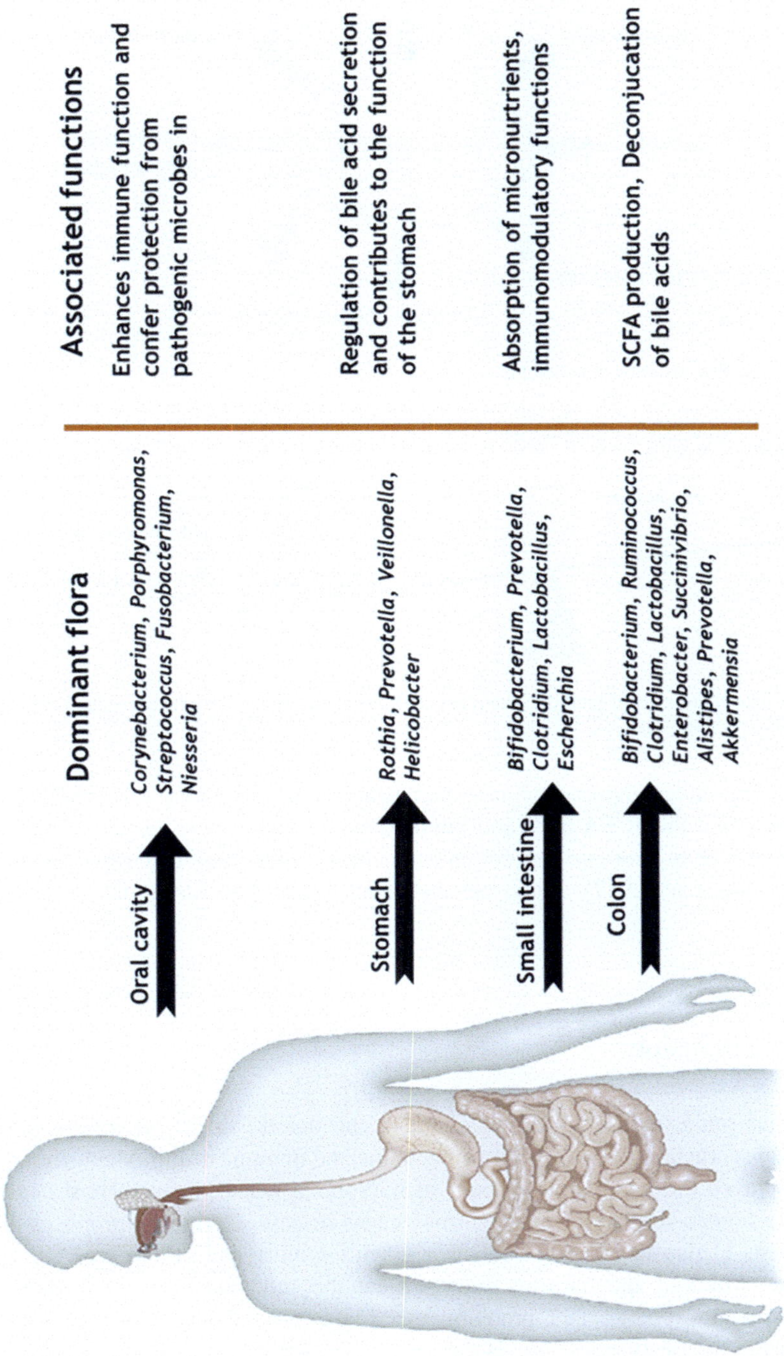

Fig. 13.1 Dominant bacterial genera and their associated function of the organs of a human gastrointestinal tract

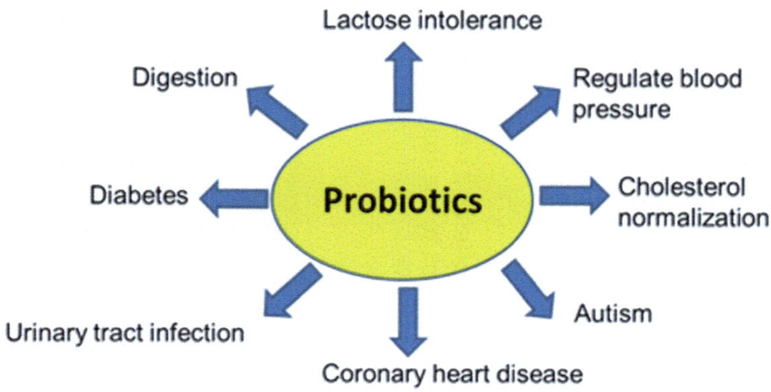

Fig. 13.2 Health attributes of probiotics

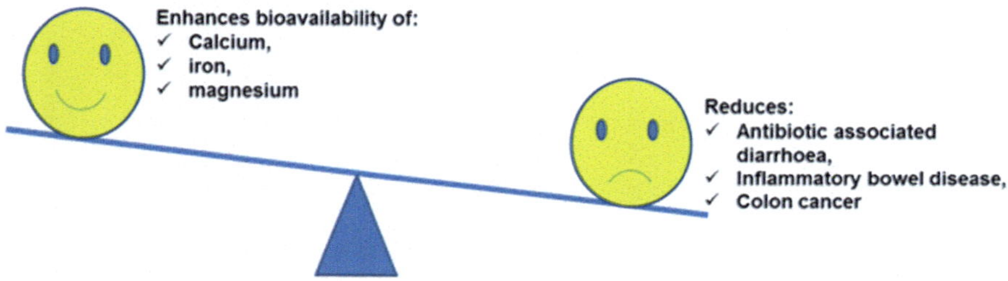

Fig. 13.3 Health benefits of prebiotics help in enrichment of selective bacteria and reduce multiple inflammatory bowel diseases

microbes to produce important metabolites such as short-chain fatty acids (SCFA). It is proposed that the SCFA have anti-inflammatory effect in host health and reduce the colonic pH to promote the colonization of *Lactobacillus* and *Bifidobacteria* (McLoughlin et al. 2017). There are enormous health benefits of the prebiotics (Fig. 13.3).

13.6 Conclusion

The human gut is harboring the uncountable and diverse microbial communities, which play a crucial role in supporting metabolic and immune health. Metagenomics have displayed the important functions of microbiota which maintain host health, and if the homeostasis is disturbed, it may cause diseases. Evidence proposes that there exists a plausible linkage between gut microbiota and etiology of chronic ailments. However, the harmful microbes can be modulated by the beneficial microbes with probiotic and prebiotic therapies, and recently, FMT is widely employed in the treatment of bowel diseases. An extensive study on the functional property of each microbe might help in targeting several metabolic diseases.

Multitude of mysteries about the compositions and functions still remain to be explicated as the microbiota differ across the populations, geography, sex, and age. Therefore, a distinct microbial profile of each population is baseline to use microbiome-based treatment of an individuals. It is imperative to spot the underlying mechanism by which the perturbations caused due to dysbiosis can be modulated by bacteria-based therapy. Additionally, by exploring microbiome functions, its interaction with other co-occurring bacteria is equally important.

References

Angelakis E et al (2015) A metagenomic investigation of the duodenal microbiota reveals links with obesity. PloS one 10:e0137784

Aroniadis OC, Brandt LJ (2013) Fecal microbiota transplantation: past, present and future. Curr Opin Gastroenterol 29:79–84

Arthur JC et al (2012) Intestinal inflammation targets cancer-inducing activity of the microbiota. Science 338:120–123

Atarashi K et al (2017) Ectopic colonization of oral bacteria in the intestine drives TH1 cell induction and inflammation. Science 358:359–365

Bäckhed F et al (2004) The gut microbiota as an environmental factor that regulates fat storage. Proc Nat Acad Sci 101:15718–15723

Cani PD, Bibiloni R, Knauf C, Waget A, Neyrinck AM, Delzenne NM, Burcelin R (2008) Changes in gut microbiota control metabolic endotoxemia-induced inflammation in high-fat diet–induced obesity and diabetes in mice. Diabetes 57:1470–1481

Carding S, Verbeke K, Vipond DT, Corfe BM, Owen LJ (2015) Dysbiosis of the gut microbiota in disease. Microb Ecol Health Dis 26:26191

Chassard C et al (2012) Functional dysbiosis within the gut microbiota of patients with constipated-irritable bowel syndrome alimentary. Pharmacol Therap 35:828–838

Cummings J, Pomare E, Branch W, Naylor C, Macfarlane G (1987) Short chain fatty acids in human large intestine, portal, hepatic and venous blood. Gut 28:1221–1227

Cummings JH, Beatty ER, Kingman SM, Bingham SA, Englyst HN (1996) Digestion and physiological properties of resistant starch in the human large bowel. Br J Nutr 75:733–747

Das S, Deb D, Adak A, Khan MR (2019) Exploring the microbiota and metabolites of traditional rice beer varieties of Assam and their functionalities 3. Biotech 9:174

David LA et al (2014) Diet rapidly and reproducibly alters the human gut microbiome. Nature 505: 559

De Groot PF, Frissen M, De Clercq N, Nieuwdorp M (2017) Fecal microbiota transplantation in metabolic syndrome: history, present and future gut microbes 8:253-267

Dewhirst FE et al (2010) The human oral microbiome. J Bacteriol 192:5002–5017

El Aidy S et al (2013) Gut bacteria–host metabolic interplay during conventionalisation of the mouse germfree colon. ISME J 7:743–755

Frank DN, Amand ALS, Feldman RA, Boedeker EC, Harpaz N, Pace NR (2007) Molecular-phylogenetic characterization of microbial community imbalances in human inflammatory bowel diseases. Proc Nat Acad Sci 104:13780–13785

Friedman ES et al (2018) Microbes vs. chemistry in the origin of the anaerobic gut lumen. Proc Nat Acad Sci 115:4170–4175

Fuentes S et al (2014) Reset of a critically disturbed microbial ecosystem: faecal transplant in recurrent Clostridium difficile infection. ISME J 8:1621–1633

Gérard P (2014) Metabolism of cholesterol and bile acids by the gut microbiota pathogens 3:14-24

Guarner F, Malagelada J-R (2003) Gut flora in health and disease. The Lancet 361:512–519

Gupta VK, Paul S, Dutta C (2017) Geography, ethnicity or subsistence-specific variations in human microbiome composition and diversity. Front Microbiol 8:1162
Hiippala K et al. (2018) The potential of gut commensals in reinforcing intestinal barrier function and alleviating inflammation. Nutrients 10(8):988
Hill C et al (2014) Expert consensus document: The International Scientific Association for Probiotics and Prebiotics consensus statement on the scope and appropriate use of the term probiotic Nature reviews. Gastroenterol Hepatol 11:506
Jiang W et al (2015) Dysbiosis gut microbiota associated with inflammation and impaired mucosal immune function in intestine of humans with non-alcoholic fatty liver disease. Sci Rep 5:1–7
Koh A, De Vadder F, Kovatcheva-Datchary P, Bäckhed F (2016) From dietary fiber to host physiology: short-chain fatty acids as key bacterial metabolites. Cell 165:1332–1345
Kovatcheva-Datchary P et al. (2015) Dietary fiber-induced improvement in glucose metabolism is associated with increased abundance of Prevotella cell metabolism 22:971-982
Kristensen NB, Bryrup T, Allin KH, Nielsen T, Hansen TH, Pedersen O (2016) Alterations in fecal microbiota composition by probiotic supplementation in healthy adults: a systematic review of randomized controlled trials. Genome Med 8:1–11
Ley RE, Turnbaugh PJ, Klein S, Gordon JI (2006) Microbial ecology: human gut microbes associated with obesity. Nature 444:1022
Li SS et al (2016) Durable coexistence of donor and recipient strains after fecal microbiota transplantation. Science 352:586–589
Lloyd-Price J, Abu-Ali G, Huttenhower C (2016) The healthy human microbiome. Genome Med 8: 1–11
Martinez-Guryn K, Leone V, Chang EB (2019) Regional diversity of the gastrointestinal microbiome. Cell Host & Microbe 26:314–324
May M, Abrams JA (2018) Emerging insights into the esophageal microbiome. Curr Treat Opt Gastroenterol 16:72–85
McLoughlin RF, Berthon BS, Jensen ME, Baines KJ, Wood LG (2017) Short-chain fatty acids, prebiotics, synbiotics, and systemic inflammation: a systematic review and meta-analysis. Am J Clin Nutr 106:930–945
Moss EL et al (2017) Long-term taxonomic and functional divergence from donor bacterial strains following fecal microbiota transplantation in immunocompromised patients. PloS One 12: e0182585
Nagpal R et al. (2012) Milk, milk products, and disease free health: an updated overview critical reviews in food science and nutrition 52:321-333
Nagpal R, Yadav H, Marotta F (2014) Gut microbiota: the next-gen frontier in preventive and therapeutic medicine? Front Med 1:15
Nardone G, Compare D (2015) The human gastric microbiota: is it time to rethink the pathogenesis of stomach diseases? United European Gastroenterol J 3:255–260
Nobel YR et al (2018) Increasing dietary fiber intake is associated with a distinct esophageal microbiome. Clin Transl Gastroenterol:9
O'Toole PW, Marchesi JR, Hill C (2017) Next-generation probiotics: the spectrum from probiotics to live biotherapeutics. Nat Microbiol 2:1–6
O'Hara AM, Shanahan F (2006) The gut flora as a forgotten organ. EMBO Rep 7:688–693
Qin J et al (2012) A metagenome-wide association study of gut microbiota in type 2 diabetes. Nature 490:55
Roberfroid M, Bornet F, Ce B, Cummings J (1995) Colonic microflora: nutrition and health. Summary and conclusions of an international Life Sciences Institute (ILSI)[Europe] workshop held in Barcelona. Spain Nutr Rev 53:127–130
Schmidt TS, Raes J, Bork P (2018) The human gut microbiome: from association to modulation. Cell 172:1198–1215
Seekatz AM, Aas J, Gessert CE, Rubin TA, Saman DM, Bakken JS, Young VB (2014) Recovery of the gut microbiome following fecal microbiota transplantation MBio:5

Shang L et al (2018) Multi-species oral biofilm promotes reconstructed human gingiva epithelial barrier function. Sci Rep 8:1–12

Sommer F, Bäckhed F (2016) Know your neighbor: Microbiota and host epithelial cells interact locally to control intestinal function and physiology. Bioessays 38:455–464

Sonnenburg JL, Bäckhed F (2016) Diet–microbiota interactions as moderators of human metabolism. Nature 535:56–64

Tap J et al (2009) Towards the human intestinal microbiota phylogenetic core. Environ Microbiol 11:2574–2584

Tropini C, Earle KA, Huang KC, Sonnenburg JL (2017) The gut microbiome: connecting spatial organization to function. Cell Host & Microbe 21:433–442

Turnbaugh PJ, Ley RE, Mahowald MA, Magrini V, Mardis ER, Gordon JI (2006) An obesity-associated gut microbiome with increased capacity for energy harvest. Nature 444:1027–1031

Van der Waaij L, Limburg P, Mesander G, Van der Waaij D (1996) In vivo IgA coating of anaerobic bacteria in human faeces. Gut 38:348–354

Van Nood E et al. (2013) Duodenal infusion of donor feces for recurrent Clostridium difficile. New Engl J Med 368:407-415

Villanueva-Millán M, Perez-Matute P, Oteo J (2015) Gut microbiota: a key player in health and disease. A review focused on obesity. J Physiol Biochem 71:509–525

Wang Z et al (2011) Gut flora metabolism of phosphatidylcholine promotes cardiovascular disease. Nature 472:57–63

Wu GD et al (2011) Linking long-term dietary patterns with gut microbial enterotypes. Science 334: 105–108

Yadav M, Verma MK, Chauhan NS (2018) A review of metabolic potential of human gut microbiome in human nutrition. Arch Microbiol 200:203–217

Yu J, Marsh S, Hu J, Feng W, Wu C (2016) The pathogenesis of nonalcoholic fatty liver disease: interplay between diet, gut microbiota, and genetic background Gastroenterology research and practice:2016

Zhang F, Luo W, Shi Y, Fan Z, Ji G (2012) Should we standardize the 1,700-year-old fecal microbiota transplantation? Am J Gastroenterol 107:1755–1756

Zhang C et al. (2015) Dietary modulation of gut microbiota contributes to alleviation of both genetic and simple obesity in children EBioMedicine 2:968-984

Chapter 14
Ecology and Abundance of Benzoate-Degrading Bacteria in Industrial Waste

Shweena Krishnani, Kunal Meena, Vineet, Deepansh Sharma, Gajender Kumar Aseri, Jagdip Singh Sohal, Neeraj Khare, and Deepti Singh

Abstract The role of microbial diversity of effluent treatment plant (ETP) for treating the higher concentrations of naturally or artificially occurring aromatic compounds such as benzoate and its salts as a result of a wide range of food and beverage industrial discharges is found to be very important for seeking the treatment of ETP, which might be disturbing the normal microflora of the ETP, resulting in a decrease in essential microbiome that might be helpful in conventional and advanced techniques of treating waste waters. Isolation, detection, and characterization of benzoate degraders are necessary to optimize it for industrial use. Aerobic and anaerobic metabolic pathways of different benzoate degraders result in different end products. This chapter communicates various benzoate-degrading microbes isolated from the wastewater and related sites analyzing the different end products of the reaction and benzoate as intermediate to some reactions, comparing the traditional and new techniques involving treatment of such sites, analyzing the risk of the loss of normal microflora, bioremediation, and the ecological study of various diversified microbes involved in benzoate degradation.

Keywords Benzoate · Bioremediation · Wastewater · Effluent treatment plant · Microbiome

14.1 Introduction

The microbial ecosystem of an industrial wastewater treatment plant illustrates an intense gathering of the diverse gene pool, which forms the possibilities of co-occurrence at an expense of toxins. Microbial communities of the treatment plant active sludge are capable of adaptation to unique and competitive habitats

---

S. Krishnani · K. Meena · Vineet · D. Sharma · G. K. Aseri · J. S. Sohal · N. Khare · D. Singh (✉)
Amity Institute of Microbial Technology, Amity University Rajasthan Jaipur, Rajasthan, India

R. P. Singh et al. (eds.), *Microbes in Microbial Communities*,
https://doi.org/10.1007/978-981-16-5617-0_14

for their survival, usually procuring the metabolic capabilities that will help them survive at such sites. For their survival, they signify discrete genetic devices to operate repetitious stress situations that cover various ecological determinants such as pH, temperature, and nutrient inaccessibility, etc. (Jadeja et al. 2019). These determinants present them as chief performers in the bioremediation and biodegradation processes that carry around the degradation of the pollutants originating via human activities within the lithosphere and hydrosphere. Bioremediation refers to when a consortia/single microbe utilizes the toxic pollutants as its sole carbon, nitrogen, and another trace of the elemental source for the conversion of the complex structural units into simple monomers aiming the reduction of toxicity and environmental pollution (Leewis et al. 2016). For the bioremediation and biodegradation of complex chemicals, unique systems have been adopted by industries such as effluent treatment plant (ETP), common effluent treatment plant (CETP), etc. An ETP is a type of wastewater treatment plant (WWTP) that provides a unique ecosystem used as a black box (Saunders et al. 2016). The notable characteristic of the microbial association of a CETP lies in the adaptation to constantly changing carbon source occurring due to chemical pollution discharges from industries. The carbon source concentration is quite dependent on the industrial production and composition, along with the activated sludge water variation covering different industries that holds the management of bioremediation and biodegradation of various organic compounds (Jadeja et al. 2019).

A WWTP is designed concerning the control of a wide range of pollutants before approaching the environment. Yet implying the most common oxygenic path toward the processing of industrial effluent, the method is not used sufficiently for its maximum potential. The lack of knowledge and shreds of evidence are the key factors for base comparing the microbial community in the activated sludge to its degradative performance (Kapley et al. 2015). Hence, it is essential to maintain as well as reduce the toxins whenever the environmental pollution level rises. Therefore, maintaining the household and industrial effluents is a key role of WWTPs. It is a well-known oxygenic plan employed for the cleaning of effluents, which is also referred to as the activated sludge process (ASP). The microbial community residing in the effluent's accounts for the bioremediation of toxicants in the activated sludge. Industrial effluents confer numerous trials among the normal microflora, besides their possession of a diverse collection of xenobiotics, its composition changes as per the production plan creating differences in the normal microflora population characterization. Certain obstacles remain united in the CETP, wherever the effluent of numerous industrial assemblies is combined before handling processes (Kapley et al. 2007). The value of ETP could turn the economics of small-scale industries (SSI), causing the process of handling of a non-viable alternative. To support such SSIs, an idea of the CETP method got implied. A CETP is a type of WWTP, which is specially intended for the treatment of effluent arising via clusters of the SSIs (Yadav et al. 2014). The effluents combined from the SSIs are brought to a common treatment plant (CTP) merged for the treatment before releasing in the atmosphere. The basic foundation defeats operational value yielding from a single unit up to its minimum while protecting the environment at its most. The last decade of research

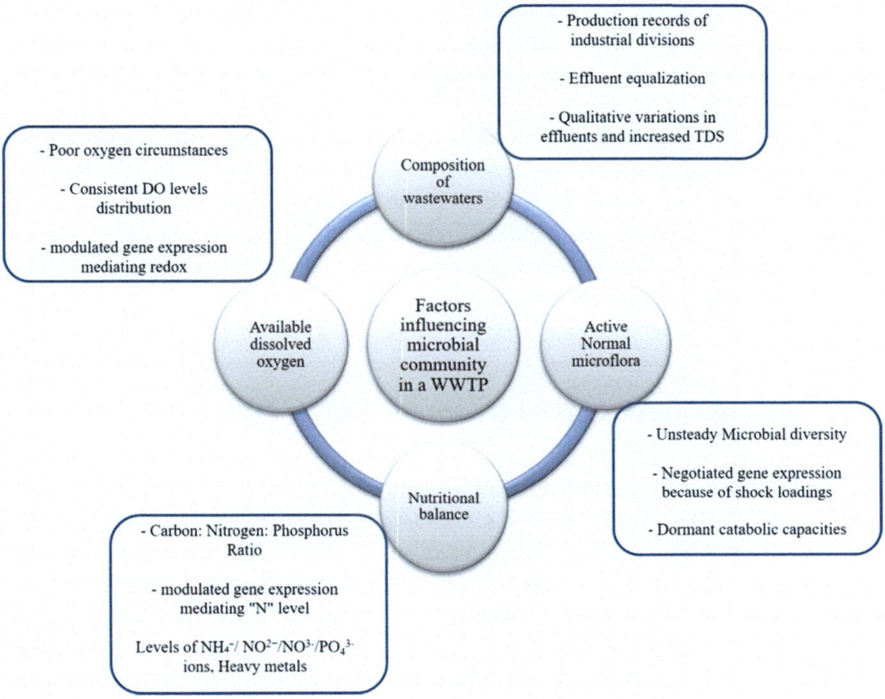

Fig. 14.1 Factors influencing normal microfloral population present in the WWTPs

for the understanding of microbial communities of these sites holds the observation of the universal utilization of activated sludge. One of the several aims is to know the naturally occurring interface that brings biodegradation in a WWT. It illustrates that the microbial culture can be observed employing simplistic molecular mechanisms like random amplification of polymorphic DNA (RAPD) or denaturing gradient gel electrophoresis (DGGE) and improved through gene quantification data that aids our learning of diminished vitality of a WWTP (Kapley et al. 2015).

Microbial communities present in the industrial treatment plants are essential for its function. In addition to that, there are several factors that influence the normal microflora of a WWTP like, (1) available dissolved oxygen that determines the reduced O_2 conditions, consistency of dissolved oxygen levels throughout the WWTP, and redox reactions taking place, (2) the composition of the wastewater that includes industrial schedules, uniform effluent distributions, enhanced TDS, and qualitative variation of effluents over time, (3) active normal microflora includes uneven microbial diversity in an effluent, dormant catabolic capacities, etc., (4) nutritional balance includes ratio of C:N:P, and levels of various nitrogen-containing compounds along with heavy metals (Fig. 14.1).

Recent investigations can communicate a variety of both direct and indirect impacts of WWTP and how the microbial population reacts toward it. These considerations include: (a) partial attenuation of microbial population downstream contribution of effluents, (b) association between overall microbial population and the three Orders proposed to track effluent in WWTP, and (c) variations in species abundance and microbial community distribution underneath effluent discharges (Price et al. 2018).

This chapter discusses about various benzoate-degrading microbes isolated from the wastewater and related sites analyzing the different end products of the reaction and benzoate as intermediate to some reactions, comparing the traditional and the new techniques involving treatment of such sites, analyzing the risk of the loss of normal microflora, bioremediation, and the ecological study of various diversified microbes involved in benzoate degradation.

14.2 Ecological Aspects of Benzoate and its Degraders

The enormous metabolic potential of microbes is crucial to degrade not only natural products but also xenobiotics because of anthropogenic activities (Singh et al. 2017; Junghare et al. 2019; Zhuang et al. 2019). For instance, phthalate (3-carboxybenzoate) is a derivative of benzoate that frequently pollutes the surroundings and has lately become subject to microbial degradation (Sawers 2018). Beyond the compounds of NO^{3-}, Fe^{-} or SO_4^{2-} dependent oxidation, syntrophic fermentation of phthalates ($C_8H_4O_4^{-2}$) is a necessary method as a broad spectrum regarding tangible and engineered conditions. It is operated through a synergistic approach in bacteria capable of fermentation along with H_2-utilizing anoxygenic microbes. Therefore, syntrophic anoxygenic microbes imply essential parts of the anoxygenic carbon cycle occurring globally (McInerney et al. 2009). The *Sphingomonas aromaticivorans* fermenting isophthalate and the *Desulfovibrio* sp. consuming H_2 are SI sulfate reducer that represents syntrophic model arrangement exhibiting the biochemical reactions in strictly anoxygenic microbial systems (Nobu et al. 2017). Hence, this is a useful example for understanding the microbial vital adaption of their metabolism in order to serve a variety of differential substrates carried out by anthropogenic activities.

Partial elimination of naturally occurring pollutants by the time of wastewater treatment is one of the foremost ways to interject micro-pollutants within particular environmental conditions since both of them are the monitored aggregates and structures that are recognized as unique toxins (Montes-Grajales et al. 2017). One of these accumulations of toxicants involves UV filters. Normally used UV filters imply 2-ethylhexyl 4-(dimethylamino) benzoate (ODPABA) plus 2-ethylhexyl 4-methoxycinnamate (EHMC). Because of their lipophilic nature, certain aggregates aggregate in effluents by direct accumulation. Organic UV filters are utilized for the protection of the skin from the damaging sun rays. Whereas synthetic UV filters go straight into the external conditions due to bathing, cleaning clothes, training water festivities, and obliquely by the community or industrial effluents. Those aggregates are identified as pollutants in effluents (Ekpeghere et al. 2016) at levels of ng L^{-1} and µg L^{-1} also within tap water systems (Da Silva et al. 2015). In cosmetic industries, these serve as an ingredient of various products such as shampoos, lotions, and different individual shield commodities, and they have also been found in industrial effluents (Biel-Maeso et al. 2019).

Different types of pollutants are imported into the ecosystem at a minimum concentration and are stable to support conversions besides forming pollutants that can injure life. The accumulation of the toxicants involves pharmaceuticals, pesticides, chemicals, and nonchemical UV filters. Their consistencies in effluents differ from ng L^{-1} to mg L^{-1} (Fagervold et al. 2019; Mackuľak et al. 2019). Generally, utilized UV filters such as EHMC and ODPABA used as an ingredient in beautifiers, though they occupy the surroundings with industrial effluents. It has been observed that in industrial effluents entering WWTP, the EHMC is seen at a concentration scale of 120–1134 ng L^{-1} (Ekpeghere et al. 2016). The variation in the intensity of UV filters separated from effluents implies that the process of WWTP carried out is crucial when assessing the rate of removal of nonbiodegradable micro-pollutants. The elimination of EHMC and ODPABA from WWTP basically advances by the adsorption of these toxicants in the residue. The UV filter concentration in the activated sludge improves with an increase of the activated sludge treatment depending on the duration of its process (Gackowska and Studziński 2020).

14.3 Ecological Characteristics and the Risk of Loss of Microbial Community

Ecological differentiation among bacterial strains is developed by molecular and metabolic heterogeneity. Though, comparing genotypes over environmental habitats persists in a larger complication. Phenotypic variations are compared with various competitive co-cultures and their respective geographic indications, symbolizing connections among intraspecific heterogeneity, microbiological intercommunications, and biological geography (Koch et al. 2020; Singh et al. 2020). Metabolic diversity is an initial supervisor of ecological differentiation in bacterial species, developing evolutionary approaches, consequently the habitat extent of related taxa (Larkin and Martiny 2017; Subhashini et al. 2017). With the rising figures of genes sequenced yet, valuable working heterogeneity was detected between closely related species (Farrant et al. 2016), with suggestions to bacterial species concepts. The heterogeneity is examined by questioning the pangenome of a taxon organization for genetic alterations with biological associations (Arevalo et al. 2018). Ecological differentiation in a species principally correlates to a pair of elastic genetic divisions: (1) the associate genome, and (2) the novel genome. The modern classification methods such as 16S rRNA sequencing and core-genome phylogenies are commonly used to identify the benzoate degrading bacteria (Table 14.1). Acknowledging the community organization of *Marinobacter* spp. with phototrophs (Gärdes et al. 2011), the group might furthermore facilitate the cyanobacterial biodegradation of ecologically more suitable aromatics, e.g., cinnamate, benzoate derivatives

Table 14.1 Bacterial communities involved in benzoate degradation

S. No.	Name of bacteria	Benzoate derivative	Conc. (mM)	Characteristics of bacteria	References
1.	*Desulfoprunum benzoelyticum*	Benzoate	0.98	Anaerobic, mesophilic, sulfate reducers	Junghare and Schink (2015)
2.	*Sporotomaculum syntrophicum*	Benzoate	5	Anaerobic, mesophilic, syntrophic benzoate degraders	Qiu et al. (2003)
3.	*Methanospirillum hungatei*	Benzoate	1	Gram negative, weakly motile, and a strict anaerobe	Qiu et al. (2004)
4.	*Desulfovibrio* strain G-11	Sodium benzoate	7	Syntrophic, benzoate degraders; anaerobic, motile, gram negative, rod shaped	Hopkins et al. (1995)
5.	*Methanospirillum hungatei* and *Desulfovibrio* sp.	Sodium benzoate	13.8	Anaerobic, motile, gram negative, rod shaped	Mountfort and Bryant (1982a, b)
6.	*Desulfotomaculumthermobenzoicum* sp.	Benzoate	10	Thermophilic sulfate-reducing bacterium, spore forming, rod-shaped, slightly motile, and gram positive	Tasaki et al. (1991)

(Żyszka-Haberecht et al. 2019). In addition to plasmids, ecological differentiation also correlates diverse capabilities of microbial synergies (Kastman et al. 2016). Ecological differentiation relates to the ability to equate production at the population level. Further, the co-culturing systems could address the coexistence of microbial species in more complex ecological situations, for example, pioneer–scavenger relations while polysaccharide degradation (Hehemann et al. 2016).

The nitrifying bacteria normally synchronize along with heterotrophic bacteria in the WWTP bioreactor. This synergism shed the light covering the ecology of nitrifying effluents to enhance our knowledge of the biotransformation and biodegradation method happening within nitrifying effluents. Furthermore, the works are designed to investigate the opportunity to utilize the autotrophic nitrifying microbial diversity to remove the emerging organic micro-pollutants like drinking water treatment or wastewater tertiary treatment (Sun et al. 2019).

14.4 Functionality of Microbial Community Involved in Benzoate Degradation Sites

Degradation of dangerous pollutants arbitrated by microbes is broadly observed as an efficient approach to bioremediation. Generally, aromatic compounds are organic molecules comprised of one or more aromatic rings, particularly benzene rings, and are the most concerning ecological pollutants that harshly endanger the environment and individual health because of their widespread and determined traits and bioaccumulation through the food web (Yang et al. 2020a, b).

14.4.1 Mechanism of Oxygenic and an-Oxygenic Benzoate Degradation

Absence of the terminal electron acceptor or hydrogen utilizing partners favors an-oxygenic pathways. In order to degrade benzoate, anaerobes imply reduction pathway (Dutton and Evans 1969) and will undergo fermentation due to the absence of oxygen such that the benzoate is converted into acetate, carbon dioxide, and cyclohexane carboxyl-ate resulting in a thermodynamically stable reaction which in turn favors free energy (more positive energy) for substrate utilization (Elshahed and Mcinerney 2001). About-face is the oxygenic route that follows ortho-cleavage pathway where catechol and protocatechuate are the initial transitional compounds. Catechol and procatechuate are the substrates of an enzyme dioxygenases that directly attack the aromatic ring between hydroxyl groups. This leads to the formation of 3-ketoadipate, leading it to conversion of the final products (Fig. 14.2), i.e., acetyl-CoA and succinyl-CoA (Valderrama et al. 2012).

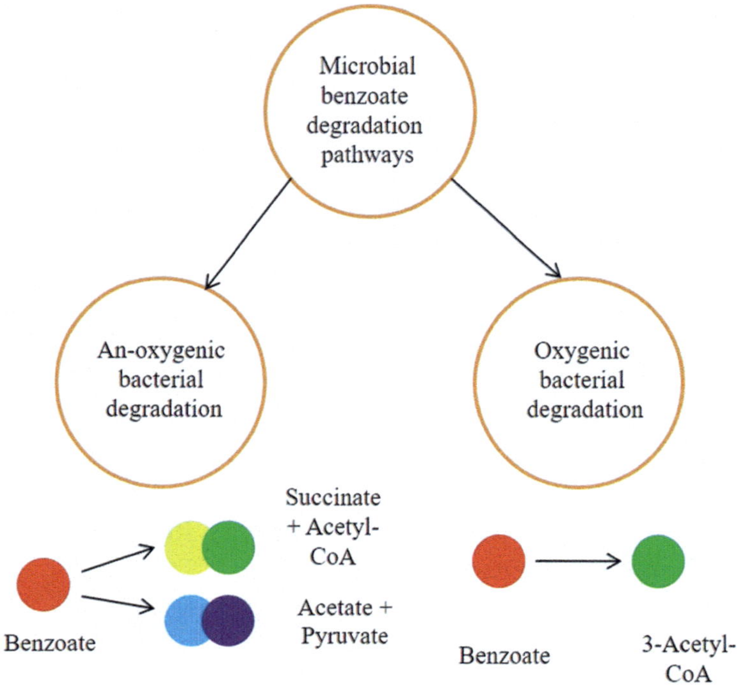

Fig. 14.2 Types of microbial degradation pathways and the corresponding end products of benzoate degradation

14.4.2 Anoxygenic Pathways: Reduction Pathways Implied by Anaerobes

The metabolism of benzoate can be achieved by the reduction pathway by anaerobes in the presence of sunlight by the reduction to cyclohex-1ene-1carboxylate supported with the hydration to 2-hydroxycyclohexanecarboxylate and dehydrogenation to 2-oxocyclohexanecarboxylate (Fig. 14.3). Moreover, hydration can result in ring-fission and the generation of pimelate occurs (Dutton and Evans 1969). The most common intermediate formed in the aromatic metabolism of aromatic compounds is either benzoate or benzyl-CoA. Ring cleavage formed due to the reduction of the aromatic ring as there is no molecular oxygen is present. Benzyl-CoA is activated by benzoate.

De-aromatization of the benzene ring befalls as 2-electron by benzoyl-CoA reduction into cyclohex-diene-1-carboxyl-CoA. Two metabolic pathways are observed for the oxidation of cyclohex-diene-1-carboxyl-CoA to 3-hydroxy pimeloyl-CoA; one that involves cyclohex-1-ene-1-carboxyl-CoA as an intermediate and another that requires 6-hydroxy-cyclohex-1-ene-1-carboxyl-CoA (Harwood and Gibson 1997).

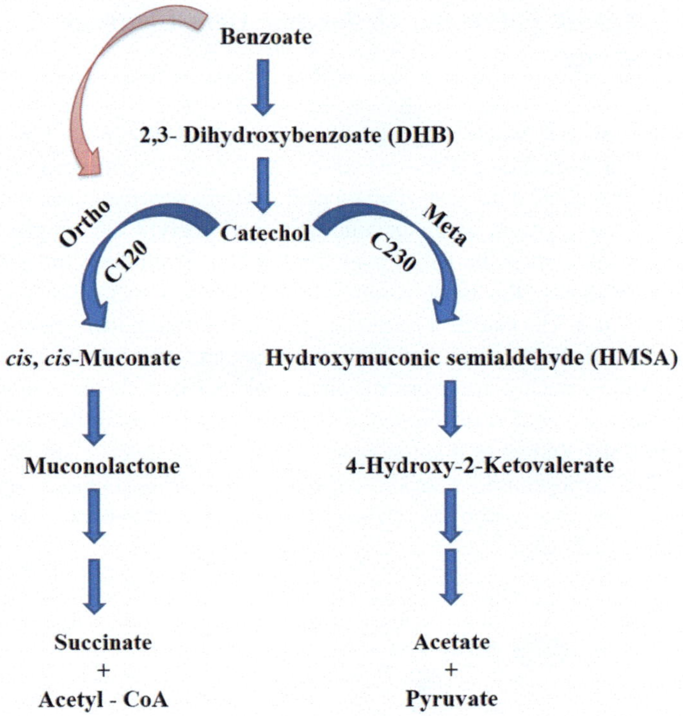

Fig. 14.3 Anaerobic benzoate degradation pathway

Phototrophic bacterium, e.g., *Rhodopseudomonas palustris*, *Thauera aromatic* K172, and *Azoarcusevansii* are known for an-oxygenic degradation.

In the environment rich in methanogens, syntrophic bacteria (e.g., *Syntrophus buswellii*) can degrade benzoate to methanogenic substrates as acetate, CO_2, H_2, and possibly formate.

$$benzoate^- + 7H_2O \rightarrow 3\ acetate^- + 3H^+ + HCO_3^- + 3H_2$$

As per the above equation, benzoate oxidation including the acetate and hydrogen is a thermodynamically unfavorable reaction that executes the metabolism of benzoate by syntrophic bacteria dependent on H_2^- using microbes. This implies that even at low levels of hydrogen levels, accumulated acetate may inhibit benzoate degradation (Hopkins et al. 1995).

14.4.3 Oxygenic Pathways: Ortho and Meta-Cleavage Pathway

The pathway describes the biodegradation of aromatic compounds like benzoate, which leads to the formation of the most commonly known intermediates that denote catechol, protocatechuic acid, and gentisic acid. Certain intermediates are additionally converted into pyruvic acid, succinic acid, and acetyl-CoA, implying a ring fission mechanism to enter the Krebs cycle (Fig. 14.4). From these three dihydroxy aromatic intermediates, the most commonly encountered metabolite before ring cleavage is catechol. The catechol formation proceeds through incorporating molecular oxygen in its respective aromatic precursor, and the initial substrates are funneled into the catechol spectrum of single-ring benzene to three-ring phenanthrene. Once the catechol is developed, it is deteriorated by both the meta and the ortho, which is also known as β-ketoadipate, ring cleavage mechanism via enzymes catechol 2,3-dioxygenase, and catechol 1,2-dioxygenase, generating 2-hydroxymuconic semialdehyde (HMSA) and cis,cis-muconate, sequentially (Valderrama et al. 2012).

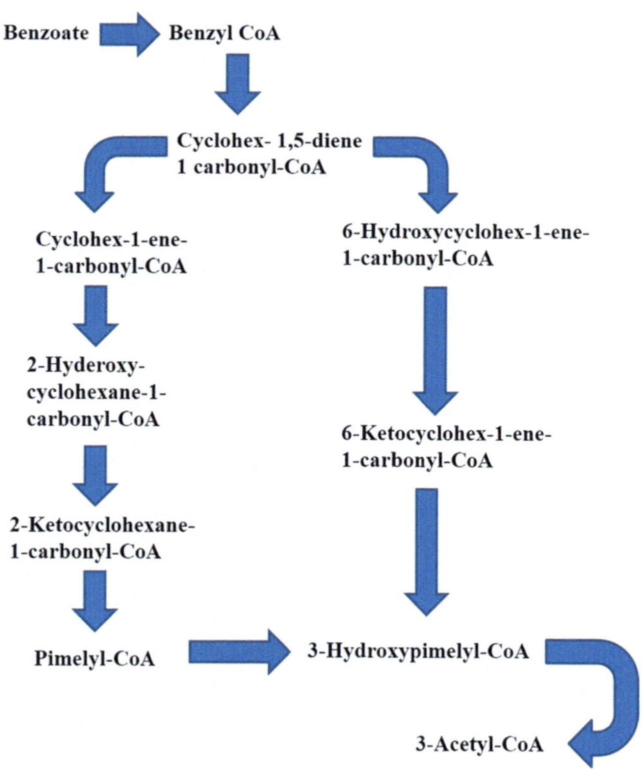

Fig. 14.4 Aerobic benzoate degradation pathway

14.5 Application of Benzoate Degrading Bacteria in Various Industrial Sectors

Various benzoate degrading bacteria are involved in broad applications in the anoxygenic and oxygenic treatment of residential and industrial effluents (Table 14.2). Benzoate is one of the fundamental intermediates in the degradation of many naturally or synthetically produced aromatic compounds. Benzoate and related aromatic compounds are widely used in industries, for instance, in the production of terephthalic acid (1,4-benzenedicarboxylic acid), and some of these chemicals are often discharged as pollutants in industrial wastewater. Since these compounds may be difficult to degrade under anoxic conditions, wastewaters are generally treated aerobically (Lau 1977). In the past decades, investigations have been conducted for the demonstration of naive aromatic compounds such as phenol and benzoate that can be efficiently degraded by anaerobic mechanisms (Li 1995; Fang et al. 1996). All new-generation anaerobic biological treatment systems employ dense populations of anaerobic microorganisms as sludge aggregates, which convert complex organic substances to methane and carbon dioxide. In the commercial and dynamic perspectives, anaerobic (methanogenic) mechanisms have frequently been offered to treat those effluents, and as an outcome, >10 large-scale anoxygenic bioreactors are currently in operation or under development for the treatment of phthalate isomer-containing effluents (Fang et al. 1996). Isolation of the benzoate organism extends the range of known syntrophs to those that degrade benzoate. The non-syntrophic bacteria might also participate in the degradation of benzoate in these environments (Mountfort et al. 1984). Over the past few decades,

Table 14.2 Various isolated benzoate degrading bacteria and their functional applications

S. No.	Organism	Functional application	References
1.	*Desulfoprunum benzoelyticum*	Anaerobic treatment of domestic and industrial wastewaters	Junghare and Schink (2015)
2.	*Desulfovibrio*	Complex effluents consist of man-made compounds and/or compounds resistant to biodegradation	Hopkins et al. (1995)
3.	*Methanospirillum hungatei*	Used for the treatment of effluents through the producers of terephthalic acid	Qiu et al. (2003)
4.	*Methanospirillum hungatei Desulfovibrio*	Biotransformations of aromatics to innocuous end-products such as methane and carbon dioxide	Mountfort and Bryant (1982a, b)
5.	*Sporotomaculum syntrophicum*	These have the capacity to ferment benzoate to their pure culture, formation of cyclohexane, carboxylate, and acetate as their products	Qiu et al. (2003)
6.	*Thauera* sp.	In situ bioremediation processes with varying oxygen level at several polluted sites and w.r.t time	Yoshifumi et al. (2004)
7.	*Pseudomonas, Mesorhizobium*	Halobenzoate-degrading denitrifying bacteria	Song et al. (2000)

several anaerobic digestion technologies, such as up-flow anaerobic sludge blanket processes, have been developed and demonstrated to be successful for the treatment of wastewater, e.g., from the petrochemical industries. The methanogenic bacteria (anaerobic fermentation technology) have been extensively utilized for the treatment of community and industrial sludge and effluents in addition to various syntrophically fermenting benzoate degrading bacteria, such as *Syntrophus entianae* (Szewzyk and Schink 1989) and *S. buswellii* (Mountfort et al. 1984). A large number of anoxygenic mechanisms are improved, and their applicability is now extending to low-strength effluents, to sludge and effluent under extreme temperature circumstances, and to more complex wastewaters containing man-made compounds and/or compounds resistant to biodegradation.

Wastewaters with high concentrations of phthalate isomers (ortho-, meta-, and para-benzene dicarboxylic acid) are one of the complications in industrial wastewaters that are now being challenged by anoxygenic processes. Phthalate isomers that are initially man-made compounds are being created in huge quantities for the manufacturing of polyester resins, plastic bottles, plasticizers, polyester fibers, and additional petroleum-based products across the globe and are consequently eluted in the effluents produced by the analogous industries (Macarie et al. 1992; Cheng et al. 1997). Since most industrial effluents comprise aromatic compounds with huge concentrations of sulfate, sulfate reduction displays a significant process following oxygen-deficient situations, contributing to biogeochemical sulfur cycling in different ecological scenarios (Castro et al. 2000).

14.6 Conclusion and Summary

Bioremediation and biodegradation of toxic aromatic compounds as a result of both aerobic and anaerobic processes in WWTPs, ETP, CETP results in the conversion to lesser toxic substance are the sole purpose and aim focusing on the reduction of pollution in industrial treatment plants. This is required for the normal functioning of industrial sludge system as well as growth of normal microflora operating the same. Regardless of the common substrate, the conversion depends on the presence and absence of the O_2. Catechol is observed to be an important intermediate in the process, and the overall intermediates lead them to merge to TCA cycle with the role of different genes at every step. Various organisms are isolated, characterized, and identified at their threshold concentrations along with their morphological characteristics. Various kinetics studies show the oxidation of benzoate which converts them to methanogenic substances like acetate. Further omics approaches and next-generation sequencing techniques can be used for the further research.

References

Arevalo P, Van Insberghe D, Polz, MF (2018) A reverse ecology framework for bacteria and archaea. In Population genomics: microorganisms (pp 77-96) Springer, Cham

Biel-Maeso M, Corada-Fernández C, Lara-Martín PA (2019) Removal of personal care products (PCPs) in wastewater and sludge treatment and their occurrence in receiving soils. Water Res 150:129–139

Castro HF, Williams NH, Ogram A (2000) Phylogeny of sulfate-reducing bacteria. FEMS Microbiol Ecol 31(1):1–9

Cheng SS, Ho CY, Wu JH (1997) Pilot study of UASB process treating PTA manufacturing waste water. Water Sci Technol 36(6–7):73–82

Da Silva CP, Emídio ES, de Marchi MRR (2015) The occurrence of UV filters in natural and drinking water in São Paulo state (Brazil). Environ Sci Pollut Res 22(24):19706–19715

Dutton PL, Evans WC (1969) The metabolism of aromatic compounds by *Rhodopseudomonas palustris* a new, reductive, method of aromatic ring metabolism. Biochem J 113(3):525–536

Ekpeghere KI, Kim UJ, Sung-Hee O, Kim HY, Oh JE (2016) Distribution and seasonal occurrence of UV filters in rivers and wastewater treatment plants in Korea. Sci Total Environ 542:121–128

Elshahed MS, McInerney MJ (2001) Benzoate fermentation by the anaerobic bacterium Syntrophusaciditrophicus in the absence of hydrogen-using microorganisms. Appl Environ Microbiol 67(12):5520–5525

Fagervold SK, Rodrigues AS, Rohée C, Roe R, Bourrain M, Stien D, Lebaron P (2019) Occurrence and environmental distribution of 5 UV filters during the summer season in different water bodies. Water Air Soil Pollut 230(7):1–13

Fang HH, Chen T, Li YY, Chui HK (1996) Degradation of phenol in wastewater in an upflow anaerobic sludge blanket reactor. Water Res 30(6):1353–1360

Farrant GK, Doré H, Cornejo-Castillo FM, Partensky F, Ratin M, Ostrowski M, Garczarek L (2016) Delineating ecologically significant taxonomic units from global patterns of marine picocyanobacteria. Proc Natl Acad Sci 113(24):E3365–E3374

Gackowska A, Studziński W (2020) Effect of activated sludge on the degradation of 2-Ethylhexyl 4-Methoxycinnamate and 2-Ethylhexyl 4-(Dimethylamino) benzoate in waste water. Water Air Soil Pollut 231(4):1–12

Gärdes A, Iversen MH, Grossart HP, Passow U, Ullrich MS (2011) Diatom-associated bacteria are required for aggregation of *Thalassiosira weissflogii*. ISME J 5(3):436–445

Harwood CS, Gibson J (1997) Shedding light on anaerobic benzene ring degradation: a process unique to prokaryotes? J Bacteriol 179(2):301

Hehemann JH, Arevalo P, Datta MS, Yu X, Corzett CH, Henschel A, Polz MF (2016) Adaptive radiation by waves of gene transfer leads to fine-scale resource partitioning in marine microbes. Nat Commun 7(1):1–10

Hopkins BT, McInerney MJ, Warikoo V (1995) Evidence for anaerobic syntrophic benzoate degradation threshold and isolation of the syntrophic benzoate degrader. Appl Environ Microbiol 61(2):526–530

Jadeja NB, Purohit HJ, Kapley A (2019) Decoding microbial community intelligence through metagenomics for efficient wastewater treatment. Funct Integr Genomics 19(6):839–851

Junghare M, Schink B (2015) *Desulfoprunum benzoelyticum* gen nov, sp nov, a gram-stain-negative, benzoate-degrading, sulfate-reducing bacterium isolated from a wastewater treatment plant. Int J Syst Evol Microbiol 65(1):77–84

Junghare M, Spiteller D, Schink B (2019) Anaerobic degradation of xenobiotic isophthalate by the fermenting bacterium *Syntrophorhabdus aromaticivorans*. ISME J 13(5):1252–1268

Kapley A, Prasa S, Purohit HJ (2007) Changes in microbial diversity in fed-batch reactor operation with wastewater containing nitroaromatic residues. Bioresour Technol 98(13):2479–2484

Kapley A, Liu R, Jadeja NB, Zhang Y, Yang M, Purohit HJ (2015) Shifts in microbial community and its correlation with degradative efficiency in a wastewater treatment plant. Appl Biochem Biotechnol 176(8):2131–2143

Kastman EK, Kamelamela N, Norville JW, Cosetta CM, Dutton RJ, Wolfe BE (2016) Biotic interactions shape the ecological distributions of *staphylococcus* species. MBio,7(5)

Koch H, Germscheid N, Freese HM, Noriega-Ortega B, Lücking D, Berger M, Wietz M (2020) Genomic, metabolic and phenotypic variability shapes ecological differentiation and intraspecies interactions of *Alteromonas macleodii*. Sci Rep 10(1):1–14

Larkin AA, Martiny AC (2017) Microdiversity shapes the traits, niche space, and biogeography of microbial taxa. Environ Microbiol Rep 9(2):55–70

Lau CM (1977, January) Staging aeration for high-efficiency treatment of aromatic acids plant wastewater. In proc Ind waste Conf; (United States) (no CONF-7705185-)

Leewis MC, Uhlik O, Leigh MB (2016) Synergistic processing of biphenyl and benzoate: carbon flow through the bacterial community in polychlorinated-biphenyl-contaminated soil. Sci Rep 6 (1):1–12

Li YY, P Fang HH, Chen T, Chui, HK (1995) UASB treatment of wastewater containing concentrated benzoate. J Environ Eng, 121(10): 748–751

Macarie H, Noyola A, Guyot JP (1992) Anaerobic treatment of a petrochemical wastewater from a terephthalic acid plant. Water Sci Technol 25(7):223–235

Mackuľak T, Černanský S, Fehér M, Birošová L, Gál M (2019) Pharmaceuticals, drugs, and resistant microorganisms—environmental impact on population health. Curr Opin Environ Sci Healt 9:40–48

McInerney MJ, Sieber JR, Gunsalus RP (2009) Syntrophy in anaerobic global carbon cycles. Curr Opin Biotechnol 20(6):623–632

Montes-Grajales D, Fennix-Agudelo M, Miranda-Castro W (2017) Occurrence of personal care products as emerging chemicals of concern in water resources: A review. Sci Total Environ 595:601–614

Mountfort DO, Bryant MP (1982a) Isolation and characterization of an anaerobic syntrophic benzoate-degrading bacterium from sewage sludge. Arch Microbiol 133(4):249–256

Mountfort DO, Bryant MP (1982b) Isolation and characterization of an anaerobic syntrophic benzoate-degrading bacterium from sewage sludge. Arch Microbiol 133(4):249–256

Mountfort DO, Brulla WJ, Krumholz LR, Bryant MP (1984) *Syntrophus buswellii* gen nov, sp nov: a benzoate catabolizer from methanogenic ecosystems. Int J Syst Evol Microbiol 34(2):216–217

Nobu MK, Narihiro T, Liu M, Kuroda K, Mei R, Liu WT (2017) Thermodynamically diverse syntrophic aromatic compound catabolism. Environ Microbiol 19(11):4576–4586

Price JR, Ledford SH, Ryan MO, Toran L, Sales CM (2018) Wastewater treatment plant effluent introduces recoverable shifts in microbial community composition in receiving streams. Sci Total Environ 613:1104–1116

Qiu YL, Sekiguchi Y, Imachi H, Kamagata Y, Tseng IC, Cheng SS, Harada H (2003) *Sporotomaculum syntrophicum* sp nov, a novel anaerobic, syntrophic benzoate-degrading bacterium isolated from methanogenic sludge treating wastewater from terephthalate manufacturing. Arch Microbiol 179(4):242–249

Qiu YL, Sekiguchi Y, Imachi H, Kamagata Y, Tseng IC, Cheng SS, Harada H (2004) Identification and isolation of anaerobic, syntrophic phthalate isomer-degrading microbes from methanogenic sludges treating wastewater from terephthalate manufacturing. Appl Environ Microbiol 70 (3):1617–1626

Saunders AM, Albertsen M, Vollertsen J, Nielsen PH (2016) The activated sludge ecosystem contains a core community of abundant organisms. ISME J 10(1):11–20

Sawers RG (2018) O-phthalate derived from plastics' plasticizers and a bacterium's solution to its anaerobic degradation. Mol Microbiol 108(6):595–600

Singh RP, Manchanda G, Li ZF, Rai AR (2017) Insight of proteomics and genomics in environmental bioremediation. In: Bhakta JN (ed) Handbook of research on inventive bioremediation techniques. IGI Global, Hershey. https://doi.org/10.4018/978-1-5225-2325-3

Singh R, Manchanda G, Maurya I, Wei Y (2020) Microbial versatility in varied environments. Springer, Singapore. https://doi.org/10.1007/978-981-15-3028-9

Song B, Palleroni NJ, Häggblom MM (2000) Isolation and characterization of diverse halobenzoate-degrading denitrifying bacteria from soils and sediments. Appl Environ Microbiol 66(8):3446–3453

Subhashini DV, Singh RP, Manchanda G (2017) OMICS approaches: tools to unravel microbial systems. Directorate of Knowledge Management in Agriculture, Indian Council of Agricultural Research. ISBN: 9788171641703. https://books.google.co.in/books?id=vSaLtAEACAAJ

Sun H, Narihiro T, Ma X, Zhang XX, Ren H, Ye L (2019) Diverse aromatic-degrading bacteria present in a highly enriched autotrophic nitrifying sludge. Sci Total Environ 666:245–251

Szewzyk U, Schink B (1989) Degradation of hydroquinone, gentisate, and benzoate by a fermenting bacterium in pure or defined mixed culture. Arch Microbiol 151:541–545

Tasaki M, Kamagata Y, Nakamura K, Mikami E (1991) Isolation and characterization of a thermophilic benzoate-degrading, sulfate-reducing bacterium, *Desulfotomaculum thermobenzoicum* sp nov. Arch Microbiol 155(4):348–352

Valderrama JA, Durante-Rodríguez G, Blázquez B, García JL, Carmona M, Díaz E (2012) Bacterial degradation of benzoate: cross-regulation between aerobic and anaerobic pathways. J Biol Chem 287(13):10494–10508

Yadav TC, Khardenavis AA, Kapley A (2014) Shifts in microbial community in response to dissolved oxygen levels in activated sludge. Bioresour Technol 165:257–264

Yang F, Feng H, Massey IY, Huang F, Guo J, Zhang X (2020a) Genome-wide analysis reveals genetic potential for aromatic compounds biodegradation of Sphingopyxis. BioMed Res Int

Yang Y, Singh RP, Song D, Chen Q, Zheng X, Zhang C, Zhang M, Li Y (2020b) Synergistic effect of *pseudomonas putida* II-2 and *Achromobacter* sp. QC36 for the effective biodegradation of the herbicide quinclorac. Ecotoxicol Environ Safety 188:109826

Yoshifumi S, Yasuyoshi S, Hiroshi U, Yasumitsu U, Akira H, Hideaki Y, Nobuo K (2004) Aerobic and anaerobic toluene degradation by a newly isolated denitrifying bacterium, *Thauera* sp strain DNT-1. Appl Environ Microbiol 70(3):1385–1392

Zhuang L, Tang Z, Ma J, Yu Z, Wang Y, Tang J (2019) Enhanced anaerobic biodegradation of benzoate under sulfate-reducing conditions with conductive iron-oxides in sediment of Pearl River Estuary. Front Microbiol 10:374

Żyszka-Haberecht B, Niemczyk E, Lipok J (2019) Metabolic relation of cyanobacteria to aromatic compounds. Appl Microbiol Biotechnol 103(3):1167–1178

Chapter 15
Metallotolerant Microorganisms and Microbe-Assisted Phytoremediation for a Sustainable Clean Environment

Dina Barman and Dhruva Kumar Jha

Abstract Both natural and anthropogenic activities have upsurged the accumulation of heavy metals in the environment. These pollutants affect the natural ecosystems, and on entering the food chain, they become hazardous to public health. In the polluted soil, where survival of plants and microbes is difficult, metallotolerant microbes can thrive by tolerating the toxic effects of heavy metals. For that, they use diverse survival mechanisms which also assist them to perform bioremediation. In comparison to conventional and physical methods of conversion of the toxic effect of metals to its non-toxic form, bioremediation is a more effective method for retrieving the metal-contaminated environments and convert the degraded area into green covers. Considering the importance, this book chapter sheds light on the mechanism, which encourages the metallotolerant microbes thriving in these metal-rich environments and performs bioremediation.

Keywords Soil · Heavy metals · Metallotolerant microbe · Bioremediation · Microbe-assisted phytoremediation

15.1 Introduction

Land degradation is among the most imperative problems facing the world today. Approximately, one-third of the earth's land surface is degraded, affecting more than 2.6 billion people. Degradation of land is mainly caused by the accumulation of elevated level of heavy metals released due to various geological and anthropogenic activities including mining, industrial emissions, fertilizer erosion from agricultural run-off, sewage, and municipal wastes (Sharma and Nagpal 2020; Romaniuk et al. 2018). It is estimated that heavy metals or metalloids have affected approximately five million sites around the globe (Liu et al. 2018). Various reports are claiming that

D. Barman · D. K. Jha (✉)
Microbial Ecology Laboratory, Department of Botany, Gauhati University, Guwahati, India
e-mail: dhruvakjha@gauhati.ac.in

R. P. Singh et al. (eds.), *Microbes in Microbial Communities*,
https://doi.org/10.1007/978-981-16-5617-0_15

the high content of heavy metals converts fertile land to degraded one in many parts of the world (Sharma and Nagpal 2020). In India, approximately 55% of the geographical area is degraded, and out of which, mining activities have degraded approximately 0.8 mha (MOEF 2001). Most of these heavy metals are generally nondegradable, and the persistent nature of these heavy metals for a longer period in aquatic and terrestrial ecosystems consequently creates harsh conditions for plant growth and development. This is responsible for the conversion of the green landscape of an area into degraded land (Sarma and Barik 2011). Among various heavy metals, Co, Cu, Fe, Mn, Mo, Ni, V, and Zn are required in minute quantities by organisms, but it becomes harmful to organisms with their presence in excessive amounts. There are some other heavy metals like Pb, Cd, Hg, and As which do not have any beneficial effect and regarded as the major threats to organisms (Chibuike and Obiora 2014; Singh et al. 2020a; Barman et al. 2020). The United States Agency for Toxic Substances and Disease Registry (ATSDR) has also listed As, Pb, Hg, and Cd as the major threat to human health (Wood et al. 2016). These heavy metals reduce plant growth by reducing photosynthetic activities, essential enzyme activities, and mineral nutrition (Ojuederie and Babalola 2017; Sivakumar 2016). This issue has attracted worldwide attention as heavy metals enter the food chain and cause detrimental impacts on human health. Heavy metals also enhance the production of reactive oxygen species (ROS) which causes a harmful effect on cells (Ojuederie and Babalola 2017).

Hence, it is imperative to remediate metal-contaminated soil. The treatment of soil using conventional methods including chemical precipitation, electrochemical treatment, and ion exchange is extremely expensive and adversely affects biological activity, soil structure, and fertility (Gupta et al. 2016). In contrast to conventional methods, bioremediation is increasingly gaining importance due to its low cost, simplicity, and better efficiency (Wei et al. 2014; Singh et al. 2017). Bioremediation was first commercially used to clean up the Sun Oil pipeline spill in Ambler, Pennsylvania during 1972 (National Research Council 1993). Since then, bioremediation has received increasing recognition for remediation of a contaminated site like Exxon Valdez and Mega Borg oil spills, Alaskan Oil Spills, and the Iraq–Kuwait war and its consequences (Shannon and Unterman 1993; Pritchard and Costa 1991). The Environmental Protection Agency in 1992 reported 240 cases of bioremediation in the United States (Alexander 1999). Despite the overwhelming advantages, the exact mechanisms by which microbes exist in such a type of environment and decontaminant pollutants are not precisely known.

Under metal stress conditions, some of the soil microorganisms (metallotolerant microorganisms) have developed certain mechanisms to avoid the toxicity arising due to the presence of an array of heavy metals. These mechanisms include an extracellular barrier, efflux of toxic ions from cells, incorporation of heavy metals into complexes by metal-binding proteins, enzymatic transformations of metals, bioaccumulation of the metal ions inside the cell actively or passively, etc. (Romaniuk et al. 2018). They can survive and detoxify heavy metals in polluted soil by expressing different metal-resistant genes (Crupper et al. 1999; Borremans et al. 2001; Yang et al. 2019). Microbes also facilitate bioremediation on interacting with plants termed as microbe-assisted phytoremediation where microbes enhance the process of phytoremediation, as well as increase the growth and biomass of the hyperaccumulating plant at the polluted sites (Tirry et al. 2018). Microbes facilitate the bioavailability of heavy metals to plant by acidification, releasing chelating substances, and changing the redox potentials (Whiting et al. 2001). Besides, microbes facilitate plant growth in heavy metal-contaminated soils by phosphorus solubilization and N_2 fixation and by producing siderophores, phytohormones, antibiotics, and antifungal metabolites. They can also alleviate the ethylene-mediated stress on synthesizing 1-aminocyclopropane-1-carboxylate (ACC) deaminase which can improve plant stress tolerance to metals (Ahemad 2019). Therefore, these beneficial microbial strains can be used as biofertilizers that significantly enhance phytoremediation as well as the growth of plants in heavy metal-contaminated soils (Ahemad 2019).

Further, there are different environmental factors that greatly influence the process of bioremediation, i.e., concentration of contaminants, availability of nutrients, characteristics of soil of the contaminated site. Studies have implied that these factors control the efficiency of bioremediation by various mechanisms. Recently, the importance of genetically engineered microorganisms (GEMs) to remediate contaminated site has increased due to their efficient genetic makeup. But still, the application of GEMs in metal-contaminated site has been limited to laboratory trial only because of regulatory risk and ecological concerns. They also hamper the indigenous population of microbes due to their uncontrolled propagation and horizontal gene transfer. Hence, it is essential to construct the life cycle of GEMs and allowing their death as soon as the pollution level is decreased to minimize their detrimental effects on the native population.

Considering the global significance of bioremediation of heavy metal contaminated sites, it is necessary to critically analyze the various strategies adopted by microbes to survive in metal-contaminated environments and the speculative mechanisms underlying detoxification and/or removal of toxicity from the contaminated site. Additionally, the role of omics and multi-omics approaches in bioremediation also needs to be delineated. Moreover, we also analyzed different relevant published data on the contribution of microbes to remediate the heavy metal contaminated environments.

15.2 Effects of Heavy Metals

Heavy metals are ions with partially or filled *d*-orbital having an atomic weight ranging between 63.5 u and 200.6 u, specific gravity of greater than 5. The physicochemical properties like pH, organic matter, clay contents, inorganic anions, and cations of soil get changed due to the presence of heavy metals (Sarma and Barik 2011; Lauwerys et al. 2007). The toxic effects of heavy metals also change the population size, diversity, and activities of soil microbiota, which in due course affect the soil enzymatic activities, recycling of plant nutrients, and ultimately hamper plant growth (Karaca et al. 2010; Wang et al. 2007). It is interesting to note that plants growing in metal contaminated soils show abnormalities in their biochemistry and physiology (Chibuike and Obiora 2014). For example, the presence of arsenic (As) in the soil leads to decreasing seed germination, reduction of seedling height, leaf area, and declining production of dry matter in *Oryza sativa*. Arsenic (As) also causes chlorosis, wilting, and stunted growth in *Brassica napus* while it inhibits the rate of transpiration of *Avena sativa* seedlings. Similarly, the presence of Pb in soil results in stunted growth, reduced germination percentage, protein content, and biomass of *Zea mays*, and inhibited ribulose-1,5-bisphosphate carboxylase/oxygenase activity that affected CO_2 fixation in *Avena sativa* (Chibuike and Obiora 2014). These effects are attributed to the inhibition of vital metabolic processes of plants like photosynthesis, water absorption, mitosis that sometimes lead to the death of the affected plants (Shun-hong et al. 2009). It is worth mentioning that, due to mining activities, generally soil become polluted not only with one heavy metal but with a combination of heavy metals which results in more harmful effects to plants (Chibuike and Obiora 2014). It was observed that the combination of Pb and Cu at high (1000 mg/kg each) and low concentrations (500 mg/kg) in soil cause fast death of the leaves and stems of *Lythrum salicaria* (Nicholls and Mal 2003). The uptake of heavy metals by plants and its consequent accumulation along the food chain also caused depletion of essential nutrients in the body that further resulted in cancer in humans, decreasing immunological defenses, intrauterine growth retardation, and disabilities associated with malnutrition (Ojuederie and Babalola 2017).

15.3 Bioremediation

Bioremediation is the eco-friendly, efficient technique to remove heavy metals from the contaminated site (Dixit et al. 2015). Bioremediation is of two main types, i.e., in situ or ex situ. In situ bioremediation involves a process where the indigenous microorganisms are stimulated to degrade heavy metals on supplying nutrients and oxygen with negligible or not interfering the soil structure. This technique has been successfully used to treat metal-contaminated site and is found to be less expensive and superior than ex situ bioremediation (Roy et al. 2015).

The in situ bioremediation process can be enhanced by chemotaxis, and the formation of biosurfactants or biofilm. Chemotaxis is a phenomenon that guides microbes to move toward or away in response to a chemical stimulus which helps in decontamination of pollutants (Ahmad et al. 2020). This behavior is not only useful for nutritional requirements but also required for their interaction with the environment. Microbes generally move toward a chemical when they utilize it for their growth and move away from a chemical when it is toxic. Microbes also form biofilm or biosurfactants to survive in metal-contaminated environments and thus enhance their bioremediation potential. It has been reported that *Pseudomonas* sp. produces biofilm to tolerate the toxicity of cadmium ion, and *Rhodotorula mucilaginosa* produces biofilm to remove toxicity of heavy metals (Tarekegn et al. 2020; Chien et al. 2013). In situ bioremediation can also be enhanced by improving native microorganisms by genetic engineering.

Ex situ bioremediation involves the transfer of contaminated pollutants from the original site to a different location for the treatment depending on the type of pollutants, cost of treatment, degree of pollution, and geology of the polluted site (Ojuederie and Babalola 2017). Based on the physical condition of the pollutant, ex situ bioremediation is of two types, i.e., solid-phase bioremediation and semi-solid-phase bioremediation. Solid-phase bioremediation includes biopile, landfarming, and composting. Landfarming is the technique where contaminated soil is excavated from the site and transported to a prepared bed to allow aerobic degradation by autochthonous microbes. Sometimes instead of transferring contaminated soil, they are treated on that site; hence, landfarming is also regarded an in situ bioremediation technique. In composting, excavated soil is mixed with compost to allow effective growth of native isolates and to permit bioremediation of contaminated soil. Bioremediation by biopile includes piling of contaminated soil and subsequently maintaining favorable condition for native microorganisms (Pande et al. 2020). Semi-solid-phase bioremediation is performed in a sludge bioreactor where polluted soil is mixed with liquid that favors better interaction between native microorganisms and pollutants (da Silva et al. 2020).

The efficiency of bioremediation depends upon several biotic and abiotic factors (Brar et al. 2006). The microorganism capable of performing degradation is affected by the characteristics of contaminants, chemical condition of the surrounding environment, and the other indigenous microflora and fauna. The competition between degrading microorganisms with other indigenous microflora and fauna for carbon sources leads to deficient conditions of nutrients, oxygen, and ultimately hamper their growth and to perform bioremediation successfully. The condition can be overcome by the application of bioaugmentation, repeated inoculation, and pre-induction (Pande et al. 2020). Bioremediation is also affected by various abiotic factors of the contaminated site. One of the most important factors is pH which has a high impact on biological activity (Singh et al. 2016a, b). Generally, bioremediation rate increased in the pH range 6.5–8.5, and it is hampered above and below this. Another important factor is temperature, and 30–40 °C is optimum for biodegradation. It has been observed that degradation of the contaminant is affected by very low or very high temperatures. The water-holding capacity of soil also affects the bioremediation process. Water is essential for the transportation of nutrients into microbes, oxygen exchange, and ejection of metabolic waste which directly influence its cell growth and efficiency to perform bioremediation. However, an excessive amount of water in soil prevents oxygen exchange and thus hamper bioremediation. Moreover, an adequate amount of nutrients are required for the growth of cells and their efficiency of biodegradation. Generally, metal-contaminated site deficiency of nutrients hampers the process of biodegradation, and it can be overcome by adding the nutrients in their useable form (Pande et al. 2020).

Though bioremediation has advantages over conventional techniques like less expensive method, it can be done on site, can permanently eliminate waste, and has more public acceptance (Boopathy 2000); however, the process of bioremediation is linked with some limitations like site-specificity where bioremediation approaches that are successful at one site may not be fruitful in other sites. Second, the microbe-mediated bioremediation process may fail in the field even it is successful under lab condition. Third, the uncertain mechanism of microorganisms is inhabiting in contaminated environments (Malla et al. 2018). Therefore, it is important to gather knowledge on the strategy used by microorganisms to grow in contaminated environments and subsequently perform bioremediation.

15.3.1 Microbial Strategies to Strive in Metal-Contaminated Environment and Underlying Mechanism

Most of the heavy metals disrupt the cell membrane of microorganisms, but the one capable of bioremediation is generally adapted to a range of resistance mechanisms through which they can utilize various toxic compounds as a source of energy for their growth and development and/or convert them into nontoxic products (Wei et al. 2014; Brar et al. 2006). Metallotolerant microbes tolerate the toxicity of heavy metals and perform bioremediation by different mechanisms like exclusion by permeability barrier, effluxing metal ions, oxidizing metals, enzymatic conversion of metals, intracellular and extracellular metal sequestration, producing metal chelators like metallothioneins and biosurfactants (Igiri et al. 2018).

Microbes can block the entry of heavy metals into the cell by using their extracellular membrane, i.e., plasma membrane, cell wall, and capsule. The extracellular surfaces are negatively charged which adsorb the positively charged heavy metals onto the binding sites of the cell wall by electrostatic interaction, ion exchange, precipitation, redox process, and surface complexation (Ayangbenro and Babalola 2017; Diep et al. 2018) (Fig. 15.1). On binding the heavy metals with the cell surface, microbes reduce their toxicity by transforming them from one oxidation state to another and thus prohibit the transportation of metal ions into the cytoplasm (Ayangbenro and Babalola 2017; Singh et al. 2020b). The phenomenon of uptake of heavy metals through surface complexation to the extracellular surface of microorganisms is termed as biosorption (Diep et al. 2018). The capacity of biosorption is influenced by three factors: (1) characteristics of the metal ion like an ionic ray, atomic weight, valence; (2) conditions of the environment such as pH, temperature, ionic strength, contact time, biomass concentration; and (3) the nature of the biosorbent (Perpetuo et al. 2011). The method of biosorption is of two types, i.e., metabolism-independent biosorption and metabolism-dependent biosorption. Metabolism-dependent biosorption mainly takes place within viable cells where metabolism occurs. Here metals get transported across the cell membrane and yield intracellular accumulation. However, metabolism-independent biosorption is mainly occurring on the exterior of cells and is a relatively rapid and reversible process (Perpetuo et al. 2011). If heavy metals enter into cytoplasm of the cell,

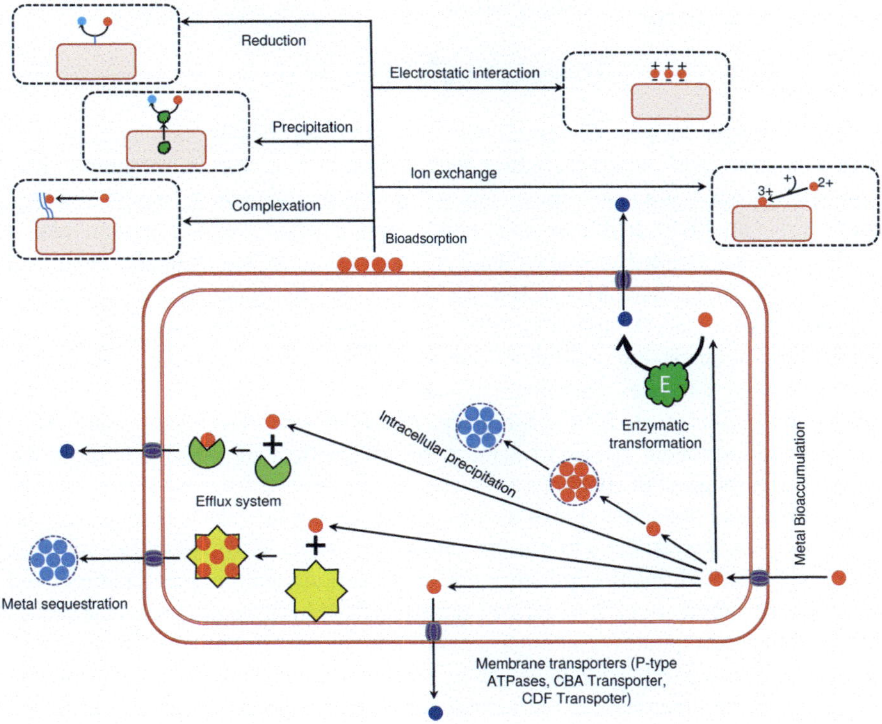

Fig. 15.1 A generalized illustration of different mechanisms involved in tolerance to toxic metals in bacteria

metallotolerant microbes efflux metal ions from the cytoplasm using three different proteins, i.e., resistance nodulation-cell division (RND superfamily) proteins, cation diffusion facilitators (CDF family), and P-type ATPases (Nies 2003).

Biotransformation is another mechanism by which microbes can detoxify the toxic effects of heavy metals. It includes oxidation, reduction, methylation, and alkylation or by synthesizing and producing metal-binding proteins such as metallothioneins (MTs) (Valls and de Lorenzo 2002) (Fig. 15.1). For example, *Alcaligenes faecalis* becomes resistant to toxic effects of arsenite [As(III)] on oxidizing arsenite to arsenate [As(V)] (Valls and de Lorenzo 2002).

$$\frac{1}{2}\ H_3AsO_3 + \frac{1}{4}\ O_2 \rightarrow \frac{1}{4} H_2AsO_4^- + \frac{1}{4} HAsO_4^{2-} + \frac{3}{4}\ H^+$$

Iron oxidizing bacteria reduce Fe(III) to Fe(II) abiotically; mercury (Hg^{2+}) into less toxic and volatile mercury (Hg°) by mercury reductase enzyme (Lloyd 2003; Valls and de Lorenzo 2002).

$$Hg^{2+} + NADPH \xrightarrow{\text{mercury reductase}} Hg^0 + H^+ + NADP^+$$

$$Fe_2O_3 + 3C \rightarrow 2FeO + 3CO$$

Some bacteria such as Clostridia, Methanogens, and Sulfate-reducing bacteria methylate a range of metals including lead, cadmium, tin, arsenic, selenium, tellurium, and mercury; as a result, the metals get transformed into their volatile dimethyl form (Igiri et al. 2018). In the process of alkylation, an alkyl group other than methyl group is directly bonded to metals through a carbon atom, for example, $As(C_2H_5)(CH_3)_2$, $As(C_2H_5)_3$, $As(C_2H_5)_2(CH_3)$, $Sb(C_2H_5)_3$ and by which it can tolerate the toxic effects of metals (Krupp et al. 1996). Microbes can also remove the toxicity of metals by synthesizing metallothioneins (MTs). For example, *Rhizobium leguminosarum* becomes cadmium resistant by sequestering cadmium ions by glutathione (Lima et al. 2006). Similarly, *Pseudomonas aeruginosa* strain WI-1 having metallothionein (BmtA) tolerates the toxic effect of lead by intracellular sequestration (Naik et al. 2012).

Microbes also precipitate lethal metal compounds intracellularly and/or extracellularly and thus convert them to less toxic form (Igiri et al. 2018). Metal precipitation is mainly achieved by dissimilatory metal reduction, sulfide precipitation, and phosphate precipitation (Valls and de Lorenzo 2002). In dissimilatory metal reduction, microbes precipitate metals such as uranium, selenium, chromium, technetium, and gold which is unrelated to its intake by microbial catalyst (Valls and de Lorenzo 2002). In sulfide precipitation, sulfur-reducing bacteria (SRB) precipitate metal [U(VI), Cr(VI), Tc(VI), Pd(II), and As(V)] in the form of metal sulfide on producing hydrogen sulfide (Igiri et al. 2018).

$$U(VI) + H_2S \rightarrow US_2$$

Similarly, some of the bacteria including *Vibrio harveyi, Citrobacter* sp. precipitate metal ions by producing highly insoluble metal phosphates (Valls and de Lorenzo 2002).

$$U(VI) + H_3PO_4 \rightarrow UP_2O_8$$

$$As(V) + H_3PO_4 \rightarrow As_3(PO_4)_5$$

Additionally, microbes like *Ralstonia metallidurans*, *Pseudomonas aeruginosa,* and *Alcaligenes eutrophus* detoxify toxic metals by forming metal-siderophore complexes. Siderophores are low-molecular-weight chelating agents having a strong affinity for ferric iron and thus produce Fe(III)-siderophore complexes. They also possess an affinity for other non-iron metals, e.g., copper, manganese, molybdenum, vanadium, zinc, which stimulate microbes to produce zincophores, chalkophores (copper-binding metallophores), etc. that can detoxify heavy metals. Microbes including various bacteria and yeast produce biosurfactants like rhamnolipids, lipopolysaccharides, exocellular polymeric surfactants in the form of polysaccharides, proteins, lipoproteins by which they can solubilize and precipitate different heavy metals such as Cd, Pb, and Zn (Mosa et al. 2016; Valls and de Lorenzo 2002).

Microbes survive in a metal-contaminated niche by expressing metal-resistant genes generally associated with plasmids (Dave et al. 2020). There are certain operons, i.e., *cad* operon, *czc* operon, *ncc* operon, *mer* operon, *cop* operon, *aox*/*ars* operon present in the plasmid of microbes by which they can tolerate the toxicity of Cd, Zn, Ni, Hg, Cu, and As metals, respectively (Dave et al. 2020). The *cad* operon and *czc* operon are generally found in *Staphylococcus* sp. and *Pseudomonas aeruginosa,* respectively, by which the bacteria confer Cd resistance (Das et al. 2016). In a study, it was shown that *Ralstonia metallidurans* ch34 can resist Cu, Co, and Zn by *czc* operon (Dave et al. 2020). Similarly, *chrA* gene can encode the chromate reductase protein present in *Arthrobacter aurescens*, *Bacillus atrophaeus*, *Pseudomonas putida*, *Rhodococcus erythropolis* by which they can transform toxic Cr(VI) to the non-toxic Cr(III) with co-factors NADH or NADPH (Das et al. 2016). Lead is another toxic metal, and the *pbr* operon (lead resistance operon) found in the endogenous pMOL30 megaplasmid confers resistance to lead. The operon consists of one regulatory gene (*pbrR*), and many structural genes *pbrT*, *pbrA*, *pbrB*, *pbrC*, *pbrD* help microbes to resist lead. In the presence of lead toxicity, transcription of *pbrABCD* operon from *pbrA* promoter is induced which is regulated by *pbrR* (Borremans et al. 2001). Interestingly another gene *pbrU* was discovered in *Ralstonia metallidurans* by Monchy et al. (2007), which gets induced in the presence of lead. Microbes can also resist the toxicity of mercury by expressing two different operons, i.e., narrow-spectrum *mer* operon and the broad-spectrum *mer* operon (Silver and Phung 2013). The narrow-spectrum *mer* operon found on the transposons Tn5037 consists of the genes *merR*, *merT*, *merC*, *merF*, *merP*, and *merD*. The operon gets induced in the presence of Hg^{2+} that provides resistance to the metal. Similarly, the broad-spectrum *mer* operon contains the genes *merE*, *merG*, and *merB* in addition to the genes present in narrow-spectrum *mer* operon that protect from organic mercury (Barkay et al. 2003).

Microbes also occupy and adapt themselves in contaminated niche by horizontal gene transfer (HGT). The genes encoding bioremediation transfer through the action of conjugative plasmids, transposable elements, and "integrative and conjugative transposons." An interesting example of horizontal gene transfer is that the *pheBA* operon encodes enzymes involved in phenol catabolism which are originated from the *Pseudomonas* sp. EST1001. The operon was transferred to *P. putida* PaW85 by conjugation and released into river water contaminated with phenolics, originating from a fire in an oil shale mine for bioremediation. After 6 years, though the *P. putida* PaW85 was absent in that river water nonetheless the operon was detected in nine *Pseudomonas* strains in the watershed (Perpetuo et al. 2011). Another conjugative plasmid, i.e., IncP-specific plasmid sequences that are present in heavy metal contaminated soil gets mobilized to bacteria and offers resistance capacity of bacteria to survive in that environment by HGT (Ansari et al. 2008). Smalla et al. (2006) detected the abundance of IncP-1β plasmids and mercury-resistance genes in mercury-polluted river sediments which were further detected in bacterial communities of that area indicating the role of HGT of IncP-1β plasmid.

15.3.2 Diversity of Metallotolerant Microorganisms

Several metal-tolerant microorganisms including bacteria, fungi, and algae have been used to remediate heavy metal–contaminated environments. Among the microorganisms, bacteria belonging to *Firmicutes*, *Proteobacteria*, and *Actinobacteria* play an important role in bioremediation due to their size, ubiquity, and ability to grow under controlled conditions as well as to their flexibility to varied environmental conditions (Igiri et al. 2018). They not only detoxify heavy metals in contaminated soils but also promote the growth and development of plants (Mishra et al. 2017). For the past few years, several articles have been published based on the use of bacteria for bioremediation purposes. Alboghobeish et al. (2014) isolated nickel-resistant bacteria from industrial waste waters belonging to *Cupriavidus* sp. ATHA3, *Klebsiella oxytoca* ATHA6, and *Methylobacterium* sp. ATHA7 which were found to remediate the Ni-polluted waste water and sewage. Bacteria can also successfully survive in mixed culture; hence, consortia of cultures can also be used for biosorption of metals and are found to more appropriate for field application (Igiri et al. 2018) (Table 15.1).

Fungi are also used as biosorbents for the removal of heavy metals. Both active and dead fungal cells play an important role in the adhesion of inorganic chemicals. Active fungal cells of *Saccharomyces cerevisiae*, *Aspergillus parasitica,* and *Cephalosporium aphidicola* were reported to detoxify Zn(II), Cd(II), and Pb (II) (Ayangbenro and Babalola 2017). White-rot fungi like *Phanerochaete chrysosporium*, *Trametes versicolor*, *Bjerkandera adjusta,* and *Pleurotus* sp. also transform a variety of organic pollutants by various ligninolytic enzymes. Marine fungi use enzymes to tolerate high concentrations of heavy metals like Pb and Cu (Deshmukh et al. 2016). The dead fungal biomass can also detoxify the toxic effect of metals. For that, the non-living biomass of *Rhizopus oryzae* and *Saccharomyces cerevisiae* use adsorption mechanism to convert toxic Cr(VI) to less toxic or non-toxic Cr(III) where anionic chromate ion binds to the cationic amines of the cell wall. However, the dead biomass of *Aspergillus niger* can reduce Cr(VI) to Cr (III) through a redox reaction (Park et al. 2005) (Table 15.1).

Algae are also used for bioremediation of heavy metal polluted effluent where living algae are found to be more complex than non-living algae. Living algae absorb heavy metal ions during the growth phase, and it is considered to be an intracellular process; however, the process of sorption illustrates large variations based on their growth phase. Along with this, the growth of algae is also affected by several environmental factors that directly influence biosorption. In contrast, non-living algal cells absorb metal ions on the surface of the cell membrane, and it is considered an extracellular process (Zeraatkar et al. 2016). For example, Tuzen et al. (2009) investigated the potentiality of *Ulothrix cylindricum* in the removal of arsenic ion

Table 15.1 Heavy metal detoxification from metal-contaminated sites by various microorganisms

Microbial species	Microbial class	Bioremediate toxicity of metal	Mechanism used	References
Sargassum fluitans	Phaeophyceae	Au	Biosorption	Niu and Volesky (2000)
Bacillus subtilis	Bacilli			
Penicillium chrysogenum	Eurotiomycetes			
Pilayella littoralis	Phaeophyceae	Al, Cd, Co, Cr, Cu, Fe, Ni, Zn	Biosorption	Carrilho and Gilbert (2000)
Penicillium canescens	Eurotiomycetes	Pb, Cd, Hg, As	Biosorption	Say et al. (2003)
Ecklonia maxima	Phaeophyceae	Cu, Pb, Cd	Biosorption	Feng and Aldrich (2004)
Gigartina salicornia	Florideophyceae	Cd	Biosorption	Hashim and Chu (2004)
Sargassum baccularia	Phaeophyceae			
Oscillatoria angustissima	Cyanophyceae	Cu, Co, Zn	Biosorption	Mohapatra and Gupta (2005)
Ulva reticulata	Ulvophyceae	Cu, Co, Ni	Biosorption	Vijayaraghavan et al. (2005)
Chlorella miniata	Trebouxiophyceae	Cr	Biosorption	Han et al. (2006)
Spirogyra sp.	Zygnematophyceae	Cr	Biosorption	Bishnoi et al. (2007)
Ceramium virgatum	Florideophyceae	Cd	Biosorption	Sari and Tuzen (2008)
Pseudomonas veronii	Pseudomonadaceae	Cd, Zn, Cu	Biosorption	Vullo et al. (2008)
Ulothrix cylindricum	Ulvophyceae	As	Biosorption	Tuzen et al. (2009)
Cladophora hutchinsiae	Cladophoraceae	Se	Biosorption	Tuzen and Sari (2010)
Aspergillus versicolor	Eurotiomycetes	Cr, Ni, Cu	Bioaccumulation	Tastan et al. (2010)
Cystoseira stricta	Phaeophyceae	Pb	Biosorption	Iddou et al. (2011)
Aspergillus fumigatus	Eurotiomycetes	Pb	Biosorption	Ramasamy et al. (2011)
Kocuria flava	Actinomycetia	Cu	Precipitation	Achal et al. (2011)
Burkholderia dabaoshanensis	Betaproteobacteria	Cd	Biosorption	Zhu et al. (2012)
Bacillus cereus	Bacilli	Cr	Enzyme-mediated	Dong et al. (2013)
Acinetobacter sp.	Gammaproteobacteria	Cr	Biosorption	Bhattacharya et al. (2014)
Spirulina platensis	Cyanophyceae	Cu	Biosorption	Anastopoulos and Kyzas (2015)
Spirulina maxima	Cyanophyceae	Cr	Bioaccumulation	Singh et al. (2016a, b)

(continued)

Table 15.1 (continued)

Microbial species	Microbial class	Bioremediate toxicity of metal	Mechanism used	References
Cladophora sp.	Ulvophyceae	Pb, Cu	–	Ojuederie and Babalola (2017)
Spirogyra sp.	Zygnematophyceae	Pb, Cu		
Hydrodictylon sp.	Chlorophyceae	As		
Oedogonium sp.	Chlorophyceae	As		
Rhizoclonium sp.	Ulvophyceae	As		
Aspergillus fumigatus	Eurotiomycetes	Pb		
Rhizopus oryzae MPRO	Mucoromycetes	Cr		
Saccharomyces cerevisiae	Saccharomycetes	Pb, Cd		
Bacillus cereus strain XMCr-6	Bacilli	Cr		
Kocuria flava	Actinomycetia	Cu		
Sporosarcina ginsengisoli	Bacilli	As		
Enterobacter cloacae B2-DHA	Gammaproteobacteria	Cr		
Gemella sp.	Bacilli	Pb, Cr, Cd	Plasmid mediated	Marzan et al. (2017)
Micrococcus sp.	Actinomycetia	Pb, Cr, Cd	Plasmid mediated	
Hafnia sp.	Gammaproteobacteria	Cd	Plasmid mediated	
Bacillus sp.	Bacilli	Cr	Reduction	Ontanon et al. (2018)
Aspergillus niger	Eurotiomycetes	Cr, Hg, Pb, Co	Biosorption	Acosta-Rodriguez et al. (2018)
Pseudomonas fluorescens	Pseudomonadaceae	Cr	Biodegradation	Kalaimurugan et al. (2020)
Bacillus safensis	Bacilli	Cr	Biodegradation	
Lactobacillus plantarum	Bacilli	Ni, Cr	Biosorption	Ameen et al. (2020)
Pseudomonas aeruginosa	Pseudomonadaceae	Cd, Pb	–	Oziegbe et al. (2021)
Klebsiella edwardsii	Gammaproteobacteria	Cd, Pb		

(As III), *Ulva lactuca* in the detoxification of Cd(II) and Pb(II) (Sari and Tuzen 2008) (Table 15.1).

15.4 Role of Plants in Bioremediation

Phytoremediation is another cost-effective and eco-friendly remediation method where plants are used to remove contaminants in the environment. This approach can also minimize the threat of dispersion of contaminant and protects the original ecotype (Awa and Hadibarata 2020). Phytoremediation can convert degraded land to be used for the cultivation of crops; hence, it has economic value also (Awa and Hadibarata 2020). To degrade organic contaminants, plants use mechanisms like phytoextraction, phytostabilization, rhizodegradation, rhizofiltration, phytodegradation, and phytovolatilization while phytostabilization, rhizofiltration, phytoaccumulation, and phytovolatilization are used to degrade inorganic contaminants (Tangahu et al. 2011). Phytoextraction involves the uptake and movement of heavy metals from soil to above-ground parts of the plants via roots. It removes metals like nickel (Ni), zinc (Zn), and copper (Cu) (Ojuederie and Babalola 2017). Like phytoextraction, phytofiltration also involves the accumulation of metal contaminants by the use of roots of plants (rhizofiltration), seedlings (blastofiltration), or excised plant shoots (caulofiltration) from aqueous wastes. Rhizofiltration mainly aims to clean extracted groundwater, surface water, and wastewater with low concentrations of contaminants (Sharma and Pandey 2014). Phytostabilization involves the absorption of heavy metals on plant roots or retention within the rhizosphere that rendering them harmless and prevent these pollutants from spreading in the environment (Ojuederie and Babalola 2017). Phytovolatilization, on the other hand, deals with the conversion of soil contaminants to their volatile form by plants and associated rhizosphere microorganisms and their consequent release into the atmosphere. Degradation of organic contaminants using plant enzymes such as nitroreductases and dehalogenases is called Phytodegradation while phytostimulation deals with the addition of microbial activity to degrade organic contaminants by exudates from plant roots (Ojuederie and Babalola 2017).

15.4.1 Limitations of Phytoremediation

Although phytoremediation is a promising approach to remediate metal-contaminated soil or water, this method suffers from some limitations. The method of phytoremediation applies only to low or moderately contaminated soils where the plant produces a significant amount of biomass. In highly contaminated soil, the toxic effects of contaminates hinder plant metabolism on reducing the biochemical process that is essential for the degradation and/or uptake of the contaminants. Second, the selection of plants for phytoremediation is very important especially concerning root depth and age (Chirakkara et al. 2016). Generally, the roots of herbaceous species may reach up to 1 m, bushes from 1 to 3 m, and trees up to 10 m. It is reported that phytoremediation is more successful in the top 50 cm^{-1} m layer (Cameselle et al. 2013). The growth of plants is influenced by climatic and hydrologic conditions (Tangahu et al. 2011), and their physiological activities depend on their age. Usually, the roots of a young plant absorb more ions than their older counterparts. The third limitation is related to the uptake and translocation of metals. The metals must be in bioavailable form, and if the metal is tightly bound to the organic portion of the soil, sometimes it may not available to plants. Additionally, the method is slow in comparison to other remediation technologies, and it may take more than 1 year of treatment (Chirakkara et al. 2016).

15.5 Microbe-Assisted Phytoremediation

To overcome the limitations of phytoremediation, recently, microbe-assisted phytoremediation has been used by many researchers (Rathore et al. 2019; Yamaji et al. 2016; Phieler et al. 2015). The metal-tolerant plant growth-promoting microorganisms (MT-PGPMs) have the potential to enhance the biomass production of plants and better tolerance of plants to heavy metals and help in revegetation and restoration of fertility of the metal-contaminated areas (Abou-Shanab et al. 2006). The microbiome can improve the process of phytoremediation through (1) proton (H^+) release that mediated change in soil pH or formation of organo-metal complexes; (2) binding compounds present in the cell (e.g., organic acids, phytochelatins, and amino acids); (3) influencing redox potential through enzyme-mediated transfer; and (4) enhancing microbial activity in the rhizosphere (Sessitsch et al. 2013; Singh et al. 2011). Further, MT-PGPMs induce plant growth directly by secreting enzymes, plant growth-promoting substances, and solubilization of nutrients (Ma et al. 2013). It is reported that by inoculating the effective isolates to the roots of the growing plants, heavy metal accumulation of inoculated plants increased from 66 to 135% in roots and 22 to 64% in the above-ground parts (Anwar et al. 2012).

15.5.1 Mechanisms Behind the Microbe-Assisted Phytoremediation

The plants growing in metal-contaminated areas attract the beneficial metal-tolerant microorganisms to form plant–microbe inter-relationship for better phytoremediation. For that plant releases signals or root exudates (chemotaxis) to their adjoining soil microorganisms (Bulgarelli et al. 2013). As a result, the microbes develop symbiotic/mutualistic associations with plants and live as endophytes or free-living rhizospheric microbes. Microbes release protons (H^+) and enzymes which help in acidification and electron transfer in the rhizosphere and thus enhance the bioavailability of metal to plants (Ma et al. 2016). MT-PGPMs alter the soil pH by releasing organic acids including gluconic acid, oxalic acid, and malic acid which form complex with insoluble heavy metals and make it soluble and consequently available to plants and microbes (Mishra et al. 2017). In this connection, Kim et al. (2010) have reported that translocation and bioaccumulation of metals are significantly enhanced by citric and oxalic acid, suggesting that these acids could be used as natural chelating agents for better phytoextraction. The release of metal chelators like metallothione, phytochelatin from plant root exudates and MT-PGPMs also contribute to the detoxification of heavy metals. MT-PGPMs release phytohormones such as indoleacetic acid (IAA), cytokinins, gibberellins, abscisic acid that govern the hormonal balance in plants as a response to metal stress (Ma et al. 2016; Ullah et al. 2015). MT-PGPMs produce ACC deaminase enzyme that hydrolyzes ACC which is the immediate precursor of the hormone ethylene in plants to ammonia and α-ketobutyrate and thus reduce the metal stress on lowering the level of ethylene inside the plants (Glick 2014). There is another mechanism adopted by MT-PGPMs under metal stress conditions to enhance plant growth through the production of antimicrobial enzymes (Saima et al. 2013), and polysaccharides (Naseem and Bano 2014) (Table 15.2, Fig. 15.2).

These play a major role to overcome the negative impact of both biotic (fungi or harmful insects) and abiotic stresses (waterlogging, drought, salt stress, and metals toxicity). Hence, MT-PGPMs can speed up phytoremediation and promote plant growth and development by resorting to any one or more of the above mechanisms. For that reason, MT-PGPMs can be effectively utilized in metal-contaminated environments for the phytoremediation. For instance, experiments assessed by Becerra-Castro et al. (2011) have shown that inoculation of Ni-resistant rhizosphere bacteria *Arthrobacter nitroguajacolicus* in Ni hyperaccumulator *Alyssum serpyllifolium* subsp. *lusitanicum* increases the higher translocation and concentration of Ni in the shoot. Similarly, on inoculating *Psychrobacter* sp., SRS8 in *Ricinus communis* and *Helianthus annuus* was found to enhance the phytoextraction and growth of the plants in Ni-contaminated soils (Sessitsch et al. 2013).

Arbuscular mycorrhizal fungal (AMF) colonization in the plant roots also increases heavy metal tolerance capacity of plants in metal-contaminated soils by depositing metals within cortical cells, binding metals to the cell wall or mycelium as well as sequestering them in their vacuole or other organelles, on releasing heat-shock protein and glutathione, precipitating or chelating metals in the soil matrix by producing glycoprotein or making phosphate-metal complexes inside the hyphae, and reducing the strength of metals by heightened root and shoot growth (Emamverdian et al. 2015; Manchanda et al. 2017). In addition to increasing heavy metal tolerance capacity, AMF improves plant growth by different mechanisms through releasing growth-promoting substances, hormones, improving systemic resistance, synergistic interaction with other soil microorganisms, increasing formation and stabilization of soil aggregates (Yao et al. 2005). Interaction of mycorrhizal inoculation (*Glomus mosseae*) with maize growing in HM

Table 15.2 Combination of hyperaccumulator plants and metal-tolerant microbes applied in microbial-aided phytoremediation of metal overburdened soil

Microbial species	Plant species	Bioremediate toxicity of metal	Effect of microbes on plants	References
Variovorax paradoxus	*Brassica juncea*	Cd	Stimulate root elongation	Belimov et al. (2005)
Rhodococcus sp.				
Flavobacterium sp.				
Pseudomonas sp. LK9	*Solanum nigrum*	Cd	Increases uptake of Cd in shoot and root	Sheng et al. (2008)
Enterobacter aerogenes	*Brassica juncea*	Ni, Cr	Strains enhance plant biomass, protein, and chlorophyll content	Kumar et al. (2009)
Rahnella aquatilis				
Microbacterium sp.	*B. napus*	Cu	Increases the root length	He et al. (2010)
Pseudomonas chloraraphis				
Arthrobacter sp.				
Pseudomonas aeruginosa	*Cicer arietinum*	Cr	Enhances dry matter accumulation, symbiotic attributes, grain yield, and protein content of chickpea	Oves et al. (2013)
Enterobacter ludwigii	*Helianthus annuus*	Co, Pb, Zn	Enhances dry matter accumulation	Arunakumara et al. (2014)
Rahnella sp.	*Amaranthus* sp.	Cd	Significant increase in dry weight was observed with various Cd concentrations	Yuan et al. (2014)
Klebsiella oxytoca	*Helianthus annuus*	Co, Pb, Zn	Increases uptake and translocation from root to shoot	Arunakumara et al. (2015)
S. acidiscabies	*Sorghum bicolor*	Cd, Co, Ni, Sr	Increases the phytoextraction	Phieler et al. (2015)
S. tendae				
Rhizophagus irregularis				
Phialocephala fortinii	*Clethra barbinervis*	Cu, Ni, Zn, Cd, Pb	Enhancement, promotion of nutrient uptake	Yamaji et al. (2016)
Rhizodermea veluwensis				
Rhizoscyphus sp.				

(continued)

Table 15.2 (continued)

Microbial species	Plant species	Bioremediate toxicity of metal	Effect of microbes on plants	References
Sphingomonas macrogoltabidus	*Alyssum murale*	Ni	Ni mobilizer, siderophore producer, and phosphate solubilizer; increases Ni uptake in shoots by 17%	Waigi et al. (2017)
Sphingomonas sp.	*Solanum nigrum*	Cd	AA producer, displays ACCD activity; induces heavy metal tolerance to Cd, Zn, Pb, and Cu	
Ensifer adhaerens	*Betula celtiberica*	As	Enhances plant growth and better accumulation of As	Mesa et al. (2017)
Pseudomonas aeruginosa	*Brassica juncea*	Cd	Increases root and shoot biomass	Rathore et al. (2019)
Pseudomonas tolaasii ACC23	*B. napus*	Cd	Increases root and shoot growth and the Cd content in plant	
Achromobacter xylosoxidans	*B. juncea*	Pb, Cu	Increases root and shoot length, fresh and dry weight and improves metal uptake	
Microbacterium sp. G16	*B. napus*	Pb	Increases root elongation of inoculated rape seedlings and total Pb accumulation	
Pseudomonas fluorescens G10	*B. napus*	Pb	Increases root elongation of inoculated rape seedlings and total Pb accumulation	
Pseudomonas sp. RJ10	*B. napus*	Cd	Increases uptake of Cd by plant, enhances shoot and root dry weight	
Bacillus sp. RJ16	*B. napus*	Cd	Increases uptake of Cd by plant, enhances shoot and root dry weight	
Azotobacter chroococcum	*B. juncea*	Pb, Zn, Cu	Increases the removal of Pb, Zn, Cu	
Bacillus subtilis SJ-101	*B. juncea*	Ni	Increases the accumulation of Ni by 1.5-fold and increased plant biomass	
Acaulospora sp.	*Ricinus communis*	Pb	Phytostabilization to ameliorate Pb pollution and decreasing its ecological risk	Gonzalez-Chavez et al. (2019)
Funneliformis mosseae				
Gigaspora gigantea				
Serratia sp.	*Zea mays*	Zn	Zn toxicity reduced and enhanced the plant growth parameters	Jain et al. (2020)

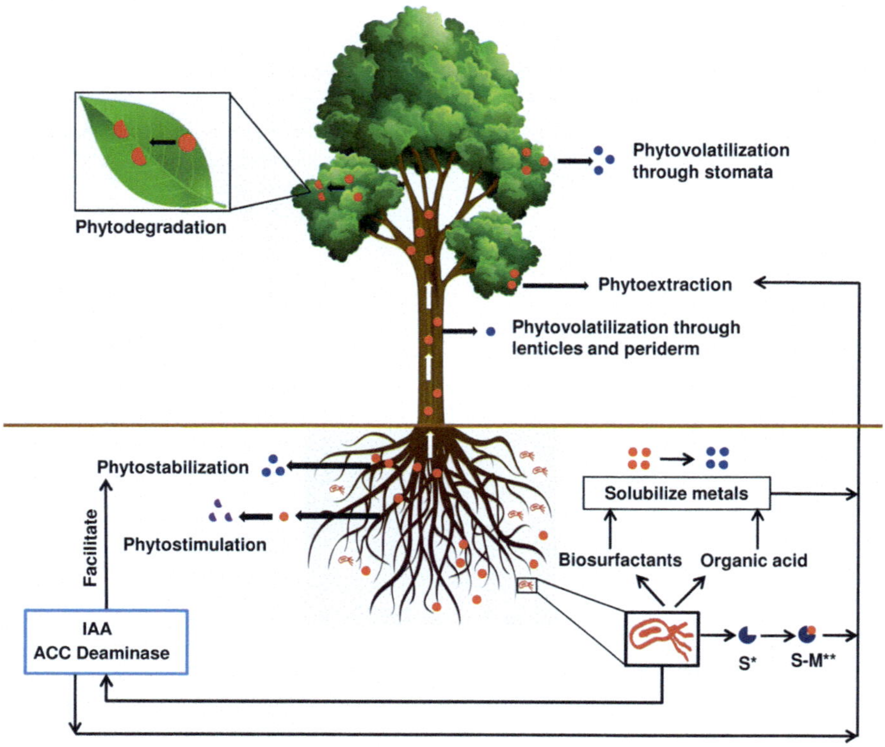

Fig. 15.2 Mechanisms of remediation of heavy metal (HM)-contaminated soil by microbial-aided phytoremediation

contaminated soil showed limiting the metal uptake capacity of the host plant on decreasing the availability of excessive Zn, Cu, and Pb (Huang et al. 2005). AMF colonization influences the production and augmentation of micronutrient uptake capacity of plants grown in heavy metal contaminated soil (Kaewdoung et al. 2016). Oxalate crystals produced by various mycorrhizal fungi (*Fomitopsis* cf. *meliae* and *Ganoderma* aff. *steyaertanum*) are also used to detoxify heavy metals by transforming them to lesser toxic forms such as copper sulfate into copper oxalate hydrate, lead nitrate into lead oxalate, cadmium sulfate into cadmium oxalate trihydrate (Kaewdoung et al. 2016).

15.6 Omics Approaches to Expedite the Remediation Process

Isolation and characterization of the microbial community responsible for bioremediation are imperative; however, with these culture-dependent methods, only 0.1–1% of the soil microbial community can be isolated, leaving more than 99% of microbes either uncultivable or difficult to culture. To overcome these limitations, a range of molecular techniques have been devised to explore the microorganisms responsible for bioremediation (Gupta et al. 2020; Subhashini et al. 2017). It includes fluorescence in situ hybridization technique (FISH), microbial lipid analysis, quantitative PCR, microradiography, stable isotope probing, clone library method, DNA microarray, and different genetic fingerprinting techniques like temperature gradient gel electrophoresis (TGGE), denaturing gradient gel electrophoresis (DGGE), single-stranded conformation polymorphism (SSCP), random amplified polymeric DNA (RAPD), terminal restriction fragment length polymorphism (T-RFLP), ribosomal intergenic spacer analysis (RISA), amplified ribosomal DNA restriction analysis (ARDRA), and length heterogeneity PCR (LH-PCR). All of these methods are based on isolation of lipids, proteins, nucleic acids targeting to amplify genes 16S rRNA, ITS, and 18S rRNA from soil (Gupta et al. 2020). Using these techniques, diversity and variation of the microbial community in contaminated soil in comparison to healthy soil can be analyzed (Panigrahi et al. 2019; Schloter et al. 2018; Malla et al. 2018; Margesin et al. 2011; Yang et al. 2020), but these techniques are unable to provide information about the mechanism involving in the remediation process.

Advanced omics strategies like metagenomics, metaproteomics, metatranscriptomics, and metabolomics provide a comprehensive and profound understanding of the underlying mechanism and adaptation strategy in microbial and plant cells in response to metal stress and thus unlimitedness in their implementation in the remediation of contaminated land (Gupta et al. 2020). Metagenomics provides us to understand not only to explore true diversity of microbes present in diverse environments but also to furnish remarkable information about the genes (*cadB*, *chrA*, *copAB*, *pbrA*, *merA*, *NiCoT*, etc.) responsible to adapt in metal-rich soil on tolerating metal toxicity, so that they can be used for bioremediation (Malla et al. 2018). In that direction, since the last few years, the genome of many metallotolerant bacteria such as *Enterobacter cloacae* B2-DHA isolated from the Hazaribagh tannery areas in Bangladesh, *Geobacillus thermodenitrificans* NG80-2 isolated from a deep oil reservoir in Northern China, *Halomonas zincidurans* strain B6 T isolated from a deep-sea heavy metal-rich sediment from the South Atlantic Mid-Ocean Ridge, *P. putida* ATH-4 isolated from soil sediments at the "Prat" Chilean military base located in Greenwich Island, Antarctica has been sequenced which provides information on the presence of heavy metal resistance genes to survive in the metal-rich environment (Barman et al. 2020). Thus metagenomics-based bioremediation approach is one of the effective tools for the removal of metal toxicity from the environment (Malla et al. 2018).

In response to metal stress, different stress response systems get activated within a given environment, and metatranscriptomics has provided a valuable insight into these gene expressions. Hence, metatranscriptomics is of immense importance for research related to environmental remediation. It was observed that on exposure to high Ni concentration to *Sphingobium*, approximately 118 genes are differentially expressed among which 90 were found to be upregulated (Volpicella et al. 2017). Transcriptome analysis of *E. coli* and *B. subtilis* showed that three membrane stress-related regulons, i.e., *cpxRA*, *rpoE,* and *basRS* get activated in response to metal stress (Hobman et al. 2007). Metaproteomics is suitable to reveal the qualitative and quantitative changes of proteofingerprints in response to metal stress. It reveals the change of physiological profiles in microbes and/or plants that undergo bioremediation. Commonly SDS-PAGE (1D), two-dimensional gel electrophoresis (2-DE), and two-dimensional difference gel electrophoresis (2-D DIGE) isobaric tags for relative and absolute quantitation (iTRAQ) are used by researchers to get information about the change of expression of the protein in response to metal toxicity (Zivkovic et al. 2018; Zhai et al. 2017; Bar et al. 2007). Combining the above-stated tools with mass spectroscopy and de novo sequencing helps to identify the proteins that get expressed on exposure to metals (Lacerda et al. 2007). The changes of proteomics profile in plants on inoculation with plant growth-promoting bacteria (PGPB) for microbe-assisted phytoremediation can also be detected by the metaproteomics approach (Li et al. 2014). However, metaproteomics offers better results in combination with other "omics" approaches. For example, Dore et al. (2015) utilized "omic" approaches with a combination of liquid chromatography and mass spectrometry techniques to identify proteins and extracellular enzymes and analyze fungal responses under various environmental conditions.

Metabolomics is the new entries to the "omics" family that provides information about the cellular metabolic architecture in response to metal stress (Booth et al. 2015). Since microbe and/or plants synthesize several metabolites to adapt metal stress condition, identification and quantification of these metabolites provide a better understanding of the functional role of these metabolites in the microbe and/or plant cells and the underlying mechanism involved in bioremediation (Malla et al. 2018). An example of this is the metabolomics profiling of *P. pseudoalcaligenes* KF707. It was observed that the strain displayed variation in levels of several metabolites with and without tellurite (Tremaroli et al. 2009). Wang et al. (2015) explore the metabolite profiling of radish roots on exposure to lead (Pb) and cadmium (Cd) stress. Results indicate that a large number of metabolites like sugars, amino acids, and organic acids alter in response to metal stress. The metabolite profiling of maize inoculated with PGPB also provide a better understanding of the upregulation of photosynthesis, hormone biosynthesis, and tricarboxylic acid cycle metabolites in maize that help the host plant to remediate metal-contaminated land as well as better growth and development of the plant in metal-contaminated land (Li et al. 2014).

15.7 Use of Genetically Modified Organisms (GMOs) in Bioremediation

GMOs mean "'any living organism that possesses a novel combination of genetic material obtained through the use of modern biotechnology.'" Microbes and/or plants can be genetically modified by recombinant DNA technology to yield a product having a special feature that has received great attention in bioremediation (Gupta and Singh 2017). Despite that, the use of GMOs in field conditions is restricted due to the associated issues of a biological system such as their reach to the contaminants, activity, competition, and most widespread contaminated sites; hence, it is largely limited in the laboratory (Gupta and Singh 2017). The requirement for the development of GMOs for bioremediation of contaminated sites involves four principal approaches. These include modification of enzyme affinity and specificity, construction and regulation of specific pathways, development of bioprocess for remediation and its monitoring and control, use and applications of biosensors for chemical sensing, toxicity reduction, and endpoint analysis (Gupta and Singh 2017). For instance, Dash and Das (2015) constructed a transgenic bacterium *Bacillus cereus* BW-03 (pPW-05) with the introduction of *merA* encoding mercuric reductase from *Bacillus thuringiensis* PW-05 in the other mercury-resistant marine bacterium *B. cereus* BW-03 (pPW-05) for better bioremediation. It was observed that the *Bacillus cereus* BW-03 (pPW-05) improves the mercury removal efficiency in comparison to the parent strains in situ. The strain also found to survive under varied conditions of pH, salinity, and mercury concentration which increase its possibility to use for bioremediation in the mercury-contaminated field. Arsenic is one of the highly toxic metals in oxidized forms, and its bioremediation is mainly associated with volatilization. Though various indigenous microflora have been reported to volatilize arsenic, the efficiency of volatilization was found to be increased by genetically modified microorganisms. Studies have reported that cloning and expression of arsenite S-adenosyl methionine methyltransferase gene (*arsM*) of *Sphingomonas desiccabilis* and *Bacillus idriensis* increase the release of methylated arsenic gas tenfold more than the wild strain (Yang 2010). Further, the introduction of microbial metal resistance genes in hyperaccumulating plants like *Arabidopsis thaliana*, *Brassica juncea*, *Populus angustifolia*, and *Nicotiana tabacum* has been found to enhance metal transformation and accumulation efficiency as compared to wild plant species. For example, the introduction of *merA* and *merB* from bacteria in *Arabidopsis thaliana* leads to an increase in the tolerance capacity of the plant as well as the better conversion of toxic mercury into its less toxic form (Bizily et al. 2000). In another study, it was observed that transformation and overexpression of the phytochelatin synthase (*TaPCS1*) gene in *Nicotiana* resulted in a better tolerance capacity of the plant toward lead (Gisbert et al. 2003).

15.8 Conclusion and Prospects

From the above thorough and critical discussion, it is evident that remediation technologies using microorganisms are more feasible to decontaminate the metal-polluted site with great economical and ecological relevance. Toward a much deeper perceptive and understanding of the microbial and microbe-assisted phytoremediation, it was observed that they employ different mechanisms to survive in the metal-contaminated site and subsequently performing bioremediation. And various omic-approaches provide a significant advantage to understand the mechanisms involved in bioremediation pathways. From the recent research articles, it is evident that MT-PGPR is an effective and sustainable measure for the reclamation of metal-polluted soils. However, in the future, the contribution of genes about Phyto beneficial traits and the occurrence of preferential symbiosis needs to be studied in-depth to harness the benefit of plant–microbe interactions. Additionally, the application of these potential microorganisms as bioinoculants to be explored for better productivity and remediating the metal-contaminated site. Hence, further research is needed to develop novel bioinoculants to tackle the threat of metal-contaminated sites. Additionally, different biotechnological approaches provide an avenue to develop the designed microbes to improve the bioremediation potentiality and better productivity under stress conditions, but in connection with regulatory risk assessment, the field application of GMOs is still restricted. Hence, further improvements in GMOs in terms of their survival, completion with an indigenous population, and chemotaxis toward the pollutants along with structural genes associated with bioremediation of contaminants should also be considered while developing GMOs for environmental cleanup.

Acknowledgments The first author is gratefully acknowledged to DBT-RA Program in Biotechnology and Life Sciences for financial support.

References

Abou-Shanab RAI, Angle JS, Chaney RL (2006) Bacterial inoculants affecting nickel uptake by *Alyssum murale* from low, moderate and high Ni soils. Soil Biol Biochem 38:2882–2889

Achal V, Pan X, Zhang D (2011) Remediation of copper-contaminated soil by *Kocuria flava* CR1, based on microbially induce calcite precipitation. Ecol Eng 37:1601–1605

Acosta-Rodriguez I, Cárdenas-González JF, Pérez ASR, Oviedo JT, Martínez-Juárez VM (2018) Bioremoval of different heavy metals by the resistant fungal strain *Aspergillus niger*. Bioinorg Chem Appl 2018(3457196):1–7

Ahemad M (2019) Remediation of metalliferous soils through the heavy metal resistant plant growth promoting bacteria: paradigms and prospects. Arab J Chem 12:1365–1377

Ahmad F, Zhu D, Sun J (2020) Bacterial chemotaxis: a way forward to aromatic compounds biodegradation. Environ Sci Eur 32:52

Alboghobeish H, Tahmourespour A, Doudi M (2014) The study of nickel resistant bacteria (NiRB) isolated from wastewaters polluted with different industrial sources. J Environ Health Sci Eng 12:44

Alexander M (1999) Biodegradation and bioremediation, 2nd edn. Academic, New York
Ameen FA, Hamdan AM, El-Naggar MY (2020) Assessment of the heavy metal bioremediation efficiency of the novel marine lactic acid bacterium, *Lctobacillus plantarum* MF042018. Sci Rep 10:314
Anastopoulos I, Kyzas GZ (2015) Progress in batch biosorption of heavy metals onto algae. J Mol Liq 209:77–86
Ansari MI, Grohmann E, Malik A (2008) Conjugative plasmids in multi-resistant bacterial isolates from Indian soil. J Appl Microbiol 104(6):1774–1781
Anwar HM, Pukclai P, da Silva JAT, Fujita M (2012) Molecular mechanism of heavy metal toxicity and tolerance in plants: central role of glutathione in detoxification of reactive oxygen species and methylglyoxal and in heavy metal chelation. J Bot 2012:1–37
Arunakumara KKIU, Walpola BC, Song JS, Shin M, Lee C, Yoon M (2014) Phytoextraction of heavy metals induced by bioaugmentation of a phosphate solubilizing bacterium. Korean J Environ Agric 33:220–230
Arunakumara KKIU, Walpola BC, Yoon MH (2015) Bioaugmentation-assisted phytoextraction of Co, Pb and Zn: an assessment with a phosphate-solubilizing bacterium isolated from metal-contaminated mines of Boryeong area in South Korea. Biotechnol Agron Soc Environ 19(2):143–152
Awa S, Hadibarata T (2020) Removal of heavy metals in contaminated soil by phytoremediation mechanism: a review. Water Air Soil Pollut 231:47
Ayangbenro AS, Babalola OO (2017) A new strategy for heavy metal polluted environments: a review of microbial biosorbents. Int J Environ Res Public Health 14(1):94
Bar C, Patil R, Doshi J, Kulkarni MJ, Gade WN (2007) Characterization of the proteins of bacterial strain isolated from contaminated site involved in heavy metal resistance—a proteomic approach. J Biotechnol 128:444–451
Barkay T, Miller SM, Summers AO (2003) Bacterial mercury resistance from atoms to ecosystems. FEMS Microbiol Rev 27:355–384
Barman D, Jha DK, Bhattacharjee K (2020) Metalotolerant bacteria: insights into bacteria thriving in metal contaminated areas. In: Singh RP, Manchanda G, Maurya IK, Wei Y (eds) Microbial versatility in varied environments. Springer Nature, Singapore, pp 135–164
Becerra-Castro C, Prieto-Fernández A, Alvarez-Lopez V, Monterroso C, Cabello-Conejo MI, Acea MJ, Kidd PS (2011) Nickel solubilizing capacity and characterization of rhizobacteria isolated from hyperaccumulating and non-hyperaccumulating subspecies of *Alyssum serpyllifolium*. Int J Phytoremediation 13:229–244
Belimov AA, Hontzeas N, Safronova VI, Demchinskaya SV, Piluzza G, Bullitta S, Glick BR (2005) Cadmium-tolerant plant growth-promoting bacteria associated with the roots of Indian mustard (*Brassica juncea* L. Czern.). Soil Biol Biochem 37:241–250
Bhattacharya A, Gupta A, Kaur A, Malik D (2014) Efficacy of *Acinetobacter* sp. B9 for simultaneous removal of phenol and hexavalent chromium from co-contaminated system. Appl Microbiol Biotechnol 98(23):9829–9841
Bishnoi NR, Kumar R, Kumar S, Rani S (2007) Biosorption of Cr(III) from aqueous solution using algal biomass *Spirogyra* spp. J Hazard Mater 145(1–2):142–147
Bizily SP, Rugh CL, Meagher RB (2000) Phytodetoxification of hazardous organomercurials by genetically engineered plants. Nat Biotechnol 18:213–217
Boopathy R (2000) Factors limiting bioremediation technologies. Bioresour Technol 74:63–67
Booth SC, Weljie AM, Turner RJ (2015) Metabolomics reveals differences of metal toxicity in cultures of *Pseudomonas pseudoalcaligenes* KF707 grown on different carbon sources. Front Microbiol 6:827
Borremans B, Hobman JL, Provoost A, Brown NL, van der Lelie D (2001) Cloning and functional analysis of the pbr lead resistance determinant of *Ralstonia metallidurans* CH34. J Bacteriol 183:5651–5658
Brar SK, Verma M, Surampalli R, Misra K, Tyagi R, Meunier N, Blais J (2006) Bioremediation of hazardous wastes—a review. Pract period Hazard toxic Radioact Waste Manag 10:59–72

Bulgarelli D, Schlaeppi K, Spaepen S, VerLorenvan Themaat E, Schulze-Lefert P (2013) Structure and functions of the bacterial microbiota of plants. Annu Rev Plant Biol 64:807–838

Cameselle C, Chirakkara RA, Reddy KR (2013) Electrokinetic-enhanced phytoremediation of soils: status and opportunities. Chemosphere 93(4):626–636

Carrilho EN, Gilbert TR (2000) Assessing metal sorption on the marine alga *Pilayella littoralis*. J Environ Mon 2:410–415

Chibuike GU, Obiora SC (2014) Heavy metal polluted soils: effect on plants and bioremediation methods. Appl Environ Soil Sci 2014(752708):1–12

Chien CC, Lin BC, Wu CH (2013) Biofilm formation and heavy metal resistance by an environmental *Pseudomonas* sp. Biochem Eng J 78:132–137

Chirakkara RA, Cameselle C, Reddy KR (2016) Assessing the applicability of phytoremediation of soils with mixed organic and heavy metal contaminants. Rev Environ Sci Biotechnol 15:299–326

Crupper SS, Worrell V, Stewart GC, Iandolo JJ (1999) Cloning and expression of *cadD*, a new cadmium resistance gene of *Staphylococcus aureus*. J Bacteriol 181:4071–4075

Das U, Dash HR, Chakraborty J (2016) Genetic basis and importance of metal resistant genes in bacteria for bioremediation of contaminated environments with toxic metal pollutants. Appl Microbiol Biotechnol 100:2967–2984

Dash HR, Das S (2015) Bioremediation of inorganic mercury through volatilization and biosorption by transgenic *Bacillus cereus* BW-03(pPW-05). Int Biodeterior Biodegradation 103:179–185

Dave D, Sarma S, Parmar P, Shukla A, Goswami D, Shukla A, Saraf M (2020) Microbes as a boon for the bane of heavy metals. Environ Sustain 3:233–255

Deshmukh R, Khardenavis AA, Purohit HJ (2016) Diverse metabolic capacities of fungi for bioremediation. Indian J Microbiol 56(3):247–264

Diep P, Mahadevan R, Yakunin AF (2018) Heavy metal removal by bioaccumulation using genetically engineered microorganisms. Front Bioeng Biotechnol 6:157

Dixit R, Wasiullah MD et al (2015) Bioremediation of heavy metals from soil and aquatic environment: an overview of principles and criteria of fundamental processes. Sustainability 7:2189–2212

Dong G, Wang Y, Gong L, Wang M, Wang H, He N, Zheng Y, Li Q (2013) Formation of soluble Cr (III) end-products and nanoparticles during Cr(VI) reduction by *Bacillus cereus* strain XMCr-6. Biochem Eng J 70:166–172

Dore J, Perraud M, Dieryckx C et al (2015) Comparative genomics, proteomics and transcriptomics give new insight into the exoproteome of the basidiomycete *Hebeloma cylindrosporum* and its involvement in ectomycorrhizal symbiosis. New Phytol 208:1169–1187

Emamverdian A, Ding Y, Mokhberdoran F, Xie Y (2015) Heavy metal stress and some mechanisms of plant defense response. Sci World J 2015(756120):1–18

Feng D, Aldrich C (2004) Adsorption of heavy metals by biomaterials derived from marine alga *Ecklonia maxima*. Hydrometallurgy 73:1–10

Gisbert C, Ros R, Haro AD, Walker DJ (2003) A plant genetically modified that accumulates Pb is especially promising for phytoremediation. Biochem Biophys Res Commun 2:440–445

Glick BR (2014) Bacteria with ACC deaminase can promote plant growth and help to feed the world. Microbiol Res 169:30–39

Gonzalez-Chavez MCA, Carrillo-Gonzalez R, Cuellar-Sanchez A et al (2019) Phytoremediation assisted by mycorrhizal fungi of a Mexican defunct lead-acid battery recycling site. Sci Total Environ 650:3134–3144

Gupta R, Singh RL (2017) Genetically modified organisms (GMOs) and environment. In: Singh RL (ed) Principles and applications of environmental biotechnology for a sustainable future. Springer, Singapore, pp 425–465

Gupta A, Joia J, Sood A, Sood R, Sidhu C, Kaur G (2016) Microbes as potential tool for remediation of heavy metals: a review. J Microb Biochem Technol 8:364–372

Gupta K, Biswas R, Sarkar A (2020) Advancement of omics: prospects for bioremediation of contaminated soils. In: Shah MP (ed) Microbial bioremediation & biodegradation. Springer Nature, Singapore, pp 113–142

Han X, Wong YS, Tam NFY (2006) Surface complexation mechanism and modeling in Cr(III) biosorption by a microalgal isolate, *Chlorella miniata*. J Colloid Interface Sci 303(2):365–371

Hashim MA, Chu KM (2004) Biosorption of cadmium by brown, green and red seaweeds. Chem Eng J 97:249–255

He LY, Zhang YF, Ma HY, Su LN, Chen ZJ et al (2010) Characterization of copper-resistant bacteria and assessment of bacterial communities in rhizosphere soils of copper-tolerant plants. Appl Soil Ecol 44:49–55

Hobman JL, Yamamoto K, Oshima T (2007) Transcriptiomic responses of bacterial cells to sub-lethal metal ion stress. In: Nies DH, Silver S (eds) Molecular microbiology of heavy metals, microbiology monographs, vol 6. Springer, Berlin, pp 73–115

Huang Y, Tao S, Chen YJ (2005) The role of arbuscular mycorrhiza on change of heavy metal speciation in rhizosphere of maize in wastewater irrigated agriculture soil. J Environ Sci 17(2):276–280

Iddou A, Hadj Youcef M, Aziz A, Ouali MS (2011) Biosorptive removal of lead (II) ions from aqueous solutions using Cystoseira stricta biomass: study of the surface modification effect. J Saudi Chem Soc 15:83–88

Igiri BE, Okoduwa SIR, Idoko GO, Akabuogu EP, Adeyi AO, Ejiogu IK (2018) Toxicity and bioremediation of heavy metals contaminated ecosystem from tannery wastewater: a review. J Toxicol 2018(2568038):1–16

Jain D, Kour R, Bhojiya AA et al (2020) Zinc tolerant plant growth promoting bacteria alleviates phytotoxic effects of zinc on maize through zinc immobilization. Sci Rep 10:13865

Kaewdoung B, Sutjaritvorakul T, Gadd GM, Whalley AJS, Sihanonth P (2016) Heavy metal tolerance and biotransformation of toxic metal compounds by new isolates of wood-rotting fungi from Thailand. Geomicrobiol J 33:283–288

Kalaimurugan D, Balamuralikrishnan B, Durairaj K et al (2020) Isolation and characterization of heavy-metal-resistant bacteria and their applications in environmental bioremediation. Int J Environ Sci Technol 17:1455–1462

Karaca A, Cetin SC, Turgay OC, Kizilkaya R (2010) Effects of heavy metals on soil enzyme activities. In: Soil heavy metals, soil biology. Springer, Berlin, pp 237–262

Kim S, Lim H, Lee I (2010) Enhanced heavy metal phytoextraction by *Echinochloa crus-galli* using root exudates. J Biosci Bioeng 109:47–50

Krupp EM, Grumping R, Furchtbar URR, Hirner AV (1996) Speciation of metals and metalloids in sediments with LTGC/ICP-MS. Fresenius J Anal Chem 354:546–549

Kumar KV, Srivastava S, Singh N, Behl HM (2009) Role of metal resistant plat growth promoting bacteria in ameliorating fly ash to the growth of *Brassica Juncea*. J Hazard Mater 170:51–57

Lacerda CMR, Choe LH, Reardon KF (2007) Metaproteomic analysis of a bacterial community response to cadmium exposure. J Proteome Res 6:1145–1152

Lauwerys R, Haufroid V, Hoet P, Lison D (2007) Toxicologie industrielle et in-toxications professionnelles, 5th edn. Elsevier-Masson, Paris

Li K, Pidatala VR, Shaik R, Datta R, Ramakrishna W (2014) Integrated metabolomic and proteomic approaches dissect the effect of metal-resistant bacteria on maize biomass and copper uptake. Environ Sci Technol 48:1184–1193

Lima AIG, Corticeiro SC, Figueira EMAP (2006) Glutathione-mediated cadmium sequestration in *Rhizobium leguminosarum*. Enzym Microb Technol 39(4):763–769

Liu X, Jiang J, Yan Y et al (2018) Distribution and risk assessment of metals in water, sediments, and wild fish from Jinjiang River in Chengdu, China. Chemosphere 196:45–52

Lloyd JR (2003) Microbial reduction of metals and radionuclides. FEMS Microbiol Rev 27(2–3):411–425

Ma Y, Rajkumar M, Luo Y, Freitas H (2013) Phytoextraction of heavy metal polluted soils using *Sedum plumbizincicola* inoculated with metal mobilizing *Phyllobacterium myrsinacearum* RC6b. Chemosphere 93:1386–1392

Ma Y, Oliveira RS, Freitas H, Zhang C (2016) Biochemical and molecular mechanisms of plant-microbe-metal interactions: relevance for phytoremediation. Front Plant Sci 7:918

Malla MA, Dubey A, Yadav S, Kumar A, Hashem A, Abd Allah EF (2018) Understanding and designing the strategies for the microbe-mediated remediation of environmental contaminants using omics approaches. Front Microbiol 9:1132

Manchanda G, Singh RP, Li ZF, Zhang JJ (2017) Mycorrhiza: creating good spaces for interactions. In: Mycorrhiza-function, diversity, state of the art. Springer, Cham, pp 39–60

Margesin R, Płaza GA, Kasenbacher S (2011) Characterization of bacterial communities at heavy-metal-contaminated sites. Chemosphere 82(11):1583–1588

Marzan LW, Hossain M, Mina SA, Akter Y, Chowdhury AMMA (2017) Isolation and biochemical characterization of heavy-metal resistant bacteria from tannery effluent in Chittagong city, Bangladesh: bioremediation viewpoint. Egypt J Aquat Res 43:65–74

Mesa V, Navazas A, González-Gil R, González A, Weyens N, Lauga B, Gallego JLR et al (2017) Use of endophytic and rhizosphere bacteria to improve phytoremediation of arsenic-contaminated industrial soils by autochthonous *Betula celtiberica*. Appl Environ Microbiol 83(8):e03411–e03416

Mishra J, Singh R, Arora NK (2017) Alleviation of heavy metal stress in plants and remediation of soil by rhizosphere microorganisms. Front Microbiol 8:1706

MoEF (2001) Land degradation. State of the environment, Government of India, New Delhi, pp 61–75

Mohapatra H, Gupta R (2005) Concurrent sorption of Zn (II), Cu (II) and Co (II) by *Oscillatoria angustissima* as a function of pH in binary and ternary metal solutions. Bioresour Technol 96(12):1387–1398

Monchy S, Benotmane MA, Janssen P, Vallaeys T, Taghavi S, van der Lelie D, Mergeay M (2007) Plasmids pMOL28and pMOL30 of *Cupriavidus metallidurans* are specialized in the maximal viable response to heavy metals. J Bacteriol 189:7417–7425

Mosa KA, Saadoun I, Kumar K, Helmy M, Dhankher OP (2016) Potential biotechnological strategies for the cleanup of heavy metals and metalloids. Front Plant Sci 7:303

Naik MM, Pandey A, Dubey SK (2012) *Pseudomonas aeruginosa* strain WI-1 from Mandovi estuary possesses metallothionein to alleviate lead toxicity and promotes plant growth. Ecotoxicol Environ Saf 79:129–133

National Research Council (1993) The current practice of bioremediation. In: National Research Council (ed) In situ bioremediation: when does it work?. The National Academies Press, Washington, DC, pp 47–62

Naseem H, Bano A (2014) Role of plant growth-promoting rhizobacteria and their exopolysaccharide in drought tolerance of maize. J Plant Interact 9:689–701

Nicholls AM, Mal TK (2003) Effects of lead and copper exposure on growth of an invasive weed, *Lythrum salicaria* L. (*Purple loosestrife*). Ohio J Sci 103(5):129–133

Nies DH (2003) Efflux-mediated heavy metal resistance in prokaryotes. FEMS Microbiol Rev 27: 313–339

Niu H, Volesky B (2000) Gold-cyanide biosorption with L-cysteine. J Chem Technol Biotechnol 75:436–442

Ojuederie OB, Babalola OO (2017) Microbial and plant-assisted bioremediation of heavy metal polluted environments: a review. Int J Environ Res Public Health 14(12):1504

Ontanon OM, Fernandez M, Agostini E, González PS (2018) Identification of the main mechanisms involved in the tolerance and bioremediation of Cr(VI) by *Bacillus* sp. SFC 500-1E. Environ Sci Pollut Res 25(16):16111–16120

Oves M, Khan MS, Zaidi A (2013) Chromium reducing and plant growth promoting nivel strain *Pseudomonas aeruginosa* OSG41 enhances chickpea growth in chromium amended soils. Eur J Soil Biol 56:72–83

Oziegbe O, Oluduro AO, Oziegbe EJ, Ahuekwe EF, Olorunsola SJ (2021) Assessment of heavy metal bioremediation potential of bacterial isolates from landfill soils. Saudi J Biol Sci 28(7):3948–3956. https://doi.org/10.1016/j.sjbs.2021.03.072

Pande V, Pandey SC, Sati D et al (2020) Bioremediation: an emerging effective approach towards environment restoration. Environ Sustain 3:91–103

Panigrahi S, Velraj P, Rao TS (2019) Functional microbial diversity in contaminated environment and application in bioremediation. In: Microbial diversity in the genomic era. Academic, New York, pp 359–385

Park D, Yun YS, Jo JH, Park JM (2005) Mechanism of hexavalent chromium removal by dead fungal biomass of *Aspergillus niger*. Water Res 39(4):533–540

Perpetuo EA, Souza CB, Nascimento CAO (2011) Engineering bacteria for bioremediation. In: Carpi A (ed) Progress in molecular and environmental bioengineering: from analysis and modeling to technology applications. IntechOpen, London, pp 605–632

Phieler R, Merten D, Roth M, Büchel G, Kothe E (2015) Phytoremediation using microbially mediated metal accumulation in *Sorghum bicolour*. Environ Sci Pollut Res 22:19408–19416

Pritchard PH, Costa CF (1991) EPA's Alaska oil spill bioremediation project, part 5. Environ Sci Technol 25(3):372–379

Ramasamy RK, Congeevaram S, Thamaraiselvi K (2011) Evaluation of isolated fungal strain from e-waste recycling facility for effective sorption of toxic heavy metals Pb(II) ions and fungal protein molecular characterization—a Mycoremediation approach. Asian J Exp Biol 2(2):342–347

Rathore SS, Shekhawat K, Dass A, Kandpal BK, Singh VK (2019) Phytoremediation mechanism in Indian mustard (*Brassica juncea*) and its enhancement through agronomic interventions. Proc Natl Acad Sci, India, Sect B Biol Sci 89(2):419–427

Romaniuk K, Ciok A, Decewicz P et al (2018) Insight into heavy metal resistome of soil psychrotolerant bacteria originating from King George Island (Antarctica). Polar Biol 41: 1319–1333

Roy M, Giri AK, Dutta S, Mukherjee P (2015) Integrated phytobial remediation for sustainable management of arsenic in soil and water. Environ Int 75:180–198

Saima K, Roohi M, Ahmad IZ (2013) Isolation of novel chitinolytic bacteria and production optimization of extracellular chitinases. J Genet Eng Biotechnol 11:39–46

Sari A, Tuzen M (2008) Biosorption of cadmium(II) from aqueous solution by red algae (*Ceramium virgatum*): equilibrium, kinetic and thermodynamic studies. J Hazard Mater 157(2–3):448–454

Sarma K, Barik SK (2011) Coal mining impact on vegetation of the Nokrek biosphere reserve, Meghalaya, India. Biodiversity 12(3):154–164

Say R, Yimaz N, Denizli A (2003) Removal of heavy metal ions using the fungus *Penicillium canescens*. Adsorpt Sci Technol 21:643–650

Schloter M, Nannipieri P, Sorensen SJ, van Elsas JD (2018) Microbial indicators for soil quality. Biol Fertil Soils 54:1–10

Sessitsch A, Kuffner M, Kidd P, Vangronsveld J, Wenzel WW, Fallmann K, Puschenreiter M (2013) The role of plant-associated bacteria in the mobilization and phytoextraction of trace elements in contaminated soils. Soil Biol Biochem 60:182–194

Shannon MJ, Unterman R (1993) Evaluating bioremediation: distinguishing fact from fiction. Annu Rev Microbial 47:715–738

Sharma A, Nagpal AK (2020) Contamination of vegetables with heavy metals across the globe: hampering food security goal. J Food Sci Technol 57(2):391–403

Sharma P, Pandey S (2014) Status of phytoremediation in world scenario. Int J Environ Bioremediat Biodegrad 2(4):178–191

Sheng XF, Xia JJ, Jiang CY, He LY, Qian M (2008) Characterization of heavy metal-resistant endophytic bacteria from rape *Brassica napus* roots and their potential in promoting the growth and lead accumulation of rape. Environ Pollut 156:1164–1170

Shun-hong H, Bing P, Zhi-hui Y, Li-yuan C, Li-cheng Z (2009) Chromium accumulation, microorganism population and enzyme activities in soils around chromium-containing slag heap of steel alloy factory. Trans Nonferrous Met Soc China 19:241–248

da Silva IGS, de Almeida FCG, da Rochae Silva NMP, Casazza AA, Converti A, Sarubbo LA (2020) Soil bioremediation: overview of technologies and trends. Energies 13(18):4664

Silver S, Phung LT (2013) Bacterial mercury resistance proteins. In: Kretsinger RH, Uversky VN, Permiakov EA (eds) Encyclopedia of metalloproteins. Springer-Verlag, New York, pp 209–217

Singh RP, Singh RN, Srivastava AK, Kumar S, Dubey RC et al (2011) Structural analysis and 3D-modelling of fur protein from *Bradyrhizobium japonicum*. J Appl Sci Environ Sani 6(3):357–366

Singh RP, Manchanda G, Singh RN, Srivastava AK, Dubey RC (2016a) Selection of alkalotolerant and symbiotically efficient chickpea nodulating rhizobia from north-west indo Gangetic Plains. J Basic Microbiol 56:14–25. https://doi.org/10.1002/jobm.201500267

Singh N, Raghubanshi A, Upadhyay A, Rai U (2016b) Arsenic and other heavy metal accumulation in plants and algae growing naturally in contaminated area of West Bengal, India. Ecotoxicol Environ Saf 130:224–233

Singh RP, Manchanda G, Li ZF, Rai AR (2017) Insight of proteomics and genomics in environmental bioremediation. In: Bhakta JN (ed) Handbook of research on inventive bioremediation techniques. IGI Global, Hershey. https://doi.org/10.4018/978-1-5225-2325-3

Singh RP, Anwar MN, Singh D, Bahuguna V, Manchanda G, Yang Y (2020a) Deciphering the key factors for heavy metal resistance in gram-negative bacteria. In: Singh R, Manchanda G, Maurya I, Wei Y (eds) Microbial versatility in varied environments. Springer, Singapore. https://doi.org/10.1007/978-981-15-3028-9_7

Singh R, Manchanda G, Maurya I, Wei Y (eds) (2020b) Microbial versatility in varied environments. Springer, Singapore. https://doi.org/10.1007/978-981-15-3028-9

Sivakumar D (2016) Biosorption of hexavalent chromium in a tannery industry waste water using fungi species. Glob J Environ Sci Manag 2:105–124

Smalla K, Haines AS, Jones K, Krogerrecklenfort E, Heuer H, Schloter M, Thomas CM (2006) Increased abundance of IncP-1β plasmids and mercury resistance genes in mercury polluted river sediments first discovery of IncP-1β plasmids with a complex mer transposon as sole accessory element. Appl Environ Microbiol 72:7253–7259

Subhashini DV, Singh RP, Manchanda G (2017) OMICS approaches: tools to unravel microbial systems. Directorate of Knowledge Management in Agriculture, Indian Council of Agricultural Research, New Delhi. ISBN: 97881 71641703. https://books.google.co.in/books?id=vSaLtAEACAAJ

Tangahu BV, Abdullah SRS, Basri H, Idris M, Anuar N, Mukhlisin M (2011) A review on heavy metals (As, Pb, and Hg) uptake by plants through phytoremediation. Int J Chem Eng 2011(939161):1–31

Tarekegn MM, Salilih FZ, Ishetu AI (2020) Microbes used as a tool for bioremediation of heavy metal from the environment. Cogent food agric 6:1

Tastan BE, Ertugrul S, Donmez G (2010) Effective bioremoval of reactive dye and heavy metals by *Aspergillus versicolor*. Bioresour Technol 101(3):870–876

Tirry N, Joutey NT, Sayel H, Aziza K, Bahafid W et al (2018) Screening of plant growth promoting traits in heavy metals resistant bacteria: prospects in phytoremediation. J Genet Eng Biotechnol 16(2):613–619

Tremaroli V, Workentine ML, Weljie AM et al (2009) Metabolomic investigation of the bacterial response to a metal challenge. Appl Environ Microbiol 75:719–728

Tuzen M, Sari A (2010) Biosorption of selenium from aqueous solution by green algae (*Cladophora hutchinsiae*) biomass: equilibrium, thermodynamic and kinetic studies. Chem Eng J 158:200–206

Tuzen M, Sari A, Mendil D, Uluozlu OD, Soylak M, Dogan M (2009) Characterization of biosorption process of As(III) on green algae *Ulothrix cylindricum*. J Hazard Mater 165(1–3):566–572

Ullah A, Heng S, Munis MFH, Fhad S, Yang X (2015) Phytoremediation of heavy metals assisted by plant growth promoting (PGP) bacteria: a review. Environ Exp Bot 117:28–40
Valls M, de Lorenzo V (2002) Exploiting the genetic and biochemical capacities of bacteria for the remediation of heavy metal pollution. FEMS Microbiol Rev 26(4):327–338
Vijayaraghavan K, Jegan J, Palanivenu K, Velan M (2005) Biosorption of copper, cobalt and nickel by marine green alga *Ulva reticulate* in a packed column. Chemosphere 60:419–426
Volpicella M, Leoni C, Manzari C et al (2017) Transcriptomic analysis of nickel exposure in *Sphingobium* sp. ba1 cells using RNA-seq. Sci Rep 7:8262
Vullo DL, Ceretti HM, Hughes EA, Ramyrez S, Zalts A (2008) Cadmium, zinc and copper biosorption mediated by *Pseudomonas veronii* 2E. Bioresour Technol 99:5574–5581
Waigi MG, Sun K, Gao Y (2017) *Sphingomonads* in microbe-assisted phytoremediation: tackling soil pollution. Trends Biotechnol 35(9):883–899
Wang YP, Shi JY, Wang H, Li Q, Chen XC, Chen YX (2007) The influence of soil heavy metals pollution on soil microbial biomass, enzyme activity, and community composition near a copper smelters. Ecotoxicol Environ Saf 67:75–81
Wang Y, Xu L, Shen H et al (2015) Metabolomic analysis with GC-MS to reveal potential metabolites and biological pathways involved in Pb & Cd stress response of radish roots. Sci Rep 5:18296
Wei W, Liu X, Sun P, Wang X, Zhu H, Hong M, Mao ZW, Zhao J (2014) Simple whole-cell biodetection and bioremediation of heavy metals based on an engineered lead-specific operon. Environ Sci Technol 48:3363–3371
Whiting SN, de Souza MP, Terry N (2001) Rhizosphere bacteria mobilize Zn for hyperaccumulation by *Thlaspi caerulescens*. Environ Sci Technol 35(15):3144–3150
Wood JL, Liu W, Tang C, Franks AE (2016) Microorganisms in heavy metal bioremediation: strategies for applying microbial-community engineering to remediate soils. AIMS Bioeng 3(2):211–229
Yamaji K, Watanabe Y, Masuya H, Shigeto A, Yui H, Haruma T (2016) Root fungal endophytes enhance heavy-metal stress tolerance of *Clethra barbinervis* growing naturally at mining sites via growth enhancement, promotion of nutrient uptake and decrease of heavy-metal concentration. PLoS One 11(12):e0169089
Yang CXL (2010) Construction of a genetically engineered microorganism with high tolerance to arsenite and strong arsenite oxidative ability. J Environ Sci Health A Tox Hazard Subst Environ Eng 45:732–737
Yang Y, Singh RP, Song D, Chen Q, Zheng X, Zhang C, Zhang M, Li Y (2019) Synergistic effect of *Pseudomonas putida* II-2 and *Achromobacter* sp. QC36 for the effective biodegradation of the herbicide quinclorac. Ecotoxicol Environ Safety 188:109826
Yang Y, Liu L, Singh RP, Meng C, Ma S, Jing C, Li Y, Zhang C (2020) Nodule and root zone microbiota of salt-tolerant wild soybean in coastal sand and saline-alkali soil. Front Microbiol 11:2178. https://doi.org/10.3389/fmicb.2020.523142
Yao Q, Zhu HH, Chen JZ (2005) Growth responses and endogenous IAA and iPAs changes of litchi (*Litchi chinensis* Sonn.) seedlings induced by arbuscular mycorrhizal fungal inoculation. Sci Hortic 105:145–151
Yuan M, He H, Xiao L, Zhong T, Liu H et al (2014) Enhancement of Cd phytoextraction by two *Amaranthus* species with endophytic *Rahnella* sp. JN27. Chemosphere 103:99–104

Zeraatkar AK, Ahmadzadeh H, Talebi AF, Moheimani NR, McHenry MP (2016) Potential use of algae for heavy metal bioremediation, a critical review. J Environ Manag 181:817–831
Zhai Q, Xiao Y, Zhao J, Tian F, Zhang H, Narbad A, Wei C (2017) Identification of key proteins and pathways in cadmium tolerance of *Lactobacillus plantarum* strains by proteomic analysis. Sci Rep 7(1182):1–17
Zhu H, Guo J, Chen M, Feng G, Yao Q (2012) *Burkholderia dabaoshanensis* sp. nov., a heavy-metal-tolerant bacteria isolated from Dabaoshan mining area soil in China. PLoS One 7(12): e50225
Zivkovic LI, Rikalovic M, Cvijovic GG, Kazazic S, Vrvic M, Brceski I, Beskoski V, Loncarevic B, Gopcevic K, Karadzic I (2018) Cadmium specific proteomic responses of a highly resistant *Pseudomonas aeruginosa* san ai. RSC Adv 8:10549–10560

Chapter 16
Actinobacterial Abundance and Interrelationships in Ecosystems of Northwest Africa

Noureddine Bouras, Amine Yekkour, Slim Smaoui, Lotfi Mellouli, and Mustapha Barakate

This book chapter is dedicated to the late Dr. Nasserdine Sabaou (1956–2019) for his genuine love and service to microbiology. He was a superb researcher and professor at the École Normale Supérieure de Kouba, Alger (Algeria) *and the ex-head of the* Laboratoire de Biologie des Systèmes Microbiens (LBSM)*. He published many papers on rare Saharan* Actinobacteria *and their metabolites.*

Abstract *Actinobacteria* are of special interest because of their versatile metabolic activities. In nature, they play a crucial role in the decomposition of environmental pollutants and organic compounds. The goal of this chapter was to review the

N. Bouras (✉)
Laboratoire de Biologie des Systèmes Microbiens (LBSM), Ecole Normale Supérieure de Kouba, Alger, Algeria

Département de Biologie, Faculté des Sciences de la Nature et de la Vie et Sciences de la Terre, Université de Ghardaia, Ghardaïa, Algeria

A. Yekkour
Laboratoire de Biologie des Systèmes Microbiens (LBSM), Ecole Normale Supérieure de Kouba, Alger, Algeria

Institut National de la Recherhce Agronomique d'Algérie (INRAA), Alger, Algeria

R. P. Singh et al. (eds.), *Microbes in Microbial Communities*,
https://doi.org/10.1007/978-981-16-5617-0_16

research on actinobacterial diversity reported in three countries in Northwest Africa: Morocco, Algeria, and Tunisia. During this investigation, a total of 30 new species within the phylum *Actinobacteria* were investigated in Algeria: *Saccharothrix* (six new species), *Actinopolyspora* (five new species), *Streptosporangium* (three new species), *Actinomadura* (two new species), and one new species from each following genera: *Actinoalloteichus*, *Actinokineospora*, *Actinophytocola*, *Glycomyces*, *Mycobacterium*, *Mycolicibacter*, *Nocardiopsis*, *Planomonospora*, *Prauserella*, *Saccharopolyspora* (reclassified to *Salinifilum*), *Streptomonospora*, and *Streptomyces*. Furthermore, two new species belonging to two new genera have been reported: *Mzabimyces* has been reclassified to *Halopolyspora* and *Bounagaea*. Eleven new species: *Geodermatophilus* (four new species), *Blastococcus* (three new species), *Streptomyces* (two new species), *Nocardia* (one new species), and *Frankia* (one new species) have been described from Tunisian ecosystems. From Moroccan ecosystems, three new species of *Streptomyces* have been isolated. To the best of our knowledge, no new actinobacterial species have been reported in the remaining Maghrebian geographical area. Maghrebian efforts to isolate *Actinobacteria* have yielded 46 new taxa, 44 new species, and 2 new genera.

Keywords *Actinobacteria* · Maghreb · Taxonomy · Biodiversity · Saharan ecosystem

16.1 Introduction

The *Actinobacteria* represent one of the largest taxonomic phyla in the *Bacteria* domain. These Gram-positive bacteria are found in all varieties of aquatic and terrestrial ecosystems. Multiple lifestyles occur in the group such as plant commensals, nitrogen-fixing symbionts, and animal, plant, and even human pathogens.

A review on the biodiversity and distribution of *Actinobacteria* in Northwest African environments reveals the scant knowledge of these topics. To the best of our knowledge, this is the first overview on the biodiversity and bioprospecting of new taxa of *Actinobacteria* from Northwest African ecosystems.

S. Smaoui · L. Mellouli
Laboratory of Microorganisms and Enzymatic Biotechnology and Biomolecules, Center of Biotechnology of Sfax, University of Sfax, Sfax, Tunisia

M. Barakate
Laboratory of Microbial Biotechnologies, Agrosciences & Environment, Faculty of Sciences Semlalia, Cadi Ayyad University, Marrakech, Morocco

Mohammed IV Polytechnic University (UM6P), AGBT, AgroBiosciences, Benguerir, Morocco

16.1.1 Actinobacteria

In soil microbiology, it is necessary to distinguish between Actinomycetes (morphological term) and *Actinobacteria* (phylogenetic term). Actinomycetes constitute a diverse group of bacteria which are able to form branching filaments in at least one stage of their development, and the majority (but not all) of them belongs to the *Actinobacteria* phylum. *Actinobacteria* are a group with high guanine and cytosine content in their genomes, and most of them (but not all) are mycelial bacteria (Actinomycetes). Advances in the molecular studies revealed that the possession of branched hyphae should not automatically place bacterial genera within the *Actinobacteria* phylum. For example, the genus *Thermoactinomyces*, classified in the *Firmicutes* phylum (characterized by genomes with low guanine and cytosine content), is a Gram-positive endospore-forming bacterium that produce a mycelium that is non-septate and slenderer by comparison to micro-fungi. Consequently, *Thermoactinomyces* is an Actinomycete, but belongs to the *Firmicutes* and not the *Actinobacteria* phylum. However, other genera such as *Arthrobacter*, *Micrococcus*, and *Mycobacterium* belong to *Actinobacteria*, but are not Actinomycetes (because unable to form branching hyphae). On the other hand, *Sterptomyces* and *Nocardiopsis* are considered at the same time to be both Actinomycetes and *Actinobacteria*.

Actinobacteria is one of the largest phyla within the domain *Bacteria* (Ait Barka et al. 2015). Actinobacterial strains occupy all the niches of the biosphere, are known for their ubiquitous presence, and are recovered from a wide range of aquatic and terrestrial environments and also from living organisms such as plants, animals, and insects. Within this phylum are some of the most well-known producers of secondary metabolites, most notably the *Streptomyces* genus (Raja and Prabakaran 2011; Singh et al. 2019). The members of this phylum produce almost 70% of all antibiotics used currently, as well as many other bioactive compounds such as antimicrobial (antibacterial and antifungal), anticancer, anthelmintic, anti-inflammatory, antitumor, antioxidant, herbicide and immunosuppressive agents, and plant-growth-promoting substances and also plant disease biocontrol agents and plant growth promoting agents. Consequently, these bacteria are of major importance for biotechnology, medicine, industry, and agriculture.

Two of the main strategies to search for new taxa as sources of bioactive compounds is the isolation of rare *Actinobacteria* also named NSA (non-*Streptomyces Actinobacteria*) from different ecosystems including endophyte species associated with plants and algae and also the isolation of extremophilic *Actinobacteria* (whatever the genus) from harsh environments. Harsh conditions such as high or low temperature, high or low pH, high radiation, and high salt concentration are able to affect the metabolite profile of the extremophilic *Actinobacteria* and increase the probability to obtain new bioactive compounds. These harsh conditions can be imitated experimentally, during incubation, by using different approaches such as thermic treatment, and also applied physical (high temperature, high pH, etc.) and chemical conditions (addition of NaCl, phenol, incorporation of antibacterial agents, etc.). Furthermore, selective culture media (using some specific compounds such as chitin or humic acid as sole carbon source, and sometimes as specific nitrogen sources too) could also be used. Many scientists combined both strategies and isolated rare extremophilic actinobacterial strains from the less or underexplored and uncommon ecosystems such as arid and semi-arid habitats, Saharan soil, and palm groves (Meklat et al. 2012). It is important to say that the most dominant genus of *Actinobacteria* is always *Streptomyces*, even from exotic habitats; however, the percentage of uncommon and rare *Actinobacteria* is higher in these specific ecosystems.

16.2 Northwest African Environments

Northwest Africa, also known as Maghreb, which means both the western land and the place where the sun sets. The majority of the Maghrebian land is desert (North African desert also called Sahara). The variations in precipitation, temperature, humidity, elevation, soil type, and distance from the Mediterranean Sea or Pacific Ocean make Northwest Africa a diversified region. With an area of more than 6 million km^2, the Maghreb has a spectacular climatic diversity going from snow-capped mountains such as high summits of Toubkal (4167 m) and Chélia (2328 m) in the northern regions overlooking the Mediterranean Sea to the hottest desert in the world in the south. These climatic characteristics significantly affect the biodiversity of all living organisms, including bacteria.

Based on the Köppen–Geiger climate classification systems, the Maghreb is divided into eight distinct ecoregions. Two climatic systems: *BWh* ("hot desert climate" also named "arid, desert, hot") and *BSh* ("hot semi-arid climate" also named "arid, steppe, hot") are found in all Maghrebian countries. Three climatic systems are present in Morocco, Algeria, and Tunisia: *Csa* ("hot-summer Mediterranean climate" also named "temperate, dry summer, hot summer"); *BSk* ("cold semi-arid climate" also named "arid, steppe, cold") and *BWk* ("cold desert climate" also named "arid, desert, cold"). The climatic system *Csb* ("warm-summer Mediterranean climate" also named "cool-summer Mediterranean climate" or "temperate, dry summer, warm summer") is found only in Algeria and Morocco. In addition, the climatic system *Dsb* ("warm humid continental climate" also named "cold, dry summer, warm summer") is present only in Morocco.

16.3 Culture Media and Cultivation Conditions

Many culture media were employed in the isolation of new taxa from Maghrebian countries. Generally, the culture medium was supplemented with the antifungal agent actidione (50 mg/L) to inhibit the growth of micro-fungi. Occasionally, the addition of antibacterial agents to reduce or inhibit the growth of Gram-negative bacteria was also employed. It is necessary to mention that the addition of the antibacterial agents could also reduce or inhibit the growth of some species or strains of *Actinobacteria*. For isolation of halophilic bacteria, the addition of NaCl was used. The following culture media were used for isolating *Actinobacteria* from Northwest Africa:

- Chitin-vitamin agar medium described by Hayakawa and Nonomura (1987) consisting of (per liter of deionized water): 2 g chitin, 0.35 g K_2HPO_4, 0.15 g KH_2PO_4, 0.2 g $MgSO_4{\cdot}7H_2O$, 0.3 g NaCl, 0.02 g $CaCO_3$, 10 mg $FeSO_4{\cdot}7H_2O$, 1 mg $ZnSO_4{\cdot}7H_2O$, 1 mg $MnCl_2{\cdot}4H_2O$, and 18 g agar. The pH was adjusted to 7.2 prior to autoclaving. The B vitamins including: thiamine hydrochloride, riboflavin, niacin, pyridoxine hydrochloride, inositol, calcium pantothenate, *p*-aminobenzoic acid (0.5 mg/L of each), and biotin (0.25 mg/L) were added to the autoclaved medium. This medium has been used to isolate 12 Maghrebian taxa: 3 species of *Streptosporangium* (*S. algeriense*, *S. becharense*, and *S. saharense*), 2 species of *Saccharothrix* (*S. tamanrassetensis* and *S. ghardaiensis*), 2 species of *Actinomadura* (*A. algeriensis* and *A. adrarensis*), *Actinokineospora mzabensis*, *Actinophytocola algeriensis*, *Prauserella isguenensis*, *Planomonospora algeriensis*, and *Halopolyspora algeriensis*.
- Humic acid-vitamin agar medium reported by Hayakawa and Nonomura (1987) composed of (per liter of deionized water): 1 g humic acid, 0.5 g Na_2HPO_4, 0.05 g $MgSO_4{\cdot}7H_2O$, 1.7 g KCl, 0.01 g $FeSO_4{\cdot}7H_2O$, 0.02 g $CaCO_3$, 18 g agar. The pH was adjusted to 7.2 before autoclaving. This culture medium has been used in the isolation of nine Maghrebian taxa: three species of *Actinopolyspora* (*A. algeriensis*, *A. mzabensis*, and *A. saharensis*), four species of *Saccharothrix* (*S. algeriensis*, *S. hoggarensis*, *S. saharensis*, and *S. isguenensis*), *Actinoalloteichus hoggarensis*, and *Salinifilum ghardaiensis*.
- Luedemann medium (DSMZ medium 877) reported by Luedemann (1968), consisting of (per liter of deionized water): 0.5 g yeast extract, 1.5 g malt extract broth, 1 g soluble starch, 1.0 g glucose, 0.2 g $CaCO_3$, 0.5 g NaCl, 15 g agar. The pH was adjusted to 8.6 before autoclaving. This culture medium has been used to isolate four species of *Geodermatophilus* (*G. aquaeductus*, *G. bullaregiensis*, *G. pulveris*, and *G. sabuli*) and three species of *Blastococcus* (*B. capsensis*, *B. colisei*, and *B. xanthinilyticus*).
- Culture medium R2A agar (DSMZ medium 830) reported by Reasoner et al. (1979), consisting of (per liter of deionized water): 0.5 g yeast extract, 0.5 g proteose peptone, 0.5 g casamino acids, 0.5 g glucose, 0.5 g soluble starch, 0.3 g sodium pyruvate, 0.3 g K_2HPO_4, 0.05 g $MgSO_4{\cdot}7H_2O$, and 15 g agar (the pH was adjusted to 7.2 before autoclaving) was also used to isolate *G. pulveris* and *B. xanthinilyticus*.
- Complex agar medium reported by Chun et al. (2000), composed of (per liter of deionized water): 7.5 g peptone, 10 g yeast extract, 3 g sodium citrate, 10 g $MgSO_4{\cdot}7H_2O$, 2 g KCl, 1 mL 4.98% $FeSO_4{\cdot}7H_2O$, and 20 g agar (pH = 7.2). This medium was used to isolate two species of *Actinopolyspora* (*A. biskrensis*

and *A. righensis*), *Nocardiopsis algeriensis*, *Planomonospora algeriensis*, and *Bounagaea algeriensis*.

- Modified DPM (Defined Propionate Minimal) medium reported by Baker and O'Keefer (1984) containing per liter of distilled water: 0.5 g sodium propionate, 0.5 g sodium succinate, 0.5 g K_2HPO_4, 1 g $MgSO_4{\cdot}7H_2O$, 0.1 g $CaCl_2$, 0.01 g $FeSO_4{\cdot}7H_2O$, 0.0055 g Na_2EDTA, 0.0074 g H_3BO_3, 0.005 g $MnCl_2{\cdot}4H_2O$, 0.0032 g $ZnSO_4{\cdot}7H_2O$, 0.0004 g $CuSO_4{\cdot}5H_2O$, 0.00014 g Na_2MoO_4, and 0.000045 g $CoCl_2$. This medium was used to isolate two new species: *Frankia elaeagni* and *Nocardia casuarinae*.
- Löwenstein–Jensen growth medium has been used to isolate *Mycolicibacter algericus* as reported by Sahraoui et al. (2011).
- Middlebrook 7H10 agar plus OADC (oleic acid, bovine albumin, sodium chloride, dextrose, catalase) supplemented with 40,000 U/L polymyxin B, 16 g/L nalidixic acid, 4 g/L amphotericin B, and 20 g/L vancomycin has been used to isolate "*Mycobacterium icosiumassiliensis*" as reported by Djouadi et al. (2016).
- On the other hand, six culture media were used to isolate *Streptomyces* species from Maghrebian regions:
- International *Streptomyces* Project medium 2 (ISP 2), composed of (per liter of deionized water): 4 g yeast extract, 10 g malt extract, 4 g glucose, and 20 g agar (Shirling and Gottlieb 1966).
- Modified ISP2 (DSMZ medium 65 = GYM *Streptomyces* medium), composed of (per liter of deionized water): 4 g glucose, 4 g yeast extract, 10 g malt extract, 2 g $CaCO_3$, and 12 g agar. The pH was adjusted to 7.2 prior to autoclaving.
- International *Streptomyces* Project medium 3 (ISP 3), composed of (per liter of deionized water): 20 g oatmeal, 1 mL of trace salts solution, and 18 g agar. Trace salt solution was composed of (per 100 mL of deionized water): 0.1 g $FeSO_4{\cdot}7H_2O$, 0.1 g $MnCl_2{\cdot}4H_2O$, 0.1 g $ZnSO_4{\cdot}7H_2O$ (Shirling and Gottlieb 1966).
- International *Streptomyces* Project medium 4 (ISP 4) reported by Shirling and Gottlieb (1966). This medium consisted of (per liter of deionized water): 10 g starch, 1 g K_2HPO_4, 1 g $MgSO_4{\cdot}7H_2O$, 1 g NaCl, 2 g $(NH_4)_2SO_4$, 2 g $CaCO_3$, 1 mL of trace salt solution (as indicated in ISP3 medium), and 20 g agar. The pH was adjusted to 8 before autoclaving.
- Actinomycetes isolation agar (Olson's medium) reported by Olson (1968) which consisted of (per liter of deionized water): 5% glycerol, 0.2% sodium caseinate, 0.01% L-asparagine, 0.4% sodium propionate, 0.05% K_2HPO_4, 0.0001% $FeSO_4$, and 1.5% agar.
- Soil extract agar which reported by Ouhdouch et al. (2001) and Barakate et al. (2002).

16.4 Biodiversity of New *Actinobacteria* Taxa

Table 16.1 shows the list of new species of *Actinobacteria* isolated from Northwest African environments, as gathered from the scientific literature. Exploration of such unique habitats has led to the discovery of 44 novel taxa. The identification of 42 new species and 2 new genera (including 2 new species) indicates that these environments constitute an ecological niche with a solid potential for the study of biodiversity. The conditions of high salinity in certain niches, and the stability of the microbial communities over the years, make these habitats of great interest and a target for in-depth study, to obtain an understanding not only of the processes forming natural actinobacterial communities and of their role in the interaction with biotic and abiotic elements but also of the biodiversity of such ecosystems.

The number of new species of *Actinobacteria* from Algeria was 30; 11 from Tunisia and 3 from Morocco, as indicated in Table 16.1. Until now, no reports are available regarding the discovery of new species of *Actinobacteria* in the remaining geographical area of Maghreb. In fact, it is not possible to talk about the biogeography of prokaryotes, for example, *Saccharothrix algeriensis*'s name does not indicate that this species is living only in Algeria, it only means that this species was reported for the first time in Algeria.

It is necessary to mention that all the new taxa of strictly halophilic *Actinobacteria* discovered from Algeria, such as different species of *Actinopolyspora*, *Prauserella*, *Streptomonospora*, *Salinifilum* (*Basonym*: *Saccharopolyspora*) and *Halopolyspora* (*Basonym*: *Mzabimyces*) which require at least 7% NaCl for growth (Table 16.2), were isolated from non-saline soil.

It seems that these strictly halophilic bacteria remain attached to the salt granules although the soil is not saline. This phenomenon could be explained by the complexity of any soil sample which could include multiple microenvironments, some very saline, others less salty, and even non-saline ones allowing all these groups to occur in the same sample.

Generally, the new species of strictly halophilic bacteria reported in the literature were isolated from hypersaline areas, and the strictly alkaliphilic bacteria were isolated from alkaliphilic zones, etc. However, the strategy of the Sabaou laboratory's research is the isolation of halophiles from non-saline soil and the alkaliphiles from non-alkaline soil, etc. This original idea allowed this research team to isolate many new taxa of *Actinobacteria* never reported before.

A total of 30 new taxa (68%) were discovered from the ecoregion *BWh*: "hot desert climate" also named "arid, desert, hot," which represents almost 80% of the total area of Northwest Africa. The number of actinobacterial species reported from the ecoregion *Csa*: "hot-summer Mediterranean climate" also named "temperate, dry summer, hot summer" was 10. Furthermore, two new taxa were isolated from the both ecoregions *BSk*: "cold semi-arid climate" also named "arid, steppe, cold" and *BSh* "hot semi-arid climate" also named "arid, steppe, hot." Until now, no studies on the isolation of *Actinobacteria* from other ecoregions (*BWk*, *Csb*, and *Dsb*) have been published.

Table 16.1 List of new species of *Actinobacteria* isolated from different countries in Northwest Africa

Order	Family	Genus	Species/etymology	Country
Actinomycetales	*Actinopolysporaceae*	*Actinopolyspora*	*A. algeriensis* sp. nov. (Meklat et al. 2012) al.ger.i.en'sis. N.L. fem. adj. *algeriensis*, pertaining to Algeria, the source of the soil from which the type strain was isolated	Algeria
			A. biskrensis sp. nov. (Saker et al. 2015a) bis.kren'sis, N.L. fem. adj. *biskrensis* pertaining to Biskra, where the type strain was isolated	Algeria
			A. mzabensis sp. nov. (Meklat et al. 2013a) m.za.ben'sis. N.L. fem. adj. *mzabensis* pertaining to Mzab, the source of the soil from which the type strain was isolated	Algeria
			A. righensis sp. nov. (Meklat et al. 2013b) ri.ghen'sis, N.L. fem. adj. *righensis* pertaining to Oued Righ, where the type strain was isolated	Algeria
			A. saharensis sp. nov. (Meklat et al. 2013c) sa.ha.ren'sis, N.L. fem. adj. *saharensis* pertaining to Sahara, where the type strain was isolated	Algeria
		Mzabimyces	*M. algeriensis* sp. nov. gen. nov. (Saker et al. 2014): *Mzabimycetaceae* M.za.bi.my'ces. N.L. n. *Mzab*, an arid region named Mzab in the south of Algeria, from where the microorganism was isolated; Gr. masc. n. *mukês* fungus; N.L. masc. n. *Mzabimyces*, a fungus of Mzab region al.ge.ri.en'sis. N.L. masc. adj. *algeriensis*, pertaining to Algeria, the source of the soil from which the type strain was isolated →Correct name: *Halopolyspora algeriensis* comb. nov. (Lai et al. 2017): *Actinopolysporaceae* Ha.lo.po.ly.spo'ra. Gr. n. *hals* salt; Gr. adj. *polus* many; Gr. fem. n. *spora* seed and in biology 191 a spore; N.L. fem. n. *Halopolyspora*, the saline	Algeria

			many-spore bacteria al.ge.ri.en'sis. N.L. fem. n. *algeriensis*, pertaining to Algeria, the source of the soil from which the type strain was isolated	
Frankiales	*Frankiaceae*	*Frankia*	*F. elaeagni* sp. nov. (Nouioui et al. 2016) e. lae.ag'ni. N.L. gen. fem. n. *elaeagni* of *Elaeagnus*, referring to the source of the isolate	Tunisia
Geodermatophilales	*Geodermatophilaceae*	*Blastococcus*	*B. capsensis* sp. nov. (Hezbri et al. 2016b) caps.en'sis, N.L. masc. adj. *capsensis* of or belonging to Capsa, the archeological city now known as Gafsa located in southeast Tunisia, where the bacterium was collected	Tunisia
			B. colisei sp. nov. (Hezbri et al. 2017) co.li.se'i. N.L. gen. neut. n. *colisei* of the coliseum Thysdrus, site of isolation of the type strain	Tunisia
			B. xanthinilyticus sp. nov. (Hezbri et al. 2018) xan.thi.ni.ly'ti.cus. N.L. n. *xanthinum* xanthine; N.L. adj. *lyticus* (from Gr. adj. *lytikos*) able to dissolve; N.L. masc. adj. *xanthinilyticus* hydrolysing xanthine.	Tunisia
		Geodermatophilus	*G. aquaeductus* sp. nov. (Hezbri et al. 2015a) a.quae.duc'tus. L. gen. n. *aquaeductus*, of an aqueduct, pertaining to the Roman aqueducts that were used to carry water, referring to the location where the type strain of the species was sampled	Tunisia
			G. bullaregiensis sp. nov. (Hezbri et al. 2015b) bul.la.re.gi.en'sis. L. n. Bulla Regia, a Roman town in Northern Africa, today northwestern Tunisia; N.L. masc. adj. *bullaregiensis*, derived from Bulla Regia referring to the origin of isolation	Tunisia

(continued)

Table 16.1 (continued)

Order	Family	Genus	Species/etymology	Country
			G. pulveris sp. nov. (Hezbri et al. 2016a) pul've.ris L. gen. n. *pulveris* of dust, referring to the isolation source	Tunisia
			G. sabuli sp. nov. (Hezbri et al. 2015c) sa'bu.li. L. gen. neut. n. *sabuli* of sand, referring to the origin of isolation of the species	Tunisia
Glycomycetales	*Glycomycetaceae*	*Glycomyces*	*G. algeriensis* sp. nov. (Labeda and Kroppenstedt 2004) al.ger.i.en'sis. N.L. masc. adj. *algeriensis* from Algeria, the place of origin of the type strain	Algeria
Mycobacteriales	*Mycobacteriaceae*	*Mycobacterium*	*M. algericum* sp. nov. (Sahraoui et al. 2011) al.ge'ri.cum. N.L. neut. adj. *algericum* pertaining to Algeria, the country where the strain was first isolated →Correct name: *Mycolicibacter algericus* comb. nov. (Gupta et al. 2018) My.co.li.ci.bac'ter. N.L. n. *acidum mycolicum*, mycolic acid; N.L. masc. n. *bacter*, rod; N.L. masc. n. *Mycolicibacter*, a genus of mycolic acid containing rod-shaped bacteria al.ge'ri.cus. N.L. masc. adj. *algericus*, of or pertaining to Algeria, the country where the strain was first isolated	Algeria
			"*M. icosiumassiliensis*" sp. nov. (Djouadi et al. 2016) → Not validly published i.co.si.u.ma.si.li.en'sis; L. neut. n. *icosiumassiliensis*, from the combination of Icosium, the Latin name of Algiers where the strain was first isolated and Massilia, the Latin name of Marseille, where the strain was described	Algeria
	Nocardiaceae	*Nocardia*	*N. casuarinae* sp. nov. (Ghodhbane-Gtari et al. 2014) ca.su.a.ri'nae. N.L. gen. N. casuarinae of the *casuarina*, referring to the isolation of the type strain from *C. glauca*	Tunisia

Pseudonocardiales	*Pseudonocardiaceae*	*Actinoalloteichus*	*A. hoggarensis* sp. nov. (Boudjelal et al. 2015) hog.gar.en'sis. N.L. masc. adj. *hoggarensis* pertaining to Hoggar, the source of the soil from which the type strain was isolated	Algeria
		Actinokineospora	*A. mzabensis* sp. nov. (Aouiche et al. 2015) mzab.en'sis. N. L. fem. adj. *mzabensis* pertaining to Mzab, the source of the soil from which the type strain was isolated	Algeria
		Actinophytocola	*A. algeriensis* sp. nov. (Bouznada et al. 2016a) al.ge.ri.en'sis. N.L. masc. adj. *algeriensis* of Algeria, where the type strain originated	Algeria
		Bounagaea	*B. algeriensis* sp. nov. gen. nov. (Meklat et al. 2015) bou.na.ga'ea, N.L. gen. masc. n. *bounagaea* of Bounaga, named in honor of the memory and untimely death of our late Professor Djilali Bounaga (1931–1980) of the Centre National de Recherche sur les Zones Arides (CNRZA), and also in honor of our Professor Nicole Bounaga-Riveill (CNRZA and URZA) for their broad contributions in teaching and training of researchers in the field of biology in Algeria al.ge.ri.en'sis. N.L. fem. adj. *algeriensis*, pertaining to Algeria, the source of the soil from which the type strain was isolated	Algeria
		Prauserella	*P. isguenensis* sp. nov. (Saker et al. 2015b) is.guen.en'sis. N.L. fem. adj. *isguenensis* pertaining to Béni-Isguen, the source of the soil from which the type strain was isolated	Algeria
		Saccharopolyspora	*S. ghardaiensis* sp. nov. (Meklat et al. 2014a) ghar.da.i.en'sis. N.L. fem. adj. *ghardaiensis*, pertaining to Ghardaïa, the source of the soil from which the type strain was isolated →Correct name: *Salinifilum ghardaiensis* comb. nov. (Moshtaghi Nikou et al. 2017) Sa.li.ni.fi'lum. L. pl. n. *salinae*, salt source; L. neut. n. *filum*, thread; N.L. neut. n. *Salinifilum*, a thread from a salt source ghar.da.i.en'sis. N.L. fem. adj. *ghardaiensis*, pertaining to Ghardaïa, the source of the soil from which the type strain was isolated	Algeria

(continued)

Table 16.1 (continued)

Order	Family	Genus	Species/etymology	Country
		Saccharothrix	*S. algeriensis* sp. nov. (Zitouni et al. 2004) al.ger.i.en'sis. N.L. fem. adj. *algeriensis* of Algeria, where the type strain originated	Algeria
			S. ghardaiensis sp. nov. (Bouznada et al. 2017) ghar.da.i.en'sis, N.L. fem. adj. *ghardaiensis* pertaining to Ghardaïa, Mzab, the source of the soil from which the type strain was isolated	Algeria
			S. hoggarensis sp. nov. (Boubetra et al. 2013a) hog.gar.en'sis. N.L. fem. adj. *hoggarensis* pertaining to Hoggar, the source of the soil from which the type strain was isolated	Algeria
			S. isguenensis sp. nov. (Bouznada et al. 2016b) is.guen.en'sis. N.L. fem. adj. *isguenensis* pertaining to Béni-Isguen (Ghardaïa, Mzab), the source of the soil from which the type strain was isolated	Algeria
			S. saharensis sp. nov. (Boubetra et al. 2013b) sa.har.en'sis. N.L. fem. adj. *saharensis* referring to the Sahara, where the type strain was isolated), from which the type strain was isolated	Algeria
			S. tamanrassetensis sp. nov. (Boubetra et al. 2015) ta.man.ras.set.en'sis. N.L. fem. adj. *tamanrassetensis* referring to Tamanrasset, where the type strain was isolated	Algeria
Streptomycetales	*Streptomycetaceae*	*Streptomyces*	*S. marokkonensis* sp. nov. (Bouizgarne et al. 2009) ma.rok.ko.nen'sis. N.L. masc. adj. *marokkonensis* pertaining to Marokko, the Dutch name of Morocco, from where the type strain was isolated	Morocco
			"*S. massilialgeriensis*" sp. nov. (Djaballah et al. 2018) → Not validly published ma.sil.io.al.ge.ri.en'sis. L. gen. masc. n. *massilioalgeriensis*, a combination of massili for Massilia, the Roman name of Marseille where the type	Algeria

			strain was identified and sequenced, and algeriensis, for Algeria, the country in North Africa, where the strain was isolated	
			S. thinghirensis sp. nov. (Loqman et al. 2009) thin.ghi.ren'sis. N.L. masc. adj. *thinghirensis* of Thinghir, named after the town in Southern Morocco where the strain was isolated	Morocco
			"*S. tunisialbus*" sp. nov. (Ayed et al. 2018) → Not validly published tu.ni.si.al'bus. N.L. fem. n. *Tunisia*, Tunisia, the country where the strain was isolated; L. adj. *albus*, white, referring to the white color of spores; N.L. fem. adj. *tunisialbus*, an organism with white spores from Tunisia	Tunisia
			S. tunisiensis sp. nov. (Slama et al. 2014) tu.ni.si.en'sis. N.L. mas. adj. *tunisiensis* pertaining to Tunisie, the French name of Tunisia where the strain was isolated	Tunisia
			S. youssoufiensis sp. nov. (Hamdali et al. 2011) yous.sou.fi.en'sis: N.L. gen. masc. n. *youssoufiensis* of Youssoufia, named after the rock phosphate mine town in Morocco where the strain was isolated	Morocco
Streptosporangiales	*Nocardiopsaceae*	*Nocardiopsis*	*N. algeriensis* sp. nov. (Bouras et al. 2015) al.ger.i.en'sis. N.L. fem. adj. *algeriensis* referring to Algeria, the country from where the type strain was isolated	Algeria
		Streptomonospora	*S. algeriensis* sp. nov. (Meklat et al. 2014b) al.ger.i.en'sis. N.L. fem. adj. *algeriensis*, pertaining to Algeria, the source of the soil from which the type strain was isolated	Algeria
	Streptosporangiaceae	*Planomonospora*	*P. algeriensis* sp. nov. (Chaabane Chaouch et al. 2017) al.ge.ri.en'sis. N.L. fem. adj. *algeriensis* pertaining to Algeria, the source of the soil from which the type strain was isolated	Algeria
		Streptosporangium	*S. algeriense* sp. nov. (Boubetra et al. 2016) al.ge.ri.en'se. N.L. neut. adj., *algeriense* pertaining to Algeria, from where the type strain was isolated	Algeria

(continued)

Table 16.1 (continued)

Order	Family	Genus	Species/etymology	Country
			S. becharense sp. nov. (Chaabane Chaouch et al. 2016b) be.char.en'se. N.L. neut. adj. *becharense* pertaining to Béchar, south-west Algeria, the source of the soil from which the type strain was isolated	Algeria
			S. saharense sp. nov. (Chaabane Chaouch et al. 2016a) sa.har.en'se. N.L. neut. adj. *saharense* pertaining to Sahara, the source of the soil from which the type strain was isolated	Algeria
	Thermomonosporaceae	*Actinomadura*	*A. adrarensis* sp. nov. (Lahoum et al. 2016b) ad.rar. en'sis. N.L. fem. adj. *adrarensis* pertaining to Adrar, the source of the soil from which the type strain was isolated	Algeria
			A. algeriensis sp. nov. (Lahoum et al. 2016a) al.ger.i.en'sis. N.L. fem. adj. *algeriensis* referring to Algeria, the country from where the type strain was isolated	Algeria

Table 16.2 Molecular and physiological characteristics of new species of *Actinobacteria* isolated from different countries in Northwest Africa

Species/type strains/ 16S rRNA gene/risk group	Source/climate classification system	Isolation conditions	16S rRNA similarity (DDH) with the neighboring species	Temperature range (°C) (optimal)	pH range (optimal)	NaCl range (optimal)	Major menaquinones (>10%)	Major fatty acids (>10%)
Actinopolyspora algeriensis **Type strains:** DSM 45476^T, CCUG 62415^T. **16S rRNA gene:** HQ918195. **Risk group:** 1	Saharan soil from Bamendil palm grove, Ouargla, South of Algeria (*BWh*)	Humic acid-vitamin agar supplemented with 20% (w/v) NaCl, incubation at 30 °C for 25 days	98.5% (43.6%) with *A. halophila* DSM 43834^T	25–40 (30)	5–8 (7)	7–32 (15–25)	MK-9(H_4), MK-10(H_4)	anteiso-$C_{17:0}$, iso-$C_{15:0}$, anteiso-$C_{15:0}$, iso-$C_{16:0}$
Actinopolyspora biskrensis **Type strains:** DSM 46684^T, CECT 8576 ^{T}. **16S rRNA gene:** KJ504178. **Risk group:** 1	Saharan soil from Biskra, North-East of Algeria (*BWh*)	Complex medium agar supplemented with 15% (w/v) NaCl, incubation at 30 °C for 25 days	99.2% (57.2%) with *A. saharensis* DSM 45459^T, 99.1% (68.4%) with *A. halophila* DSM 43834^T and 99.0% (45.6%) with *A. algeriensis* DSM 45476^T	20–37 (30)	6–8 (7)	10–30 (15–20)	MK-9(H_4), MK-10(H_4)	anteiso-$C_{17:0}$, iso-$C_{15:0}$, iso-$C_{17:0}$
Actinopolyspora mzabensis **Type strains:** DSM 45460^T, CCUG 62965^T. **16S rRNA gene:** HQ918202. **Risk group:** 1	Saharan soil from Ntissa palm grove, Béni-isguen, Ghardaïa, Mzab, South of Algeria (*BWh*)	Humic acid-vitamin agar supplemented with 20% (w/v) NaCl, incubation at 30 °C for 25 days	98.0% (44.5%) with *A. erythraea* DSM 45583^T and 97.6% (42.35%) with *A. alba* DSM 45004^T	25–40 (30)	5–8 (7)	7–32 (10–28)	MK-10(H_4), MK-9(H_4)	anteiso-$C_{17:0}$, iso-$C_{16\ 0}$, iso-$C_{15:0}$
Actinopolyspora righensis **Type strains:**	Saharan soil from Djamâa, Oued Righ,	Complex medium agar supplemented	97.8% with *A. xinjiangensis* TRM 40136^T, 97.7% with	20–40 (28–32)	5–8 (6–7)	10–30 (15–25)	MK-10(H_4), MK-9(H_4)	anteiso-$C_{17:0}$, iso-$C_{17:0}$,

(continued)

Table 16.2 (continued)

Species/type strains/ 16S rRNA gene/risk group	Source/climate classification system	Isolation conditions	16S rRNA similarity (DDH) with the neighboring species	Temperature range (°C) (optimal)	pH range (optimal)	NaCl range (optimal)	Major menaquinones (>10%)	Major fatty acids (>10%)
DSM 45501^T, CCUG 63368^T, MTCC 11562^T. **16S rRNA gene:** HQ918196. **Risk group:** 1	El-Oued, South of Algeria (*BWh*)	with 20% (w/v) NaCl, incubation at 30 °C for 35 days	*A. erythraea* DSM 45583^T and 97.5% with *A. alba* DSM 45004^T					iso-$C_{15:0}$, iso-$C_{16:0}$
Actinopolyspora saharensis **Type strains:** DSM 45459^T, CCUG 62966^T. **16S rRNA gene:** HQ918198. **Risk group:** 1	Saharan soil from El-Oued, South of Algeria (*BWh*)	Humic acid-vitamin agar supplemented with 20% (w/v) NaCl, incubation at 30 °C for 28 days	98.8% (30.5%) with *A. algeriensis* DSM 45476^T, 98.5% (55.1%) with *A. halophila* DSM 43834^T and 97.2% (31.5%) *A. mortivallis* DSM 44261^T	20–35 (28–32)	5–8 (6–7)	10–30 (15–25)	MK-10(H_4), MK-9(H_4)	anteiso-$C_{17:0}$, iso-$C_{15:0}$, iso-$C_{16:0}$
Halopolyspora algeriensis (*Basonym*: *Mzabimyces algeriensis*) **Type strains:** DSM 46680^T, MTCC 12101^T. **16S rRNA gene:** KJ574202. **Risk group:** 1	Saharan soil from Ahbas palm grove, Béni-isguen, Mzab, Ghardaïa, South of Algeria (*BWh*)	Chitin-vitamin agar supplemented with 15% (w/v) NaCl, incubation at 30 °C for 25 days	92.02% with *S. qijiaojingensis* YIM 91168^T	20–45 (30)	5–9 (7)	7–30 (10–20)	MK-9(H_4)	iso-$C_{15:0}$, iso-$C_{16:0}$, iso-$C_{17:0}$, anteiso-$C_{17:0}$
Frankia elaeagni **Type strains:**	Fragments from *Elaeagnus*	Modified DPM, incubation at	The 16S rRNA gene sequence identity for	(28)	(6.3–6.8)	ND		

DSM 46783^T, CECT 9031^T. **16S rRNA gene:** KX674377. **Risk group:** 1	*angustifolia*, Gafsa, Tunisia (*BWh*)	30 °C for 3 weeks	the strains ranged from 98.1 to 98.9% (DDH well below 70%)				MK-9(H$_4$), MK-9(H$_6$), MK-9(H$_8$)	iso-C$_{16:0}$, C$_{16:0}$, C$_{17:1}$ $\omega 8c$
Blastococcus capsensis **Type strains:** DSM 46835^T, CECT 8876^T. **16S rRNA gene:** LN626274. **Risk group:** 1	From an archeological limestone sample (the wall rock of a Roman pool) in Gafsa, Tunisia (*BWh*)	Luedemann medium, incubation at 28 °C for 2 weeks	99.4 % (48.6%) with *B. saxobsidens* DSM 44509^T	20–37 (30)	5.5–9 (6–8)	0–3	MK-9(H$_4$), MK-9	C$_{17:1}$ $\omega 8c$, C$_{16:1}$ $\omega 7c$, iso-C$_{15:0}$, iso-C$_{16:0}$
Blastococcus colisei **Type strains:** DSM 46837^T, CECT 8823^T. **16S rRNA gene:** LN626273. **Risk group:** 1	Limestone from a wall rock of ruins (amphitheater) of the Coliseum Thysdrus monument, El-Jem, Mahdia, Tunisia (*BSh*)	Luedemann medium, incubation at 28 °C for 2 weeks	98.3% with *B. jejuensis* DSM 19597^T	10–40 (30)	6–10 (6.5–8)	0–4	MK-9(H$_4$), MK-9(H$_2$), MK-9	iso-C$_{16:0}$, C$_{18:1}$ $\omega 9c$, C$_{17:1}$ $\omega 8c$, iso-H-C$_{16:1}$
Blastococcus xanthinilyticus **Type strains:** DSM 46842^T, CECT 8884^T, **16S rRNA gene:** LN626275. **Risk group:** 1	From marble dust collected from Bulla Regia monuments, Jendouba, northern Tunisia (*Csa*)	Luedemann medium and R2A, incubation at 28 °C for 2 weeks	99.5% (38.0%) with *B. saxobsidens* DSM 44509^T and 99.3% (11.5%) with *B. capsensis* DSM 46835^T	10–45 (28–35)	6–10 (6.5–8)	0	MK-9(H$_4$), MK-8(H$_4$), MK-9(H$_2$)	iso-C$_{16:0}$, C$_{17:1}$ $\omega 8c$, iso-C$_{15:0}$, C$_{18:1}$ $\omega 9c$

(continued)

Table 16.2 (continued)

Species/type strains/ 16S rRNA gene/risk group	Source/climate classification system	Isolation conditions	16S rRNA similarity (DDH) with the neighboring species	Temperature range (°C) (optimal)	pH range (optimal)	NaCl range (optimal)	Major menaquinones (>10%)	Major fatty acids (>10%)
Geodermatophilus aquaeductus **Type strains:** DSM 46834^T, CECT 8822^T, **16S rRNA gene:** LN626272. **Risk group:** 1	From the surface of an altered calcarenite stone found in the ruins of a Roman aqueduct, the aqueduct of Hadrian, northern Tunisia (*Csa*)	Luedemann medium, incubation at 28 °C for 20 days	99.6% (39.9%) with *G. amargosae* DSM 46136^T, 99.6% (33.9%) with *G. normandii* DSM 45417^T, 99.2% (27.0%) with *G. saharensis* DSM 45423^T, and 99.1% (13.2%) with *G. tzadiensis* DSM 45416^T	10–45 (25–35)	5–10.5 (6–9.5)	0–3	MK-9(H_4)	iso-$C_{16:0}$, iso-$C_{15:0}$, iso-H-$C_{16:0}$
Geodermatophilus sabuli **Type strains:** DSM 46844^T, CECT 8820^T. **16S rRNA gene:** LN626269. **Risk group:** 1	From a limestone grooves collected from Ong Jmal, Tozeur, Sahara desert of Tunisia (*BWh*)	Luedemann medium	98.3% with G. ruber DSM 45317^T	25–45 (35–40)	6.5–10.5 (6.5–8)	0–4	MK-9(H_4), MK-9(H_6)	iso$C_{16:0,}$ iso-$C_{15:0}$, anteiso-$C_{15:0}$, anteiso-$C_{17:0}$
Geodermatophilus bullaregiensis **Type strains:** DSM 46841^T, CECT 8821^T. **16S rRNA gene:** LN626271. **Risk group:** 1	From crevices of a marble rock surface located in the ruin of Bulla Regia monument, located in north western Tunisia (*Csa*)	Luedemann medium, incubation at 28 °C for 10 days	98.3% with *G. saharensis* DSM 45423^T	10–40 (25–35)	6.5–10.5 (7–8.5)	0–4	MK9(H_4), MK-9(H_0)	iso-$C_{16:0}$, iso-$C_{15:0}$

Geodermatophilus pulveris **Type strains:** DSM 46839^T, CECT 9003^T. **16S rRNA gene:** LN626270. **Risk group:** 1	From desert limestone dust from the Tunisian part of the Grand Erg Oriental in the Sahara desert (*BWh*)	Luedemann medium and R2A, incubation at 28 °C for 3 weeks	99.1% (33.7%) with *G. poikilotrophus* DSM 44209^T and 98.7% (36.2%) with *G. siccatus* DSM 45419^T	10–40 (30–35)	5.5–11 (7–10)	0–2 (0)	MK-9(H_4)	iso-$C_{16:0}$, $C_{16:1}$ $\omega 7c$
Glycomyces algeriensis **Type strains:** DSM 44727^T, NRRL B-16327^T, JCM 14910^T, NBRC 103888^T. **16S rRNA gene:** AY462044. **Risk group:** 1	From soil collected in a potato field in Oran, Algeria (*Csa*)	–	99.6% (40%) with *G. rutgersensis* IFO 14488^T	15–37	7–10	0–5	MK-10, MK-11, MK-12	anteiso-$C_{15:0}$, anteiso-$C_{17:0}$, iso-$C_{15:0}$, anteiso-$C_{17:0}$ iso-$C_{16:0}$
Mycolicibacter algericus (*Basonym*: *Mycobacterium algericum*) **Type strains:** DSM 45454^T, CIP 110121^T, **16S rRNA gene:** GU564404. **Risk group:** 1	From a goat pulmonary lesion from slaughterhouse in Souk El-Tenine, Bejaïa, North of Algeria (*Csa*)	Löwenstein–Jensen growth medium, incubation at 37 °C for 35 days	98.0% with *M. senuense* 05-832^T, 98.7% with *M. terrae* ATCC 15755^T and 97.5% with *M. arupense* AR30097^T	25–40 (37–40)	–	–	–	–
"*Mycobacterium icosiumassiliensis*" **Type strains:**	Water lake surface of Réghaïa,	Modified Middlebrook 7H10 agar,	99.4% with *M. arupense* AR-30097^T and 99.1%	28–42 (35)	–	0–3	–	–

(continued)

Table 16.2 (continued)

Species/type strains/ 16S rRNA gene/risk group	Source/climate classification system	Isolation conditions	16S rRNA similarity (DDH) with the neighboring species	Temperature range (°C) (optimal)	pH range (optimal)	NaCl range (optimal)	Major menaquinones (>10%)	Major fatty acids (>10%)
DSM 100711^T, CSUR P1561^T. **16S rRNA gene:** KT592291. **Risk group:** 1	Algiers, North of Algeria (*Csa*)	incubation at 35 °C for 1 month	() with *M. nonchromogenicum* ATCC 19530^T					
Nocardia casuarinae **Type strains:** DSM 45978^T, CECT 8469^T. **16S rRNA gene:** KF924767. **Risk group:** 1	From a root nodule of *Casuarina glauca* (filao) plants, growing in Zerniza arboreatum located in Sejnane, Bizerte Tunisia (*Csa*)	Liquid DPM medium at 30 °C for 1 month	93.33% with *N. seriolae* JCM3360^T, 96.53% with *N. aobensis* DSM 44805^T, 96.54% with *N. transvalensis* DSM 43405^T and 96.55% with *N. elegans* IMMIB N-402^T	15–32 (28)	6–10 (7)	0–4	MK-8(H_4c)	10-methyl $C_{18:0}$, $C_{17:1}$ *ω9c*, $C_{16:1}$ *ω7c*, $C_{16:0}$
Actinoalloteichus hoggarensis **Type strains:** DSM 45943^T, CECT 8639^T. **16S rRNA gene:** KJ504175. **Risk group:** 1	Saharan soil sample from Hoggar, Tamanrasset, southern Algeria (*BWh*)	Humic acid-vitamin agar, incubation at 30 °C for 21 days	99.3% (26.5%) with *A. hymeniacidonis* DSM 45092^T and 98.7% (28.0%) with *A. nanshanensis* DSM 45655^T	20–37 (30)	6–9 (7)	3–7 (5)	MK-9(H_4), MK-10(H_4).	iso-$C_{15:0}$, iso-$C_{16:0}$, anteiso-$C_{15:0}$ $C_{17:1}$ *ω9c*
Actinokineospora mzabensis **Type strains:** DSM 45961^T,	Saharan soil from Ntissa palm grove, Béni-isguen,	Chitin-vitamin agar supplemented with rifampicin	96.2% with *A. inagensis* DSM 44258^T and 97.8% with	20–35 (25–30)	5–9 (5–7)	0–2	MK-9(H_4)	iso-$C_{16:0}$

CECT 8578^T. **16S rRNA gene:** KJ504177. **Risk group:** 1	Mzab, Ghardaïa, South of Algeria (*BWh*)	(25 mg/L), incubation at 30 °C for 14 days	*A. baliensis* NBRC 104211^T					
Actinophytocola algeriensis **Type strains:** DSM 46746^T, CECT 8960^T. **16S rRNA gene:** KU242749. **Risk group:** 1	Saharan soil from Berriane palm grove, Ghardaïa, Mzab, southern Algeria (*BWh*)	Chitin-vitamin agar, incubation at 30 °C for 6 weeks	98.5% (23.7%) with *A. gilvus* DSM 45828^T, 98.0% (17.9%) with *A. corallina* DSM 45659^T and 97.5% (32.9%) with *A. timorensis* DSM 45660^T	20–40	6–9	0–4 (0–2)	MK-9(H_4)	iso-$C_{16:0}$
Bounagaea algeriensis **Type strains:** DSM 45966^T, CECT 8470^T. **16S rRNA gene:** KF981441. **Risk group:** 1	Saharan soil from El-Goléa, Ghardaïa, South of Algeria (*BWh*)	Complex medium agar supplemented with 20% (w/v) NaCl, incubation at 30 °C for 21 days	93.0% with *Mzabimyces algeriensis* DSM 46680^T, 91.2% with *Saccharopolyspora ghardaiensis* DSM 45606^T, 90.8% with *Halopolyspora alba* DSM 45976^T and 90.0% with *Actinopolyspora mortivallis* DSM 44261^T	20–40 (28–35)	5–8 (6–7)	10–35 (15–25)	MK-9(H_4), MK-9(H_2), MK-10(H_2)	anteiso-$C_{17:0}$, iso-$C_{15:0}$, iso-$C_{17:0}$
Prauserella isguenensis **Type strains:** DSM 46664^T, CECT 8577^T. **16S rRNA gene:**	Saharan soil from Ahbas, Béni-Isguen, Ghardaïa, Mzab, Algeria (*BWh*)	Chitin-vitamin agar supplemented with 15% NaCl, incubation at 30 °C for 6 weeks	98.9% (43.6%) with *P. flava* DSM 45265^T, 98.8% (65.5%) with *P. alba* DSM 44590^T, 98.6% (40.6%) with *P. aidingensis* DSM 45266T, 98.5%	20–45 (30–37)	5–9 (7)	5–25 (7–15)	MK-9(H_4)	iso-$C_{16:0}$

(continued)

Table 16.2 (continued)

Species/type strains/ 16S rRNA gene/risk group	Source/climate classification system	Isolation conditions	16S rRNA similarity (DDH) with the neighboring species	Temperature range (°C) (optimal)	pH range (optimal)	NaCl range (optimal)	Major menaquinones (>10%)	Major fatty acids (>10%)
KJ504173. **Risk group:** 1			(27.9%) with *P. salsuginis* DSM 45264^T and 98.4% (45.0%) with *P. sediminis* DSM 45267^T					
Salinifilum ghardaiensis (*Basonym*: *Saccharopolyspora ghardaiensis*) **Type strains:** DSM 45606^T, CECT 8304^T, CCUG 63370^T. **16S rRNA gene:** KC427277. **Risk group:** 1	Saharan soil from Chaâbet Ntissa, Béni-isguen, Ghardaïa, Mzab, South of Algeria (*BWh*)	Humic acid-vitamin agar supplemented with 20% (w/v) NaCl, incubation at 30 °C for 45 days	92.1–94.3% with other members of the genus *Saccharopolyspora*	25–45 (30–35)	5–8 (6–7)	7–32 (15–25)	MK-9(H_4)	anteiso-$C_{17:0}$, iso-$C_{15:0}$, iso-$C_{17:0}$, iso *w*9*c*-$C_{17:1}$, (*cis*9 iso-$C_{17:1}$)
Saccharothrix algeriensis **Type strains:** DSM 44581^T, NRRL B-24137^T, JCM 13242^T, NBRC 101915^T, **16S rRNA gene:** AY054972. **Risk group:** 1	Saharan soil from the palm grove of Adrar oasis, Algeria (*BWh*)	Humic acid-vitamin agar supplemented with streptomycin sulfate (10 μg/mL), incubation at 30 °C for 14 days	98.8% (55.9%) with *S. australiensis* DSM 43800^T	18–45	5–9	0–4	MK-9(II_4)	iso-$C_{16:0}$, iso-H-$C_{16:0}$ (iso-$C_{16:0}$ H), iso-2-hydroxy-$C_{16:0}$, iso-$C_{15:0}$

Saccharothrix hoggarensis **Type strains:** DSM 45457^T, CCUG 60214^T. **16S rRNA gene:** HQ399564. **Risk group:** 1	Saharan soil in the Hoggar, Tamanrasset, South of Algeria (*BWh*)	Humic acid-vitamin agar, incubation at 30 °C for 14 days	98.9% (16.05%) with *S. longispora* DSM 43749^T, 98.4% (22.0%) with *S. xinjiangensis* DSM 44896^T and 98.2% (50.05%) with *S. texasensis* DSM 44231^T	20–45 (30)	6–9 (7)	0–5	MK-9(H_4), MK-7(H_4)	iso-$C_{16:0}$, iso-$C_{15:0}$
Saccharothrix saharensis **Type strains:** DSM 45456^T, CCUG 60213^T. **16S rRNA gene:** FJ379333. **Risk group:** 1	From a soil sample collected from Adrar palm grove, southern Algeria (*BWh*)	Humic acid-vitamins agar, incubation at 30 °C for 14 days	99.3% (16.2%) with *S. xinjiangensis* DSM 44896^T and 98.9% (33.9%) with *S. texasensis* DSM 44231^T	15–45 (30)	7–9 (7)		MK-9(H_4), MK-7(H_4)	iso-$C_{16:0}$, iso-$C_{15:0}$
Saccharothrix tamanrassetensis **Type strains:** DSM 45947^T, CECT 8640^T. **16S rRNA gene:** KJ504176. **Risk group:** 1	Saharan soil from Tamanrasset, Hoggar, southern Algeria (*BWh*)	Chitin-vitamin agar, incubation at 30 °C for 14 days	98.9% (60.5%) with *S. australiensis* DSM 43800^T, 98.8% (36.2%) with *S. xinjiangensis* DSM 44896^T, 98.7% (24.0%) with *S. algeriensis* DSM 44581^T and 98.6% (49.3%) with *S. espanaensis* DSM 44229^T	–	–	–	MK-9(H_4), MK-10(H_4)	iso-$C_{16:0}$, iso-$C_{15:0}$
Saccharothrix ghardaiensis **Type strains:** DSM 46886^T, CECT 9046^T.	Saharan soil from Metlili, Mzab, Ghardaïa, South	Chitin-vitamin agar, incubation at 30 °C for 21 days	99.2% (17.5%) with *S. espanaensis* DSM 44229^T, 98.7% (12.5%) with *S. variisporea* DSM 43911^T and	20–40 (30)	6–9 (7)	0–1 (0)	MK-9(H_4)	iso-$C_{16:0}$, iso-$C_{15:0}$

(continued)

Table 16.2 (continued)

Species/type strains/ 16S rRNA gene/risk group	Source/climate classification system	Isolation conditions	16S rRNA similarity (DDH) with the neighboring species	Temperature range (°C) (optimal)	pH range (optimal)	NaCl range (optimal)	Major menaquinones (>10%)	Major fatty acids (>10%)
16S rRNA gene: KY021820. **Risk group:** 1	of Algeria (*BWh*)		98.6% (13.3%) with *S. texasensis* NRRL B-16134^T					
Saccharothrix isguenensis **Type strains:** DSM 46885^T, CECT 9045^T. **16S rRNA gene:** KU933253. **Risk group:** 1	Saharan soil sample from Béni-isguen, Mzab, Ghardaïa, South of Algeria (*BWh*)	Humic acid-vitamin agar, incubation at 30 °C for 21 days	99.8% (11.7%) with *S. ecbatanensis* DSM 45486^T, 99.3% (41.9%) with *S. hoggarensis* DSM 45457^T, 98.6% (29.7%) with *S. longispora* DSM 43749^T and 98.6% (34.2%) with *S. yanglingensis* DSM 45665^T	20–40 (30)	6–9 (7)	0–2 (0-1)	MK-9(H_4), MK-12(H_4)	iso-$C_{16:0}$, iso-$C_{15:0}$, iso-$C_{17:0}$, iso-$C_{17:1}$ $\omega 9c$
Streptomyces marokkonensis **Type strains:** DSM 41918^T, LMG 23016^T, JCM 17027^T. **16S rRNA gene:** AJ965470. **Risk group:** 1	A rhizosphere soil of an argan tree *Argania spinosa* L., Essaouira, south of Morocco (*BSk*)	ISP3, incubation at 28 °C for 2 weeks	99.5% (21%) with *S. rubrogriseus* DSM 41477^T, 99.5% (17%) with *S. lienomycini* DSM 41475^T and 99.3% (8%) with *S. violaceoruber* DSM 40783^T	15–40 (28–30)	4.5–10	–	–	iso-$C_{16:0}$, anteiso-$C_{17:0}$, anteiso-$C_{15:0}$, $C_{16:0}$
Streptomyces thinghirensis **Type strains:** DSM 41919^T,	Rhizosphere soil of wild healthy *Vitis vinifera* plant in	Actinomycetes isolation agar, incubation at	99.2% (4.6%) with *S. coelescens* DSM 40421^T, 99.32% (6.7%) with *S. aurantiogriseus*	28–42	5–10	0–7	–	anteiso-$C_{15:0}$, iso-$C_{16:0}$, anteiso-

CCMM B35^T. **16S rRNA gene:** FM202482. **Risk group:** 1	Thinghir, Ouarzazate, southern Morocco (*BWh*)	28 °C for 3 weeks	DSM 40138^T, 99.24% (9.3%) with *S. lienomycini* DSM 41475^T, 99.18% (10.5%) with *S. violaceolatus* DSM 40438^T and 99.65% (33.4%) with *S. marokkonensis* DSM 41918^T					$C_{17:0}$, iso-$C_{15:0}$
Streptomyces youssoufiensis **Type strains:** DSM 41920^T, CCMM B709^T. **16S rRNA gene:** FN421338. **Risk group:** 1	From a phosphate mine in Youssoufia, North of Marrakesh, southern Morocco (*BSh*)	Soil extract agar, incubation at 28 °C for 3 weeks	98.97% (28.5%) with *S. coerulescens* NBRC 12758^T, 98.72% (11.4%) with *S. ramulosus* NRRL B-2714^T, 98.43% (24%) with *S. kasugaensis* M338-M1^T and 98.66% (5.4%) with *S. varsoviensis* NRRL B-3589^T	28–42	5–10	0–7	–	iso-$C_{16:0}$, iso-$C_{16:1}$
Streptomyces tunisiensis **Type strains:** DSM 42037^T, JCM 17589^T. **16S rRNA gene:** KF697135. **Risk group:** 1	Soil from a forest in Tunis, northern Tunisia (*Csa*)	ISP 2, incubation at 28 °C for 21 days	(3.2%) with *S. griseoincarnatus* DSM 40274^T, (23.8%) with *S. variabilis* DSM 40179^T, (19.3%) with *S. labedae* DSM 41446^T and (24.1%) with *S. erythrogriseus* DSM 40116^T	20–37 (28)	(7)	0–7	MK-9(H_6), MK-9(H_8)	iso-$C_{16:0}$, anteiso-$C_{15:0}$, iso-$C_{14:0}$

(continued)

Table 16.2 (continued)

Species/type strains/ 16S rRNA gene/risk group	Source/climate classification system	Isolation conditions	16S rRNA similarity (DDH) with the neighboring species	Temperature range (°C) (optimal)	pH range (optimal)	NaCl range (optimal)	Major menaquinones (>10%)	Major fatty acids (>10%)
"Streptomyces tunisialbus" **Type strains:** DSM 105760^T, JCM 32165^T. **16S rRNA gene:** KY242372. **Risk group:** 1	*Lavandula officinalis* rhizosphere soil, collected from Park El Montazah, Tunis, Tunisia (*Csa*)	Modified ISP2, incubation at 30 °C for 21 days	99.86% (63.6%) with *S. netropsis* DSM 40259^T	20–37 (28–30)	5–11 (6–7)	0–7 (0–4)	MK-9(H_8), MK-9(H_6)	$C_{16:0}$, anteiso-$C_{15:0}$, iso-$C_{16:0}$
"Streptomyces massilialgeriensis" **Type strains:** CSUR P3927^T. **16S rRNA gene:** LT838406. **Risk group:** 1	Extremely saline soil from the dry lake (Garaet) of Ank el Djamel in Oum El Bouaghi, semi-arid region, Algeria (*Csa*)	ISP4 supplemented with 5% NaCl, incubation at 30 °C for 14 days	97.26% with *S. albus* ATCC 25426^T	(37)	(7.5)	–	–	–
Nocardiopsis algeriensis **Type strains:** DSM 45462^T, CECT 8712^T. **16S rRNA gene:** KJ470139. **Risk group:** 1	Saharan soil from Adrar, Touat, South Algeria (*BWh*)	Complex medium agar supplemented with 0.5 % (w/v) NaCl, incubation at 30 °C for 7 days	98.7% (17.9%) with *N. alba* DSM 43377^T, 98.6% (14.6%) with *N. lucentensis* DSM 44048^T, 98.6% (31.1%) with *N. aegyptia* DSM 44442^T, 98.6% (27.1%) with *N. sinuspersici* DSM 45277^T and 98.5% (14.1%) with *N. arvandica* DSM 45278^T	20–45 (25–35)	7–12 (7–10)	0–7.5 (0–5)	MK-10(H_4), MK-11(H_4)	iso-$C_{16:0}$, anteiso-$C_{15:0}$

Streptomonospora algeriensis **Type strains:** DSM 45604T, CCUG 63369^T, MTCC 11563^T. **16S rRNA gene:** HQ918204. **Risk group:** 1	From Djelfa, semiarid region, North-central of Algeria (*BSk*).	Complex medium agar supplemented with 20% (w/v) NaCl, incubation at 30 °C for 35 days	98.8% (17.1%) with *S. alba* DSM 44588^T and 98.7% (57.9%) with *S. flavalba* DSM 45155^T	20–45 (28–37)	5–8 (6–7)	7–20 (10–15)	MK-11(H_4), MK-10(H_6)	iso-$C_{16:0}$, anteiso-$C_{17:0}$
Planomonospora algeriensis **Type strains:** DSM 46752^T, CECT 9047^T. **16S rRNA gene:** KX029461. **Risk group:** 1	Saharan soil from Béni-Abbès, Béchar, Saoura, South-West of Algeria (*BWh*)	Chitin-vitamin agar supplemented with polymyxin (25 mg/L), incubation at 30 °C for 21 days	99.3% (58.4%) with *P. sphaerica* DSM 44632^T, 99.2% (70.1%) with *P. parontospora* subsp. *parontospora* DSM 43177^T and 99.0% (64.9%) with *P. parontospora* subsp. *antibiotica* DSM 43869^T	20–40 (30)	6–11 (8–9)	0–2 (0–1)	MK-9(H_2), MK-9(H_0)	$C_{17:1}$ $\omega 9c$, $C_{17:0}$
Streptosporangium algeriense **Type strains:** DSM 45455^T, CCUG 62974^T, MTCC 11561^T. **16S rRNA gene:** KP974644. **Risk group:** 1	Saharan soil from Adrar palm grove, southern Algeria (*BWh*)	Chitin-vitamin agar supplemented with polymyxin (25 mg/L), incubation at 30 °C for 21 days	99.3% (60%) with *S. jomthongense* NBRC 110047^T	(30)			MK-9(H_2), MK-9(H_4)	iso-$C_{16:0}$, $C_{16:0}$, $C_{17:1}$ $\omega 8c$, 10-methyl $C_{17:0}$
Streptosporangium becharense **Type strains:** DSM 46887^T,	Saharan soil from Béni-Abbès, Béchar, Saoura, south-	Chitin-vitamin agar supplemented with polymyxin	99.7% (58.1%) with ‘*S. subfuscum*’ DSM 46724, 98.7% (38.8%) with *S. pseudovulgare*	20–42 (35–37)	6–12 (7.5–8.5)	0–2	MK-9(H_2), MK-9(H_4)	$C_{17:1}$ $\omega 8c$, iso-$C_{16:0}$, 10-methyl

(continued)

Table 16.2 (continued)

Species/type strains/ 16S rRNA gene/risk group	Source/climate classification system	Isolation conditions	16S rRNA similarity (DDH) with the neighboring species	Temperature range (°C) (optimal)	pH range (optimal)	NaCl range (optimal)	Major menaquinones (>10%)	Major fatty acids (>10%)
CECT 8961^T. **16S rRNA gene:** KU593506. **Risk group:** 1	west of Algeria (*BWh*)	(25 mg/L), incubation at 30 °C for 28 days	DSM 43181^T, 98.6% (18.5%) with *S. fragile* DSM 43847^T and 98.5% (18.2%) with *S. sandarakinum* DSM 45763^T					$C_{17:0}$, $C_{18:1}$ *ω9c*
Streptosporangium saharense **Type strains:** DSM 46743^T, CECT 8840^T. **16S rRNA gene:** KT581983. **Risk group:** 1	Saharan saoil from Béni-isguen, Mzab, Ghardaïa, southern Algeria (*BWh*)	From a soil sample that was air dried at room temperature for 10 days before being baked at 120 °C for 1 h. Chitin-vitamin agar supplemented with penicillin (25 mg/L), incubation at 30 °C for 21 days	98.5% with *S. jomthongense* BCC 53154^T.	20–42	3–12	0–2 (0–1)	MK-9(H_4), MK-9(H_6), MK-9(H_2)	iso-$C_{16:0}$, $C_{16:0}$
Actinomadura adrarensis **Type strains:** DSM 46745^T, CECT 8842^T. **16S rRNA gene:** KU356942. **Risk group:** 1	Sandy loam Saharan soil from the palm grove of the Bouda, Adrar, southern Algeria (*BWh*)	Chitin-vitamin agar, incubation at 30 °C for 21 days	98.3% (39.8%) with *A. sputi* DSM 45233^T and 97.8% (18.7%) with *A. hallensis* DSM 45043^T	25–37 (30)	7–10 (8)	0–2 (0)	MK-9(H6), MK-9(H4), MK-9(H8)	$C_{16:0}$, 10-methyl $C_{17:0}$, $C_{17:1}$ *ω9c*, $C_{15:0}$

Actinomadura algeriensis **Type strains:** DSM 46744^T, CECT 8841^T. **16S rRNA gene:** KT259320. **Risk group:** 1	Saharan soil in the Hoggar, Tamanrasset, South of Algeria (*BWh*)	Chitin-vitamin agar, incubation at 30 °C for 21 days	98.5% (48.0%) with *A. sediminis* DSM 45500^T and 98.3% (33.2%) with *A. cremea* subsp. *cremea* DSM 43676^T	20–37	6–9	0–3	MK-9(H4), MK-9(H2), MK-9(H6)	$C_{16:0}$, $C_{18:1}$ *cis*9, iso-$C_{16:0}$, 10-methyl $C_{18:0}$

Among 42 new species and 2 new genera belonging to 11 families of *Actinobacteria* discovered in Northwest Africa, the most recovered and abundant genera: *Streptomyces* and *Saccharothrix* each had 6 new species identified, accounting for 37% of the total novel *Actinobacteria* species from Maghreb regions, followed by the genus *Actinopolyspora* with 5 species which represent 11% of the total. Moreover, four species of *Geodermatophilus*, three species of *Streptosporangium*, three species of *Blastococcus*, and two species of *Actinomadura* were discovered from Northwest Africa. In addition, one new species from other genera was described.

The taxonomic status of any bacterial strain, according to the polyphasic taxonomy, is determined by both phenotypic and genotypic characterization. A combination of chemotaxonomic analysis and other phenotypic features (tolerance tests, enzyme production, ability to metabolize carbon and nitrogen sources) together with other genetic traits of the taxon (16S rRNA phylogeny, GC content, DNA–DNA hybridization) was classically used for new *Actinobacteria* species descriptions (Carro and Nouiuoi 2017).

16.5 Conclusion

Maghrebian region though less explored possesses a great actinobacterial biodiversity. This bacterial diversity can be explored for isolation of novel rare *Actinobacteria* and characterization of new antibacterial and antifungal molecules. Hence, this biodiversity is seen as a consequence of adaptation to these taxa to harsh conditions. *Actinobacteria* from unexplored ecosystems will be novel and useful host in the future for the production of enzymes, chemicals, antibiotics with low cost.

References

Ait Barka E, Vatsa P, Sanchez L, Gaveau-Vaillant N, Jacquard C, Meier-Kolthoff JP, Klenk H-P, Clément C, Ouhdouch Y, van Wezel GP (2015) Taxonomy, physiology, and natural products of Actinobacteria. Microbiol Mol Biol Rev 80:1–43

Aouiche A, Bouras N, Mokrane S, Zitouni A, Schumann P, Sproer C, Sabaou N, Klenk HP (2015) *Actinokineospora mzabensis* sp. nov., a novel actinomycete isolated from Saharan soil. Anton Van Leeuwenhoek 107:291–296

Ayed A, Slama N, Mankai H, Bachkouel S, ElKahoui S, Tabbene O, Limam F (2018) *Streptomyces tunisialbus* sp. nov., a novel *Streptomyces* species with antimicrobial activity. Anton Van Leeuwenhoek 111:1571–1581

Baker D, O'Keefer D (1984) A modified sucrose fractionation procedure for the isolation of *Frankiae* from actinorhizal root nodules and soil samples. Plant S 78:23–28

Barakate M, Ouhdouch Y, Oufdou K, Beaulieu C (2002) Characterization of rhizospheric soil streptomycetes from Moroccan habitats and their antimicrobial activities. World J Microbiol Biotechnol 18:49–54

Boubetra D, Zitouni A, Bouras N, Mathieu F, Lebrihi A, Schumann P, Sproer C, Klenk HP, Sabaou N (2013a) *Saccharothrix hoggarensis* sp. nov., an actinomycete isolated from Saharan soil. Int J Syst Evol Microbiol 63:549–553

Boubetra D, Zitouni A, Bouras N, Mathieu F, Lebrihi A, Schumann P, Sproer C, Klenk HP, Sabaou N (2013b) *Saccharothrix saharensis* sp. nov., an actinomycete isolated from Algerian Saharan soil. Int J Syst Evol Microbiol 63:3744–3749

Boubetra D, Zitouni A, Bouras N, Schumann P, Sproer C, Klenk HP, Sabaou N (2015) *Saccharothrix tamanrassetensis* sp. nov., an actinomycete isolated from Saharan soil. Int J Syst Evol Microbiol 65:1316–1320

Boubetra D, Bouras N, Zitouni A, Schumann P, Sproer C, Sabaou N, Klenk HP (2016) *Streptosporangium algeriense* sp. nov., an actinobacterium isolated from desert soil. Int J Syst Evol Microbiol 66:1034–1038

Boudjelal F, Zitouni A, Bouras N, Schumann P, Sproer C, Sabaou N, Klenk HP (2015) *Actinoalloteichus hoggarensis* sp. nov., an actinomycete isolated from Saharan soil. Int J Syst Evol Microbiol 65:2006–2010

Bouizgarne B, Lanoot B, Loqman S, Sproer C, Klenk HP, Swings J, Ouhdouch Y (2009) *Streptomyces marokkonensis* sp. nov., isolated from rhizosphere soil of Argania spinosa L. Int J Syst Evol Microbiol 59:2857–2863

Bouras N, Meklat A, Zitouni A, Mathieu F, Schumann P, Sproer C, Sabaou N, Klenk HP (2015) *Nocardiopsis algeriensis* sp. nov., an alkalitolerant actinomycete isolated from Saharan soil. Anton Van Leeuwenhoek 107:313–320

Bouznada K, Bouras N, Schumann P, Sproer C, Sabaou N, Klenk HP (2016a) *Actinophytocola algeriensis* sp. nov., an actinobacterium isolated from Saharan soil. Int J Syst Evol Microbiol 66:2760–2765

Bouznada K, Bouras N, Mokrane S, Chaabane Chaouch F, Zitouni A, Potter G, Sproer C, Klenk HP, Sabaou N (2016b) *Saccharothrix isguenensis* sp. nov., an actinobacterium isolated from desert soil. Int J Syst Evol Microbiol 66:4785–4790

Bouznada K, Bouras N, Mokrane S, Chaabane Chaouch F, Zitouni A, Potter G, Sproer C, Klenk HP, Sabaou N (2017) *Saccharothrix ghardaiensis* sp. nov., an actinobacterium isolated from Saharan soil. Anton Van Leeuwenhoek 110:399–405

Carro L, Nouiuoi I (2017) Taxonomy and systematics of plant probiotic bacteria in the genomic era. AIMS Microbiol 3:383–412

Chaabane Chaouch F, Bouras N, Mokrane S, Zitouni A, Schumann P, Sproer C, Sabaou N, Klenk HP (2016a) *Streptosporangium saharense* sp. nov., an actinobacterium isolated from Saharan soil. Int J Syst Evol Microbiol 66:1371–1376

Chaabane Chaouch F, Bouras N, Mokrane S, Zitouni A, Schumann P, Sproer C, Sabaou N, Klenk HP (2016b) *Streptosporangium becharense* sp. nov., an actinobacterium isolated from desert soil. Int J Syst Evol Microbiol 66:2484–2490

Chaabane Chaouch F, Bouras N, Mokrane S, Bouznada K, Zitouni A, Potter G, Sproer C, Klenk HP, Sabaou N (2017) *Planomonospora algeriensis* sp. nov., an actinobacterium isolated from a Saharan soil of Algeria. Anton Van Leeuwenhoek 110:245–252

Chun J, Bae KS, Moon EY, Jung SO, Lee HK, Kim SJ (2000) *Nocardiopsis kunsanensis* sp. nov., a moderately halophilic actinomycete isolated from a saltern. Int J Syst Evol Microbiol 50:1909–1913

Djaballah CE, Kitouni M, Raoult D, Khelaifia S (2018) "Streptomyces massilialgeriensis" sp. nov., a new bacterial species isolated from an extremely saline soil collected from the dry lake of Ank el Djamel in Algeria. New Microbes New Infect 21:18–19

Djouadi LN, Levasseur A, Khalil JB, Blanc-Taileur C, Asmar S, Ghiloubi W, Nateche F, Drancourt M (2016) *Mycobacterium icosiumassiliensis* sp. nov., a new member in the *Mycobacterium terrae* complex isolated from surface water in Algeria. Curr Microbiol 73:255–264

Ghodhbane-Gtari F, Nouioui I, Salem K, Ktari A, Montero-Calasanz Mdel C, Tisa LS, Klenk HP, Gtari M (2014) *Nocardia casuarinae* sp. nov., an actinobacterial endophyte isolated from root nodules of *Casuarina glauca*. Anton Van Leeuwenhoek 105:1099–1106

Gupta RS, Lo B, Son J (2018) Phylogenomics and comparative genomic studies robustly support division of the genus *Mycobacterium* into an emended genus *Mycobacterium* and four novel genera. Front Microbiol 9:67

Hamdali H, Virolle MJ, von Jan M, Sproer C, Klenk HP, Ouhdouch Y (2011) *Streptomyces youssoufiensis* sp. nov., isolated from a Moroccan phosphate mine. Int J Syst Evol Microbiol 61:1104–1108

Hayakawa M, Nonomura H (1987) Humic acid-vitamin agar, a new medium for the selective isolation of soil actinomycetes. J Ferment Technol 65:501–509

Hezbri K, Ghodhbane-Gtari F, Del Carmen Montero-Calasanz M, Sghaier H, Rohde M, Sproer C, Schumann P, Klenk HP, Gtari M (2015a) *Geodermatophilus aquaeductus* sp. nov., isolated from the ruins of Hadrian's aqueduct. Anton Van Leeuwenhoek 108:41–50

Hezbri K, Ghodhbane-Gtari F, Del Carmen M-CM, Sghaier H, Rohde M, Schumann P, Klenk HP, Gtari M (2015b) Description of *Geodermatophilus bullaregiensis* sp. nov. Anton Van Leeuwenhoek 108:415–425

Hezbri K, Ghodhbane-Gtari F, Montero-Calasanz Mdel C, Sghaier H, Rohde M, Schumann P, Klenk HP, Gtari M (2015c) *Geodermatophilus sabuli* sp. nov., a gamma-radiation-resistant actinobacterium isolated from desert limestone. Int J Syst Evol Microbiol 65:3365–3372

Hezbri K, Ghodhbane-Gtari F, Montero-Calasanz MD, Nouioui I, Rohde M, Sproer C, Schumann P, Klenk HP, Gtari M (2016a) *Geodermatophilus pulveris* sp. nov., a gamma-radiation-resistant actinobacterium isolated from the Sahara desert. Int J Syst Evol Microbiol 66:3828–3834

Hezbri K, Louati M, Nouioui I, Gtari M, Rohde M, Sproer C, Schumann P, Klenk HP, Ghodhbane-Gtari F, Montero-Calasanz MD (2016b) *Blastococcus capsensis* sp. nov., isolated from an archaeological Roman pool and emended description of the genus *Blastococcus*, *B. aggregatus*, *B. saxobsidens*, *B. jejuensis* and *B. endophyticus*. Int J Syst Evol Microbiol 66:4864–4872

Hezbri K, Nouioui I, Rohde M, Schumann P, Gtari M, Klenk HP, Montero-Calasanz MD, Ghodhbane-Gtari F (2017) *Blastococcus colisei* sp. nov, isolated from an archaeological amphitheatre. Anton Van Leeuwenhoek 110:339–346

Hezbri K, Nouioui I, Rohde M, Sproer C, Schumann P, Gtari M, Klenk HP, Montero-Calasanz MDC, Ghodhbane-Gtari F (2018) *Blastococcus xanthinilyticus* sp. nov., isolated from monument. Int J Syst Evol Microbiol 68:1177–1183

Labeda DP, Kroppenstedt RM (2004) Emended description of the genus *Glycomyces* and description of *Glycomyces algeriensis* sp. nov., *Glycomyces arizonensis* sp. nov. and *Glycomyces lechevalierae* sp. nov. Int J Syst Evol Microbiol 54:2343–2346

Lahoum A, Bouras N, Mathieu F, Schumann P, Sproer C, Klenk HP, Sabaou N (2016a) *Actinomadura algeriensis* sp. nov., an actinobacterium isolated from Saharan soil. Anton Van Leeuwenhoek 109:159–165

Lahoum A, Bouras N, Verheecke C, Mathieu F, Schumann P, Sproer C, Klenk HP, Sabaou N (2016b) *Actinomadura adrarensis* sp. nov., an actinobacterium isolated from Saharan soil. Int J Syst Evol Microbiol 66:2724–2729

Lai H, Jiang Y, Saker R, Chen X, Bouras N, Klenk HP, Wei X, Jiang Y, Sabaou N (2017) Reclassification of *Mzabimyces algeriensis* Saker et al. 2015 as *Halopolyspora algeriensis* comb. nov. Int J Syst Evol Microbiol 67:2787–2790

Loqman S, Bouizgarne B, Barka EA, Clement C, von Jan M, Sproer C, Klenk HP, Ouhdouch Y (2009) *Streptomyces thinghirensis* sp. nov., isolated from rhizosphere soil of *Vitis vinifera*. Int J Syst Evol Microbiol 59:3063–3067

Luedemann GM (1968) Geodermatophilus, a new genus of the *Dermatophilaceae* (Actinomycetales). J Bacteriol 96:1848–1858

Meklat A, Bouras N, Zitouni A, Mathieu F, Lebrihi A, Schumann P, Sproer C, Klenk HP, Sabaou N (2012) *Actinopolyspora algeriensis* sp. nov., a novel halophilic actinomycete isolated from a Saharan soil. Extremophiles 16:771–776

Meklat A, Bouras N, Zitouni A, Mathieu F, Lebrihi A, Schumann P, Sproer C, Klenk HP, Sabaou N (2013a) *Actinopolyspora mzabensis* sp. nov., a halophilic actinomycete isolated from an Algerian Saharan soil. Int J Syst Evol Microbiol 63:3787–3792

Meklat A, Bouras N, Zitouni A, Mathieu F, Lebrihi A, Schumann P, Sproer C, Klenk HP, Sabaou N (2013b) *Actinopolyspora righensis* sp. nov., a novel halophilic actinomycete isolated from Saharan soil in Algeria. Anton Van Leeuwenhoek 104:301–307

Meklat A, Bouras N, Zitouni A, Mathieu F, Lebrihi A, Schumann P, Sproer C, Klenk HP, Sabaou N (2013c) *Actinopolyspora saharensis* sp. nov., a novel halophilic actinomycete isolated from a Saharan soil of Algeria. Anton Van Leeuwenhoek 103:771–776

Meklat A, Bouras N, Zitouni A, Sabaou N, Mathieu F, Schumann P, Sproer C, Klenk HP (2014a) *Saccharopolyspora ghardaiensis* sp. nov., an extremely halophilic actinomycete isolated from Algerian Saharan soil. J Antibiot (Tokyo) 67:299–303

Meklat A, Bouras N, Riba A, Zitouni A, Mathieu F, Rohde M, Schumann P, Sproer C, Klenk HP, Sabaou N (2014b) *Streptomonospora algeriensis* sp. nov., a halophilic actinomycete isolated from soil in Algeria. Anton Van Leeuwenhoek 106:287–292

Meklat A, Bouras N, Mokrane S, Zitouni A, Schumann P, Sproer C, Klenk HP, Sabaou N (2015) *Bounagaea algeriensis* gen. nov., sp. nov., an extremely halophilic actinobacterium isolated from a Saharan soil of Algeria. Anton Van Leeuwenhoek 108:473–482

Moshtaghi Nikou M, Ramezani M, Harirchi S, Makzoom S, Amoozegar MA, Shahzadeh Fazeli SA, Schumann P, Ventosa A (2017) *Salinifilum* gen. nov., with description of *Salinifilum proteinilyticum* sp. nov., an extremely halophilic actinomycete isolated from Meighan wetland, Iran, and reclassification of *Saccharopolyspora aidingensis* as *Salinifilum aidingensis* comb. nov. and *Saccharopolyspora ghardaiensis* as *Salinifilum ghardaiensis* comb. nov. Int J Syst Evol Microbiol 67:4221–4227

Nouioui I, Ghodhbane-Gtari F, Montero-Calasanz MD, Goker M, Meier-Kolthoff JP, Schumann P, Rohde M, Goodfellow M, Fernandez MP, Normand P et al (2016) Proposal of a type strain for *Frankia alni* (Woronin 1866) Von Tubeuf 1895, emended description of *Frankia alni*, and recognition of *Frankia casuarinae* sp. nov. and *Frankia elaeagni* sp. nov. Int J Syst Evol Microbiol 66:5201–5210

Olson EH (1968) Actinomycetes isolation agar. Difco: Supplementary literature. Difco Lab, Detroit

Ouhdouch Y, Barakate M, Finance C (2001) Actinomycetes of Moroccan habitats: isolation and screening for antifungal activities. Eur J Soil Biol 37:69–74

Raja A, Prabakaran P (2011) Actinomycetes and drug—an overview. Am J Drug Discov Dev 1:75–84

Reasoner J, Blannon JC, Geldreich EE (1979) Rapid sevenhour fecal coliform test. Appl Environ Microbiol 38:229–236

Sahraoui N, Ballif M, Zelleg S, Yousfi N, Ritter C, Friedel U, Amstutz B, Yala D, Boulahbal F, Guetarni D et al (2011) *Mycobacterium algericum* sp. nov., a novel rapidly growing species related to the *Mycobacterium terrae* complex and associated with goat lung lesions. Int J Syst Evol Microbiol 61:1870–1874

Saker R, Bouras N, Zitouni A, Ghoul M, Rohde M, Schumann P, Sproer C, Sabaou N, Klenk HP (2014) *Mzabimyces algeriensis* gen. nov., sp. nov., a halophilic filamentous actinobacterium isolated from a Saharan soil, and proposal of *Mzabimycetaceae* fam. nov. Anton Van Leeuwenhoek 106:1021–1030

Saker R, Bouras N, Meklat A, Zitouni A, Schumann P, Sproer C, Klenk HP, Sabaou N (2015a) *Actinopolyspora biskrensis* sp. nov., a novel halophilic actinomycete isolated from northern Sahara. Curr Microbiol 70:423–428

Saker R, Bouras N, Meklat A, Zitouni A, Schumann P, Sproer C, Sabaou N, Klenk HP (2015b) *Prauserella isguenensis* sp. nov., a halophilic actinomycete isolated from desert soil. Int J Syst Evol Microbiol 65:1598–1603

Shirling EB, Gottlieb D (1966) Methods for characterization of *Streptomyces* species. Int J Syst Bacteriol 16:313–340

Singh RP, Manchanda G, Maurya IK, Maheshwari NK, Tiwari PK et al (2019) *Streptomyces* from rotten wheat straw endowed the high plant growth potential traits and agro-active compounds. Biocatal Agric Biotechnol 17:507–513. https://doi.org/10.1016/j.bcab.2019.01.014

Slama N, Mankai H, Ayed A, Mezhoud K, Rauch C, Lazim H, Barkallah I, Gtari M, Limam F (2014) *Streptomyces tunisiensis* sp. nov., a novel *Streptomyces* species with antibacterial activity. Anton Van Leeuwenhoek 105:377–387

Zitouni A, Lamari L, Boudjella H, Badji B, Sabaou N, Gaouar A, Mathieu F, Lebrihi A, Labeda DP (2004) *Saccharothrix algeriensis* sp. nov., isolated from Saharan soil. Int J Syst Evol Microbiol 54:1377–1381

The manufacturer's authorised representative in the EU is Springer Nature Customer Service Centre GmbH, Europaplatz 3, 69115 Heidelberg, Germany. If you have any concerns regarding our products, please contact ProductSafety@springernature.com

Printed and bound by CPI Group (UK) Ltd, Croydon, CR0 4YY
30/09/2026
02225248-0013